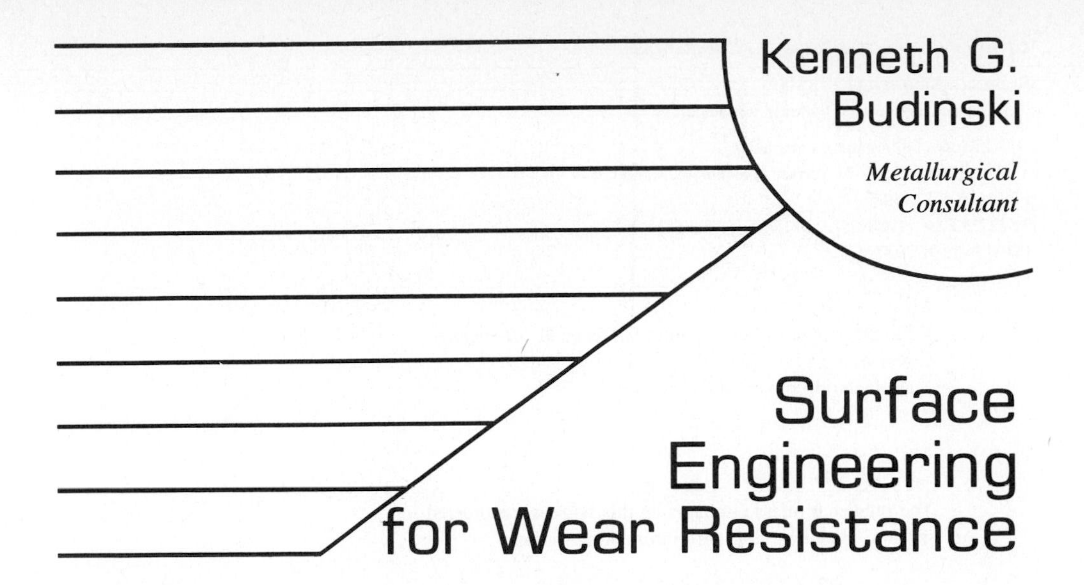

Kenneth G. Budinski

Metallurgical Consultant

Surface Engineering for Wear Resistance

Prentice Hall, Englewood Cliffs, New Jersey 07632

Library of Congress Cataloging-in-Publication Data

Budinski, Kenneth G.
Surface engineering for wear resistance.

Includes bibliographies and index.
1. Hardfacing. 2. Surface hardening. 3. Metals—
Finishing. I. Title.
TS227.3.B84 1988 671.7 88–5831
ISBN 0–13–877937–6

Editorial/production supervision: Gretchen K. Chenenko
Cover design: Ben Santora
Manufacturing buyer: Mary Noonan

The publisher offers discounts on this book when ordered in bulk quantities. For more information, write:

Special Sales/College Marketing
Prentice Hall
College Technical and Reference Division
Englewood Cliffs, New Jersey 07632

Printed in the United States of America
10 9 8 7 6 5 4 3 2 1

ISBN 0-13-877937-6

Prentice-Hall International (UK) Limited, *London*
Prentice-Hall of Australia Pty. Limited, *Sydney*
Prentice-Hall Canada Inc., *Toronto*
Prentice-Hall Hispanoamericana, S.A., *Mexico*
Prentice-Hall of India Private Limited, *New Delhi*
Prentice-Hall of Japan, Inc., *Tokyo*
Simon & Schuster Asia Pte. Ltd., *Singapore*
Editora Prentice-Hall do Brasil, Ltda., *Rio de Janeiro*

To my beloved father

Anthony L. Budinski
(1905–1987)

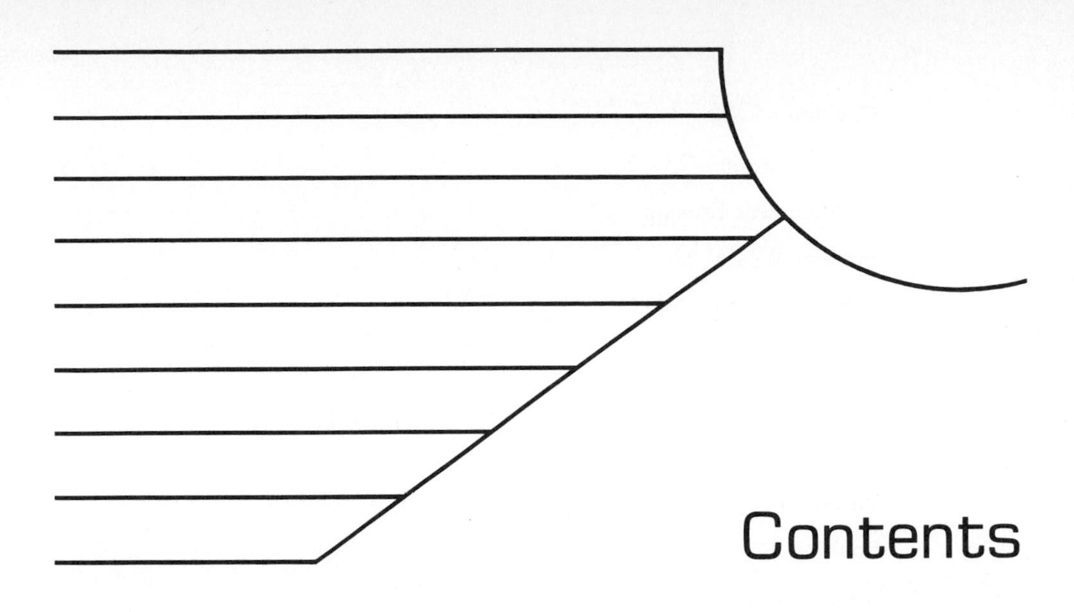

Contents

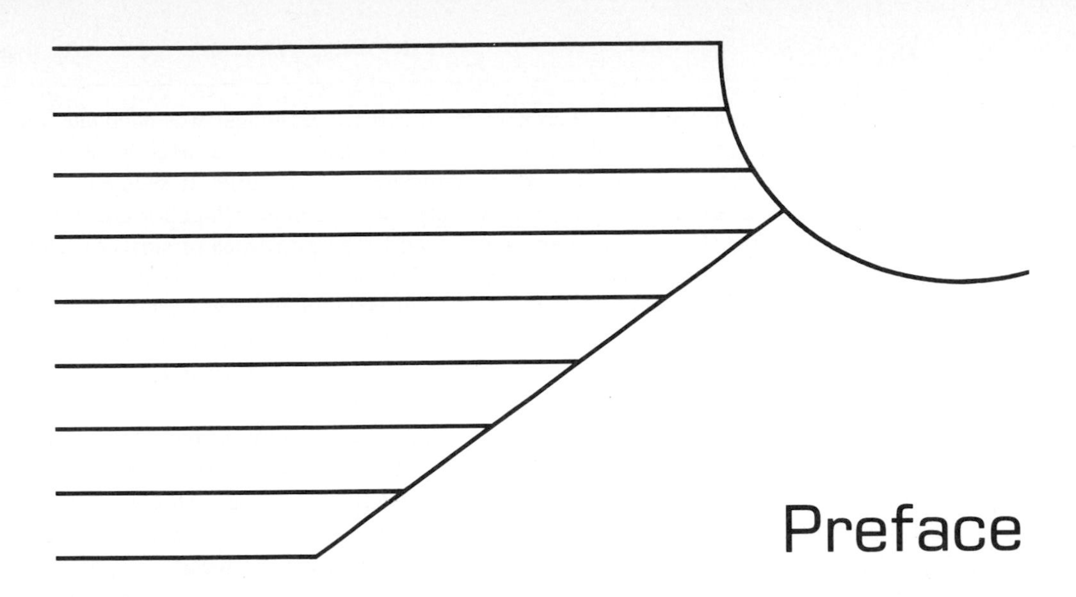

Preface

For the last ten years or so, the Welding Research Council (WRC) has been trying to legitimatize the field of hardfacing through the activities of the High Alloys Committee and its subcommittee on Hardfacing and Wear. The subcommittee has conducted a number of symposia to address the uses and benefits of hardfacing, but these gatherings were mostly attended by people who were already working in the field of hardfacing or who were experienced users of hardfacing and surface treatments. At a 1982 meeting of the Hardfacing Subcommittee, a task group was established to explore the feasibility of publishing a document to educate the general public on the subject of hardfacing. More specifically, the document would answer the application questions of potential users and promote wider acceptance in industry.

In 1983, I wrote a brief manual entitled "Guide to Hardfacing" and presented it to about a dozen of the subcommittee members for review. They graciously made helpful suggestions and the document was presented to WRC for possible publication. WRC thought that the document was too lengthy to fit into their normal publishing format (mostly interpretive reports), and it was suggested that we seek a commercial publisher. This was done, but the commercial publisher thought that it was too brief to publish as a reference text or handbook and suggested expansion. I agreed with all parties and after three more years of writing and additional reviews by WRC members, the present form emerged, not a handbook, not a manual, but a teaching and reference book on hardfacing and about 20 surface treatment and coating processes that compete with hardfacing in improving the wear resistance of surfaces.

The competitive processes include through hardening, case hardening, plating, selective hardening, and some of the newer surface technologies, such as ion implantation, thin-film coatings, and high-energy surface modifications with laser and electron beam.

The use of the original title, ''Guide to Hardfacing,'' no longer was adequate since the expansion resulted in more about other processes than about hardfacing. One reviewer suggested the use of the term surface engineering in the title. It seemed appropriate since there is a Surface Engineering Society in the United Kingdom and this society deals with all the things that people do to modify the properties of surfaces. Hence the title became *Surface Engineering for Wear Resistance*.

I feel that the expansion of the original hardfacing guide to include a wide variety of surface treatments and coatings makes it much more valuable than the original work. This is a first attempt at a very broad field, and much of what is said about these surface processes is based on 25 years of personal experience in dealing with wear problems in a very large manufacturing plant. I do not rely on hardfacing for solving wear problems, but rather I have used each of the 20 or so processes covered in this book, as well as some processes that are best left anonymous. This work is intended to serve as a reference for designers, manufacturing people, maintenance personnel, and students, in selecting processes to make surfaces more wear resistant. It can be used for courses on welding, metallurgy, wear, or material selection; it can be put in reference centers in design areas and it is hoped that it will be adopted by some venturesome professors to start courses on surface engineering. All the important surface-treating processes are described in simple terms; it is shown how they can be applied to various wear situations, and information is presented on selection and specification. An appendix lists the trade names and chemical compositions of over 800 consumables that are used in hardfacing. The overall goal of this work is to supply the reader with sufficient information to select the most appropriate and cost-effective surface to resist a particular form of wear.

A work of this nature does not come from one person. I acknowledge the contributions of my associates in the WRC, discussions with my coworkers, and suggestions from my friends and associates in the wear community. I especially thank Dr. Robert Tucker, chairman of the WRC Wear and Hardfacing Subcommittee, for his support in this effort and in keeping the WRC hardfacing activity progressing.

Kenneth G. Budinski

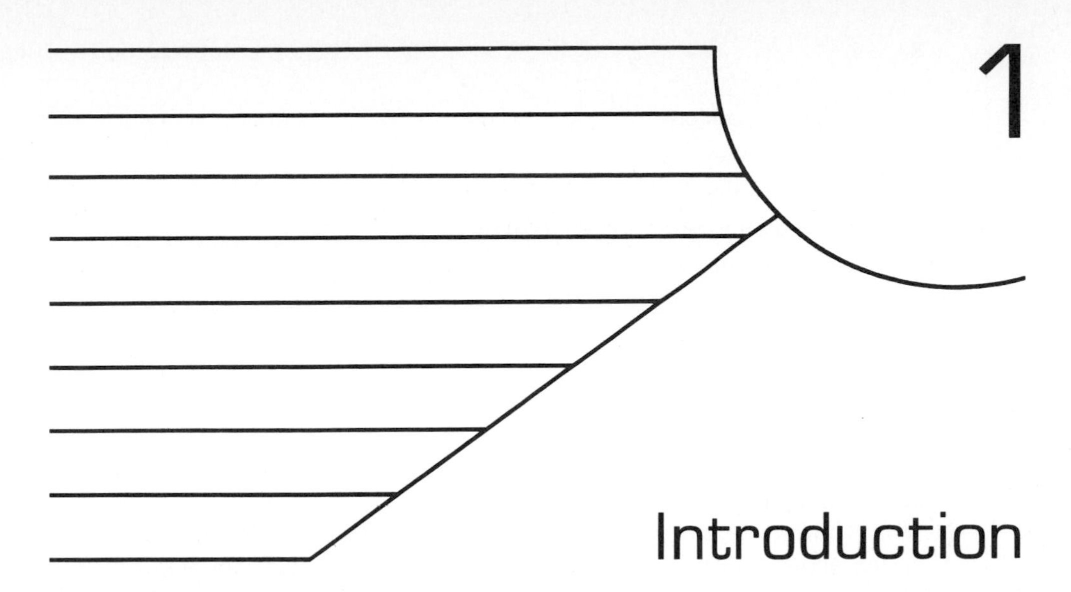

1 Introduction

This book is about ways to make surfaces wear resistant. It started out to be a guide to one family of processes to make surfaces wear resistant, hardfacing with welding techniques; but it became apparent that there was a need to show how hardfacing processes compare with the other techniques that are used to make surfaces wear resistant. The term *surface engineering* was put into the title to reflect this broader scope. Surface engineering is a relatively new term that has come into use in the last decade or so to describe multidiscipline activities aimed at tailoring the properties of surfaces of engineering materials to improve their function or serviceability. There are probably no engineers that actually call themselves surface engineers, but there are many physicists, metallurgists, chemists, and mechanical engineers that do work that can be called surface engineering. They work on coatings to protect surfaces from deterioration in their use environment, and they work on ways to alter the functional properties of surfaces. There is a surface engineering society in the United Kingdom whose journal, *Surface Engineering,* deals with subjects such as the following:

Fusion hardfacing
Weld cladding
Explosive cladding
Sputter coating
Chemical vapor deposition (CVD)
Physical vapor deposition (PVD)
Ion plating
Plasma-assisted CVD

Electrodeposition
Thermal spraying
Plastic coating
Methods for testing surface coatings
Ion implantation
Mechanical plating
Electroless plating
Selective electrodeposition
Surface hardening

This book covers most of these subjects; thus the term surface engineering was felt to be an appropriate part of the title.

The term *wear* in the title is intended to show that we will only cover the application of the above ''shopping list'' of techniques to enhancing the wear life of surfaces. Many coatings and surface treatments are used to alter the physical or chemical properties of surfaces; platings are used to provide RF shielding; PVD coatings are used to produce electrical insulation; many weld overlays are done to produce corrosion resistance; some thin-film coatings are applied to alter the optical properties of surfaces, such as the index of refraction. These are the kinds of applications that will not be included in this book. We will discuss all these subjects as they apply to the wear life of surfaces.

Thus, the purpose of this book is to present information on coatings that can be applied to surfaces and treatments to surfaces that make the surface characteristics superior to those of the bulk material. This information is intended to serve as a guide to process selection and specification for people working in design engineering or maintenance and for students aspiring to work in these areas. The objective of sharing this information, simply put, is to save money for the people and companies that must foot the bill for machines and parts that wear out. The annual cost of wear in the United States in 1985 was estimated to be $20 billion, but the cost of wear may be better illustrated by what it costs the average person. If you own an automobile costing $10,000, you know that this automobile will be worth only a small fraction of this amount after only a kilogram or so of metal is removed by wear. This material removal will gradually come from engine and suspension parts. The critical kilogram will occur after about 150,000 miles of driving. To a trucker this is only one year; to a traveling salesman this may be two years of use; but to all automobile owners, 150,000 miles of use, assuming a lifetime average speed of 30 miles per hour, is a service life of only 5000 hours. Engines on leisure vehicles and boats typically have a wear life of only 500 hours. My personal experience with lawnmower engines suggests that their wear life is less than 200 hours. These are not long service lives compared to the 100,000-hour service life that can be expected from an average well-built electric motor.

Wear costs a lot of money, and the reduction of these costs is the goal of this book. It should be the goal of every person who designs a device that employs a *tribosystem,* a system involving surfaces in relative motion or a system involving unwanted mechanical action on a surface that causes material removal.

The plan of this book is to first define wear, to discuss modes of wear, to comment on what is known and what is not known about wear mechanisms, and then to present selection and specification information on the processes that can be used to reduce the damaging effects of wear: metallic platings, conversion coatings, diffusion treatments, surface hardening of quench-hardenable steels, thin-film coatings, high-energy surface modifications, as well as some processes that do not fit into any of these categories. The last part of this book will present information on the use of welding hardfacing processes for wear resistance. Hardfacing methods are given more emphasis than the other subjects because it is felt that they have the least acceptance by designers and engineers and they probably offer more potential benefits than the processes that are more widely used. The average designer probably does not know how to properly specify a hard weld deposit or a plasma coating on a surface that will be subject to wear. The hardfacing approach may have the most favorable cost/benefit ratio, but the designers default to a more comfortable, but less effective, solution because of their lack of knowledge about hardfacing. We hope to remedy this situation.

In the remainder of this chapter, we will discuss the history of surface engineering, the common ways to alter surfaces to improve their wear life, and finally the history of hardfacing to show how it developed and to point out how it should have been a part of materials engineering all along.

HISTORY OF SURFACE ENGINEERING

Tribology, the science of friction and wear, started to appear as an engineering discipline after World War II in the United Kingdom. A few people called themselves tribologists at a major wear conference in the United States in 1968, and since that time tribology labs have gradually started to appear in the United States. At the present time there are many tribology conferences in the United States, but there are still very few people who call themselves tribologists. This is not the case in other parts of the world where the field of tribology is much more established. Duncan Dowson has superbly documented the history of tribology in a book of the same name, and the use of surface treatments to reduce wear can be traced from his account of the chronology of tribology events.

In Figure 1–1, we annotated one of Dowson's time lines to show how important events in the development of tribosystems coincide with the development of surface treatments to reduce wear. The history of tribosystems is really a tracking of the discovery of new mechanical devices. The first tribosystem was probably the use of smooth logs for rollers to make the movement of large building stones easier. Then came the wheel and the need to have bearings to support wheel journals; then came ball and roller bearings, and so on. Each new device presented a new wear problem, and the users of these new devices had to deal with these problems.

Some time in the early years of ferrous metals, metalsmiths learned how to surface harden. Wrought irons were heated in bone or other carbonaceous materials, and carbon from these materials diffused into the iron and allowed quench hardening, the start of

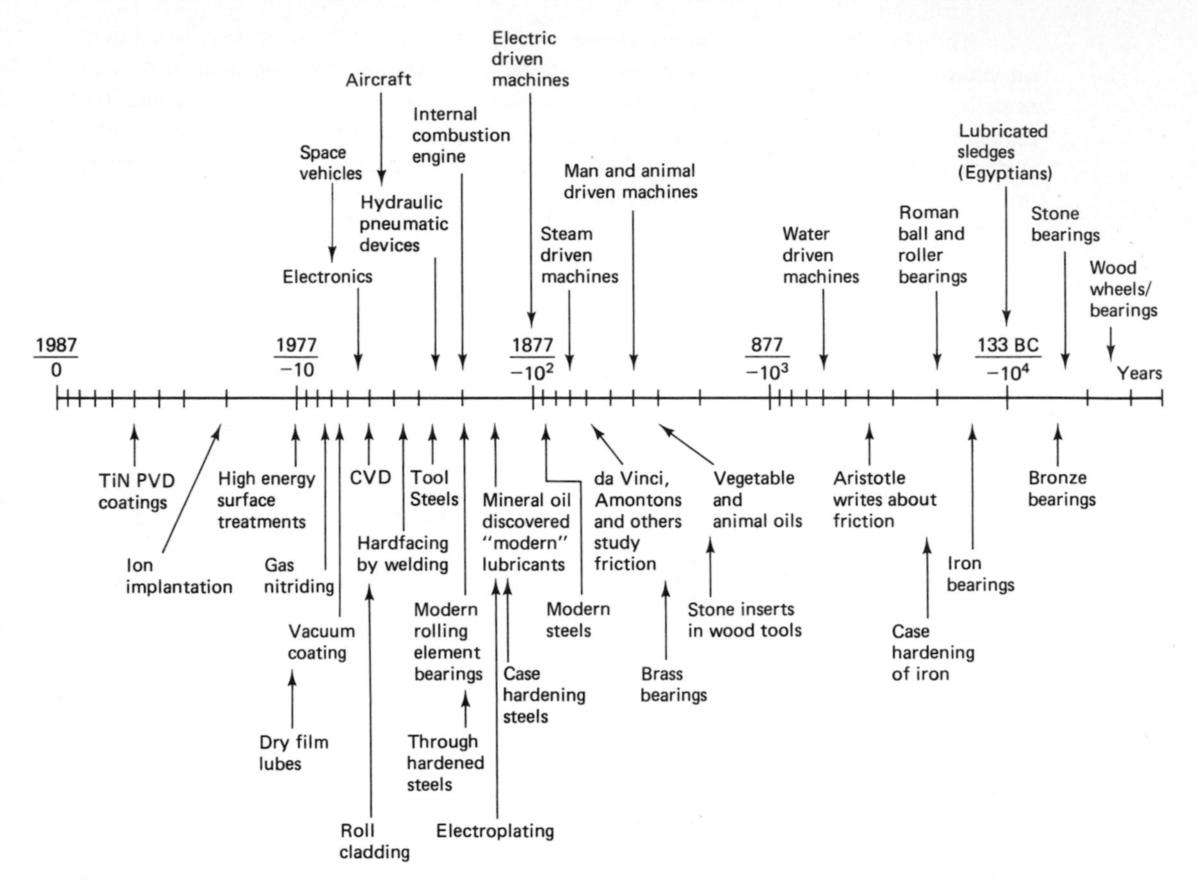

Figure 1–1 Chronology of tribosystems and techniques to deal with wear (after Dowson).

case hardening. An early form of hardfacing was practiced by Vikings; hard stones were embedded into the leading edges of wooden plows to resist soil abrasion.

Not much happened in the development of wear-resistant materials until the Industrial Revolution. The invention of the steam engine necessitated the development of lubricants and efficient bearings. Electric-driven devices required even better tribosystems, and finally the internal combustion engine required the types of bearings and lubricants that are similar in concept to the ones that are used today. Surface engineering made considerable strides in the early part of this century with the development of hard welding consumables, flame hardening, gas carburizing, nitriding, electroplating of hard deposits, roll cladding, chill casting, and many other processes. Most of the tools available to us in 1987 for treatment of surfaces have been developed since 1900.

This review in the chronology of surface treatments is intended to illustrate that surface treatments and coatings have been around for a very long time, and each new development in mechanisms creates a new set of tribological problems. Improving the properties of surfaces is the current trend for solving today's wear problems, rather than the development of new wear-resistant bulk materials. In the next generation of tribosystems, we will probably rely almost entirely on improvements to surfaces. Wear takes place at the surface and the near-surface, and it makes sense to concentrate on the surface rather than to make new bulk materials to resist wear.

TOOLS FOR DEALING WITH WEAR

Wear is defined as damage to a solid surface, most of the time in the form of gradual material removal from a surface, by the action of relative motion with a contacting substance or substances. There are many types of wear, and we will describe them in detail in Chapter 2, but there are only four main types of tribosystems that produce wear and about six basic things that are done to materials to reduce the effects of wear. The four basic tribosystems are as follows:

1. Relatively smooth solids sliding on other smooth solids
2. Hard, sharp substances sliding on softer surfaces
3. Fatigue of surfaces by repeated stressing (usually compressive)
4. Fluids with or without suspended solids in motion with respect to a solid surface

We will give names to these tribosystems and discuss them in more detail, but at this point it can be said that most wear damage occurs in tribosystems that contain one or more of these triboevents.

The six traditional techniques applied to materials to deal with wear produced in the preceding tribosystems are as follows:

1. Separate conforming surfaces with a lubricating film
2. Make the wearing surface hard
3. Make the wearing surface resistant to fracture
4. Make the eroding surface resistant to corrosion
5. Choose material couples that are resistant to interaction in sliding
6. Make the wearing surface fatigue resistant

Lubricants

A discussion of the various characteristics of lubricants is outside the scope of this book, but the role of all lubricants is simply to separate conforming surfaces so that they do not touch during relative sliding. If they do not touch, they will not damage each other. Any fluid that wets a surface will probably lubricate it. Effective long-term prevention of wear by lubricants requires a definite separating film. This is the hard part of lubricated tribosystems—arriving at a lubricant that will sustain a separating film under the service conditions. Solid film lubricants work the same as fluid lubricants. They separate conforming surfaces and the secret to their successful performance is to select a solid film that will stay in place under the service conditions. The bond of the solid film to the substrate is usually the key to keeping the film intact. Fluid lubricants have been developed to such a degree that there is probably a lubricant that will keep wear to a minimum in most continuous sliding tribosystems. Intermittent-motion sliding systems and oscillating systems are more troublesome because separating films are more difficult to form under these conditions. The thermal spray systems that are a part of hardfacing can be used to apply solid film coatings of intercalative materials such as graphite or molybdenum disulfide composites, but the entire field of solid film lubricants is not developed to the point where we can specify a particular lubricant on these types of systems and be sure that it will work. Each application probably requires some development work, and this is the extent to which we will discuss solid film lubricants.

Hardness

The modern definition of *hardness* is resistance to penetration, and this property is measured using machines that impress a ball or point into the surface and measure the degree of penetration under a prescribed load. There are far too many hardness measuring systems, but basically hardness is really measuring the flow stress of a material. Absolute hardness is the universal hardness measurement that theoretically will work on any type of material. It has units of kilograms per square millimeter (kg/mm^2) and is determined by impressing a penetrator into the surface of a material under some load; then the projected area of the penetration mark is measured and the load is divided by this area. In this book we will use absolute hardness as well as the other hardness scales. Appendix I contains some data sheets to compare these scales.

Early metalsmiths learned how to harden steels by carburizing them. They did not have penetration devices to rate the effectiveness of their hardening, but they could tell by scratching their work with stones if their hardening was effective. This technique was essentially all that was used until modern penetration techniques were developed in this century. The importance of hardness in resisting wear dates back to antiquity, and it remains probably the foremost technique for combatting many forms of wear.

Hardfacing is defined as applying with welding techniques a material with properties that are superior to those of the substrate to which it is applied. Most often the property that is made superior to that of the substrate is hardness. The bulk hardness of the weld deposit can be harder than that of the substrate or the weld may contain second phases or even hard particles that contribute to the effective hardness of the deposit. In the latter types of materials, the wear characteristics may not be proportional to the bulk hardness, but they will probably be proportional to the hardness of the matrix and a factor that considers the volume fraction of hard phases, the hardness of these hard phases, and the size and distribution of these phases.

Hard platings such as chromium have been used for many years to improve resistance to a number of forms of wear. Diffusion treatments such as carburizing and nitriding do the same, and more recently the high-energy surface treatments and vacuum coatings are used to produce a hard layer on the surfaces of materials or hard coatings. Thus, of all the various tools that are available to reduce wear, increasing the surface hardness is one of the most effective.

Fracture Resistance

Many wear processes involve fracture of material from a surface. In fact, some researchers are of the opinion that fracture is the main way that material is removed in wear processes. Whether this is true or not, there is no doubt that the fracture resistance of materials in some types of tribosystems is of major importance. For example, an examination of many knives and cutting devices that have allegedly wore out will show under microscopic examination that the cutting edge did not round over by gradual removal of material, but it became dull by chipping. Fragments of the cutting edge fractured off. When this happens, the solution is to use a material of surface treatment that is more resistant to fracture (chipping).

The usual measure of resistance to fracture is toughness. Toughness is measured by a variety of techniques, such as impact strength, fracture energy, and fracture mechanics tests. By definition, toughness is a measure of the energy required to fracture a given volume of material. The oldest way to measure this parameter is to perform a tensile test on a sample and then measure the area under the stress–strain curve. The larger the area is the greater the toughness. The most popular technique for measuring toughness is to fracture a specimen with a swinging pendulum and measure the energy absorbed by the specimen in the fracturing process. This same type of test is also done by dropping a weight on a notched specimen and measuring the energy absorbed. The newest technique to assess the fracture resistance of a material is to make a sharp fatigue crack in a specimen and then to measure the crack propagation rate in a tensile tester. The latter is a fracture mechanics test, and one fracture resistance parameter obtained by this test is the critical stress intensity factor. It is essentially the stress level at the tip of the fatigue crack at which the crack will readily propagate. The higher the critical stress intensity factor is the greater the fracture resistance.

When a wear process seems to take place primarily by a fracture mechanism, a material substitution can be made using a material that has better fracture resistance as measured by one of the preceding fracture or toughness parameters. There are even a number of wear models based on the fracture toughness parameters that are developed in the fracture mechanics types of tests. Such models are not widely used, but a person trying to solve wear problems should be cognizant of the role of toughness in wear. Sometimes hard materials can be too hard, and the tribosystem would be improved if one or both members had better toughness. Ceramic types of materials are particularly prone to wear by fracture, especially when they are sharpened to edges for cutting devices. In fact, their poor toughness will have to be overcome before these types of materials are widely used for tool materials. Hardfacing alloys are often tailored for improved fracture resistance by using hard phases in a tough matrix. The brittleness of hard chromium in some cases can limit its use in certain wear systems. The usual solution that is applied is to keep the chromium very thin. Many cemented carbides are too brittle to use for wear components, and new varieties with better fracture resistance are continually being developed. Thus, toughness and fracture resistance play a significant role in wear-resistant surfaces, and the use of hard materials to solve wear problems can lead to fracture problems that overwhelm the benefits of the hard surface. Material toughness must also be considered in material selection.

Corrosion Resistance

Some wear processes relate to the resistance of a material to reaction with its use environments. Liquid erosion, like that which occurs in pipes carrying fluids at high velocity, is resisted by surfaces with tenacious passive films. Fretting corrosion occurs when oscillatory motion between mating surfaces causes the faying surfaces to react with air or another environment. A material that does not oxidize or react in some other way can resist this form of wear. Cavitation resistance requires surfaces with extremely tenacious protective films. This is the way that the corrosion resistance of surfaces interfaces with wear processes. A part of material removal can be from mechanical action of the substance on the surface, and part of the material removal can be from dissolution of the surface from chemical effects.

The cobalt-base hardfacing alloys have corrosion characteristics that are superior to those of many 300 series stainless steels. They have significant utility in resisting liquid erosion, cavitation, slurry erosion, and some of the other corrosive forms of wear. The aluminum bronze surfacing alloys are used on marine propellers to prevent cavitation damage and to repair props that have suffered cavitation damage. The nickel-base and cobalt-base hardfacing alloys are widely used to resist chemical attack and abrasion from plastic resins in extrusion systems. Thus some hardfacing alloys have significant corrosion resistance and are adaptable to the corrosive forms of wear.

The competitive surface treatments for resisting the corrosive types of wear are relatively few. Selective hardening processes such as induction and flame hardening usually do nothing for corrosion resistance. Diffusion treatments fall into the same category. Thin-film coatings and platings have some utility, but these coatings are seldom continuous or pinhole free. They do not completely protect a surface.

The strongest competitors to hardfacing in resisting the corrosive forms of wear are stainless steels and other corrosion-resistant metals. They have the disadvantage, compared to hardfacing, of being expensive, and the whole structure of the component may have to be made from these expensive metals. With hardfacing, only the surfaces that are likely to be attacked need to be treated. Ceramic types of wear tiles can be used to prevent corrosive wear in the form of slurry erosion, but these types of materials are difficult to apply to complex surfaces.

Finally, rebuilding cements of the filled epoxy type can be used to resist, for example, slurry erosion in pumps. They can be quite competitive with hardfacing in many applications, but they have the disadvantage, compared to hardfacing, that if they are not properly applied they may not adhere in service. The same application risk can apply to some hardfacings, but there are techniques to determine bond integrity. It may not be possible to detect bond problems with repair cements prior to their being placed into service.

In summary, hardfacing processes are very competitive in resisting the forms of wear that involve corrosion—material removal by the conjoint action of wear and reaction of the worn surface with its environment.

Resistance to Adhesive Interactions

As researchers dating to antiquity have learned, it is not possible to eliminate friction between surfaces in relative motion. The same factors that caused friction contribute to wear between conforming solids. When two solids are in contact and relative motion occurs, there is a tendency for localized solid-state bonding in the real area of contact. We will discuss the prevailing theories on the mechanisms of this type of wear in Chapter 2, but there are no material combinations that are free from adhesion types of reactions. There is no agreed on theoretical way to predict the severity of adhesion tendencies between surfaces in relative motion. Attempts have been made to relate this phenomenon to surface energy considerations, melting point, and a number of other factors, but the present status is to empirically test relative compatibilities in laboratory wear tests. For example, if an application requires the best counterface to run against 416 stainless steel hardened to 34 HRC, some candidates could be recommended; but the best counterface determination would require testing at loads and speeds comparable to the intended application. Thus material selection for metal-to-metal and similar types of wear usually requires compatibility testing.

Relating this to hardfacing, laboratory compatibility tests have shown that some hardfacing alloys have the lowest tendencies for adhesive interactions; they have excellent metal-to-metal wear resistance. The cobalt-base and nickel/chromium/boron hardfacing alloys have been used for many years for applications involving metal-to-metal wear. They can be applied as fusion deposits or as plasma spray deposits. More recently, composite surfacings that contain a significant volume fraction of carbides are used for applications involving a combination of metal-to-metal and abrasive wear.

The competitors to hardfacing alloys in the area of adhesive wear are through-hardened tool steels, diffusion-hardened surfaces, selective-hardened alloy steels, and some platings. If a metal-to-metal couple is adequately lubricated in use, the hardened steels will probably perform as well as the hardfacing alloys. In boundary lubrication and in unlubricated wear, the hardfacing alloys often show superior wear characteristics. In addition, the cobalt-base alloys have corrosion characteristics that are usually superior to any of the hardened steels or surface treatments. They are even superior to the hardenable stainless steels in most corrosive environments. Thus, when a wear system involves adhesive interactions between surfaces, a number of hardfacing alloys can provide characteristics that are hard to match with through-hardened material and surface treatments.

Fatigue Resistance

Rolling element bearing, gears, cams, and similar power-transmission devices often wear by a mechanism of surface fatigue. Repeated point or line contact stresses can lead to subsurface cracks that eventually grow to produce surface pits and eventual failure of the device. Devices that are subject to repeated impacts suffer wear in a similar fashion. The materials that resist this type of wear usually have high compressive strength and a homogeneous microstructure. Hardfacing alloys are not widely used for rolling element bearings except when the rollers are too large to make by other techniques. For example, hardfacing alloys are commonly used to make the bearing surfaces for huge rotating kilns. These kilns may weigh 100 tons and they are supported on rollers. A huge ring on the kiln runs against these rollers. The wear ring on the kiln may be 15 feet in diameter, and the most cost-effective way to produce a wear-resistant surface on them is with hardfacing. Large thrust bearings for cranes and bridges may be similarly surfaced for resistance to fatigue types of loading.

Through-hardened steels, heavy case-hardened steels, and flame- or induction-hardened steels are the prime competitors to hardfacing in this area. Surface coatings, especially those with mechanical bonds to the surface, are not good candidates for this type of application. The hertzian contact stresses can lead to failure of the coating bond and subsequent spalling of the coating. Thus hardfacings are not widely used for average rolling element bearings, but fusion deposits of a variety of alloys are used for large components that are subject to rolling fatigue and cyclic impact.

The point to be made in this discussion is that wear-resistant surfaces require certain material properties, and these properties can be obtained by a variety of techniques. Sometimes a surface coating is the best way to get the desired surface properties; sometimes it is best to use a hardened steel; sometimes diffusion hardening treatments are the best choice; and sometimes hardfacing is the best process to use. As shown in Figure 1–2, there are different types of tribosystems, and the wear that occurs in these tribosystems can be addressed by coatings or by modifications to the substrate. The choice of the right method for a particular application is materials engineering, and we will concentrate on how to make this choice in the remainder of this book.

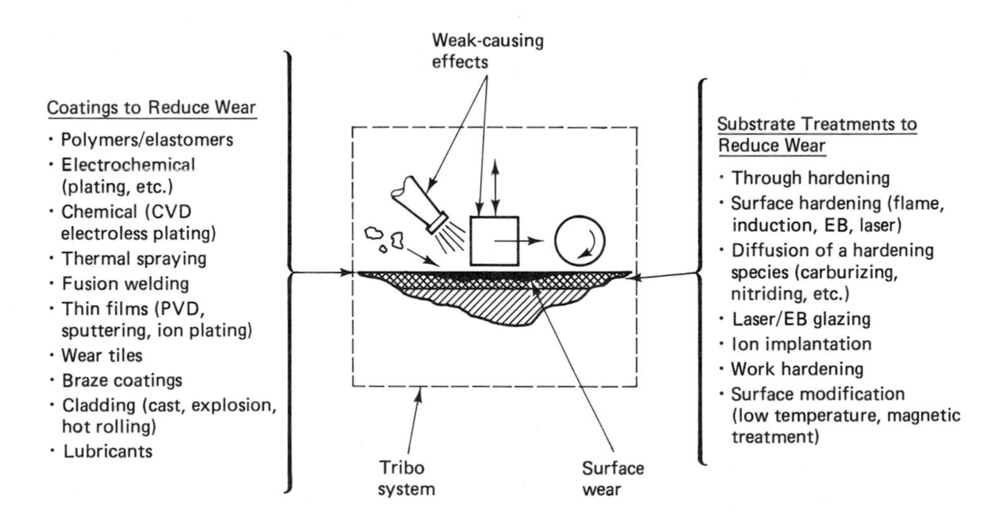

Figure 1–2 Spectrum of surface engineering processes.

CURRENT STATUS OF SURFACE ENGINEERING

In a 1987 conference at the Massachusetts Institute of Technology on surface science, one speaker made the statement that most of the important reactions that occur in materials take place within 100 angstroms of the surface. Obviously, this is an exaggeration, but he was trying to illustrate the importance of the nature of films and morphology of surfaces. A significant amount of research in chemistry is aimed at controlling catalysis, chemisorption, and other processes that take place at surfaces; in physics, thin-film coatings are being engineering to alter physical properties such as magnetism, surface conduction, and electrical characteristics. Most state-of-the-art electronic devices are made as composites of thin-film coatings. Electrical engineers rely heavily on these films for the development of mass memory devices, and material scientists are working toward the control of surfaces for corrosion control, crystal growth epitaxy, and modification of a number of use properties. Thus surface engineering is becoming more important as we evolve from users of massive, bulky devices to miniature, lightweight, fast-response, solid-state devices and devices made with a minimum of materials and a maximum of technology.

Hardfacing could be used as an important surface engineering tool. It is not used as widely as it could be because the details of its use are not as widely known, for example carburizing or plating. From the design standpoint, it is no more difficult to design for hardfacing than it is for plating and diffusion-hardening processes. There are guidelines for applying plating, such as chromium does not throw into holes, heavy deposits form nodules that are difficult to remove, hand masking is expensive and inaccurate, and plating of hardened steels can cause embrittlement of the part. There are things that have to be known about plating to properly specify it on a drawing. The designer must allow for the deposit thickness in part dimension and must specify coverage and postplating treatments. Hardfacing is no different. This process can be used by the average designer if the following things are known:

1. Application techniques
2. Available consumables
3. Properties of consumables
4. Design requirements
5. Where to get it done

Two basic categories of application processes are used in hardfacing: fusion welding and nonfusion. The important fusion welding processes are gas and arc welding. The important nonfusion processes are thermal spray techniques, such as plasma arc spraying and metallizing. The user of hardfacing must make the decision very early as to which of these categories will meet the application requirements. Fusion welding processes produce a deposit that can be almost any thickness, with a metallurgical bond to the substrate. The nonfusion processes have a mechanical bond to the substrate, very similar to that of plating. It is often the designer's responsibility to specify the hardfacing application process, so it is necessary to know something about each.

The basic categories of materials that can be applied by hardfacing processes are illustrated in Table 1–1. There are 11 of these categories, and many proprietary consumables do not fit into these categories. One problem to be addressed in using hardfacing is coping with alloy designation. The American Welding Society has developed an alloy-designation system that could be used instead of using trade names, but the acceptance of this system is not universal. Trade names for consumables are still widely used. Since there are about 1200 different trade names on the books, a user cannot be expected to be familiar with very many of them. The problem of specifying a consumable can be solved by becoming familiar with one or two consumables in each consumable category that is likely to be used. For example, scores of materials can be applied by plasma spray techniques, but if a very hard surface is desired to resist low-stress abrasion, one material, chromium oxide, works in about 90 percent of the applications. Similarly, the number of candidates for other applications can be reduced. The situation is really no different than it is with some metals and plastics. There are hundreds of copper alloys commercially available, but familiarity with about 10 alloys will suffice for 95 percent of applications. There are thousands of polymers on the market, but about 80 percent of the usage is made up of polymers from three families, polyolifins, polyvinylchlo-

TABLE 1–1 TYPES OF HARDFACINGS AVAILABLE FOR CONTROL OF SURFACE CHARACTERISTICS

Basic hardfacing material	Description	Usual application method
Tool steels	Properties comparable to common AISI tool steels	Arc welding
Martensitic steels	From moderate hardness buildup alloys to high hardness	Arc welding
High-chromium white irons	Martensitic matrix with wear-resistant alloy carbides	Arc welding
Austenitic irons	Austenitic matrix with wear-resistant alloy carbides	Arc welding
Manganese steels	Austenitic "soft" alloy that work hardens in use	Arc, gas welding
WC composites	Steel matrix containing WC particles of varying size	Arc, gas welding
Copper alloys	Brasses and bronzes	Arc, gas welding
Nickel alloys	Ni/Cr/B alloys with high hardness and hard microconstituents	Gas welding
Ceramics	Almost any type of oxide, carbide, or ceramic compound	Thermal spray
Cermets	Cobalt/WC	Thermal spray
Cobalt alloys	Co/Cr/W alloys with high hardness and hard microconstituents	Gas welding

ride, and polystyrene. Thus understanding hardfacing consumables is approached by limiting the consumables to a realistic repertoire.

The properties of hardfacing consumables vary from that of very soft tin alloys to some of the hardest ceramics. Just about any desired mechanical or physical property can be achieved because just about any material can be applied by hardfacing processes. We will concentrate on the wear characteristics of these materials in later discussions. For wear-resistance applications, the user must become familiar with the different types of wear and select a hardfacing or other process that is compatible with this form of wear. Parts do not just wear; they wear by a particular wear process. A hardfacing or other surface treatment is not wear resistant; it may be resistant to certain types of wear, but certainly not to all types of wear. Understanding wear processes is a key step in designing to minimize wear.

The design requirements for hardfacing are not many, but the successful use of the process requires that it be known how to allow for hardfacing deposits in part dimensions. One of the first requirements is to know the thickness limitations of the various processes. For example, a thickness of 0.1 in. (2500 μm) on a plasma spray deposit is completely out of line. It could possibly be done, but few companies could afford the cost. Normal thicknesses are between 3 and 10 mils (75 and 250 μm). On the other hand, a fusion weld deposit of 0.010 in. (250 μm) would be out of line. The thinnest deposit that could be made by gas welding may be 0.060 in. (1500 μm) and deposits are not uniform. It is also necessary to know how to undercut a part to accept a deposit, and on large earth-moving machinery and the like it may be necessary to specify even the arrangement of weld beads.

The last item in our list of essentials, where to get hardfacing done, pertains to the actual mechanism of getting hardfacing done once a hardfacing has been properly specified. Any welding shop can apply the consumables that are available for arc and gas welding. Most factories have a welding function. It may be a single welder who fixes broken equipment. This welder can apply hardfacing even though he or she has never heard of the term. A proper specification will tell the welder what consumable to use and where to apply it. If thermal spray equipment is not available in house, these hardfacing processes are readily available in job shops. In fact, there are usually enough job shops in a given location that there may be several shops willing to quote on a particular job. Thus any place that does some welding can do some hardfacing, and the more sophisticated techniques are readily available in specialty welding shops.

SUMMARY

In this discussion we have defined surface engineering and showed that it includes many processes, some old, some new. The common thread is that they are applied to a subsrate with service characteristics that are not up to our expectations; we want better use properties without using a more expensive substrate. In this book we will be dealing with how surfaces can be improved by surface engineering for applications that involve wear. Surface engineering also applies to altering physical and chemical properties of surfaces. The economic justification for improving wear resistance of surfaces should be obvious; there are only three reasons for a piece of machinery becoming disfunctional: (1) it wore out, (2) it corroded, or (3) some part fractured. We are addressing the first two.

We showed that surface engineering had its roots in antiquity, but present day tools for surface engineering have increased its potential to the point that it should be elevated to a standard engineering tool. It should be taught in schools just like mechanics, heat transfer, and the other applied sciences.

Our discussion of the tools for dealing with wear illustrated that dealing with wear is a manageable task. It is not as much of a science as one would like, but progress is being made. Math models and empirical relationships are still being sought for most types of wear, but there has been enough research into the basic types of wear and into the response of engineering materials in various tribosystems that sharing and application of this information will produce designs with superior service lives. For example, there are no agreed-to math models for the wear life of a high-speed-tool steel drill that is being used to drill holes in carbon steel, but it is known that a good PVD coating on the drill will let it drill twice as many holes as without the coating. Hardfacing and the other coatings that we will discuss have proven track records in many areas of wear. We will be basing our recommendations for their use on these proven successes.

Finally, we discussed the current status of surface engineering. Hardfacing is an under-utilized tool as are quite a number of the other surface engineering processes shown in Figure 1–2. Thin-film coatings are readily available, but few designers know how to properly specify them. The application of high-energy surface modification processes has been painfully slow. Ion implantation has been shown to have significant promise for many industrial applications, but once again it is not in the repertoire of surface treatments of the average designer. We could continue to cite examples of neglect in many of the other surface processes. What we are trying to illustrate is that surface engineering is the most promising way to obtain higher serviceability in components; the implementation of these processes in new designs is lagging because of unfamiliarity with these processes. Future successes in the marketplace will go to those companies who have designers proficient in surface engineering; the goal of this book is to encourage the use of surface engineering processes.

REFERENCES

DOWSON, D. *History of Tribology*. London: Longman Group Ltd., 1979.

STRATFORD, K. N., and others, eds. *Coatings and Surface Treatment for Corrosion and Wear Resistance*. Birmingham, U.K.: Institution of Corrosion Science and Technology, 1984.

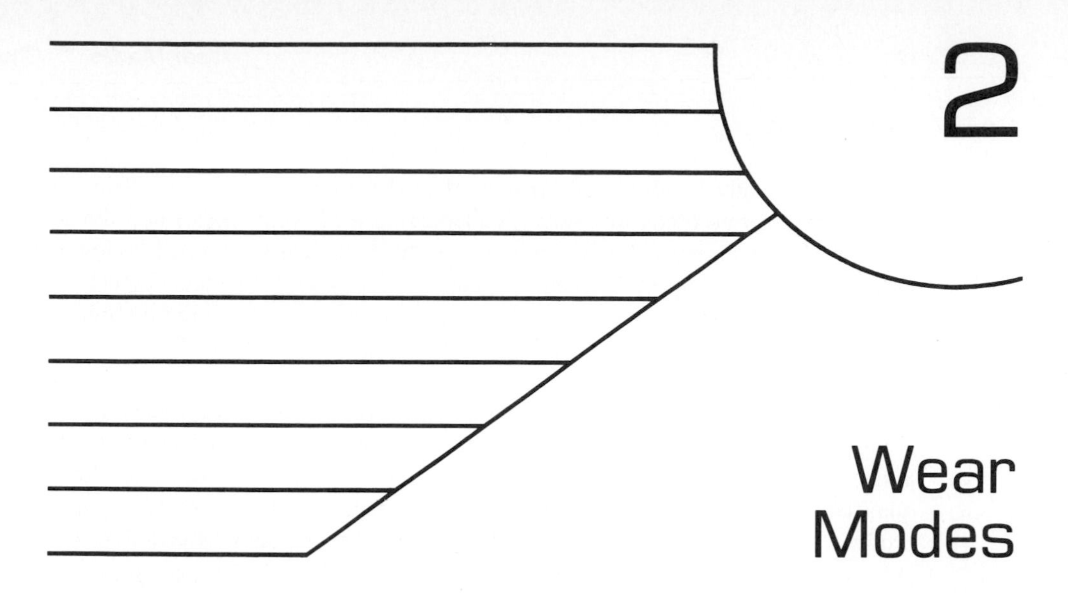

2 Wear Modes

Parts do not just wear out; they wear by various modes that are different in what they look like, different in cause, usually different in mechanism, and different in solution. Before one can address the solution of a wear problem, it is necessary to pinpoint the type or mode of wear that is causing the perceived problem. It is the purpose of this discussion to define the important wear modes, to show their manifestations, and to summarize what is currently known about the mechanisms that control these modes of wear.

DEFINITION OF WEAR

Let us start this discussion by redefining wear. *Wear* is damage to a solid surface, usually involving progressive loss of material, due to relative motion between that surface and a contacting substance or substances. One wear researcher, Zum Ghar, made this definition even simpler; he stated that there are only two ways of removing material from a surface: the material can be dissolved from a surface as in a chemical reaction or material can be fractured from a surface. As we will see when we discuss specific wear modes, most mechanisms involve some type of fracture of material from a surface. At this point, it is sufficient to keep in mind that wear is usually progressive and that it is a system effect; it is always caused by interaction between solids in relative motion or by the mechanical action of a fluid in motion with respect to a solid.

CATEGORIES OF WEAR

There are literally hundreds of terms used to describe various wear effects. The loose use of these terms tends to confuse understanding of wear modes and the solution of wear problems. Figure 2–1 is an attempt to reduce all wear processes into four categories based on commonality of mechanism. There are a limited number of specific modes in each category, and these specific modes will be described in detail in subsequent sections. At this point, we will simply describe the four categories.

Abrasion Wear produced by hard particles or protuberances forced against and moving along a solid surface. The term hard really means that the substance producing the abrasion is harder than the surface that is receiving the wear damage. An additional qualifier for abrasion processes is that the abradant usually has sharp angular edges to produce a cutting or shearing action on the solid that is being subjected to damage.

Erosion Progressive loss of original material from a solid surface due to the mechanical interaction between that surface and a fluid or impinging fluid stream. The fluids may

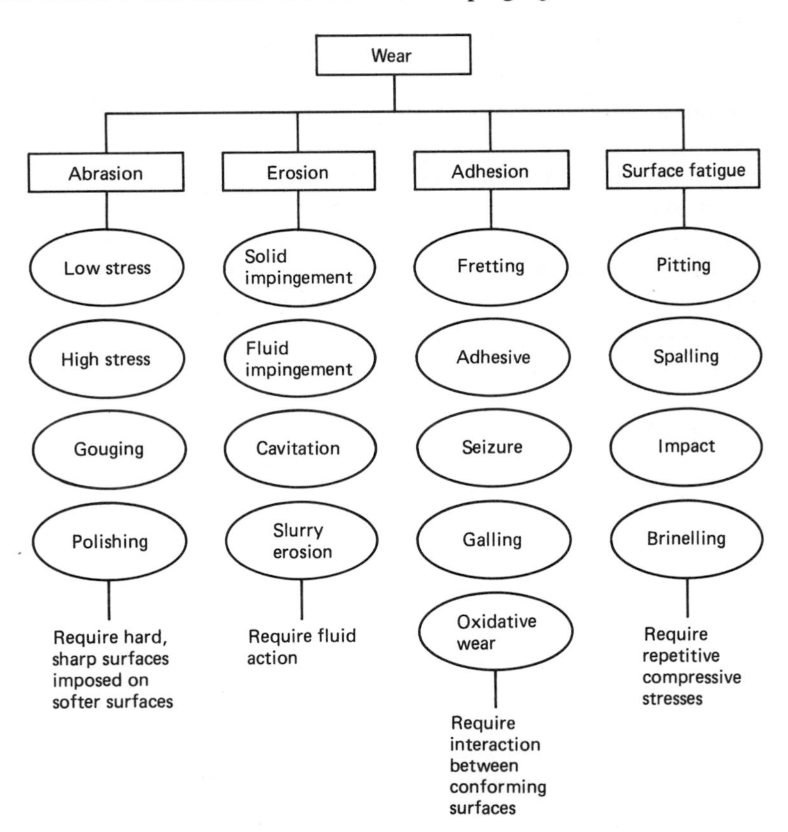

Figure 2–1 Basic categories of wear and modes of wear.

be multicomponent and they may contain solids. When the fluids are capable of chemically reacting with the solid surface in the wear system, material removal may be due to the concurrent processes of mechanical action and corrosion.

Adhesive Wear Progressive loss of material from solid surfaces in relative motion that is at least initiated by localized bonding between these surfaces. Whenever two solids experience relative motion, the friction force that tends to resist this motion occurs due to adhesion between the two surfaces. In adhesive wear, bonding between contacting surface features eventually results in fracturing of material from one or both of the interacting surfaces. If the bond to one surface is stronger than the bond to the other, transfer of material may occur. If surface features are fractured from both surfaces, wear debris is formed. The problem with using the term adhesive wear is that, after the initiation step, the surfaces are usually separated by wear particles, and adhesion between the members of the sliding couple may no longer occur. In many instances, the wear debris is abrasive and the mechanism of material removal becomes abrasion. For this reason, terms such as metal-to-metal wear can be more appropriate than the term adhesive wear.

Surface Fatigue Fracture of material from a solid surface caused by the cyclic stresses produced by repeated rolling or sliding on a surface. The most common example of a system that would be prone to this form of wear is a track that is subjected to repeated travel of a ball or roller. In this instance, material removal would occur by subsurface cracking. A pit will be produced when the subsurface crack progresses to the surface.

In the remainder of this section we will define the specific modes of wear that fit into these four basic categories. Examples will be presented of systems that are prone to these modes. The processes that are used to produce wear-resistant surfaces, hardfacing and other surface treatments covered in this book, exist to combat these types of wear. An understanding of wear modes is essential to selecting a surface treatment to resist wear. The purpose of this chapter is to introduce the "enemy."

LOW-STRESS ABRASION

Description: Another term commonly used for this wear mode is scratching abrasion. This is essentially the primary mechanism of damage. Surfaces subjected to low-stress abrasion show that material has been removed by hard, sharp particles or other hard, sharp surfaces plowing material out in furrows. Grinding with a surface grinder can be a controlled form of low-stress abrasion. The low-stress qualifier means that the abradant is imposed on the surface with relatively low normal forces. The criteria established by H. Avery for low-stress abrasion is that the forces must be low enough to prevent crushing of the abradant. Figure 2–2 shows a surface that was subjected to this form of wear.

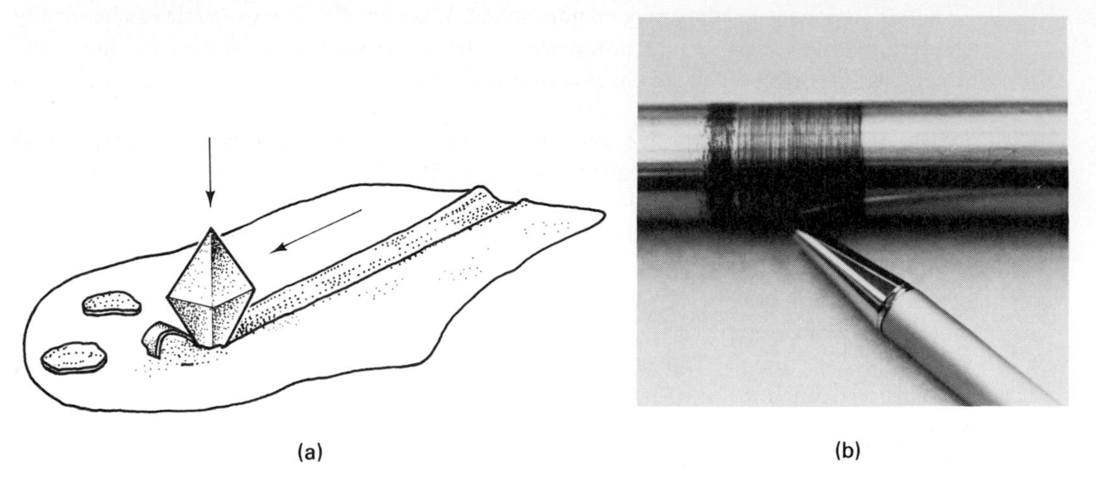

Figure 2–2 (a) Schematic of low-stress abrasion. (b) Low-stress abrasion of a shaft from hard contaminants in a plastic bushing.

The following are some additional characteristics of low-stress abrasion:

1. Abrasion rates increase with the sharpness of the abradant.
2. Abrasion rate decreases as the hardness of the surface subjected to abrasion increases.
3. Abrasion rate decreases as the size of the abradant decreases. Below a particle size of about 3 μm (0.0012 in.), scratching abrasion ceases; polishing wear commences and microchip formation no longer occurs.
4. Abrasion rate is directly proportional to the sliding distance and the load on the particles or protuberances.
5. Abrasion rates significantly increase when the hardness of the abradant is more than twice the hardness of the surface subjected to the abrasion.
6. In metals, microstructure (carbon content, carbides, hard phases, etc.) affects abrasion rates. The presence of hard microconstituents reduces wear.
7. Fixed abrasives produce more abrasion than the same abrasive used in a three-body, lapping mode. The abradant can roll in the wear interface, and microchip formation (scratching) is reduced.
8. Elastomers resist low-stress abrasion by elastically deforming when the sharp surfaces of the abradant are imposed on the surface. They often have better low-stress abrasion resistance than metals.
9. Ceramics and cermets can have effective resistance to low-stress abrasion if the ceramic is harder than the abrasive and if cermets have a significant volume fraction of a phase that is harder than the abrasive.

Examples: Particles sliding on chutes, packing running on shafting, plowing sandy soil, cutting materials containing abrasive substances, sliding systems in dirty environments, masonry-handling equipment, ash-handling equipment, mineral-handling equipment.

Applicable Hardfacings: Composite rods, high-carbon iron/chromium alloys, high speed steels, tool steels, nickel/chromium/boron alloys, cobalt-base alloys, plasma, and d-gun (detonation gun) applied ceramics and cermets.

Applicable Surface Treatments: Hard plating, case hardening, selective hardening, some heavy CVD treatments.

HIGH-STRESS ABRASION

Description: This form of abrasion is characterized by scratching, plastic deformation of surfaces, and pitting from impressed particles. Damage is almost always more severe than in low-stress abrasion. Plastic deformation of metal surfaces often occurs and deep scratches occur in the direction of motion. Most interactions of this form of abrasion with materials are the same as in low-stress abrasion (i.e., the role of hardness, the role of microstructure). Compressive strength of the surface subject to damage, however, is more important. To resist, this form of wear, it is desirable for the resisting surface to have a compressive strength greater than that of the abrasive. Figure 2–3 illustrates this form of wear.

Examples: Milling of minerals, rollers running over dirty tracks, earth-moving equipment, use of farm implements in hard soil, heavily loaded metal-to-metal sliding systems in dirty environments.

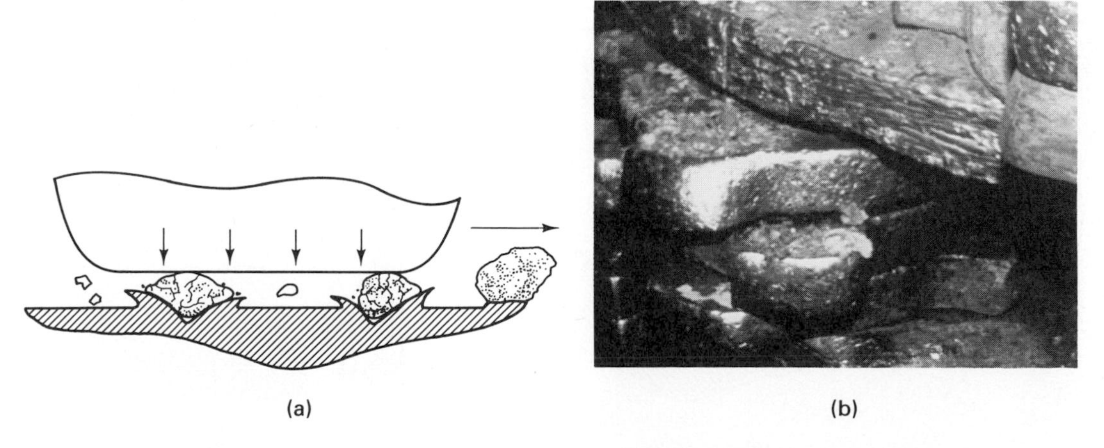

Figure 2–3 (a) Schematic of high-stress abrasion. (b) Star wheels on a refuse grinder that have been subjected to high-stress abrasion. Wheels are 2 in. (50 mm) thick.

Applicable Hardfacings: High-carbon iron/chromium alloys, high speed steels, tool steels, composites.

Applicable Surface Treatment: Heavy carburized cases, cemented carbide wear tiles, heavy flame hardening, cast white iron wear plates.

GOUGING ABRASION

Description: Gouging abrasion is material removal caused by the action of repetitive compressive loading of hard materials such as rocks against a softer surface, usually a metal. This form of abrasion is likely to be conjoint with both low- and high-stress abrasion. The most severe damage is caused by the gouging action; but if minerals are being handled, smaller particles will be present in the wear system that cause scratching abrasion, and the loading on some of the fine, abrasive particles may be high enough to cause high-stress abrasion.

The mechanism of gouging abrasion is plastic deformation coupled with chip removal. Both the deformation and chip removal are usually macroscopic in nature. Fatigue undoubtedly plays a role in this wear process. A single impact of a large rock on a metal surface will not usually produce material removal, only plastic deformation. The repetitive action of crushing rocks and the like produces material removal by fatigue action on dented surfaces; overlapping gouges can produce fracture of material in the web area between gouges. Figure 2–4 shows gouging damage on one of the cutters shown in Figure 2–3.

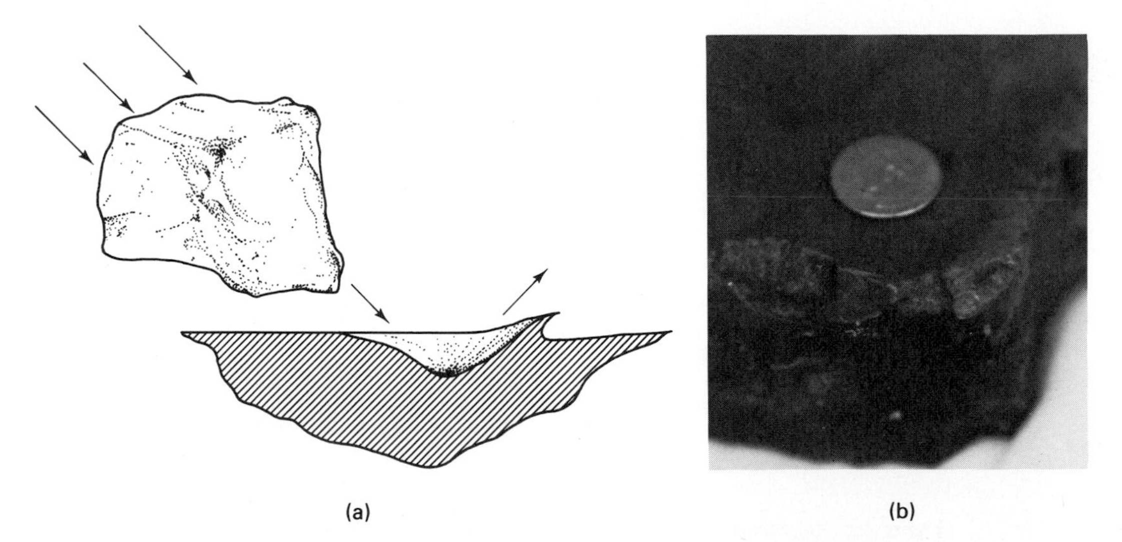

Figure 2–4 (a) Schematic of gouging abrasion. (b) Gouging damage caused by grinding of rocks.

Examples: Hammermill hammers, gyratory crusher parts, ball mill parts, jaw crushers, earth movers in rocky strata, agricultural equipment in rocky soil.

Applicable hardfacings: High-carbon iron/chromium alloys, manganese steels, composites.

Applicable Surface Treatments: High-strength-low-alloy, and manganese steel wear plates.

POLISHING WEAR

Description: Polishing wear is unintentional progressive removal of material from a surface by the action of rubbing from other solids under such conditions that material is removed without visible scratching, fracture, or plastic deformation of the surface. The dictionary definition of polishing is to smooth or brighten. Surfaces that have been subjected to polishing wear are usually smoothed or brightened, but this smoothing or brightening requires material removal and it can cause a loss of serviceability in some parts. Dimensions can be changed enough to make a part unusable. Figure 2–5 shows an example of a part worn by polishing.

The mechanism of polishing wear is not agreed on, but it was put in the category of abrasive wear because some investigators have documented that polishing produces microchip removal (like that in low-stress abrasion) when a surface is acted on by hard particles that are larger than about 3 μm (0.00012 in.). When surfaces are subjected to polishing by smaller particles, scratches and microchips are no longer observed. The mechanism of material removal under this latter condition is the part of polishing that

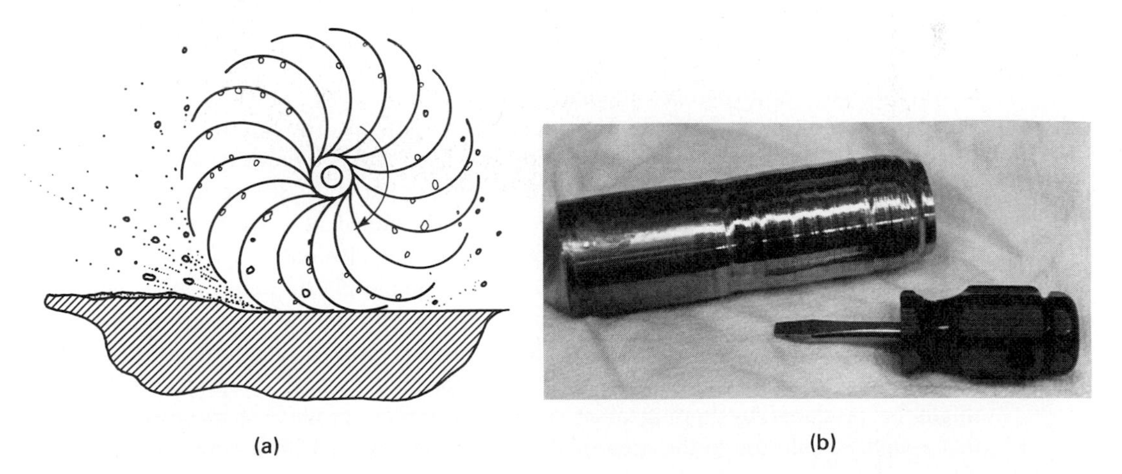

(a) (b)

Figure 2–5 (a) Polishing metal removal with a buffing wheel. Fine particles of abrasive are carried by the buff and each particle removes a minute quantity of material. (b) Pump sleeve polished by the action of inorganic fibers in packing that ran against the sleeve.

is under dispute. If chip removal does not occur, how is material removed? Rabinowicz proposes a mechanism of molecular removal. Atoms or molecules are individually removed from the surfaces by the rubbing counterface. There is no direct evidence to support this theory, but from the practical standpoint it is obvious that polishing can occur by repeated rubbing of almost anything. Steel handrails on well-used stairs often are polished from the rubbing of people's hands. In this instance, there is no abrasive on most people's hands, yet material is still removed. The likely mechanism in such cases may be something akin to Rabinowicz's proposal. In any case, polishing wear is a significant wear process with an uncertain mechanism: but in cases where hard substances are part of the wear system, the material rules of low-stress abrasion apply.

Examples: With the increased use of magnetic media for data collection, polishing wear has become very important. Recording disks and tapes contain abrasive minerals (usually iron oxides) less than 3 μm (0.00012 in) in diameter. Almost anything that they run against is subject to polishing wear. Other examples of this wear process are stair treads, lens-grinding equipment, conveying systems for filled plastics, fans exhausting fine particulates, mixing devices for grains and fine solids.

Applicable Hardfacings: High-carbon/chromium alloys, cobalt-base alloys, nickel/chromium/boron alloys, high speed steels and tool steels, thermal sprayed ceramics and cermets.

Applicable Surface Treatments: Hard platings, thin-film hard compounds, case hardening, selective hardening, wear tiles (carbides, ceramics), hard CVD treatments.

SOLID PARTICLE IMPINGEMENT

Description: Solid particle impingement is a form of erosion produced by a continuing succession of impacts from solid particles on a surface. The impacting particles are smaller than the solid subjected to the erosion, and if all the impacts are superimposed on the same spot, the term repeated impact is used. The most severe form of this wear process is a abrasive blasting nozzle continuously aimed at a single spot on a solid surface. Figure 2–6 shows an example of solid particle erosion.

The mechanism of surface damage can be simple plastic deformation, with each particle forming a small crater, or it can be microchip removal. Nonrepeating impacts usually produce craters with a depth that is usually only a small fraction (about 10 percent) of the equivalent diameter of the impinging particle. Repeated impacts, as might be produced by impingement of a sand blaster, produce a macroscopic crater that roughly conforms to the area of the impinging stream with a wavy contour and a depth that varies essentially with the velocity profile of the impinging stream. The center of the crater coincides with the center of the stream; the bottom is rounded and the crater slopes smoothly to the original surface at the outside diameter of the stream.

Figure 2–6 (a) Schematic of solid particle erosion. (b) Erosion of a wearback from a pipe carrying fly ash. Note hole and wavy surface.

The following are some of the factors that apply to solid particle impingement when repeated impact conditions exist:

1. Material removal rate (w) is proportional to particle type, size (m, mass), velocity (v), particle flux, and fluence, and it is inversely proportional to the hardness (H) of the impingement surface:

$$w = \frac{mv^2}{H}k$$

where k is a function of the impingement angle and the nature of the particles. (Sometimes an exponent is applied to the hardness term.)

2. Hard, sharp particles produce the highest material removal rates.
3. Ductile materials such as soft metals show a maximum erosion rate with an impingement angle in the range of 15 to 30 degrees.
4. Brittle materials such as ceramics and fully hardened tool steels show a maximum erosion rate with an impingement angle of about 90 degrees. Material removal is usually by microspalling/fracture.
5. The velocity exponent in the erosion rate equation can be as high as 6 depending on the system (medium, target material).
6. The degree of erosion damage decreases as the equivalent diameter of the particles decrease. Particles less than about 20 μm (0.0008 in) produce very low erosion rates and the material removal rate becomes almost nil when particles are less than 1 μm (0.00004 in.).
7. Under conditions of steady-state repeated impact with very hard particles (>1000 HV), the erosion rate of a hard tool steel (700 HV) will be comparable to the erosion rate of a soft steel (300 HV).

8. Hard microconstituents in a target material with a metal matrix have little effect on reducing erosion rates.
9. Hard ceramics and cermets are quite resistant to repeated impact erosion if the ceramic is harder than the impinging particles and if the impinging particles are small enough not to cause spalling.
10. Elastomers are very resistant to solid particle erosion except at normal incidence angles.

Examples: Fans in dirty environments, abrasive blasting, conveyance of solid particles in fluid streams, aircraft operating in sand or dirt, cyclone separators, air-blast comminution equipment, exhaust systems carrying particulates.

Applicable hardfacings: High-carbon iron/chromium alloys, high speed steels, thermal sprayed ceramics (only for fine particles, 50 μm (0.0002 in) and low-angle incidence).

Applicable Surface Treatments: Carbide and ceramic wear tiles.

CAVITATION

Description: Cavitation erosion is progressive loss of material from a solid due to the action of bubbles in a liquid collapsing near the solid surface. When bubbles collapse in a liquid, the liquid surrounding a bubble rushes in to fill the void. This action can create tiny liquid jets that can have sufficient mechanical action to cause material removal and/or plastic deformation of surfaces. In materials that rely on passive films for corrosion resistance, the mechanism of material removal may be repeated removal of these films coupled with corrosion of the surface that has become active by removal of its protective film. An example of cavitation erosion is shown in Figure 2–7.

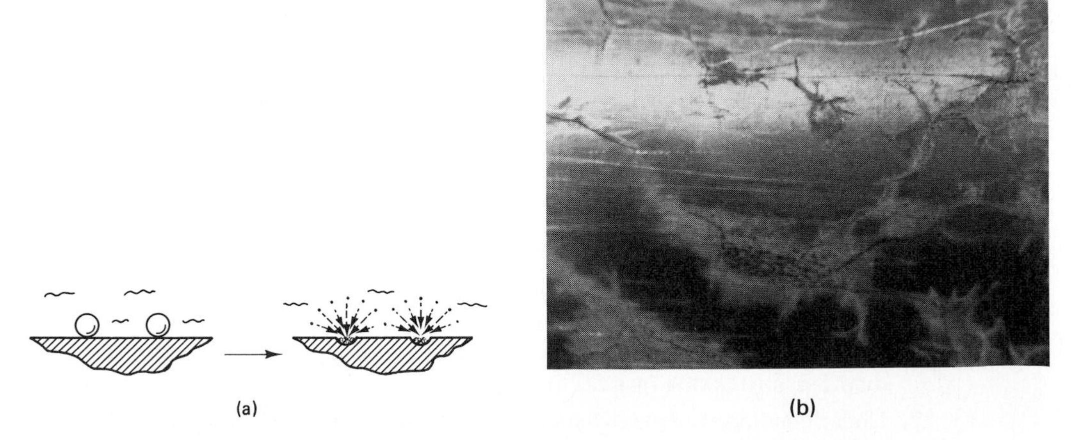

Figure 2–7 (a) Schematic of cavitation. (b) Cavitation on a stainless steel tank. An ultrasonic agitation device was attached to the other side of this section of the tank (magnification 1×).

Cavitation in liquids is governed by various laws of fluid mechanics, but cavitation is possible in any pump or propeller device. For cavitation to cause erosion damage, it is necessary that the bubbles collapse at or near the surface of the solid. Cavitation damage starts slowly; an incubation period usually exists. The rate of damage then increases until a steady-state erosion rate is achieved.

Metals that have a high tensile strength and a very tenacious passive film are the most resistant to cavitation. Plastics and ceramics that do not rely on passive films for corrosion resistance are in general resistant to this form of erosion.

Examples: Ship propellers, pipelines, pumps, mixing devices, ultrasonic agitators.

Applicable Hardfacings: Aluminum bronzes, cobalt-base alloys (fusion deposits), high work-hardening stainless steels.

Applicable Surface Treatments: Corrosion-resistant platings, ion-implanted surfaces of some metals, ceramic tiles, filled epoxy rebuilding cements.

SLURRY EROSION

Description: Slurry erosion is progressive loss of material from a solid surface by the action of a mixture of solid particles in a liquid (slurry) in motion with respect to the solid surface. If the solid surface is capable of corroding in the fluid portion of the slurry, the slurry erosion will contain a corrosion component. For example, a sand/water slurry in a steel pipe can cause damage by the abrasion of protective films; this in turn allows the steel to be corroded by the water. Without the abrasive in the slurry, the water would produce negligible damage. Conversely, a sand/oil slurry in the same pipe would be subject only to damage from the abrasive. The oil would not corrode the steel. Figure 2–8 shows an example of slurry erosion.

A slurry by definition is a physical mixture of solid particles and a liquid or multiphase fluid (gas and liquid). The particles must be in suspension in the liquid; thus a mixture of crusher run limestone in water would not be considered to be a slurry, nor would average drinking water, which may contain minute solid particles. Pumpable slurries that are prone to causing slurry erosion usually contain particles in the range of 10 μm to several millimeters (0.0004 to 0.1 in.), but particles can be larger if the fluid properties will allow the particles to stay in suspension.

The abrasivity of a slurry depends on the volume fraction of the solid phase, the size angularity, and hardness of the particles that comprise this phase, the slurry velocity, and the angle of impingement of the slurry on a surface or the load that is used to impose a slurry on a surface. The volume fraction of solid particles required to form a slurry is not governed by any rigid numbers, but most pumpable slurries contain at least 10 percent solids. The effect of particle concentration on erosion rate depends on the system; some slurries show a direct relationship between particle concentration and erosion, while other slurries can show exponential increases in erosion with a doubling

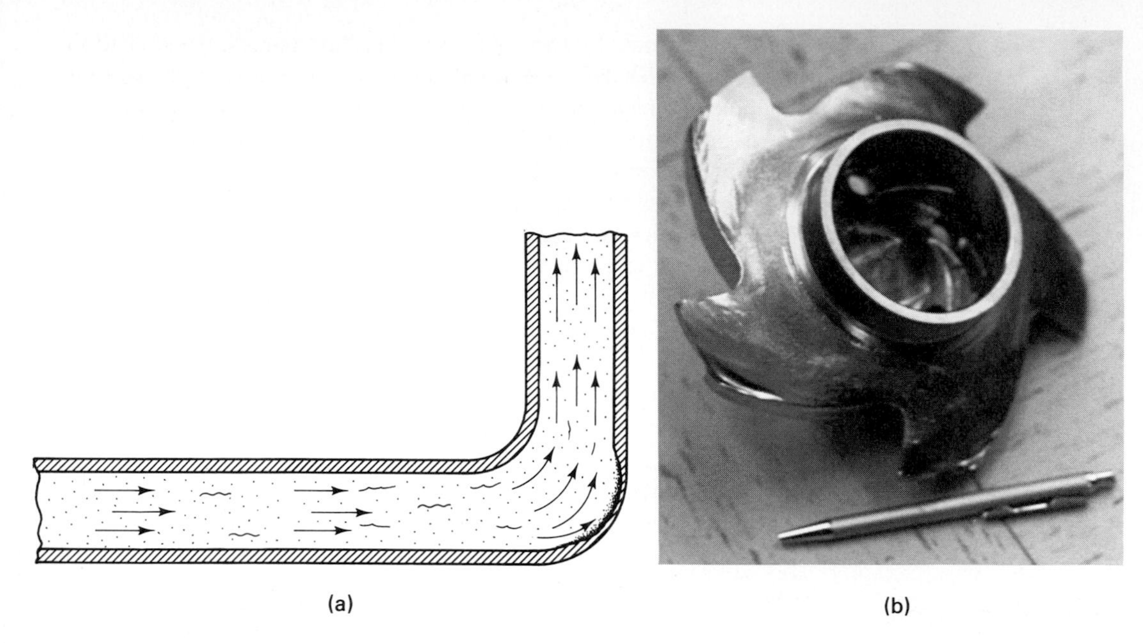

Figure 2–8 (a) Schematic of slurry erosion. (b) Pump impeller showing erosion damage from pumping a slurry of silica and water.

of particle loading. The liquid corrosivity and the nature of the substrate are the important factors in determining the effect of particle loading. In noncorrosive fluids, a direct correlation usually occurs.

A similar situation occurs concerning the effect of particle velocity and mass. The role of these factors is a function of the specific system. In noncorrosive fluids, the role of these factors is like their role in solid particle impingement.

If a slurry is comprised of hard particles in a corrosive fluid, the erosion resistance of a material will correlate with its corrosion resistance. Hard metals will offer improved erosion resistance in slurries with low corrosivity, but with high velocities even hard metals have limited erosion resistance. In such cases, it is typical to expect high erosion rates, and the effects of such high erosion rates are dealt with by system design. As an example, it is common to use replaceable wear backs on 90 degree elbows in high-velocity slurry pipelines.

Examples: Slurry pipelines, slurry pumps, mineral flotation systems, oil well down-hole equipment, mud pumps, well pumps, agitators, cement-handling equipment.

Applicable Hardfacings: High-carbon iron/chromium alloys, cobalt-base alloys, nickel/chromium/boron alloys, plasma sprayed ceramics and cermets (with small particles under mostly parallel flow conditions).

Applicable Surface Treatments: Hard platings, ceramic and carbide wear tiles, ceramic-filled repair cements, chromized steels, cast cylinder liners, plastic-lined pipe, basalt-lined pipe.

IMPINGEMENT EROSION

Description: Impingement simply means the act of hitting or striking. Impingement erosion is progressive material removal from a solid surface by the striking action of a fluid on a solid surface. The impinging fluid does not contain particles as was the case in slurry erosion. If the impinging fluid is in the form of droplets, the material damage mechanism is similar to that described in solid particle erosion. Material removal is a function of droplet size, velocity, angle, and fluence. Figure 2–9 shows an example of liquid impingement.

When the impinging fluid is in the form of a steady stream, the erosion process essentially becomes a corrosion process. If the impinging stream has sufficient energy, it can remove protective surface films, and material removal occurs by a mechanism of dissolution. The fluid can be a liquid or a gas or a combination thereof. The resistance of a material to this type of fluid erosion is a function of the corrosion resistance of the solid in the fluid, the fluid velocity, and other factors, such as temperature, that are known to have an effect on corrosion processes.

The ideal material for resisting the erosion action of impacting droplets is a resilient material that can elastically deform under the impact stress and recover without damage. Elastomers and natural rubbers often meet this criterion. The metals that resist droplet erosion are those that have tenacious passive films and high compressive strengths. Unfortunately, few metals meet these criteria. Hardened titanium and the more corrosion-resistant cobalt-base alloys can be effective in many systems.

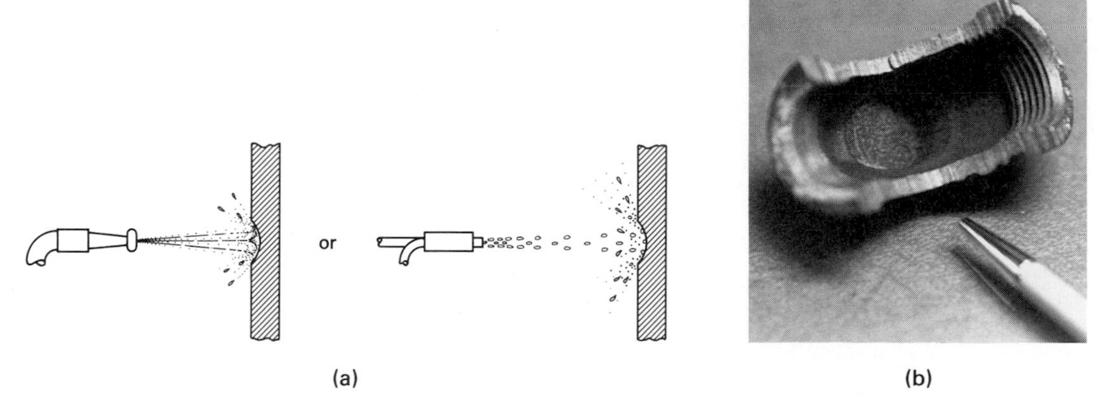

Figure 2–9 (a) Schematic of liquid impingement erosion. (b) Pipe elbow perforated by impingement from high-velocity fluid in a pipeline.

Examples: Rain impinging on aircraft, fans exhausting liquid droplets, tank inlet targets, liquid spray deflectors, hypersonic devices carrying liquids, steam turbine vanes.

Applicable Hardfacings: Cobalt-base alloys, thermal sprayed ceramics.

Applicable Surface Treatments: Ceramic and carbide wear tiles, elastomer and plastic-clad surfaces, filled repair cements, corrosion-resistant platings.

FRETTING WEAR

Description: Fretting is defined as oscillatory movement of small amplitude between two solid surfaces. It is usually tangential in nature, and typically it is unintended motion. It most often occurs between parts that are not supposed to experience relative motion. Fretting motion can lead to fretting wear or to fretting corrosion. Fretting wear initiates by local adhesion of the mating surfaces. The adhesion is usually in the form of a microscopic junction, and surface damage occurs when junctions grow and fracture from one of the mating surfaces. Once the surface finish of the mating solids is altered by these microscopic events, there is a progressive pitting and gnarling of the mating surfaces. The rate of material removal is usually low, and fretting wear is ignored in many mating material systems. Fretting damage reaches a state of importance with maintenance personnel when the pitting that is a part of the surface damage causes stress concentrations that often lead to fatigue failure. Fretting damage is illustrated in Figure 2–10.

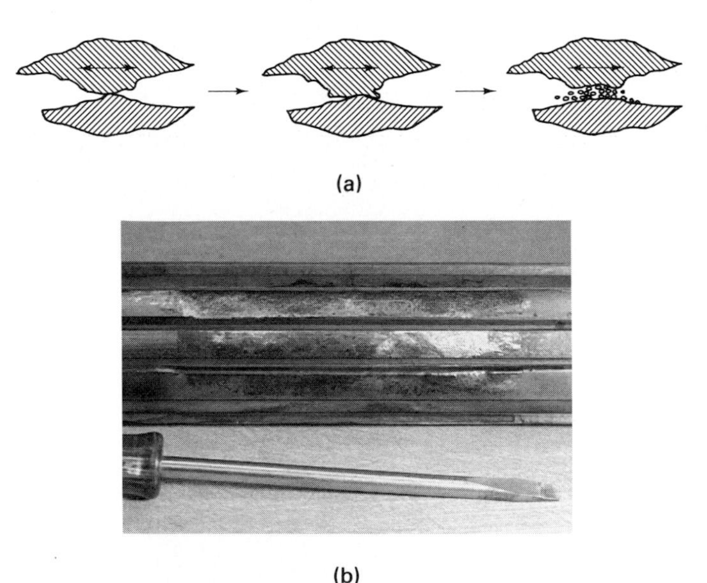

Figure 2–10 (a) Schematic of asperity interaction in fretting wear. (b) Fretting damage on a splined shaft from relative motion of a mating part.

If a mating couple experiences fretting motion in an environment that is capable of reacting with the fretted surfaces, fretting corrosion is the correct term for the mode of damage. Most fretting damage of industrial importance is of this nature. The adhered junctions fractured by the fatigue type of motion react with the ambient environment to produce an oxide or some other compound. The introduction of an oxide or some other reaction product between the faying surfaces usually increases the material removal rate by adding an abrasion component to the wear. Thus, fretting corrosion is the most common type of damage in a system that is subject to fretting motion.

The factors that affect fretting damage can be summarized as follows:

1. When ferrous metals are subjected to fretting corrosion in air, the oxidation of the surfaces produces an abrasive iron oxide (alpha) with a hardness of about 500 HV.
2. Fretting damage can occur at low and high loads, and the effect of increased load depends on the system. In some systems, the damage rate is increased; in others, it is reduced.
3. Fretting amplitudes are in the range of 10 to 300 μm. (0.0004 to 0.012 in.). The critical amplitude for accelerated damage is thought to be about 30 μm (.0012 in.). Damage in general increases with amplitude; at slip amplitudes in excess of 300 μm, reciprocating wear rather than fretting wear is said to exist.
4. Fretting damage occurs with equal propensity at low and high frequencies of oscillation. As in all wear processes, the damage increases with the total sliding distance. Frequency affects the wear rate in that the total sliding distance is larger in a system with high frequency than in the same system with low frequency. The total sliding distance statement also applies to amplitude.
5. The system environment determines if fretting wear or fretting corrosion will occur.
6. Fretting damage occurs in all material systems—metals, plastics, ceramics, and even elastomers. No material is immune.

As is implied by the preceding list of factors that affect fretting damage, damage can occur with many materials and under many conditions; the best way to minimize damage is to prevent the relative movement that is causing the fretting motion.

Examples: Gears and sheaves held on shafts with set screws, clamping faces of injection-molding cavities, bearings on shafts with a loose fit, drive-coupling components, metal parts vibrating in truck or rail transit.

Applicable Hardfacings: Cobalt-base alloys, plasma and d-gun sprayed carbides, and some ceramics.

Applicable Surface Treatments: Soft platings (gold, silver, etc.), lubricative thin-film coatings, bonded solid-film lubricant coatings.

ADHESIVE WEAR

Description: By definition, adhesive wear is due to localized bonding between contacting solid surfaces leading to material transfer between the two surfaces or loss from either surface. The origin of this form of wear is in reality the same phenomenon that is responsible for friction. If two solids are in intimate contact (at least locally), there is a tendency for these solids to bond together. The forces that produce this bonding tendency are the same as the forces that hold atoms and molecules together in solids. If two atoms are brought close to each other, there is a repulsion until the atoms get very close and the forces become an attraction. The exact mechanisms of these forces are rooted in quantum mechanics and are beyond our scope; but, in simple terms, if two solids are brought into such intimate contact that the surface atoms can interact, there will be a tendency for the surfaces to bond to each other.

In real life, all perfectly clean mating surfaces have areas on their surfaces that are in such intimate contact that there is a bonding tendency. Researchers in the field of adhesive wear model the adhesion as occurring at asperity contacts on the mating surfaces. No surface is perfectly flat or smooth, and when two solids are mated, the normal force from the loaded member is supported by the asperities and wave forms that contact each other. The real area of contact between surfaces is thought to be about 1/10,000 of the apparent area of contact. The asperities that carry the load can be under extremely high pressure. Plastic deformation of these asperities can occur, and some of the contacts will be in such intimate contact with the mating surface that there will be adhesion. Adhesive wear occurs when the junctions formed in this process plastically deform, transfer, or fracture.

Simple asperity adhesion and transfer occur very early in the wear process. Once debris is generated from microscopic asperity or junction fractures, the wear process is influenced and the predominating wear mode of wear may change. For example, when the debris oxidizes, abrasion can take over as the prevailing wear mode. Even if this does not occur, sliding surfaces can take on a scored appearance from junctions that have grown large enough that they form a protuberance on one of the surfaces that scores the other surface by plastic deformation. Adhesive wear is the initiation phase of almost all dry sliding wear systems, but as the wear progresses it becomes mixed mode. All sorts of things, such as scratching abrasion, occur. A surface subjected to adhesive wear can have many different appearances, so adhesive wear is usually used to describe the wear that occurs in unlubricated solid-to-solid sliding systems, even though adhesion was only the initiation phase. Figure 2–11 shows a gear that was not lubricated and thus the wear that occurred could be called adhesive; but the wear debris that was in the system undoubtedly caused some material removal by abrasion.

Since other forms of wear can be conjoint with adhesive wear, it is preferred to refer to this form of wear in other terms. If the solids in relative motion are metals, this form of wear can be referred to as metal-to-metal wear. Similar descriptions are appropriate for other mating couples (i.e., plastic-to-plastic, metal-to-plastic, etc.). These

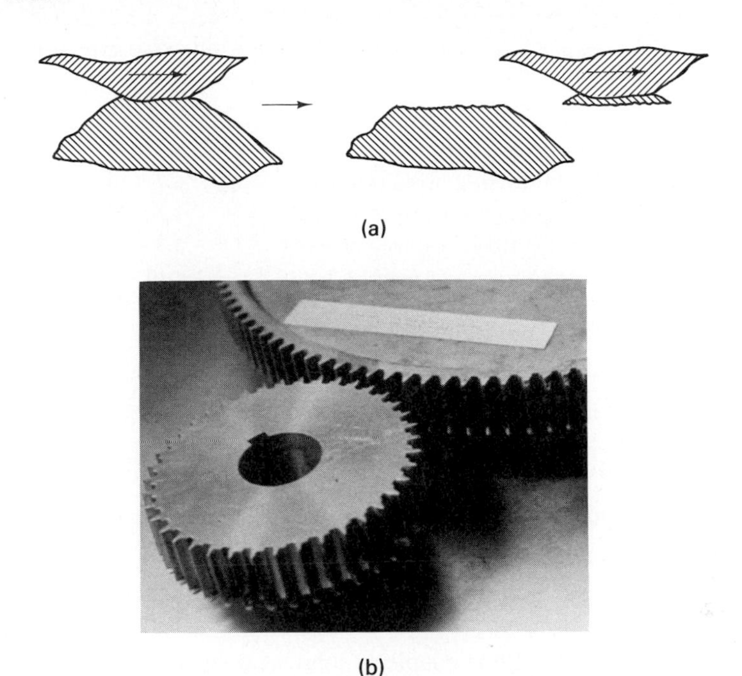

Figure 2–11 (a) Adhesion of asperities in adhesive wear. (b) Metal-to-metal wear on gear teeth (no lubrication).

latter descriptions eliminate having to deal with mechanisms. The basic Archard equation that has been applied to many wear processes essentially applies to adhesive wear:

$$W = K\frac{LP}{H}$$

where w = wear rate (volume/time or distance)

k = dimensionless wear coefficient for the system

L = sliding distance

P = normal force on the sliding member

H = penetration hardness of the softer member in the couple

Some wear coefficients for different sliding systems are given in Table 2–1. The lower the number, the lower the system wear rate is. These numbers differ with different researchers, and they should not be relied on to calculate the wear rate of any sliding system; but they serve to show the relative effect of lubrication and mixing materials in a sliding couple. Wear rates can be reduced by several orders of magnitude by lubrication; wear rate can be reduced by at least an order of magnitude in dry sliding systems by using dissimilar materials. In unlubricated systems, a metal on a nonmetal can reduce system wear by three orders of magnitude.

TABLE 2–1 WEAR COEFFICIENTS FOR SOME SLIDING SYSTEMS

	Like	Unlike
Unlubricated, metal to metal	1.2×10^{-3}	1.7×10^{-5}
Poorly lubricated, metal to metal	6.7×10^{-5}	3.3×10^{-5}
Average lubrication, metal to metal	3.3×10^{-6}	3.3×10^{-6}
Excellent lubrication, metal to metal	3.3×10^{-7}	3.3×10^{-7}
For a nonmetal on a metal: 1.7×10^{-6} to 3.3×10^{-7}		

After Rabinowicz.

Some guidelines relating to the various forms of ''adhesive'' wear are as follows:

METAL-TO-METAL WEAR

1. Avoid pure metal couples and couples of low work hardening single-phase alloys.
2. Hard metal couples produce less system wear than a soft/hard metal couple.
3. All metal-to-metal couples will be subject to adhesive wear when not lubricated; lubricate wherever possible.
4. Only use a hard/soft couple in lubricated systems where it is necessary to have a soft counterface for conformability or for embeddability. Hard/hard couples produce the lowest wear. Soft babbitt and bronze bushing are used for conformability and embeddability.

CERAMIC-TO-CERAMIC WEAR

1. Avoid self-mating.
2. Lubricate or run immersed in liquid.
3. Test for compatibility of mixed wear couples.

CERAMIC-TO-METAL WEAR

1. Use low surface roughness on the ceramic member to prevent abrasion of the metal.
2. Lubricate where possible.
3. Use hard metals (the metal will almost always sustain most of the wear).

PLASTIC-TO-METAL WEAR

1. Lubricate if possible with a fluid that will not react with either solid (some plastics are affected by oils).
2. Avoid plastics with hard inorganic fillers such as glass; they cause abrasion of the metal member.
3. If possible, use fully hardened metals or ceramics as the counterface.
4. In unlubricated systems, use a self-lubricating plastic (PTFE filled, or similar).

Examples: Face seals, O-ring seals, gears, cams, slides, ways, pistons, screws, bushings, drive chains, sprockets, actuators.

Applicable hardfacings: Cobalt-base alloys, nickel/chromium/boron alloys, thermal sprayed metals, ceramics and cermets, tool steels, iron/chromium alloys.

Applicable Surface Treatments: Hard metal platings, case hardening (all types), selective hardening, thin-film coatings, diffusion treatments that produce hard surfaces, carbide surfaces.

SEIZURE

Description: Seizure is the stopping of relative motion as the result of interfacial friction. Local solid-state welding may be a part of the mechanism of seizure. This wear process does not necessarily require progressive loss of material. Seizure can occur with no damage to either surface. As an example, an auto engine that has lost its cooling system can overheat, causing seizure by thermal expansion of the pistons in the cylinders. When allowed to cool, the engine could still be serviceable. Unfortunately, seizure usually involves local welding and significant damage to the mating surfaces. Figure 2–12 shows a protuberance on a spline. This protuberance produced seizure with the mating part. The mating part would not move even with 50 tons (45,000 kg) of encouragement. The protuberance was caused by galling.

The most common cause of seizure is loss of running clearance in a confined sliding system. The obvious solution to loss of running clearance is to calculate anticipated thermal effects on running clearances and to allow for them in the selection of a running clearance. A second common cause of seizure in these same types of systems is loss of running clearance due to buildup of wear debris. This can be dealt with by considering

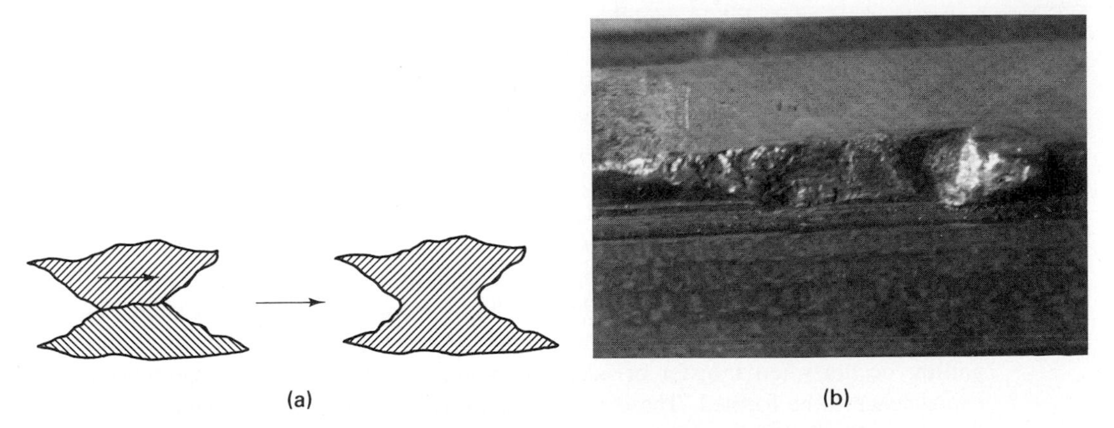

Figure 2–12 (a) Schematic of junction bonding to produce seizure. (b) Seizure of this spline was caused by galling excrescence (10×).

wear particle size in selecting running clearances. A useful rule of thumb is that the softer the materials, the larger the running clearance required. As an example, the suggested running clearance on a 1-in. (25-mm) inside diameter oil-impregnated bronze bushing is about 0.001 in. (25 μm). If the same bushing were made from plastic, the running clearance would be about 0.010 in. (0.25 mm); for a lubricated hard steel/hard steel couple, the clearance could be as low as 0.0002 in. (5 μm). Running clearances like these increase with diameter, and bushing manufacturers usually provide these data.

There are no miracle materials that resist seizure. Lubrication and proper clearances are the best solution, but some material combinations tend to have high friction; the notable ones are pure metals and most single-phase alloys. These should be avoided if seizure is to be avoided.

Examples: Hinge pins, pistons in cylinders, valves, seldom used sliding mechanisms, unlubricated sliding systems.

Applicable Hardfacings: Cobalt-base alloys, nickel/chromium/boron alloys, thermal sprayed ceramics, cermets, and tool steels.

Applicable Surface Treatments: Lubricating thin-film coatings, plating/lube co-deposits, case hardening, selective hardening.

GALLING

Description: The term galling does not have an agreed on definition. In Europe, the wear community uses the term scuffing in its place. The Organisation for Economic Co-operation and Development, Paris (OECD) defines scuffing as localized damage caused by the occurrence of solid-phase welding between sliding surfaces, without local melting. In Russia, galling is characterized by the formation of excrescences on one or both members of a sliding couple. An excrescence is metal flowed up from a surface. Galling by our definition is damage to one or both members in a solid-to-solid sliding system caused by macroscopic plastic deformation of the apparent area of contact, leading to the formation of surface excrescences that interfere with sliding. Figure 2–13 shows galling damage on a conforming couple that experienced only one rotation under a normal force of about 3000 lb (1360 kg).

The formation of excrescences is a key part of the damage mechanism. Burnishing of wear surfaces in their area of contact is normal and usually expected. It is not a damage process; in fact, soft bearing metals such as babbitt are used because of their ability to deform and conform to a harder sliding member. However, if an excrescence was formed in a babbitted surface, failure of the bearing would eventually result. Thus galling occurs when transfer or adhesion between surfaces causes a protuberance or excrescence to be formed. These surface perturbations can lead to seizure. If a couple is prone to galling, damage may occur after only one traverse of one surface on the other. In plain bearings, galling may occur when lubrication is interrupted.

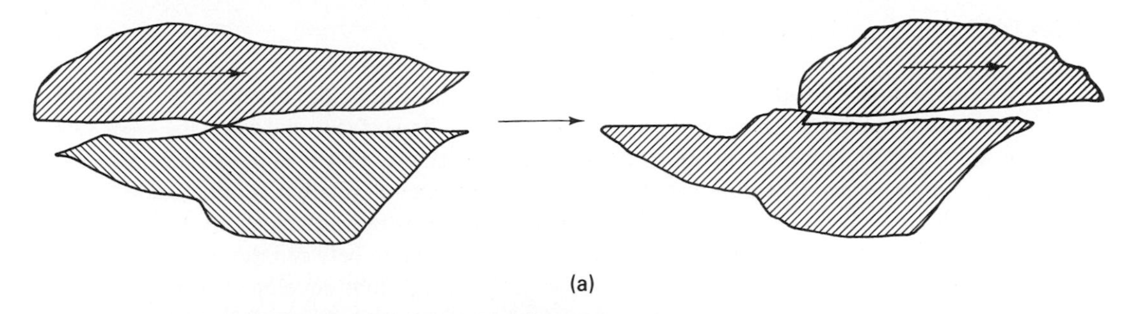

(a)

(b)

Figure 2–13 (a) Schematic of formation of an excrescence in galling. (b) Galling damage on the polished conforming surfaces of special nuts after one use.

Galling resistance is a function of the mating couple. A material does not have a galling resistance; a mating couple has a particular galling resistance. In metals, hard/hard couples often have good galling resistance. Ceramic/ceramic couples can be very prone to galling and compatibility tests are usually required. Plastic/metal couples are usually galling resistant. The galling resistance of a particular couple is not intuitively known, nor do any fundamental laws allow prediction of galling tendencies. Mating couple tests are usually required.

Examples: Fitted sliding members, plug valves, gate valves, slides, ways, gibs, heavily loaded unlubricated sliding members.

Applicable Hardfacings: Cobalt-base alloys, nickel/chromium/boron alloys, tool steels, high-hardness iron/chromium alloys.

Applicable Surface Treatments: Chromium plating, soft metal platings, hardcoating, case hardening, selective hardening, lubricating thin-film coatings, ceramic and carbide wear plates, cast hard cylinder linings.

OXIDATIVE WEAR

Description: Oxidative wear is a wear process in which sliding surfaces react with their environment to form oxide films that separate the surfaces and keep the wear rate low. This form of wear exists in lightly loaded systems, and it is sometimes referred to as mild wear. If unlubricated steels are experiencing relative motion under light load, both surfaces will develop what appears to be a rusty surface; a powder debris that also appears to be fine rust will be found to accumulate between the faying surfaces. This is oxidative wear. Figure 2–14 shows a chain link that developed oxidative wear where it connected to another link. The wear rate was very low and the wear surface was covered with red debris.

Oxidation of sliding surfaces initiates when the microscopic asperities and adhesion junctions are fractured from the surfaces. The point from which an asperity is fractured oxidizes, as does the particle that has fractured from the surface. The energy for the oxidation is supplied by the mechanical forces that are producing the relative motion of the surfaces. Oxidative wear starts with adhesion, and the rate of material removal

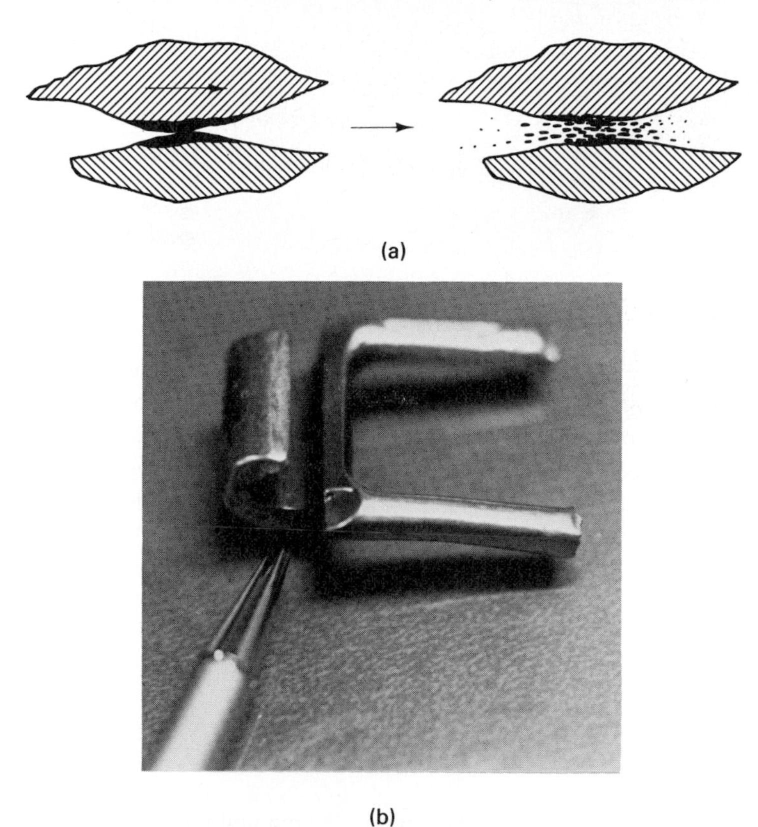

Figure 2–14 (a) Schematic of oxidative wear. (b) Oxidative wear occurred from low-speed moving with a mating chain link (dark area).

is relatively high. As oxide films form, the wear rate decreases, and eventually a steady-state wear rate is established. The oxide debris that has accumulated in the sliding interface essentially lubricates the system and keeps the wear rate low.

This form of wear is often tolerated because the material removal rates are small, but attrition is continual and serviceability will eventually be lost. Probably the most common example of oxidative wear is the hinge pin in a household door. Most homeowners ignore periodic lubrication of hinges and, since the loads and sliding speeds are low, the hinge and hinge pin are prime candidates for oxidative wear. After maybe five years, the friction may increase to the point where you are forced to take some action and put in a few drops of oil. In industrial systems, the significance of this form of wear is loss of dimension.

Almost all metals are susceptible to this form of wear. The obvious solution to oxidative wear is lubrication. When this cannot be done, hard/hard mating couples have the best resistance.

Examples: Dry sliding systems on gages and fixtures, hinge assemblies, conveyors, sliding parts on machine tools, sliding systems in furnaces and ovens, hard to lubricate components on farm implements, and earth-moving machinery.

Applicable Hardfacings: Cobalt-base alloys, nickel/chromium/boron alloys, thermal sprayed ceramics and cermets, high-carbon iron/chromium alloys, tool steels.

Applicable Surface Treatments: Soft metal platings, lubricating thin-film coatings, case-hardened surfaces, selective-hardened surfaces.

PITTING WEAR

Description: Pitting can occur in a number of wear processes. We have already mentioned that pitting can occur in cavitation and fretting. Pitting as addressed here is the removal or displacement of material by a fatigue action to form cavities in a surface. Pitting as a part of surface fatigue frequently occurs in rolling element bearings, gears, worm wheels, and cam paths. It is the prime manifestation of surface fatigue. Repeated stresses due to sliding or rolling cause subsurface cracks that grow to produce a fracture of a local area of the surface. Removal of the fractured fragment produces a pit. Figure 2–15 shows an example of pitting wear.

Pitting wear can be minimized by lowering the load, by using materials with high shear and compressive strengths, and by using homogeneous materials that do not have structure-induced stress concentrations. Materials used in rolling element bearings usually have high hardness, high compressive strength, and a structure devoid of massive microconstituents (carbides). AISI 52100 steel is the standard material for most rolling element bearings in the United States. It is hardened to about 60 to 62 HRC, and it has a microstructure of tempered martensite with a fine dispersion of submicron-size carbides. Many hardfacing alloys contain massive carbides for resistance to abrasion;

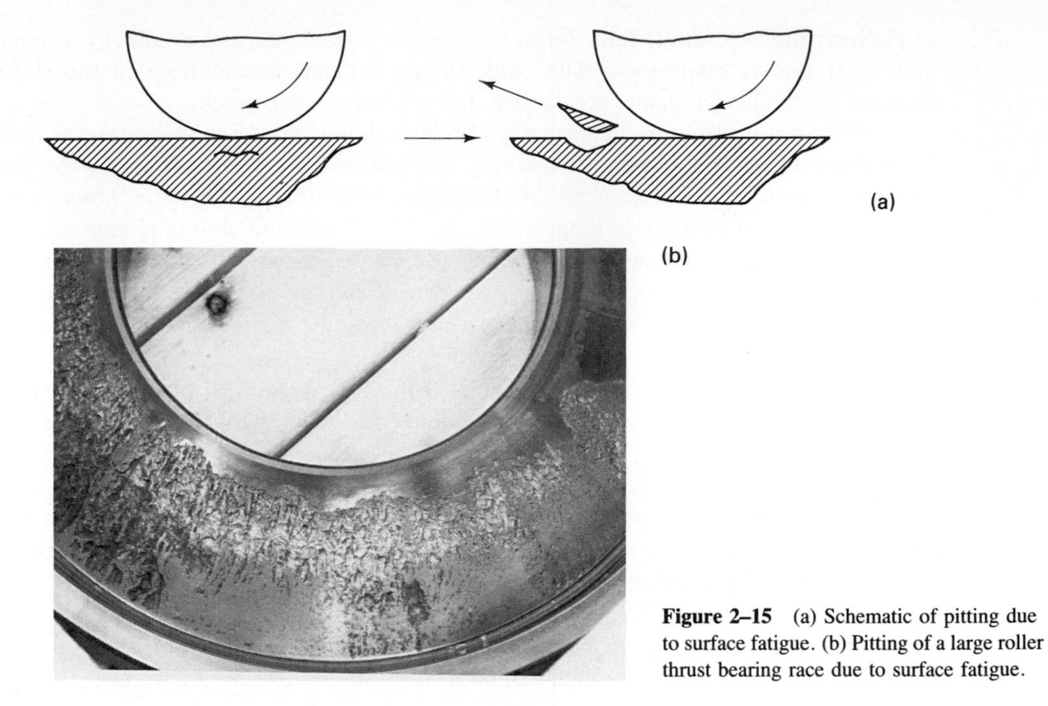

Figure 2–15 (a) Schematic of pitting due to surface fatigue. (b) Pitting of a large roller thrust bearing race due to surface fatigue.

this makes them unsuitable for surface fatigue applications. The massive carbides are stress concentrations that promote the formation of subsurface cracks.

Examples: Cam paths, gear teeth, rails and metal tires, rolling element raceways, sprockets.

Applicable Hardfacings: Lower hardness cobalt-base alloys, high speed steels, lower-hardness nickel/chromium/boron alloys, austenitic manganese steels.

Applicable Surface Treatments: Carburizing (heavy), selective hardening.

SPALLING

Description: In this form of wear, particles fracture from a surface in the form of flakes. Spalling arises from the same mechanisms as pitting. It is the result of surface fatigue, and it occurs in the same types of systems. Occasionally, wear surfaces that are subjected to rolling elements are electroplated for wear resistance. Such systems are very prone to spalling. The platings adhere to the surface with a mechanical bond and, when the shear stresses that result from hertzian loading act on the plating/basis metal interface, spalling can result. Case-hardened surfaces and thermal sprayed coatings similarly can be susceptible to spalling. Figure 2–16 shows spalling on a chromium-

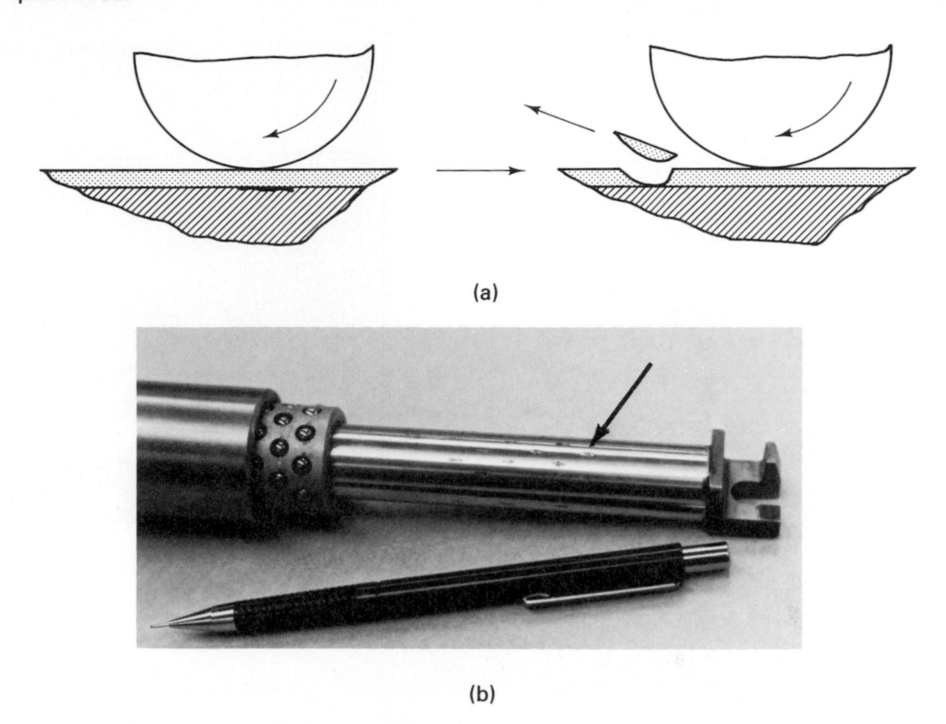

Figure 2–16 (a) Spalling of a coating from surface fatigue. (b) Spalling of plating due to surface fatigue (oscillatory movement of about 5 mm).

plated shaft that was used in a ball bushing. Thin, hard wear coatings on soft substrates are particularly prone to spalling and should be avoided in systems where hertzian or impact loading is expected.

Examples: Coated cams and gears, thin surface hardening on cams and gears, rails, plated mechanical stops, thin platings on reciprocating systems.

Applicable Hardfacings: Cobalt-base alloys, nickel/chromium/boron alloys, high speed steels, high-carbon iron/chromium alloys.

Applicable Surface Treatment: Heavy carburizing, selective hardening.

IMPACT WEAR

Description: Impact wear is material damage and removal by the action of repetitive impacting of two solid surfaces. A simple example of impact wear is the damage that occurs on the head of a high-speed riveting hammer. The hammer plastically deforms the rivet head, and there is no concern for the wear that occurs on the rivet; but the

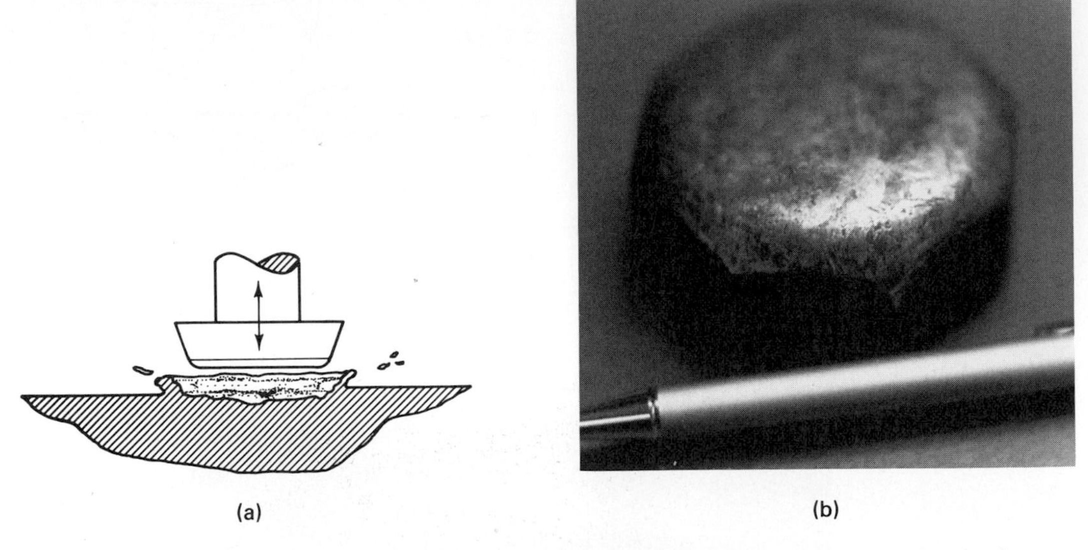

Figure 2–17 (a) Schematic of impact wear. (b) Impact wear on the striking face of a battering tool.

hammer suffers material attrition that eventually necessitates its replacement. Figure 2–17 shows mild impact wear on the head of a battering tool.

The mechanism of damage that occurs in impact wear systems somewhat depends on the nature of the impacting surfaces. For example, the damage on the end of a cold chisel from repetitive impacts of a hammer is usually plastic deformation. The striking end of the chisel eventually assumes a mushroom shape. When hard-hard combinations are subjected to repetitive impacts, the damage will be either a surface fatigue fracture, such as pitting or spalling, or fretting damage. The tangential motion required for fretting damage arises from elastic or plastic deformation of the solid surfaces from the imposed impact stresses. Surface damage from impact wear depends on the shape of the impacting surfaces, the energy of the impact, and the ability of the impacted surfaces to dissipate the absorbed energy. The metals that resist this form of wear have high hardness and good toughness. Surface hardening is often used to achieve this combination of hardness and toughness. A good carpenter's hammer will probably be made from a medium carbon steel, and the end will be induction hardened to a hardness of about 52 HRC to a depth of about 0.1 in. (2.5 mm). On the other end of the spectrum, impact wear can be minimized by making one of the impacting surfaces an elastomer that elastically deforms with the impact. The material removal is essentially eliminated on the rigid member of the system, and the damage is concentrated on the elastomer.

Examples: Hammer heads, riveting tools, pneumatic drills, mechanism stops, striking anvils.

Applicable Hardfacings: High-carbon iron/chromium alloys (50 to 55 HRC), shock-resistant tool steel rods.

Applicable Surface Treatments: Selective hardening, heavy carburizing, brazed-on cemented carbide tiles.

BRINELLING

Description: Brinelling is the wear term used to describe surface damage of solids by repeated local impact or by static overload. The origin of this term is probably from the similarity of this form of damage to a hardness indentation produced in a brincll hardness test. A very common example of this form of surface damage is local deformation (brinelling) on the faces of injection-molding cavities when the mold is inadvertently closed on a part. Even if hardened steels are used for the mold faces, a soft plastic part can produce the denting damage that is typical of brinelling. The plastic starts to extrude under the extreme clamping pressures; when it reaches a thickness of a few mils (100 μm), the friction between the plastic and the steel becomes so high that it cannot extrude any further, and the soft plastic behaves as an incompressible fluid that develops pressures sufficient to yield the hardened steel mold faces in compression. Figure 2–18 is an example of brinelling of a ball race by static overload.

Brinelling can be produced by a static overload in rolling element bearings. A small dent is produced at the contact point of each ball or roller. Brinelling by repeated impact can occur in a cam path where the follower motion is erratic; the follower may hop at some point on the path to produce brinelling or denting in that area. Thus

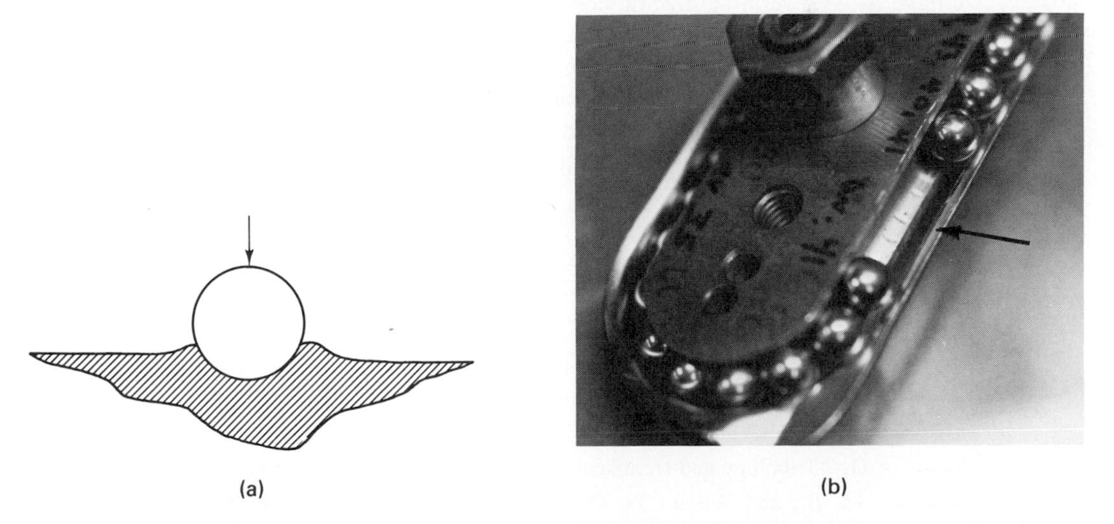

Figure 2–18 (a) Schematic of brinelling. (b) Brinelling of a bearing race by static overload.

brinelling is a form of surface damage characterized by local plastic denting or deformation of the surface. Material is not necessarily removed; it may be only displaced. Materials with very high compressive yield strengths have the best resistance to this form of wear.

Examples: Static overloads on mating surfaces: on mold faces, on wheels on rails, on rolling element bearings. Inertial overloads on cams, bearings, and mating parts (as might occur when a heavy piece of machinery is transported on a truck on a bumpy road).

Applicable Hardfacings: High-carbon iron/chromium alloys, high speed steels.

Applicable Surface Treatments: Heavy carburized, selective hardening, cemented carbide tiles.

SUMMARY

In this chapter, we have described the most agreed on modes of wear. There may be some wear manifestations that others feel should be added to this list, but our list is already longer than desired. It is our intent that the reader learn to observe a wear system with these processes in mind and mentally assess which of these processes would predominate. This information in turn is used to select a hardfacing consumable system or surface treatment that is suited to resisting this wear process. In each case, we have listed several hardfacing consumable systems and surface treatments that could be used to address a particular form of wear; in the remainder of this text, we will describe in detail hardfacing processes and their application, and we will describe the surface engineering processes that compete with hardfacing to solve wear problems. When we are done, the reader should be able to select the best system to address one of the forms of wear that we have just described.

REFERENCES

ADLER, W., ed. *STP 664: Erosion, Its Prevention and Useful Applications*. Philadelphia: American Society for Testing and Materials, 1980.

KRUSCHOV, M. M., "Principles of Abrasive Wear." *Wear,* 28:69–88, 1979.

MOORE, A. J. "Tribology and Hardfacing," *Tribology,* 8:265–270, 1976.

PETERSON, M. B., and WINER, W. O., eds. *Wear Control Handbook.* New York: American Society of Mechanical Engineers, 1980.

RABINOWICZ, E., *Friction and Wear of Materials.* New York: John Wiley & Sons, 1965.

Rigney, D. A., ed. *Fundamentals of Friction and Wear of Materials*. Metals Park Ohio: American Society for Metals, 1981.

Standard Terminology Relating to Friction and Wear, ASTM G40–82. Philadelphia: American Society for Testing and Materials, 1980.

Szeri, A. Z., ed. *Tribology*. New York: McGraw-Hill Book Co., 1979.

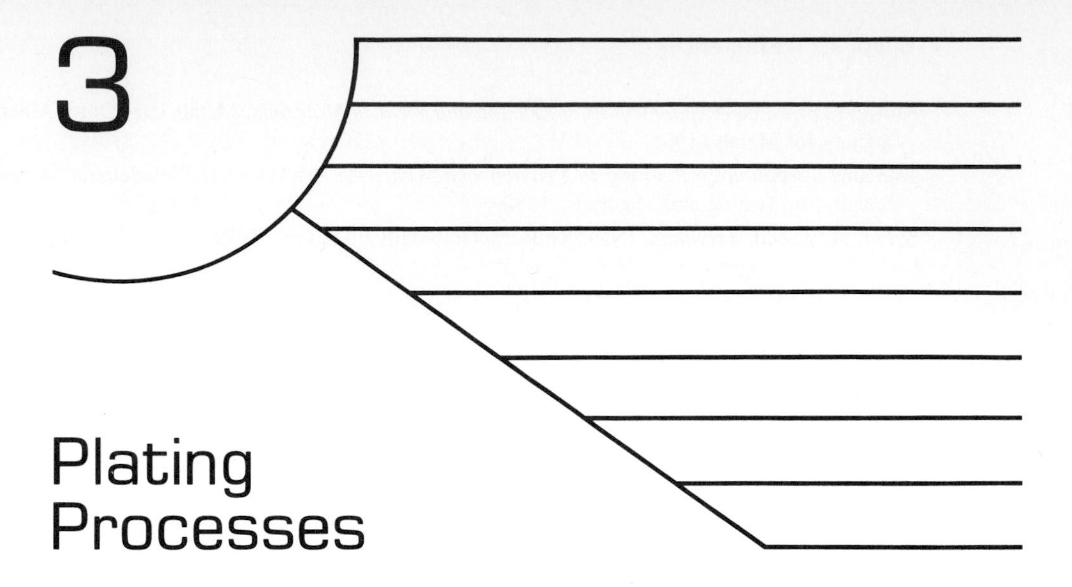

3 Plating Processes

Plating is the generic term that is widely used to describe metallic coatings that are applied by electroplating and electroless plating processes. Present-day shops and companies that apply ''platings'' usually offer a wide variety of coating services; the list usually contains the following:

Electrodeposited metals
Electroless deposition of metals
Electrochemical conversion coatings
Chemical conversion coatings
Electropolishing
Electroforming

We will describe all these processes, but the processes that are most competitive with hardfacing and other wear coatings are the metallic coatings that are applied by electrolytic and electroless plating processes. We will describe how these processes can be used to coat machine components to reduce wear and when these coatings may be the most appropriate to use.

FUNDAMENTALS OF ELECTRODEPOSITION

The fundamental aspects of the electrodeposition process impose some restrictions on the use of this process for application of wear-resistant coatings. Electrodeposition is the process of depositing a substance on an electrode immersed in an electrolyte by passing electric current through the electrolyte. The usual materials deposited are metals, and the usual electrolytes are water solutions that contain dissolved ions of the material to be deposited on the electrode. The materials that can be plated are those that can be put into solution in ion form in an electrolyte; the substrates that can be plated are those that make a suitable electrode (they must be capable of carrying current). The rate at which a material can be electrodeposited is a function of the quantity of electricity passed through the electrolyte. This is Faraday's law:

$$W = kIt$$

where W = weight of material deposited
k = electrochemical equivalent of the deposited material
I = current flowing in the cell
t = time that the process is carried on

The electrochemical equivalent of a material is the atomic weight of the element to be deposited divided by its normal valence. The significance of Faraday's law to the potential user of electrodeposited coatings is several fold. First, alloys of metals cannot be plated unless the elemental metals that comprise the alloy have the same electrochemical equivalents. Iron has an atomic weight of 55 and a valence of 2 to yield an electrochemical equivalent of about 27. Chromium has an atomic weight of 55, but with a valence of 3 (trivalent chromium) the electrochemical equivalent is only about 18. This is why stainless steels alloys of iron and chromium cannot under normal conditions be electrodeposited as an alloy. A second result of these differing electrochemical equivalents is that different metals may have different deposition rates at the same current density. Since the electrochemical equivalents of the metals are different, W, the weight of metal deposited in a given time, would be lower for chromium than it would be for iron. This is a significant factor that affects the use of electrodeposition for wear-resistant coatings; some metals deposit faster than others, and these deposition rates are controlled by the nature of the element to be plated and on the efficiency of the plating bath. Typically, chromium is deposited at a rate of about 0.3 mil (7.5 μm) per hour, and there are formidable problems to be dealt with in trying to speed up this deposition rate. Excessive current flow can cause arcing and part damage at the part/conductor contact. Sometimes plating rates are limited by the kinetics of the bath. There can be an inadequate concentration of the ions reduced from the electrolyte at the part surface (concentration polarization). Sometimes excessive plating rates cause excessive gas evolution at the surfaces to be plated, and the electrolyte contact is interrupted. In essence, the time restraints on electrodeposition rates are difficult to overcome, and normal electrodeposition processes have deposition rates that are usually in the range of 1 to several mils (25 to 75 μm) per hour.

Hydrogen Embrittlement

Another aspect of the electrochemistry of plating that can affect the user is hydrogen evolution at the electrodes. The actual mechanism of electrodeposition is a reduction of the ions of the species to be deposited from ion form to elemental form. Using the example of plating copper from a copper sulfate solution, when copper sulfate crystals are put into water, they dissolve and form copper ions and sulfate ions. As is shown in Figure 3–1, when an anode and the substrate to be plated (cathode) are electrically coupled in the copper sulfate solution, the copper ions in the solution are reduced to metallic form by gaining electrons from the substrate. At the cathode, additional copper ions are produced by the action of the copper anode being dissolved in the electrolyte. The exact electrode reactions that occur in electroplating operations are a function of the plating solutions and the electrode material, but a phenomenon that is common to most plating operations is reduction of hydrogen ions from the electrolyte to atomic hydrogen, which diffuses into the substrate. Just as copper ions are reduced to metallic copper by taking on two electrons from the cathode, hydrogen ions present in any water-based electrolyte can pick up an electron from the electrode and form atomic hydrogen. If the substrate to be plated is a ferrous material, absorbed atomic hydrogen acts as a strengthening agent, which can cause embrittlement of the part. This is a very real consideration in the plating of hardened steel parts. Unless the parts are heat treated after plating to remove absorbed hydrogen, they will be exceptionally brittle and essentially unusable for any application that involves even elastic strains on the part.

Plating Adhesion

What makes electrodeposits adhere to a surface? There are various opinions and varying supporting data for these opinions, but most people in the field agree with the van der Waals type of bonding theory. This theory states that when two atoms are brought together there is a repulsion at large distances (these large distances may be only angstroms), but as the nuclei of the two atoms overcome some critical separation distance, there will be a net attraction of the two atoms. These are the primary types of forces

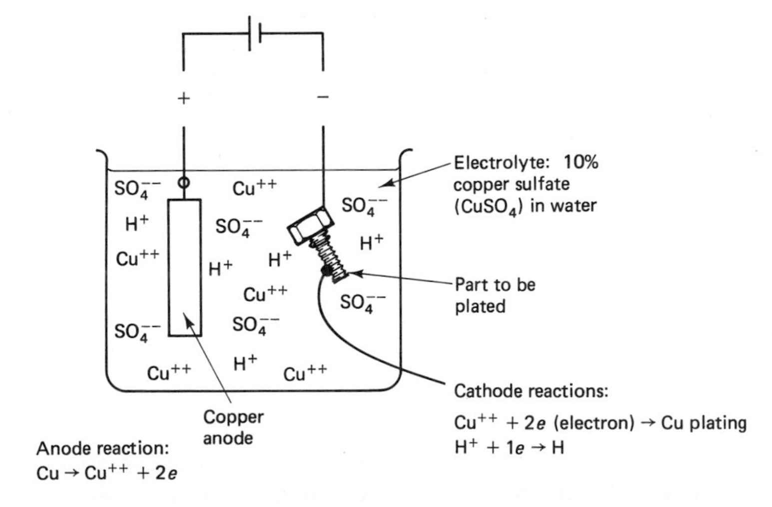

Figure 3–1 Electrode reactions in plating of copper from a copper sulfate bath.

that control adhesion of electrodeposited metallic coatings on metallic substrates. The key to getting atoms close enough to achieve this type of bonding is to have an atomically clean surface on the substrate to be plated. An atomically clean surface is usually obtained by pickling or by electrocleaning the parts to be plated prior to initiation of the plating operation. If the substrate is not atomically clean, the plating will have a poor bond.

The surface finish of the substrate to be plated is not overly important, but it is common practice to have a finish on the substrate similar to that desired after plating. The electrodeposited coating usually ends up with a surface texture that is the same as the starting substrate surface texture, with geometric leveling occurring as the deposit gets thicker.

The strength of adhesion of electrodeposited coatings on metals can be measured by a number of techniques, but the highest bond strengths reported for normal plating operations are less than 100 ksi (689 MPa) in tensile shear. Some platings can have no adhesion even though they appear to be perfectly adhered. For example, many electrodeposits do not adhere to aluminum because it is so difficult to get an atomically clean surface on aluminum (there is usually an oxide surface on aluminum). However, it is common practice to electroplate nickel or copper on aluminum rolls. The plated surfaces are perfectly serviceable but the platings are not adhered. Many platings shrink as they build up on the surface. When a cylinder is plated, the deposit essentially becomes a shrink fitted sleeve on the cylinder. The fit is so good that it would appear to be perfectly adhered even with ultrasonic inspection techniques. The point to be made is that electrodeposited metals can have very good bond to the substrate, but that bond will never be the same as a fusion bond, and poor bonds can go undetected unless techniques are used to test the actual bond strength.

OTHER PLATING PROCESSES

Electroless Plating

Electroless plating is the trade-accepted term for deposition of a metallic coating on a metallic substrate by immersion in a suitable bath without the use of electricity; thus the term electroless. The true description of this process is autocatalytic plating. The deposit is achieved by the catalytic reduction of metal ions in the plating bath. We will discuss the use of this process for wear coatings in the subsequent section on the use of these coatings for wear reduction.

Electrochemical Conversion Coatings

Conversion coatings are coatings formed by reaction of a surface with a specified environment. Electrochemical conversion coatings are formed by making the part to be coated one electrode in an electrochemical cell containing an electrolyte that is formulated to produce a particular coating result. An example of a simple type of chemical conversion coating is the oxide coating that occurs on many metals in air. A freshly abraded surface on copper will turn dark after sitting in room air for a few hours. The coating that is formed is a copper oxide, and the coating was formed by reaction of the copper with oxygen in the air.

The most common electrochemical conversion coating in use is anodizing of aluminum. When aluminum is made the anode in an electrochemical cell containing a suitable electrolyte (usually chromic or sulfuric acid), the surface of the aluminum will electrochemically react with the electrolyte to form an aluminum oxide coating. The coating is not plated on; it comes from the reaction and partial dissolution of the surface by the electrolyte. Anodizing has a role in wear prevention, and we will discuss this in a subsequent section on hardcoating.

Chemical Conversion Coatings

We have already defined what a chemical conversion coating is; the chemical conversion coatings that are available in most plating shops are (1) chromate, (2) phosphate, and (3) oxide.

Chromate chemical conversion coatings are most commonly applied to nonferrous metals to enhance their atmospheric corrosion resistance. These coatings are usually applied from proprietary solutions by dipping, spray, or brush application. The coating thickness and properties vary considerably, but all are very thin, usually less than 0.0001 in. (2.5 μm), and they are not meant to be wear resistant.

Phosphate chemical conversion coatings are usually applied to steels, and there are a number of types for specific applications. Zinc phosphate chemical conversion coatings are used to assist paint adhesion on steel surfaces to provide a modicum of atmospheric corrosion resistance. The black surfaces on some auto dashboards are black zinc phosphate conversion coatings. Iron phosphate coatings are used for the same types of things. Manganese phosphate conversion coatings are usually black, and they tend to be the thickest of the conversion coatings. Their use in machine parts is quite widespread. These coatings are soft and porous. When they are immersed in oil, it is absorbed into the coating, and the oil-impregnated coating is used to minimize atmospheric rusting. An additional and more important benefit of the manganese phosphate coating is that it assists break-in wear on parts that are used in sliding systems.

Oxide chemical conversion coatings can be applied to most metals; most of these coatings are black and the major purpose for using them is to obtain the black color. The ''bluing'' on gun barrels is usually a black oxide chemical conversion coating. They do not offer the corrosion protection of the other conversion coatings, and they have essentially no utility in wear systems. Their use is predominately restricted to decorative applications.

In general, chemical conversion coatings are not competitive in properties with plated coatings for wear applications. None of them are thicker than about 0.0001 in. (2.5 μm) and none of them are hard. The manganese phosphate coatings are very useful for reducing part damage during break-in. These coatings should be considered for carbon and low-alloy steel wear components that could benefit from added protection during break-in. They should be applied after the parts are hardened, and they should not be used on high-alloy steels and tool steels. The latter materials may be subject to dimensional loss during the coating operation.

Electropolishing

Electropolishing is the process of lowering the surface roughness of a part by electrochemically dissolving some of the part's surface. This is usually done by making the part to be polished the anode in an electrochemical cell containing special, sometimes proprietary electrolytes. The cathode is usually some nonreacting material such as titanium, and the current and cell voltage are adjusted such that surface removal is uniform. This is a difficult process to control. If all conditions are not perfect, pitting rather than polishing can occur. An additional risk is excessive dimension loss. However, when all conditions are perfect, this process can be used to polish surfaces on parts that are too complex in shape to polish by conventional techniques. It is one of the few processes that are economical to use on the inside diameters of pipes and small deep vessels.

Electropolishing has little use in the area of wear prevention. Occasionally, it is used to polish threaded parts where poor surface finish on one or both members is producing unacceptable wear.

Electroforming

This is the process of electrodepositing a material on a removable mandrel to make a part. For example, a thin nickel sleeve can be electroformed by electroplating nickel on an aluminum cylinder. The plating will not adhere because of the passive surface on the aluminum, and the plated sleeve can be removed by shrinking the aluminum mandrel with a cold treatment. This plating process normally has no utility in the area of wear-resistant coatings.

Summary

In this discussion, we have described some of the basics of various plating processes. The processes that are useful in solving wear problems will be discussed further; we will try to show where platings are more cost effective than the surface coatings and present sufficient details so that they can be used in design.

ELECTRODEPOSITION FROM PLATING BATHS

A number of factors must be weighed when bath plating is considered for producing a wear-resistant surface. By the term *bath electrodeposition,* we mean conventional plating where the part to be plated is made an electrode in an electrochemical cell and the part will be immersed in the plating cell or bath. The first question that a prospective user of this process might ask is if the part will fit in available tanks. If outside vendors are used, one can assume that tanks are available that are at least 2 ft (0.6 m) cubed. Parts that have at least one dimension that is over 5 ft (1.5 m) may limit the size capability of many plating shops, and available tank capacity should be explored before the decision is made to use a plating. Some plating shops routinely plate extruder screws that are 40 ft (12 m) long, but they are few and far between.

If it has been determined that a part can be plated, the next question is what metal to use to plate the part. Theoretically, any metal can be plated on any metal, but only 16 metals are plated in typical plating shops: copper, iron, nickel, chromium, silver, zinc, cadmium, gold, rhodium, tin, brass, osmium, cobalt, platinum, indium, and lead. There are various types of some of these: hard nickel, soft nickel, black chromium, black nickel, soft iron, and hard iron are examples. From the wear standpoint, the candidate platings are primarily hard nickel, hard chromium, and the precious metals. We will give some wear-resistance data on these platings in the concluding section of this discussion, but hard chromium is the most widely used immersion plating for wear applications.

Assuming that an available plating will do the intended job, the next thing to do is to consider if the part design is suitable for bath plating. Electroplating has two idiosyncracies that must be dealt with: edge buildup and throwing power of the bath.

Edge buildup means that the deposit thickness is not uniform. The thickness at sharp edges will be approximately twice the thickness away from edges (Figure 3–2). Some plating baths are less susceptible to this phenomenon, but chromium and nickel are susceptible. Edge buildup is caused by higher current densities on corners and on any male protrusions. The current density is higher and the plating rate is correspondingly higher. This problem can be dealt with to some extent by design. As shown in Figure 3–3, edges can be rounded and reentrant corners can be made less troublesome by the use of generous fillets. Edge buildup is not normally a problem if the plating thickness is kept below 0.0001 in. (2.5 μm). Edge buildup still occurs, but the amount is usually too small to be detected.

The problem with throwing power simply means that essentially no plating will go into fastener holes, corners, and keyways without special techniques. This phenomenon is caused by current flow patterns in the bath, and the special technique required to solve this problem is to place auxiliary anode electrodes into the holes or recesses that would not otherwise plate. The use of auxiliary electrodes is expensive and should be avoided wherever possible.

Probably the biggest consideration in using bath plating for a part is the thickness

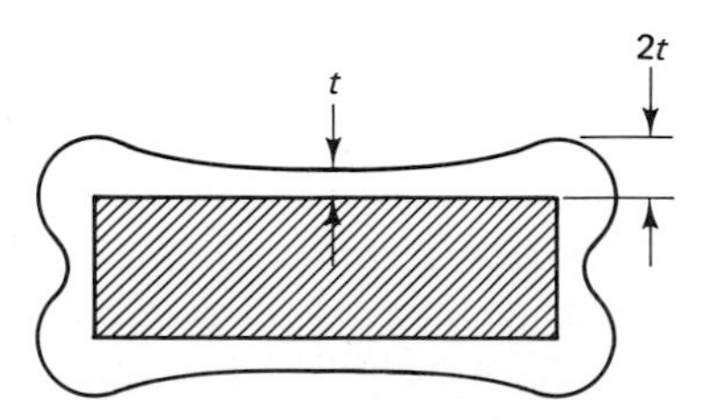

Figure 3–2 Edge buildup from immersion electroplating.

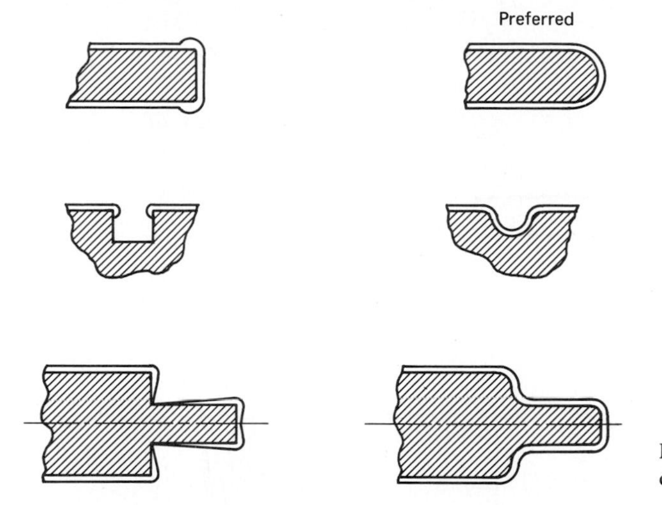

Figure 3–3 Designing for better electrodeposition.

of the deposit. If it is felt that the plating must be thicker than 0.0001 in. (2.5 μm), then most precision parts will require grinding of the deposit to get around the edge buildup problems. Figure 3–4 shows some typical parts that might be plated for wear resistance. If a thin, 0.0001-in. (2.5 μm) deposit is deemed adequate, all these parts can be plated all over. Thicker deposits will usually require grinding after plating and masking to limit the plating to functional surfaces. Masking is done with special chemical-resistant lacquers or with special plater's tape. This is a hand operation and it should be minimized. If an electrodeposit is to be ground after plating, any plating thickness can be specified, but the user should keep in mind that plating operations are slow and excessive thickness is expensive. It is usually not economical to use electrodeposited coatings in finished thicknesses greater than 0.003 in. (75 μm) unless the plating is being applied to restore dimensions. Thus a proper drawing specification would show the following:

1. Type of plating
2. Areas to be plated
3. Finished plating thickness
4. Postplating heat treatment (if needed)

Postplating heat treatment is used to remove hydrogen pickup from the plating operation. If any hardened steel (over 40 HRC) is electroplated, it must be given a postplating heat treatment. Typically, this treatment consists of baking for at least 4 hours at a temperature in the range of 375° to 600°F (190° to 315°C) as soon as possible after plating. This bake will not completely remove hydrogen effects, but without this treatment the brittleness is so great that the part would be unusable.

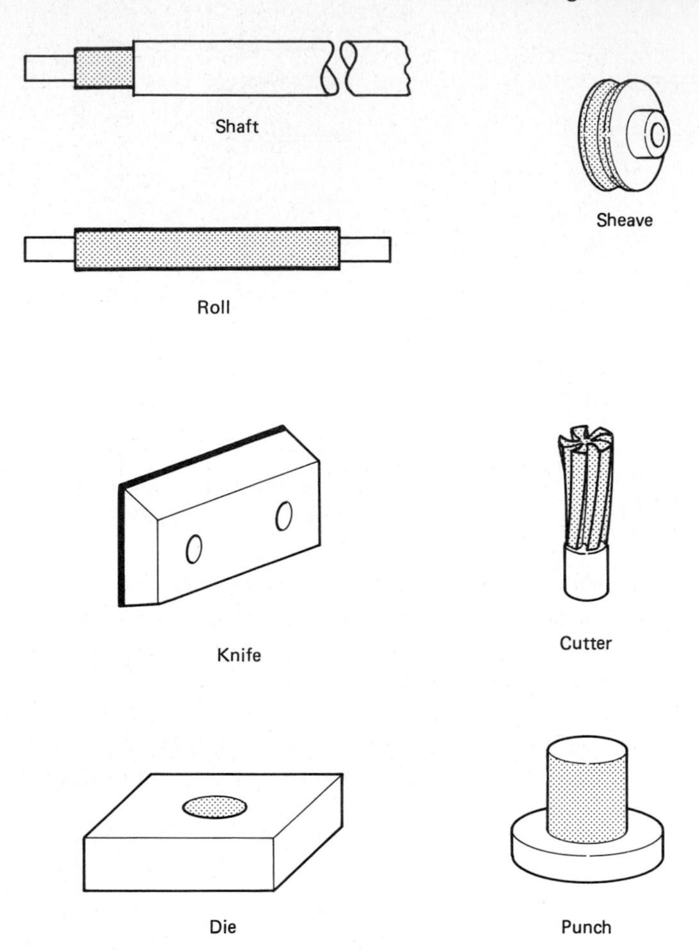

Figure 3–4 Typical parts that can be immersion electroplated for wear resistance.

ELECTROLESS PLATING

Electroless plating baths have been developed for copper, silver, nickel, and a number of other materials, but the system with the most importance for wear applications is the nickel/phosphorus system. Coating is accomplished by immersing the part in an aqueous solution containing metal salts, a reducing agent, and other chemicals that control pH and reaction rates. When a suitable substrate is put in the bath, it acts as a catalyst or aid to cause the nickel ions in solution to be reduced by the reducing agent. The ions are not picking up electrons from the cathode as in electroplating. The reducing agent is causing the metal ion reduction, and the nickel coating on the part continues to act as the catalyst as the plating process proceeds. This is why this plating process is called autocatalytic plating. The plating does not stop when the catalytic surface is covered. Once covered, the nickel becomes the catalyst for the reduction process. The reaction in a nickel sulfate bath with a sodium hypophosphite reducing agent is

$$NiSO_4 + NaH_2PO_2 + H_2O \xrightarrow[\text{catalyst}]{\text{heat}} \text{Ni plating} + NaHPO_3 + H_2SO_4$$

The best catalysts for this type of plating are iron, nickel, cobalt, and palladium, but most metal substrates will respond to electroless nickel plating with the exception of lead, tin, cadmium, zinc, antimony, and bismuth. Good bond strengths have been demonstrated even on aluminum. The heat shown in the reaction equation comes from the plating bath. Electroless nickel baths usually run at a temperature near the boiling point of water. The finished nickel coating is not pure nickel, but contains phosphorus inclusions. The phosphorus content can be as high as 13 percent. Deposited electroless nickel has a hardness of about 43 HRC, and the hardness can be increased by age hardening heat treatments at temperatures in the range of 550° to 750°F (288° to 400°C). A typical heat treatment of 2 hours at 600°F (315°C) will increase the coating hardness to around 60 HRC.

Electroless nickel deposition rates are relatively slow compared to electrodeposits. Plating rates are typically about 0.0004 in. (10 μm) per hour. The normal thickness limit used is 0.002 in. (50 μm). Some plating companies will put on coatings as thick as 0.006 in. (150 μm), but coatings this thick may be prohibitively expensive. The biggest advantage of electroless nickel over electrodeposited coatings is that there are no throwing power and corner buildup problems like those that occur in electrodeposition. All wetted surfaces are uniformly plated. The plating goes into deep holes, keyways, and reentrant corners. Because of the slow plating speeds and uniform deposition rate, it is possible to plate up to 0.002 in. (50 μm) of coating on precision parts without grinding after plating.

Another advantage of electroless nickel plating is that it is possible to make composite platings. If particles of an inert substance (a substance that will not react with the chemicals in the bath) are put into suspension in the plating bath, the particles will become part of the deposit. Figure 3–5 is a photomicrograph of a diamond-filled electroless nickel deposit. Composite coatings are available with many fillers, but some of the more common fillers besides diamond are PTFE (Teflon®) and silicon carbide. Up to 50 volume percent of the coating can be one of these materials. Inclusion of the diamond and silicon carbide is done to enhance abrasion resistance, and the PTFE addition is made to produce a self-lubricating surface. The PTFE additions are usually in the form of submicrometer-size particles, but the diamond and SiC particles can be 10 μm or larger in diameter.

Electroless nickel coatings are widely used for wear reduction, with and without the particle additions. Almost any plating shop can apply the conventional electroless nickel; the composite coatings are mostly proprietary and can only be done in certain shops.

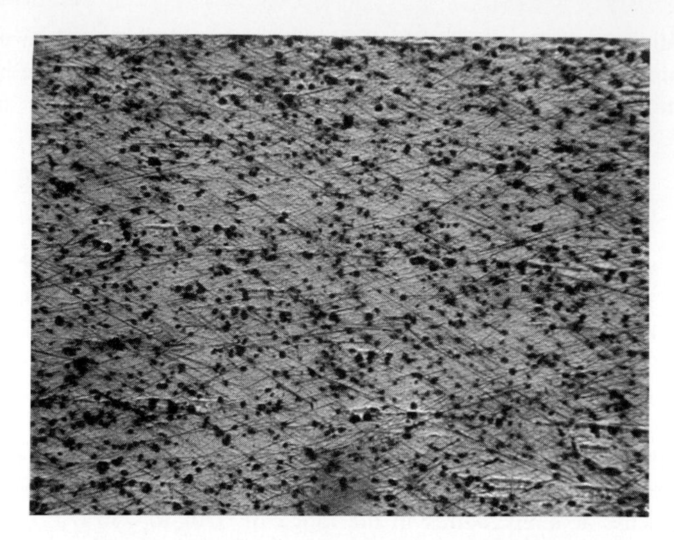

Figure 3–5 Composite plating of 15-μm (0.0006 in.) diamond particles in nickel (50×).

METALLIDING

Metalliding is a patented* electrodeposition process for electrodepositing materials at elevated temperatures in a fused salt electrolyte. This process is not widely used, but it offers some unique advantages over conventional electrodeposition processes. It may solve a problem that cannot be solved by conventional materials and coatings.

A schematic of the process is shown in Figure 3–6. The theory of this plating operation is the same as conventional electrodeposition. An anode material dissolves in the electrolyte; ions of the anode material migrate to the cathode, where they are converted from ion form to elemental form by transfer of electrons from the cathode. Two unique aspects of this process are (1) elements that cannot be plated by conventional processes may plate in metalliding, and (2) if the deposition rate is controlled to match the diffusion rate of the plated species in the substrate, the substrate will end up with a coating that is diffused into the substrate rather than deposited on the surface.

Figure 3–6 shows some of the elements that have been successfully plated and some of the substrates that have successfully accepted these coatings. One inventor of the process once stated that he could plate anything on anything. There are a number of reasons why this plating process works on elements that are not normally plated, but one of the most significant reasons is the use of molten fluoride salts as the electrolyte. These salts are very aggressive solvents and will dissolve most materials. This means that protective films on the anode will be dissolved and an atomically clean surface will be obtained. Similarly, the anode material will dissolve in the electrolyte so that ions of the materials to be deposited are available for transfer to the cathode.

The coatings can be diffused into the cathode substrate because the molten salt electrolytes are used in the temperature range of about 1000° to 2100°F (538° to 1150°C).

* Patent originally assigned to General Electric Co.

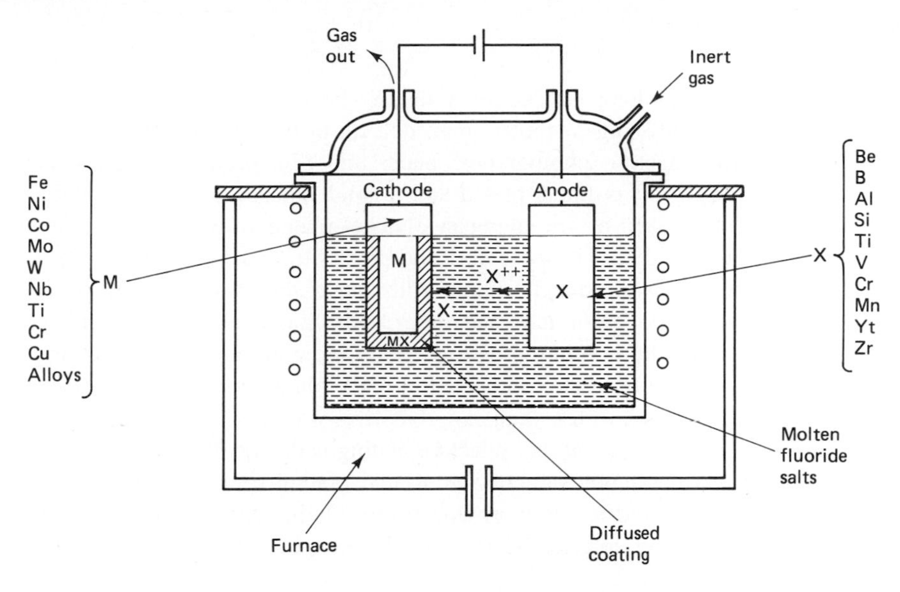

Figure 3–6 Schematic of metalliding bath.

Diffusion of many species is easy in many host materials at the higher end of this temperature range. The diffusion of the plated species into the substrate can have another important effect. The diffused species may chemically combine with the substrate material to form a compound with properties different from those of the substrate or plated material. For example, boron can be plated onto and diffused into iron-base alloys to form a surface of iron boride that has a hardness over 1000 HK, harder than any pure metal. Coating can be applied to the surface if the deposition rate is faster than the diffusion rate. The thickness of the coating or diffused layer can be as thin as 0.001 in. (25 μm) or many mils in thickness.

Since this sounds like a great process, why doesn't everybody use it? The answer is that the process is simple in concept, but the problems of controlling the purity of the salt electrolytes and the plating atmosphere are formidable. Exacting process controls are essential. Similarly, it is difficult to maintain any molten salt bath, and large salt baths are exponentially more difficult to maintain.

In summary, metalliding is a unique electrodeposition process for applying elements that cannot normally be electrodeposited on substrates that cannot normally be plated. This process is expensive and it can only be done by a few companies. However, it may be the only process that will produce a coating that will provide suitable service life in some special applications. It can produce very hard surfaces. It can produce a thick coating with no surface buildup. It can plate substrates that defy plating by conventional techniques. For example, a common application of this process is to apply tantalum to chemical-handling equipment, valves, pump parts, and the like. These parts would be too expensive to make from solid tantalum.

SELECTIVE PLATING

For probably as long as electroplating has been used, platers have practiced methods of their own devising to repair small defects in plated parts. Many parts, such as rolls used to manufacture photographic paper and film products, require perfect surfaces. When a large roll is finish plated and a small pit opens up in grinding, there are two choices: start over or try to repair. Nickel-plated rolls were sometimes repaired by drilling out the defect in the nickel and press fitting a nickel plug into the drilled hole. This does not work on chromium rolls since the nickel would polish in relief and show. Repairs to chromium and other rolls that could not be repaired by plugging were attempted by building a plating solution dam around the defect and plating with a small power supply in an improvised in-place plating tank. The results of this technique were failures more often than not, but this was the start of what is now known as selective plating. Present-day selective plating is the localized electrodeposition of metals by the use of an anode saturated in special plating solutions. A rectified ac (dc) power supply is connected to the work and to the plating anode, and plating is accomplished by motion of the solution-bearing anode on the work or by movement of the work with respect to the anode.

The earliest version of selective plating of this type came into use in the early 1950s. It was called brush plating; the crudest versions of this process used a brush and a car battery as the power supply. Later versions used cloth-covered graphite electrodes. The electrode (anode) was dunked in a series of solutions, and the area to be plated was swabbed with the wetted electrode. Plating times with early selective plating units were very slow. It was not uncommon to take an hour to deposit a plating of 0.0001 in. (2.5 μm) over an area of only 4 in.^{2} (25 cm^2).

Modern Selective Plating

The selective plating in use today is still similar in principle to brush plating, with the exception that today's equipment, solutions, and techniques allow the process to work as originally conceived. Plating rates can be as high as 2 mils (50 μm) per minute; over 75 metals and alloys can be applied, and the deposits can have adhesion and use properties that are comparable to those of bath electrodeposits. The biggest advantage of this type of plating is that the part to be plated does not have to be immersed in a plating tank. Solutions and anode contact are localized. Massive machine parts can be locally plated for wear or restoring of mismachining. Parts that cannot be removed from machines without excessive downtime can be rebuilt in place. Large plated rolls can have nicks and dings repaired. Conversely, very small parts can have small controlled amounts of metal locally applied.

The changes in selective plating that have made modern techniques successful are several:

1. Plating solutions have been developed with high metal ion concentrations (for fast plating speeds).

2. Special cleaners and activators have been developed along with exacting procedures for their use.
3. Plating power supplies have been specially designed to perform selective plating.

Most of these developments are proprietary in nature, but a number of companies sell complete systems. A modern-day selective plating system is illustrated schematically in Figure 3–7. Figure 3–8 shows a unit in use. The plating electrode is made from high-purity graphite. Ordinary graphite contains contaminants that can harm the plating solutions. Conductor handles are fastened to the graphite electrodes (usually with quick connections). The graphite is covered with an absorbant nonconducting material. The

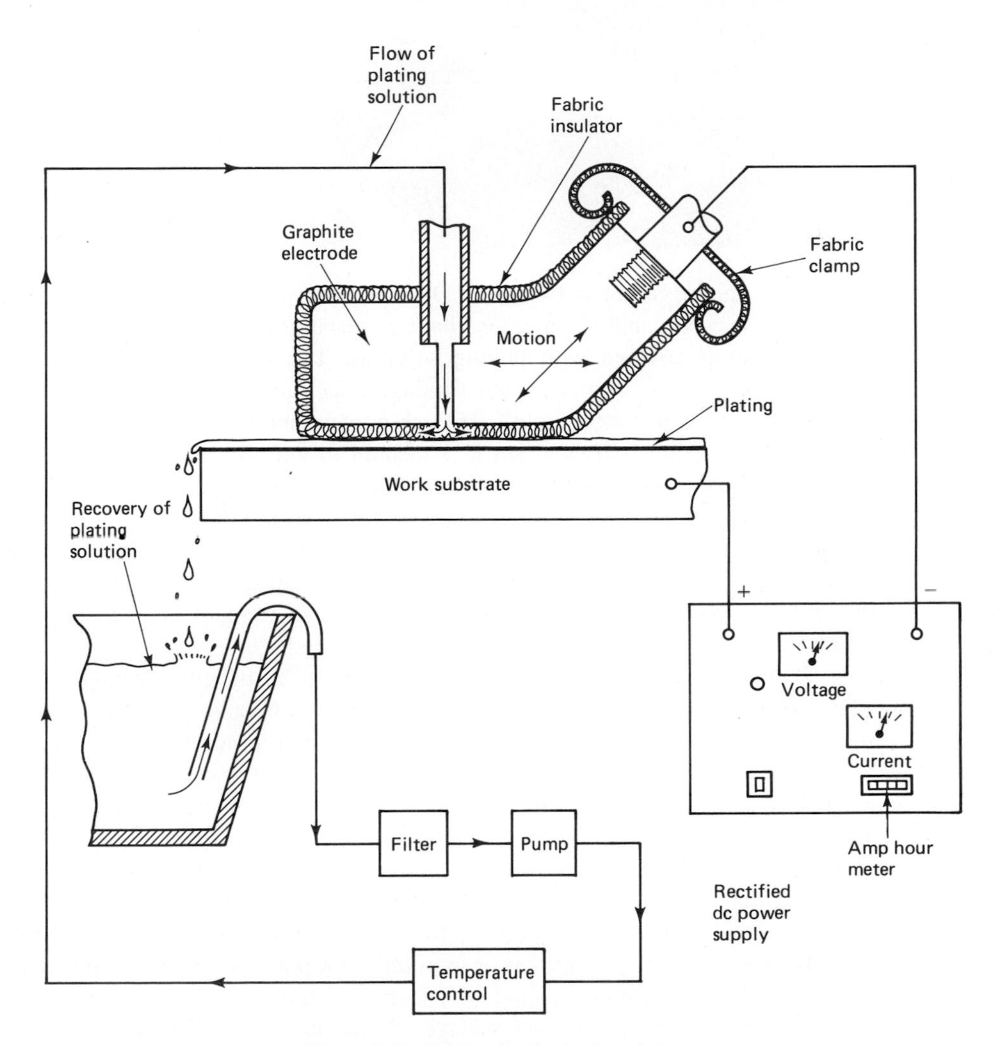

Figure 3–7 Schematic of selective plating.

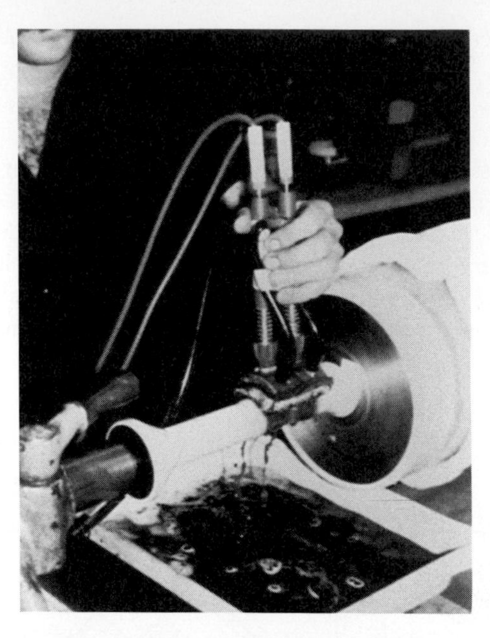

Figure 3–8 Selective plating of a roll journal.

purpose of this covering is twofold: (1) without it, the graphite electrode would short circuit to the work, and (2) the absorbant material ensures that the contact area under the electrode is covered with plating solution. The fabric electrode cover can be a sterile cotton batting, surgical finger bandage material, Dacron® fabric, or Scotchbrite. The latter is a nylon "steel wool" that can contain abrasives of various grit sizes. The Scotchbrite covering is used when it is desired to minimize corner buildup in plating; the abrasive in the Scotchbrite® will tend to abrade buildup areas and a deposit of uniform thickness can be obtained. The thickness of the electrode covering is usually about $\frac{1}{8}$ to $\frac{3}{16}$ of an in. (3 to 5 mm), but micro plating has been done with hypodermic needles covered by whipping them with ordinary Dacron thread (the solution is pumped through the needle). Nylon-covered platinum wires are also used as anodes for plating very small areas.

Plating solution can be applied to the area to be plated by dunking the covered electrode into a container of solution, but the more effective technique is to pump the solution to the electrode through PVC tubing of appropriate size. Pumping is usually done with a peristaltic pump so that the solution is not contaminated by contact with metal fittings and the like. Some solutions perform best at certain temperatures. Heating and cooling devices are used on the drip pan for the solution to control the temperature of the process. Electrodes for various part geometries are shown in Figure 3–9. It is preferred that the electrode be of sufficient size that it cover at least one-third of the area to be plated. However, covering 100 percent of the area would be ideal; this can often be done on flat surfaces, but it is usually not possible on other geometries.

The plating supplies for modern selective plating units have electronic controls to provide constant current at a set voltage. Some power supplies contain power meters

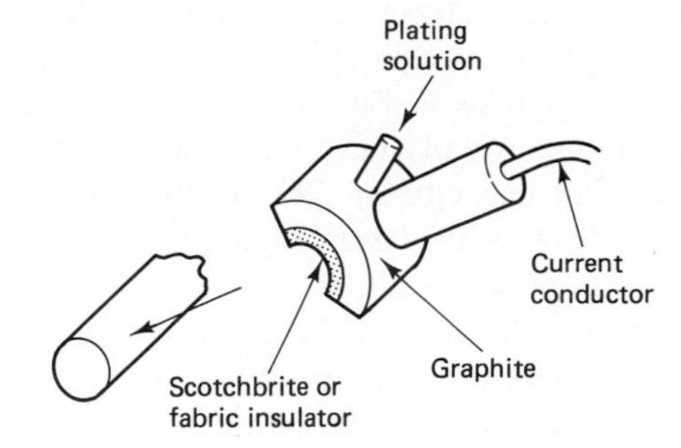

Electrode for
cylindrical ODs

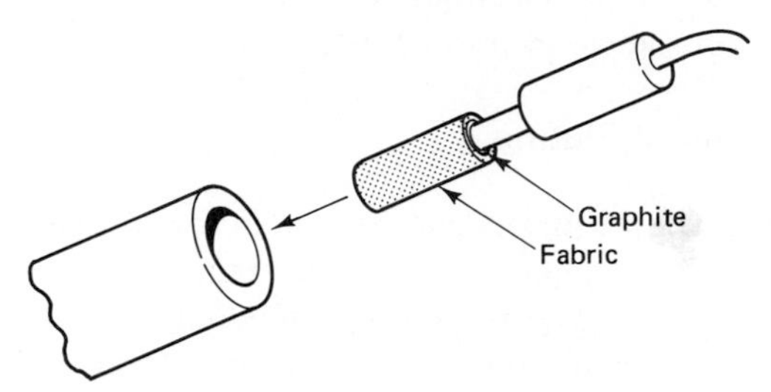

Electrode for ID bores

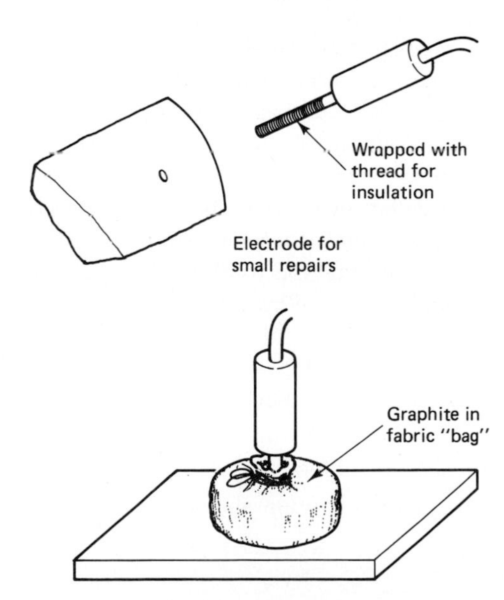

Electrode for flat surfaces

Figure 3–9 Selective plating electrode (anode) configurations.

that allow the plater to calculate the power required to plate a particular type of plating to a specified thickness. This calculated power setting is set on the power supply, and the unit will sound an alarm when the desired plating thickness is achieved. Plating voltages are usually in the range of 8 to 20 V. Power supplies are available with varying current capabilities; the smallest units usually have a current capacity of 10 A, and the largest units in use have a current capacity of 500 A. Plating current densities are normally in the range of 500 to 5000 A/ft^2, much higher than for bath electroplating.

The motion for the plating operation can be obtained by manually moving the electrode over the area to be plated, the part can be put in a rotating device such as a lathe or a rotary turntable, or the electrode can be driven by a flex shaft for plating internal diameters. Robots have been adapted to automatic selective platings. In the latter case, any desired motion of the electrode to the work can be obtained. An orbital or circular electrode is maintained on flat work, and circular shapes are rotated to achieve the desired motion. Travel speeds of the electrode over the work are usually in the range of 15 to 120 surface feet per minute (4.5 to 37 m/minute).

Suitable Substrates and Types of Plating

Selective plating can be applied to all the metals that respond to conventional immersion electrodeposition, in addition to some metals such as aluminum that are difficult to plate. Most substrates require a bonding layer or preplate (usually nickel) before putting on the desired deposit. Stainless steels, nickel/chromium alloys, chromium, and some nonferrous metals (tin, zinc, babbitt) usually require a bond coat, but this is not usually a concern of the user. The plater has detailed instructions for each type of basis metal and will apply the appropriate preplate when it is needed. The material to be plated must be clean of all rust and inorganic materials such as oils and greases. On rebuilding operations, it is usually advisable to undercut wear scars so that a uniform thickness of plating can be applied. The same thing is true for pits and scratches; they should be ground or machined so that they are open enough to allow contact with the plating electrode. If you wish to plate a defect that looks like a prick punch mark, it should be opened by grinding or other means until it resembles a ball indentation. In general, small defects should be opened up such that the width of the area to be plated is twice the depth to be plated. Plating involves the following steps:

1. Clean the surface of oils and greases (solvents, vapor degrease etc.).
2. Position surfaces to be plated to allow for electrode motion and draining of solutions to a drip tray.
3. Activate the surface (with special solutions).
4. Flood the surface with the plating solution and manipulate the electrode or part to produce the deposit.
5. Rinse and dry the part.

Preplates require the same type of procedure. Some metals require sophisticated activation steps, but the necessary procedures for most metals are well documented.

Table 3–1 is a list of some of the metal plating solutions that are available from one manufacturer of solutions. These are the same types of platings that can be applied by immersion plating. As is the case in immersion plating, there can be as many as six different solutions for putting on the same metal. Some solutions may produce lower deposit stress; some may be harder; some may be less corrosive to the substrate (acid versus alkaline). The solutions marked with the asterisks are some of the more popular maintenance solutions. Chrome cap is used as a finish plate on chromium repairs. Nickel XHB is used for heavy buildups, and the cobalt tungsten and nickel tungsten platings are used for wear surfaces over buildup plates.

TABLE 3–1 SOME SELECTIVE PLATINGS

Plating solution	Deposit hardness (HB)	Thickness limit (in.)[†]	Application comments
Babbitt, heavy build	10	<0.010	For bearings
Babbitt	10	<0.010	For bearings
Cadmium (low hydrogen)	26	<0.010	Low-hydrogen embrittlement
Cadmium, acid	17	<0.020	For corrosion resistance
Cadmium, alkaline	22	<0.020	For corrosion (replates)
*Chrome, cap	555	<0.0002	For repair of Cr plate
Chrome, neutral	555	<0.0002	Wear coating
Cobalt, machinable	375	None	For heavy buildup
Cobalt, semibright	600	None	Heavy pit repair
*Cobalt tungsten	745	<0.0002	Wear coating
Copper (high speed acid)	325	None	Rebuilding
Copper (heavy build alka.)	250	None	Rebuilding
Gold	125	None	Use for corrosion
Gold (hard)	147	None	For sliding wear
Indium	2	None	Bearing corrosion
Iron	477	<0.0005	For top coat over buildup
Iron (black)	(porous)	<0.001	Hard black surface
Lead	8	<0.010	Gasket surfaces
*Nickel (acid, high build)	555	<0.005	Buildup
Nickel, semibright	555	<0.010	Mold repair
Nickel, high speed	530	None	Buildup/wear
*Nickel, XHB	530	None	Buildup/wear
Nickel, black	(porous)	<0.0002	Black surface
*Nickel tungsten	745	<0.0005	Wear coating
Nickel cobalt	600	<0.010	Wear coating
Nickel tungsten D	555	None	Buildup
Palladium	300	<0.005	Preplate for Ag
Platinum	444	<0.0002	Electrical parts
Rhodium	568	<0.0002	Wear/electrical parts
Silver	105	<0.005	Sliding wear
Tin alkaline B	6	<0.020	Soft buildup
Zinc alkaline B	40	<0.010	Corrosion protection

From *Selection Process Instruction Manual,* Selections, Ltd., with permission.

* The more popular maintenance solutions.

[†] 0.001 inch = 25.4 μm

The cadmium (low hydrogen) solution is specially developed for plating of high-strength, heat-treated aircraft parts where hydrogen embrittlement is a service factor. This solution reportedly produces very low tendency for hydrogen embrittlement of plated parts. Unless a particular selective plating solution has documentation that it minimizes hydrogen embrittlement, the normal postplating heat treatment should be used on highly stressed hardened steels.

Although not shown in Table 3–1, there are solutions (usually phosphoric acid) for selective anodizing and hardcoating of aluminum. This process is slower than selective plating and is used mostly for repairs. These deposits can be dyed to get a color match with a previously anodized surface.

Use of Selective Plating for Repair

The most important reasons for plating in general are to protect a surface from environmental attack, to produce a wear-resistant surface, or to alter the dimensions of a surface. All these things can be accomplished with selective plating. Figure 3–10 shows some of the types of things that can be repaired with selective plating. A rule of thumb to employ to determine if selective plating should be a candidate for a repair job is that, if the desired deposit thickness is less than 0.010 in. (250 μm) and if the area of the deposit is over 1 in.^{2} (7 cm^2) and less than 300 in.^{2} (2000 cm^2), selective plating may be the most cost-effective deposition technique. There is no technical limit to the area that can be selective plated. There can be an economic limit. Selective plating was used to replate gold on the dome of the capital building in Sacramento. Anodes were 1 ft^2. Selective plating turned out to be the most cost-effective way to perform this repair.

Selective plating deposits can be as thick as 0.100 in. (2.5 mm), but usually welding is more cost effective when deposits get this thick. Conversely, welding is not usually economical when a shaft is only worn a few thousandths of an inch. Many times a repair is necessary on substrates that cannot be fusion welded without the likelihood of cracking. Selective plating and thermal spray processes are the best candidates for these types of repairs. Many times selective plating can be used to rebuild wear scars this small with a deposit accurate enough that it is not necessary to finish grind or machine the deposit. Worn shafts and journals are a natural for selective plating repair. As shown in Figure 3–11, cylindrical parts can be easily rotated in a lathe or rotary fixture. It is easy to apply and collect the solutions if done on a lathe.

Machinable deposits (less than 40 HRC) can be finished immediately after plating. This type of repair is one of the biggest applications of selective plating. Repair of mismachined parts is probably the second most popular application. Anyone who has ever machined a part to close tolerances knows that it is quite possible to miss the mark by a thousandths or two. Selective plating offers an alternative to scrapping the part.

Plating Properties

The 75 or so metals and alloys that can be applied by selective plating in general have physical properties similar to their wrought counterparts. The mechanical properties of

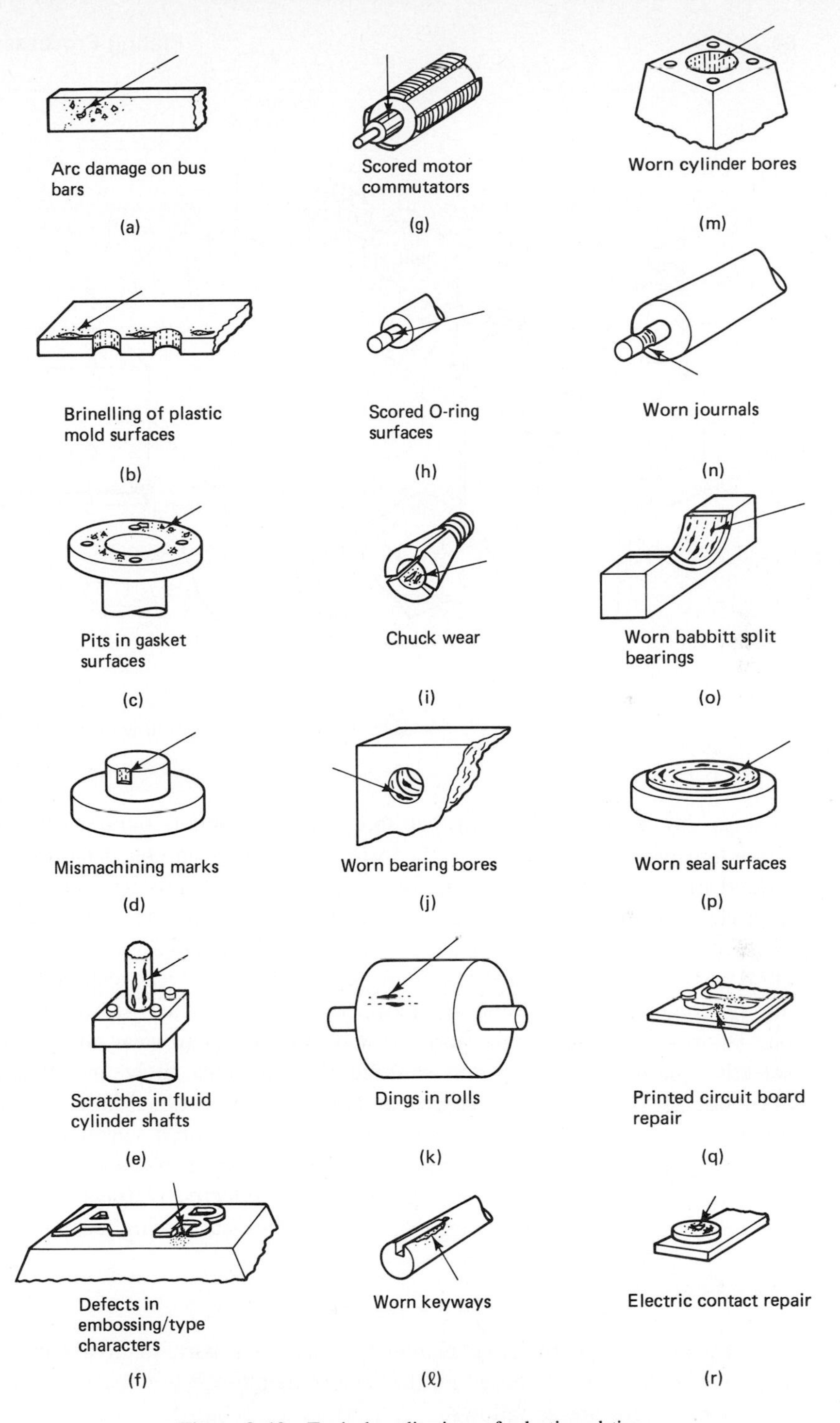

Figure 3–10 Typical applications of selective plating.

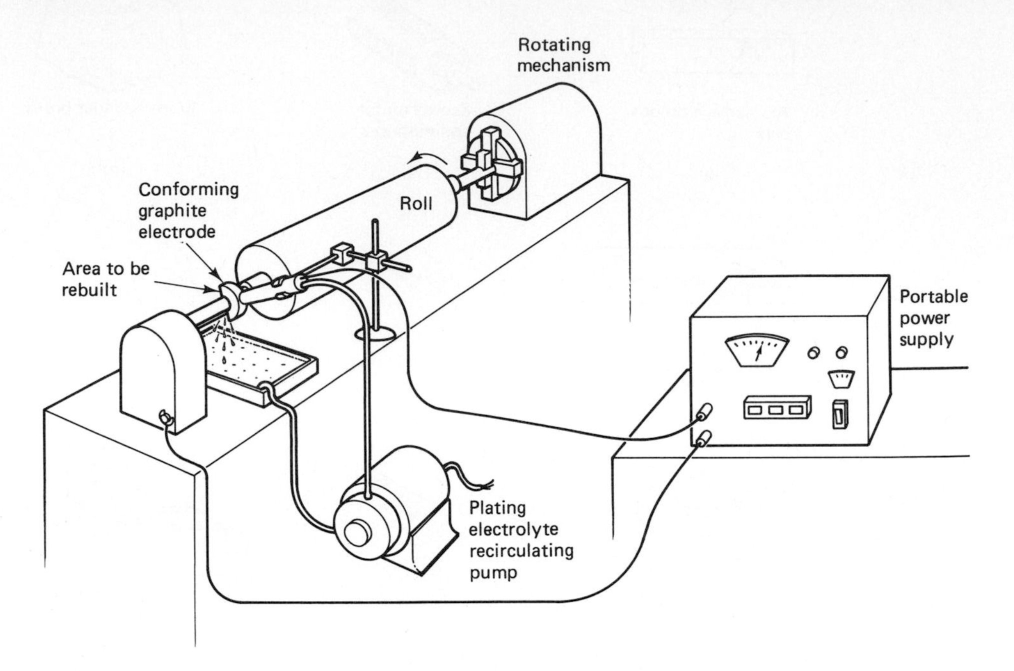

Figure 3–11 Schematic of selective plating of roll journals.

platings, however, are different than those of the wrought counterparts. Metals such as chromium are soft in their wrought form, but immersion-plated hard chromium is the hardest metal. Deposit stresses are one of the causes for this phenomenon. As of 1985, hard chromium deposits are not available by selective plating. Chromium solutions are available, but the deposit hardness is only about 53 HRC, compared with 60 to 70 HRC from bath-electroplated chromium. Some of the most wear-resistant selective platings are the cobalt and nickel/tungsten alloys. Figures 3–12 and 3–13 present test data from one laboratory comparing the abrasion and metal-to-metal wear characteristics of some selective platings to bath-plated chromium. These data show that the nickel/tungsten alloy has abrasion resistance approaching that of conventional hard chromium. Against a soft steel counterface, the system wear was lowest with a selective plating deposit of cobalt/tungsten. Against a hard counterface, cobalt/tungsten had a system wear comparable to bath-plated chromium. Thus it is possible to use selective platings for wear applications and achieve results that are comparable to conventional plating techniques.

Specification

The decision to use selective plating for coating a part, be it a repair or on a new design, should be made based on the factors mentioned previously. Some of the more important criteria are as follows:

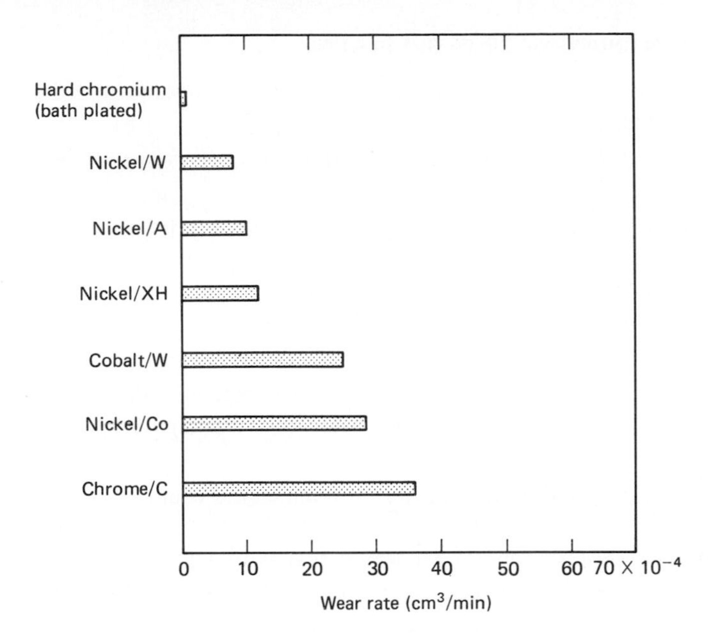

Figure 3–12 Abrasion resistance of some selective platings compared with an immersion-plated hard chromium; tested with a modified version of ASTM G 65 procedure: 5-lb (2.3 kg) normal force, 50/70 AFS silica.

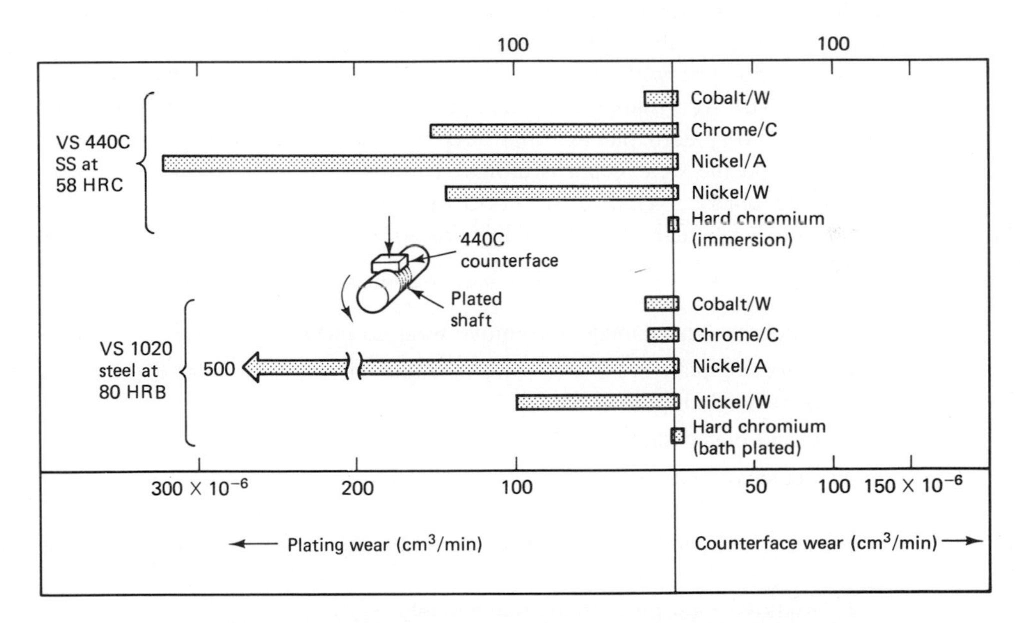

Figure 3–13 Metal-to-metal wear characteristics of some selective platings compared with immersion-plated hard chromium: block-on-ring apparatus, test rings were plated, blocks were 440 C stainless steel at 57 HRC and HRB 90 1020 steel, test speed 100 SFM (0.5 m/s), normal force 2 lb (908 g), no lubricant.

FACTORS TO CONSIDER IN CHOOSING SELECTIVE PLATING

1. Plating area should be less than 100 ft^2 (9 m^2), but it is most economical if the area to be plated is less than a few square feet.
2. Areas as small as 0.01 $in.^2$ (0.06 cm^2) can be plated without damage to surrounding surfaces.
3. There are no deposit thickness limits on many selective platings, but the economic thickness limits are:
   Hard deposits: 0.010 in. (250 μm)
   Buildup nickel deposits: 0.030 in. (750 μm)
   Buildup soft metal deposits (Cu, etc.): 0.10 in. (2.5 mm)
4. The part to be plated must have a geometry that will allow motion of the plating electrode (stylus) with respect to the area to be plated and drainage of plating solutions.
5. The part to be plated must be conductive and free of nonmetallic coatings.

If these guidelines do not pose a problem for a part that is under consideration for selective plating, the next step is to decide on the area to be plated. Plate only what needs to be done. Avoid plating over sharp corners and into reentrant corners. To decide on an appropriate thickness, consider the job that you are asking the plating to do. If you are rebuilding wear damage, it is recommended that the wear surface be made uniform. If the part is machined to remove uneven wear damage, the plating thickness that you will specify is the wear depth, plus a few thousandths of an inch (75 μm) extra for grinding allowance. If you wish to have the final surface plated with one of the very hard platings that have thickness limitations that are less than 1 mil (25 μm) and the wear depth is around 10 mils (250 μm), build up the surface with a rebuilding plating, grind to the finish dimension minus the final plate thickness, and apply the hard final plating. In this instance, the plating specification would be as follows:

1. Machine wear damage to remove wear damage.
2. Selective plate wear area with machinable nickel to wear depth plus 3 mils (75 μm) finishing allowance.
3. Grind plated surface to ××× dimension.
4. Selective plate wear area with 0.0002-in. (5-μm) hard nickel.

If the wear on a part is less than several mils, the plating specification might be:

1. Remove wear debris and contaminants.
2. Selective plate wear areas to be uniform with previously machined surfaces.

A great number of selective plating repair platings are done with no finishing after plating. This of course cannot be done if the part requires dimensional tolerances that are less than about 0.0005 in. (12 μm). In these cases, you must plate and regrind the deposit. If a particular deposit requires a preplate, the plater will do this without asking. Bonding platings are thin enough (less than 0.0001 in. or 2.5 μm) that their application should not affect subsequent plating in any way except adhesion.

In summary, selective plating is a very valuable tool for repairs and for applying metals with special characteristics in local areas. It is underused in many industries mainly because people are not familiar with the process. The information provided should serve as adequate basis for what needs to be known to use the process. There are many vendors that provide selective plating services, and the units themselves can cost as low as several thousand dollars. There is no reason why every designer and maintenance person should not have this process available for use.

HARD ANODIZING

We previously described anodizing as a form of electrochemical conversion coating. We will not discuss all electrochemical conversion coatings since the only one of these coatings that has significant utility in wear systems is hard anodizing of aluminum or, as it is also referred to, hard coating. Titanium and magnesium can be anodized, but the coatings are much thinner and softer than the coatings that can be produced on aluminum. Anodizing of aluminum is a simple operation. Anyone can anodize with a dc power supply and a solution of 10 percent sulfuric acid in water. The aluminum part is made the anode (+) in the sulfuric acid bath. The cathode can be stainless steel or lead. The current supply is turned on, the voltage is increased to about 15 V, and the current is monitored. Diffraction colors will appear on the surface of the anode. This is the coating developing. As the coating continues to increase in thickness, the current starts to decrease. The electrochemical conversion coating that is forming is aluminum oxide and it is an insulator. If the part is left in too long, the current will start to increase. The coating is being stripped. This phenomenon is the reason why there are thickness limits on conventional anodize coatings. If the plater tries to make the coating too thick, he may end up stripping the coating. Thus, hard anodizing is simply the process of building thick anodize coatings through solution and process control. One of the original hard anodizing processes utilized a solution of 15 percent sulfuric acid, not much different than the solution for conventional anodizing, but then it was learned that thick coatings could be obtained by controlling the cell parameters and by cooling the sulfuric acid bath. Conventional anodizing is performed in a bath at about 70°F (21°C) with a current density of about 15 A/ft^2 (530 A/m^2) at 15 V; hard anodizing requires a bath temperature in the range of 25° to 50°F (−4° to 10°C), a current density of about 30 A/ft^2 (1060 A/m^2) and a potential of maybe 40 V. Conventional anodize coatings are porous and clear (unless dyed), and the normal upper thickness is less than 1 mil (25 μm). Hard anodize coatings are gray to black in color and nonporous, and the normal thickness is 2 mils (50 μm).

The reported hardness of hard anodize coatings is about 1100 HK, but it is difficult to get an accurate hardness reading on these coatings because, being essentially ceramics, they tend to crack when hardness is measured by indentation techniques. It is even more difficult to get an accurate reading on the hardness of clear anodize because of the porosity. The hardness of aluminum oxide as a bulk material is about 2100 HK. Thus anodize coatings in general may be this hard if they do not contain voids that interfere with indentation hardness readings. The main point to be made is that both clear and hard anodize coatings are very hard because they are essentially aluminum oxide, a very hard ceramic.

Use of Hard Anodize Coatings

Some wear parts must be made from aluminum for various reasons, and there are limitations to hardfacing and plating of aluminum. None of the conventional fusion welding processes are applicable, and the only applicable hard platings are electroless nickel and some proprietary chromium processes. The thermal spray processes are very applicable, but they have the disadvantage of a mechanical bond to the aluminum. Hard anodizing is a way to produce a very hard surface on aluminum with an excellent bond. The coatings are essentially formed from the surface of aluminum; they are not deposited on the surface. The surface is converted into the coating. There is a volume expansion in the process, so there is a net growth of the surface. As shown in Figure 3–14, when a 2-mil (50-μm) coating is formed, 1 mil (25 μm) of the coating will be below the original part surface and 1 mil (25 μm) will be a growth (conventional anodizing is two-thirds into the surface and one-third growth). From the design standpoint, this is very important. If a user specifies a 2-mil (50-μm) hard anodize coating, he must allow for a growth of 1 mil (25 μm) per surface for coating growth. Anodizing is a relatively slow operation; thus it is possible to control coating thickness (and part growth) quite accurately. A 2-mil (50-μm) hard anodize coating may require 80 min. The coating growth rate is relatively constant; if a user wants a coating thickness of 1.5 mils (38 μm), the plater simply leaves it in the bath for 60 min.

Another factor that is quite important concerning the use of hard anodize coatings

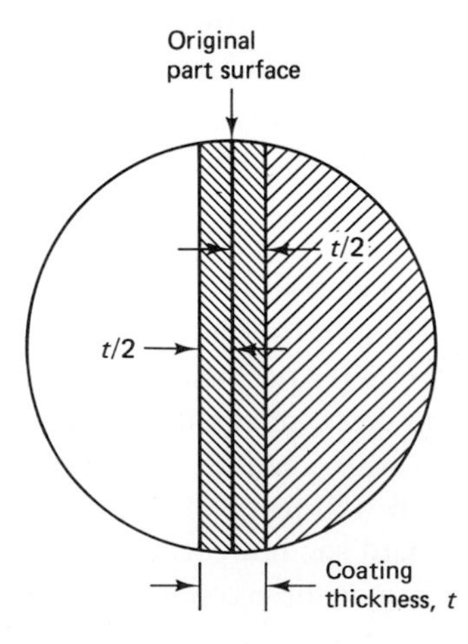

Figure 3–14 Penetration of hard coating into an aluminum substrate.

is that all aluminum alloys do not accept hard anodize coatings equally well. Hard anodize coatings on alloys with high copper or silicon content tend to be porous and not very hard. Table 3–2 lists some of the aluminum alloys that are particularly troublesome; these should be avoided.

Another factor that users should be aware of is the effect of sharp corners. Hard anodize solutions have very good throwing power; threaded fastener holes, keyways, and female corners will coat without the use of the special techniques that would be required in electrodeposition of metals. But since the coating seems to grow perpendicular to surfaces, the coating on sharp male corners tends to be brittle and prone to spalling. The solution to this problem is easy; all male corners should have a generous radius. The term "radius" is emphasized. A chamfer does not solve the corner problem; it takes a radius. The more generous the radius, the better; but a practical limit for the sharpest edge that will not cause a weak coating is a 0.030-in. (0.75-mm) radius.

Application for Wear Problems

Hard anodizing is often the lowest-cost wear-resistant coating that can be applied to aluminum. It is probably the coating that other coating candidates must compete with. If a design situation requires that an aluminum part be subject to low-stress abrasion, hard anodizing should definitely be considered. Questions should be asked such as the following: Is a 2-mil (50-μm) thickness adequate for this part? If the system in mind can tolerate 10 mils (250 μm) of wear before the part must be replaced, it may be more cost effective to use a thermal spray coating that can be applied 10 or more mils (250 μm) thick. If the part requires a conductive surface, electroplating, metallic thermal spray coatings, or hard thin-film coatings would be more appropriate wear coatings. Hard anodize coatings are dielectrics. They can be made conductive by silver impregnation, but this process is not widely available. In general, the low-stress abrasion resistance of hard anodize coatings on aluminum is comparable to most of the hard platings with the exception of hard chromium. The latter has better low-stress abrasion resistance.

For metal-to-metal wear systems, hard anodize coatings tend to wear out contacting metals (even hard metals). A hard anodized bushing would wear a steel shaft. Similarly,

TABLE 3–2 SUITABILITY OF VARIOUS ALUMINUM ALLOYS FOR HARD COATING

Preferred alloys for hard anodizing	Difficult alloys
5052	2011
5050	2017
6061	2024
6063	7075
3003	Cast and wrought alloys with
1100	Cu > 4% or
	Si > 7%

two hard anodized surfaces should not be run together. Hard anodizing tends to increase surface roughness. When the hard, rough surface of the coating rubs on a metal, it tends to have an abrasive action. When hard anodize is self-mated, the hard asperities on the hard anodize surfaces tend to cause comminution of the coatings on both surfaces. Hard anodize coatings are successfully used against hard metals and self-mated in reciprocating systems where the stroke length is sufficient to spread wear over a large area and remove trapped abrasive debris. For example, a 1-in. (25-mm) diameter hard steel shaft rotating in a hard anodized bushing that is 2 in. (50 mm) long would not work. The same system would probably work if the shaft reciprocated in the bushing with a stroke length of 6 in. (150 mm).

Metal-to-metal sliding wear problems with hard anodize coatings can be minimized by using hard anodize coatings that have been impregnated with PTFE (Teflon). There are a variety of techniques for impregnating hard anodize with PTFE. Probably all of them are proprietary, but they are readily available from coating service companies. If hard anodize coatings are contemplated for unlubricated sliding systems, the use of the PTFE-impregnated hard anodize coatings is recommended. The impregnation does not alter dimension, and the PTFE serves as the lubricant in the system.

Summary

Hard anodize coatings should definitely be considered for wear systems that require aluminum parts. Hard anodize coatings are resistant to many types of low-stress abrasion, and they can be made to work in many sliding systems. They have moderate utility in systems that involve fluid-assisted wear, slurry erosion, solid particle erosion, and liquid erosion. The coating itself is resistant to many chemicals (except caustics). Because of their brittle nature, they are not normally used for impact wear applications or for applications that involve hertzian type of loading. They can spall because of the soft substrate. They can have questionable serviceability in systems that involve fretting motion because of the abrasion tendency described in our discussion on metal-to-metal wear. Hard anodizing should be considered for any form of wear that involves aluminum parts, but the use factors that we have described should be considered.

Finally, if hard anodize is selected as the best type of coating for an application, a proper coating specification should show the following:

1. Type of aluminum
2. Areas to be coated
3. Coating thickness
4. If impregnation is desired

APPLICABILITY OF PLATINGS FOR WEAR RESISTANCE

The usual reasons for using plating are the same as those for other surface treatments and coatings:

1. Alter physical properties of a surface
2. Corrosion protection
3. Improve wear characteristics
4. Alter dimensions

There are a number of factors that a potential user of plating should keep in mind when selecting platings for one of these reasons.

Physical Properties

As mentioned previously, the physical properties of platings are usually assumed to be the same as the properties of wrought materials, but we must qualify this statement. For example, it is well known that the density and elastic modulus of chromium plating can vary considerably. Elastic modulus values have been measured to range from about 14 million to about 30 million psi (96 to 207 $\times$ 10^4 MPa). It is known that plated copper is still a better heat and electrical conductor than plated nickel or plated iron, but there is a good probability that absolute values of some of these physical properties may be different from that of their wrought counterparts. If a particular physical property of a plating is critical to the function of a part, it may be necessary to measure that physical property on the specific plating that will be used.

Mechanical Properties

It is not common practice to use platings for mechanical properties; but when this is done, it is advisable to make property measurements on the plating. The tensile strength of nickel plating deposited as an electroform can range from as low as 40 ksi (276 MPa) for a decorative nickel to 200 ksi (1380 MPa) for a 50 HRC hard nickel. In general, platings should not be used to enhance the mechanical properties of a part. The harder platings such as chromium and nickel can be brittle. When these coatings are used on mechanical components that are likely to see significant elastic strains in service, they can have an adverse effect on mechanical properties, fatigue life in particular. If hard platings on the surface crack from service strains, the coating cracks can act as stress concentrations that will promote premature failure under conditions of alternating stress. For example, it would not be recommended to hard chromium plate a machine member such as a torsion bar. Hard chromium typically contains a profuse network of cracks that are invisible to the unaided eye, but they are stress concentrations. The fatigue strength of a chromium-plated torsion bar may be only one-half that of an unplated bar.

Similarly, it is not normally recommended to use platings, even soft platings, on parts where service stresses may cause a fatigue action on the plating/substrate interface. Platings should be used with caution in design situations that involve hertzian types of contact stresses, such as on rolling elements and gears. The common mode of failure in these components is surface fatigue. Platings on fatigue-loaded surfaces may spall.

Some ball bearing manufacturers make bearings with thin chromium plating on the balls. These may work fine if used within their rated load capacity. The point to be made is that there is risk of failure of the plating/substrate bond, and platings should only be used for hertzian loaded parts if the hertzian stresses have been calculated and determined to be low enough to minimize the risk of spalling.

Another mechanical property factor that should be considered in using plating is hydrogen embrittlement. We previously mentioned that some electrodeposited coatings produce minimal embrittlement effect. But all electrodeposits on hardened steels produce some embrittling effects, and these effects are not totally removed by the heat treatments used to reduce hydrogen effects. They reduce hydrogen effects; they do not eliminate them. It is best not to use electrodeposits on hardened steels that will see high tensile stresses in use.

The machinability of platings is a mechanical property of sorts. There are a few things that a plating user should keep in mind about machining platings. The corner nodules that form with heavy electrodeposits should be removed by processes that minimize stressing of the deposit bond. If nodules are present, it is recommended that they be reduced in size with hand grinding before the part is put in a lathe or grinder to generate geometry. The intermittent cut from a lathe tool is likely to rip off nodules and fracture pieces from the deposit away from the nodules. In general, platings that have a hardness lower than 40 HRC can be finished by cutting operations. Platings harder than this should be ground. Nodules (if they are present) should be removed by selective grinding from any plating before machining.

Corrosion Resistance of Platings

Just as there are many types of wear, there are many types of corrosion, so it is difficult to make blanket statements on the corrosion resistance of platings. However, platings are seldom if ever 100 percent free of pits and defects that go through the plating to the substrate. It is advisable to avoid the use of platings for applications that involve continuous immersion of the plated surface in a substance that is capable of corroding the substrate. For example, chromium plating is very resistant to many oxidizing chemicals, but it is highly likely that the chromium will contain through-thickness cracks. If platings are to be used on a part that will be immersed in a corrodent, the substrate should also be capable of resisting attack by the corrodent.

A forté of plating is resistance to environmental corrosion—rust or oxidation resistance in room air or outdoor environments. It is beyond the scope of this book to discuss this subject in detail, but most platings used for wear and rebuilding applications also offer atmospheric corrosion protection. The more noble platings such as chromium and nickel have better chemical resistance, but they produce a galvanic couple that leads to pitting at defects. Table 3–3 presents some estimates of how long various platings will prevent rusting in a nonindustrial outdoor environment. Indoors, if the relative humidity is less than 50 percent, a 0.1-mil (2.5-μm) plating of almost any type will prevent rusting. If the relative humidity in an indoor application is over 70

TABLE 3–3 ESTIMATED ATMOSPHERIC CORROSION PROTECTION OF VARIOUS PLATINGS IN NORMAL OUTDOOR EXPOSURE (NONMARINE, INDUSTRIAL ENVIRONMENT)

Plating (mils)*	Degree of rust protection on carbon steel (years)
0.1 cadmium	1
1 cadmium	2
0.1 chromium over 0.3-mil nickel	5
1 zinc	3
5 zinc	20
0.1 nickel	3
1 nickel	5

* 1 mil ≈ 25 μm.

percent, it may be necessary to use plating thicknesses like those listed for outdoor conditions in Table 3–3.

Platings for Wear Applications

The plating processes and types of plating that find wide use in wear applications are summarized in Figure 3–15. There are so many different types of platings that selecting the proper plating to use for a particular application may seem like a very complicated task. The purpose of Figure 3–15 is to show that platings for wear applications can be as simple as a few processes and a few plating types. The coatings listed probably account for 90 percent of the platings used for wear applications. Figure 3–16 shows laboratory data on the relative abrasion resistance of some platings. Hard chromium traditionally has the best abrasion resistance in this type of wear test. The coating is applied to a flat steel sample and ground. The plated surface is held against a rotating rubber wheel and 50- to 70-mesh silica sand is flowed at a controlled rate through the wheel/sample interface. This is considered to be low-stress abrasion. Figure 3–17 shows results on the metal-to-metal wear resistance of platings against a hardened steel. Silver and gold often show the lowest system wear in these types of tests. Under unlubricated conditions, all hard-metal couples wear noticeably. Silver and gold platings plastically deform and behave like a solid film lubricant. Gold and rhodium platings are commonly used on electrical components that require resistance to wear when sliding against other metals.

One additional type of plating that finds wide application in wear systems is thin, proprietary, hard chromium platings. A number of these platings are commercially available and are often identified by trade names. Most of these are electrodeposited, and essentially they may be the same as "flash" chromium plate in thickness and hardness; but these coatings can be different from flash platings from ordinary plating shops in that the suppliers of these thin coatings have developed techniques to produce uniform

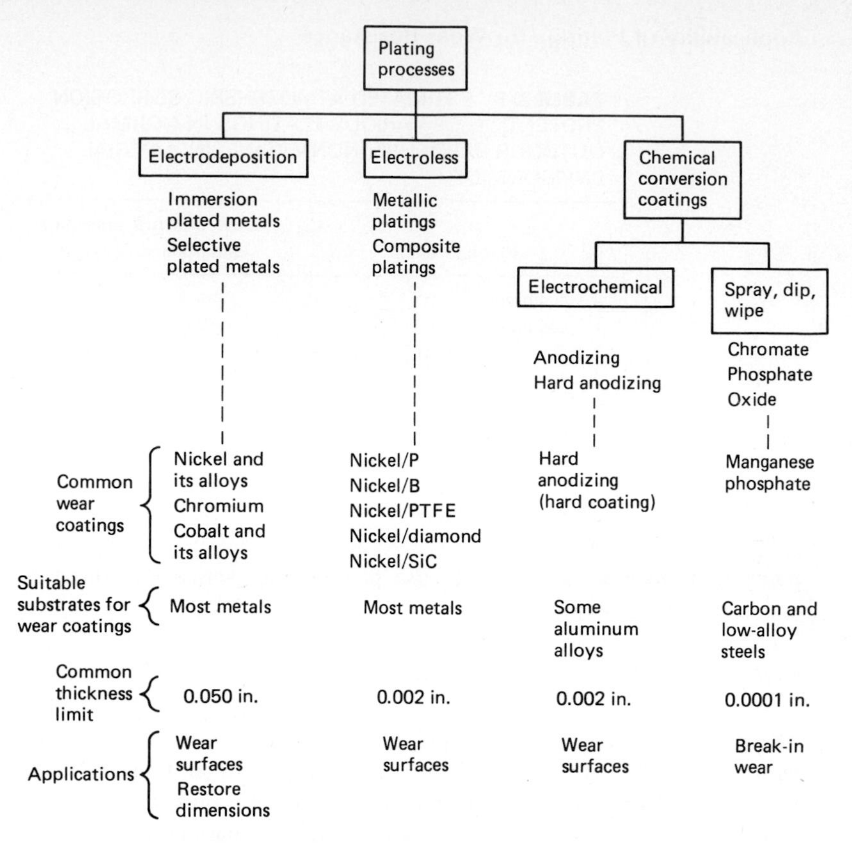

Figure 3–15 Use of plating processes for wear coatings.

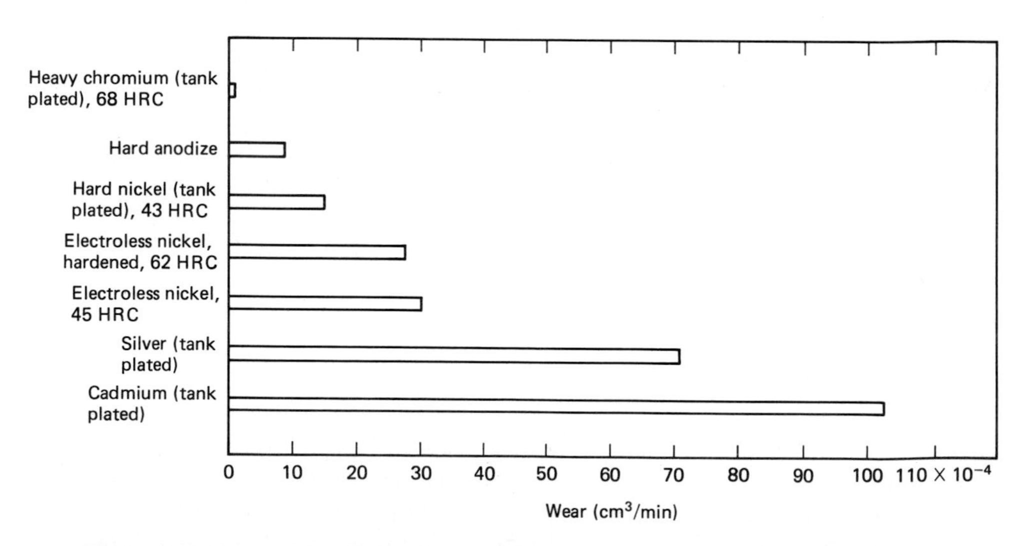

Figure 3–16 Abrasion resistance of some "wear-resistant" platings; ASTM G 65 test with 5-lb (2.3 kg) normal force.

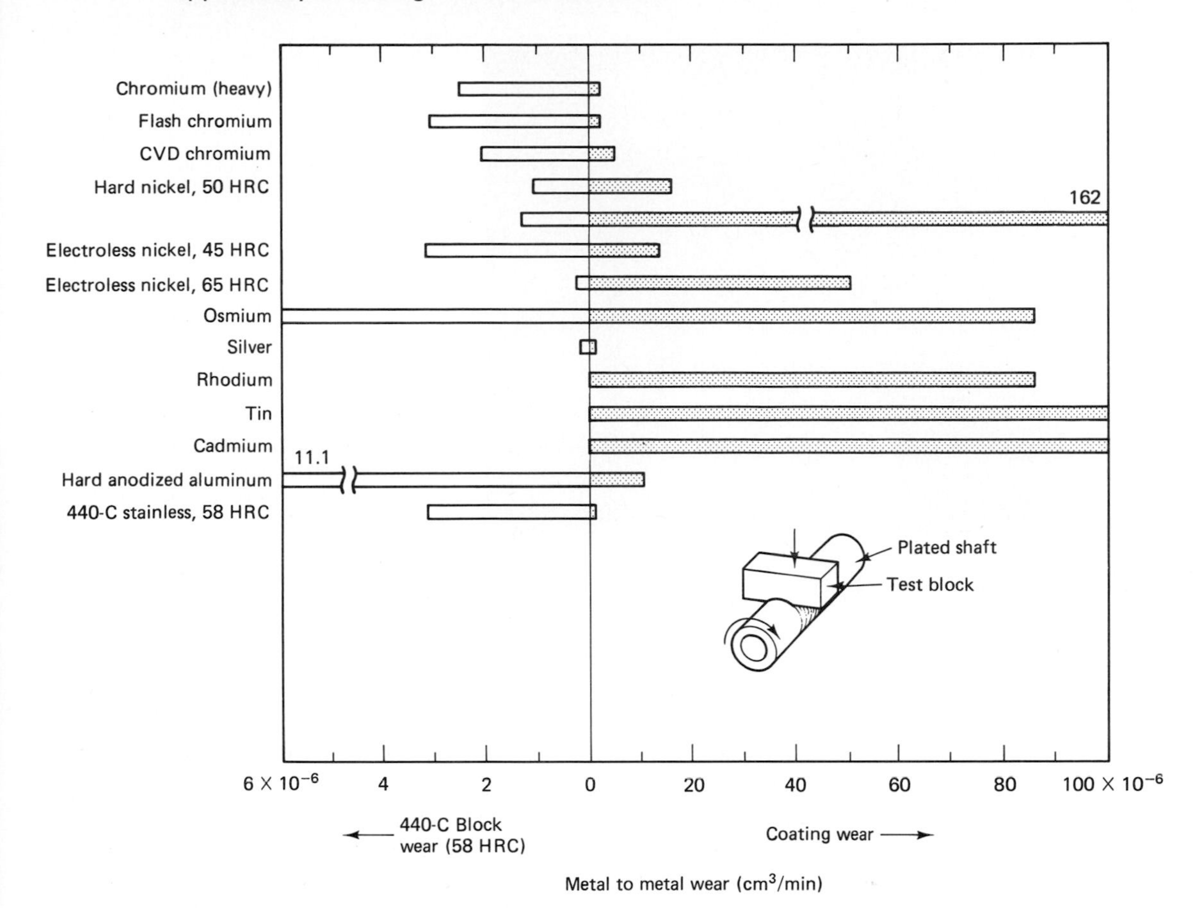

Figure 3–17 Metal-to-metal wear characteristics of some platings; sliding velocity 100 SFM (0.5 m/s), 2-lb (.9 kg) normal force.

hard deposits with excellent adhesion. Some can be applied to metals that are difficult to electroplate, such as aluminum and zinc die castings. Some can throw into holes. Some have special surface finishes. As shown in Figure 3–18, these coatings can be applied to a knife edge without significantly reducing edge sharpness. This makes these coatings very useful for applications on tools such as drills, reamers, taps, punches, and cutting devices. These coatings are normally applied at a thickness of 50 to 100 millionths of an inch (1.1 to 2.5 μm), and they often provide improved abrasion resistance compared to an uncoated tool steel at 60 HRC. For example, a new high-speed steel tap may last for ten holes. One of these thin chromium coatings may increase the useful life of the same tap to 100 holes. For this reason, many tool manufacturers sell lines of tools that are chromium plated.

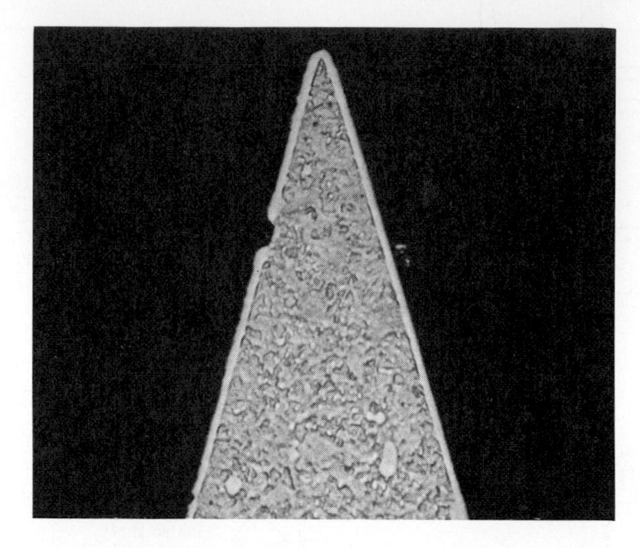

Figure 3–18 Hard chromium coating on a knife edge (800×); Tip radius 0.5 μm (0.00002 in) before plating; 2 μm (0.00008 in) after plating; plating thickness 2 μm (0.00008 in).

Use of Platings to Alter Dimensions

We have already discussed this in some detail in preceding sections, but we can summarize by restating that rebuilding is best done with the two electrodeposition processes, tank plating or selective plating. There is no technical limit to the thickness of a deposit, but the user should keep in mind that electrodeposition rates are slow compared to the deposition rates possible with welding hardfacing processes. In addition, plating labor rates are traditionally very high. Part masking can make rebuilding by electrodeposition quite expensive. It is a hand operation. When considering plating for rebuilding, it is recommended that the potential user weigh the costs by pricing all the steps that are necessary in candidate processes. Rebuilding by electrodeposition is usually most cost effective when the required buildup is less than 10 mils (250 μm) or when the part to be rebuilt is made from some material with limited weldability.

SUMMARY

Plating processes are very definitely competitive with welding hardfacing, hard vacuum coatings, and some other processes that we will discuss. They can offer lower distortion, less machining, and sometimes special properties that are not available from welding hardfacing deposits and other surface treatments. They should be considered for any application that might benefit from the application of a wear-resistant coating, but the decision on their use should be based on consideration of the selection factors that we have outlined in this chapter.

REFERENCES

DURNEY, L. J. *Electroplating Engineering Handbook,* 4th ed., New York: Van Nostrand Reinhold Co., 1984.

FELDSTEIN, N. "Composite Coatings," *Materials Engineering,* July 1981, pp. 38–41.

MAITLAND, D. W., and M. DUTSCH. "Selective Plating," *Metals Handbook,* Vol. 5, 9th ed. Metals Park, Ohio,: American Society for Metals, 1982, pp. 292–299.

RUBINSTEIN, M. "Selective Plating," *Metal Finishing,* July 1981, pp. 21–24.

Selectron Process Instruction Manual. Waterbury, Conn.: Selectrons Ltd.

SPENCER, C. F. "Electroless Nickel Plating: A Review," *Metal Finishing,* Jan. 1975, pp. 38–49.

TAPE, N. A., E. A. BAKER, and B. C. JACKSON. "Evaluation of Wear Properties of Electroless Nickel," *Plating and Surface Finishing,* Oct. 1976, pp. 30–37.

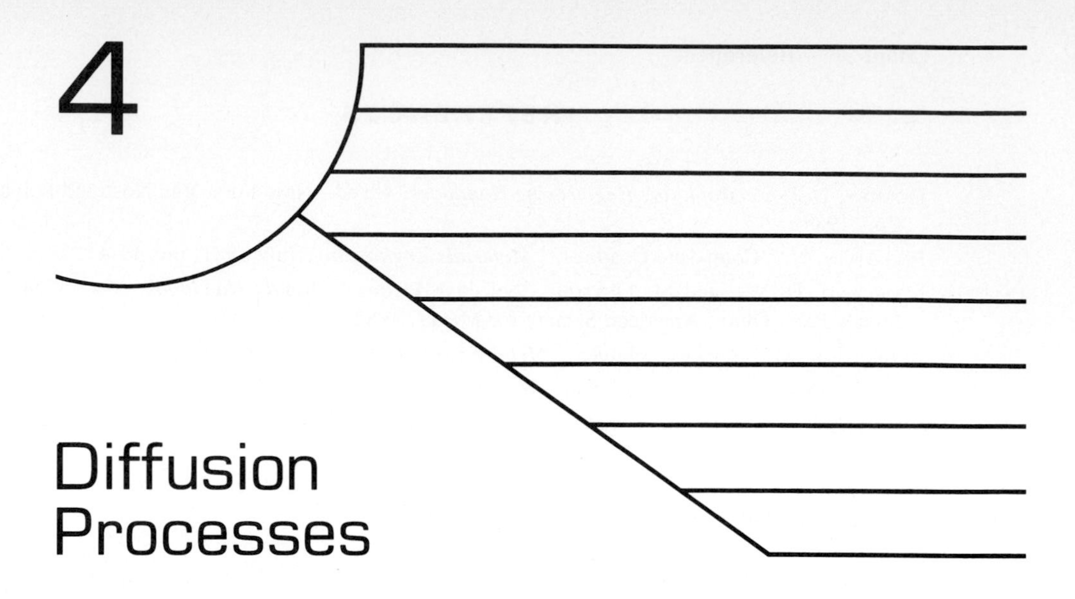

4 Diffusion Processes

In this chapter we will essentially discuss ''coatings'' that go into the substrate. There is no intentional buildup or increase in part dimensions with these processes. They cannot be used for rebuilding, but they can compete with hardfacing and surface coating in original designs. Diffusion processes by our definition are those heat-treating processes that have as their objective diffusion of some elemental species into the surface of a part to alter the surface properties. Diffusion by definition is the spontaneous movement of atoms or molecules to new sites in a material. Diffusion is easiest to see in liquids. Visualize two beakers, one containing clear water and the other containing water that has been turned black with ink. The beakers are connected at their base with a tube with a valve in the middle. When the valve is opened, the ink-colored water will start to enter the beaker containing the clear water. Within time the solution in both beakers will be the same color. Diffusion of the ink has occurred. One law that applies to diffusion processes states that for diffusion to take place there must be a concentration gradient:

$$\frac{dn}{dt} = -D\frac{dc}{dx}$$

where n = number of atoms or molecules (ink) diffusing down the concentration gradient per unit time per unit area
c = concentration of the diffusing species (number of ink molecules in the ink/water solution)
x = distance that the diffusing species (ink) travels (toward the clear beaker)
D = constant for the system in which the species is diffusing (water in our example)

that determines the rate of movement of the diffusing species. D is called the diffusion coefficient and it has units of area per unit time.

This equation is called Fick's first law of diffusion.

The diffusion coefficient contains a constant for the material in which the diffusion will occur and a term that involves absolute temperature. Thus, knowing the material and the temperature at which a diffusion process will take place, it is possible to calculate the diffusion rate of a particular species in a given host material. Diffusion processes are dependent on

1. Concentration gradient of the diffusing species (dc/dx)
2. Atomic or molecular nature of the diffusing species and the material in which the diffusion is occurring
3. Temperature at which the process will take place (D, T)

To the user of surface-hardening processes that involve diffusion, this means that, for example, if we wish to diffuse carbon into steel to increase surface hardness, we must have a concentration gradient of carbon (this is done by putting the low-carbon steel part in a carbon-rich atmosphere); the diffusion will be easy or difficult depending on the host material and the size of the atoms that we wish to diffuse in (carbon is a small atom and it diffuses easily in iron). Finally, the rate at which diffusion will occur is a function of the temperature. Diffusion is almost nonexistent at temperatures up to 800°F (425°C), but is very rapid at 1700°F (925°C). The user of diffusion processes should keep these factors in mind. Heat treaters use a wide variety of techniques to monitor the concentration gradient of the diffusing species. Sophisticated devices are used to monitor and adjust furnace atmospheres to ensure that a particular diffusion process will proceed as planned.

The complete expression for the diffusion coefficient is

$$D = de^{-Q/RT}$$

where d is a constant for a particular material with units of area/time. Q/RT is the activation energy required to make the diffusion process proceed. Q is the energy required for one atomic jump multiplied by Avogadro's number of atoms; R is Blotzmann's constant (1.38×10^{23} J/atom/K); and T is the absolute temperature. The potential user of diffusion hardening processes will not have to use any of these relationships to specify a diffusion process; they form the theoretical basis for diffusion reactions. The material constant in the preceding equation controls diffusion rate. The significance of d to the user is that diffusion is easier in some materials than others. For example, copper plating is used to mask areas on parts that are to be selectively carburized. Diffusion of carbon is difficult in copper, and it serves as a diffusion barrier.

In the remainder of this chapter, we will discuss the common and some of the not so common diffusion hardening processes. The objective in describing these processes is that they can at times be processes that are competitive with hardfacing. For example, a small 0.06 in. (1.5 mm) thick sheet metal lever may only need a hard area for a distance of 0.25 in. (10 mm) on one end. It would be possible to make this part from low-carbon steel and hardface the area that needs to be hard; it is also possible to make the piece from low-carbon steel and carburize the whole part. The process to use is the one that is most cost effective. We are discussing diffusion hardening processes to present information that can serve as the basis for the decision on the most cost-effective process.

CARBURIZING

Metallurgical Reactions

Carburizing is the process of diffusing carbon into the surface of low-carbon steels to increase the surface carbon content to sufficient levels so that the treated surface will respond to subsequent quench hardening. This process is one of the oldest hardening processes for iron alloys. Archaeologists have found wrought-iron tools from early civilizations that contained a thin, hard skin. It has been deduced that these tools were hardened by heating the tools in bones at elevated temperatures, followed by liquid quenching. Since bones are organic materials, they can satisfy the requirement of this diffusion process; they are carbonaceous and they can supply an excess of carbon at the surface to provide a concentration gradient to drive the diffusion process.

Modern-day carburizing is essentially the same as the ancient processes; the methods for heating the part and the source of carbon have changed. Carbon can be supplied by packing parts in charcoal, by organic gases, or by molten salts that contain carbon. Carburizing as used in surface hardening is mainly applicable to ferrous materials. It is possible to carburize other metals, but the effect can be detrimental and it may not form a wear-resistant, hard surface. The steels that are applicable to carburizing are the following:

1. Low-carbon steels
2. Resulfurized low-carbon steel
3. Low-carbon alloy steels
4. Low-carbon powder metal (P/M) compacts

Table 4–1 is a listing of the American Iron and Steel Institute (AISI) alloy designation numbers of some of the specific steels that are applicable. It is possible to carburize cast irons, but cast irons can all be quench hardened by other techniques and carburizing is not normally done. The metallurgical reason for carburizing is quite simple. As a review, there are essentially four common ways to harden metals:

1. Solid solution strengthening
2. Mechanical working
3. Precipitation hardening
4. Quench hardening

TABLE 4–1a CARBON STEELS SUITABLE FOR CARBURIZING

Carbon steels, nonresulfurized

AISI/SAE number	Chemical composition			
	C	Mn	P Max	S Max
Manganese 1.00 percent maximum				
1005	0.06 max	0.35 max	0.040	0.050
1006	0.08 max	0.25–0.40	0.040	0.050
1008	0.10 max	0.30–0.50	0.040	0.050
1010	0.08–0.13	0.30–0.60	0.040	0.050
1012	0.10–0.15	0.30–0.60	0.040	0.050
1013	0.11–0.16	0.50–0.80	0.040	0.050
1015	0.13–0.18	0.30–0.60	0.040	0.050
1016	0.13–0.18	0.60–0.90	0.040	0.050
1017	0.15–0.20	0.30–0.60	0.040	0.050
1018	0.15–0.20	0.60–0.90	0.040	0.050
1019	0.15–0.20	0.70–1.00	0.040	0.050
1020	0.18–0.23	0.30–0.60	0.040	0.050
1021	0.18–0.23	0.60–0.90	0.040	0.050
1022	0.18–0.23	0.70–1.00	0.040	0.050
1023	0.20–0.25	0.30–0.60	0.040	0.050
1025	0.22–0.28	0.30–0.60	0.040	0.050
1026	0.22–0.28	0.60–0.90	0.040	0.050
1029	0.25–0.31	0.60–0.90	0.040	0.050
1030	0.28–0.34	0.60–0.90	0.040	0.050
1035	0.32–0.38	0.60–0.90	0.040	0.050
1037	0.32–0.38	0.70–1.00	0.040	0.050
1038	0.35–0.42	0.60–0.90	0.040	0.050
1039	0.37–0.44	0.70–1.00	0.040	0.050
Manganese maximum over 1.00 percent				
1513	0.10–0.16	1.10–1.40	0.040	0.050
1522	0.18–0.24	1.10–1.40	0.040	0.050
1524	0.19–0.25	1.35–1.65	0.040	0.050
1526	0.22–0.29	1.10–1.40	0.040	0.050
1527	0.22–0.29	1.20–1.50	0.040	0.050

Carbon steels, resulfurized

AISI/SAE number	Chemical composition			
	C	Mn	P Max	S
1110	0.08–0.13	0.30–0.60	0.040	0.08–0.13
1117	0.14–0.20	1.00–1.30	0.040	0.08–0.13
1118	0.14–0.20	1.30–1.60	0.040	0.08–0.13
1137	0.32–0.39	1.35–1.65	0.040	0.08–0.13
1139	0.35–0.43	1.35–1.65	0.040	0.13–0.20

TABLE 4–1b ALLOY STEELS SUITABLE FOR CARBURIZING

AISI/SAE number	Chemical composition C	Mn	Ni	Cr	Mo	Other elements
1330	0.28–0.33	1.60–1.90	—	—	—	
1335	0.33–0.38	1.60–1.90	—	—	—	
4023	0.20–0.25	0.70–0.90	—	—	0.20–0.30	
4118	0.18–0.23	0.70–0.90	—	0.40–0.60	0.08–0.15	
4130	0.28–0.33	0.40–0.60	—	0.80–1.10	0.15–0.25	
4320	0.17–0.22	0.45–0.65	1.65–2.00	0.40–0.60	0.20–0.30	
4615	0.13–0.18	0.45–0.65	1.65–2.00	—	0.20–0.30	
4620	0.17–0.22	0.45–0.65	1.65–2.00	—	0.20–0.30	
4626	0.24–0.29	0.45–0.65	0.70–1.00	—	0.15–0.25	
4720	0.17–0.22	0.50–0.70	0.90–1.20	0.35–0.55	0.15–0.25	
4815	0.13–0.18	0.40–0.60	3.25–3.75	—	0.20–0.30	
4817	0.15–0.20	0.40–0.60	3.25–3.75	—	0.20–0.30	
4820	0.18–0.23	0.50–0.70	3.25–3.75	—	0.20–0.30	
5117	0.15–0.20	0.70–0.90	—	0.70–0.90	—	Si, 0.15–0.30
5130	0.28–0.33	0.70–0.90	—	0.80–1.10	—	
6118	0.16–0.21	0.50–0.70	—	0.50–0.70	—	V, 0.10–0.15
8615	0.13–0.18	0.70–0.90	0.40–0.70	0.40–0.60	0.15–0.25	
8620	0.18–0.23	0.70–0.90	0.40–0.70	0.40–0.60	0.15–0.25	
8630	0.20–0.25	0.70–0.90	0.40–0.70	0.40–0.60	0.15–0.25	
8720	0.18–0.23	0.70–0.90	0.40–0.70	0.40–0.60	0.20–0.30	
8822	0.20–0.25	0.75–1.00	0.40–0.70	0.40–0.60	0.30–0.40	
9310	0.08–0.13	0.45–0.65	3.00–3.50	1.00–1.40	0.08–0.15	

Metals can have about a dozen different types of crystal structure; some of the more common types of crystal lattices are shown in Figure 4–1. When an element (diffusing species) goes into solid solution in one of these crystal lattices, it does so by substituting for a host atom (substitutional solid solution) or by going between the sites of the host atoms (interstitial solid solution), or the diffusing species can form compounds within the host metal (Figure 4–2). Crystalline metals deform by movement of planes of atoms called dislocations. Adding impurity atoms (alloying elements) to a host metal strengthens the metal by impeding the motion of dislocations. Mechanical working strengthens metals by production of large numbers of dislocations that impede subsequent dislocation motion. Precipitation hardening strengthens metal in a fashion somewhat similar to solid solution strengthening, except in this case the motion of dislocations is impeded by the presence of precipitates, compounds that form by heating an alloy to allow agglomeration of impurity (alloy) atoms. For example, when certain types of aluminum alloys are heated (aged) at relatively low temperatures, alloy constituents such as copper form precipitates (aluminum/copper) that strain the crystal lattice and impede dislocation motion (strengthen).

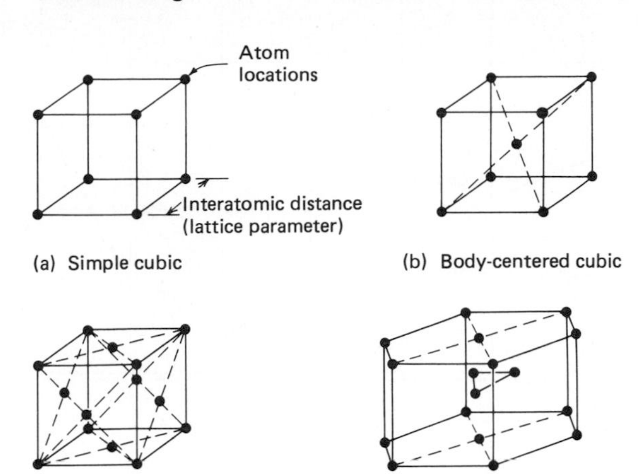

Figure 4–1 Typical crystal structures (K. G. Budinski, *Engineering Materials: Properties and Selection*, 2nd ed. Reston, VA: Reston, 1983).

The last way of hardening metals, quench hardening, is the most important. Unfortunately, it only works on a few metals, some titanium alloys, some aluminum bronzes, a few miscellaneous materials, and ferrous metals. The strengthening effect in metals other than ferrous metals is not very effective. Thus the most important quench-hardening materials are ferrous metals. The mechanism of quench hardening can be best reviewed by recalling the iron/carbon diagram. This diagram, which is shown in Figure 4–3, is

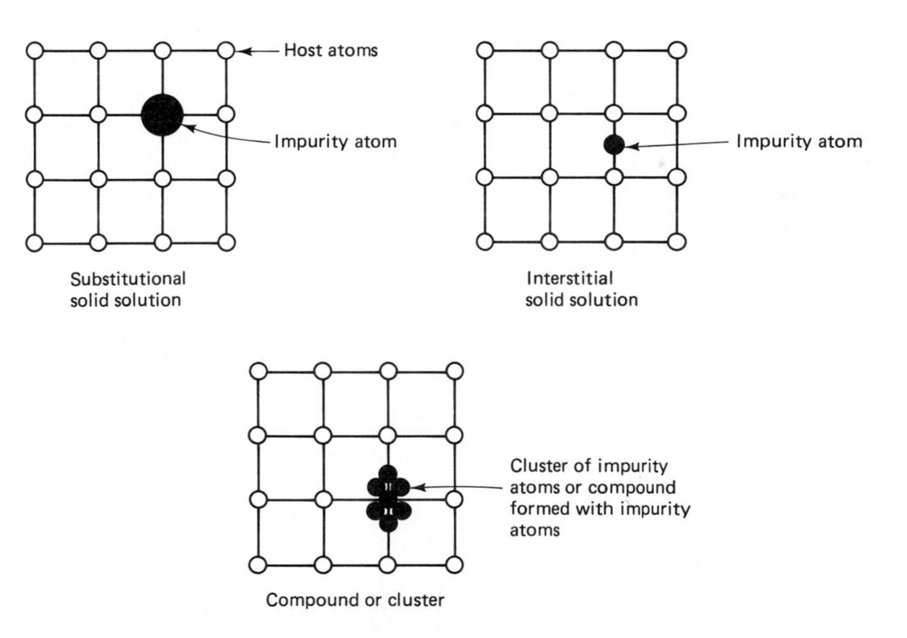

Figure 4–2 Types of solid solutions in crystalline solids.

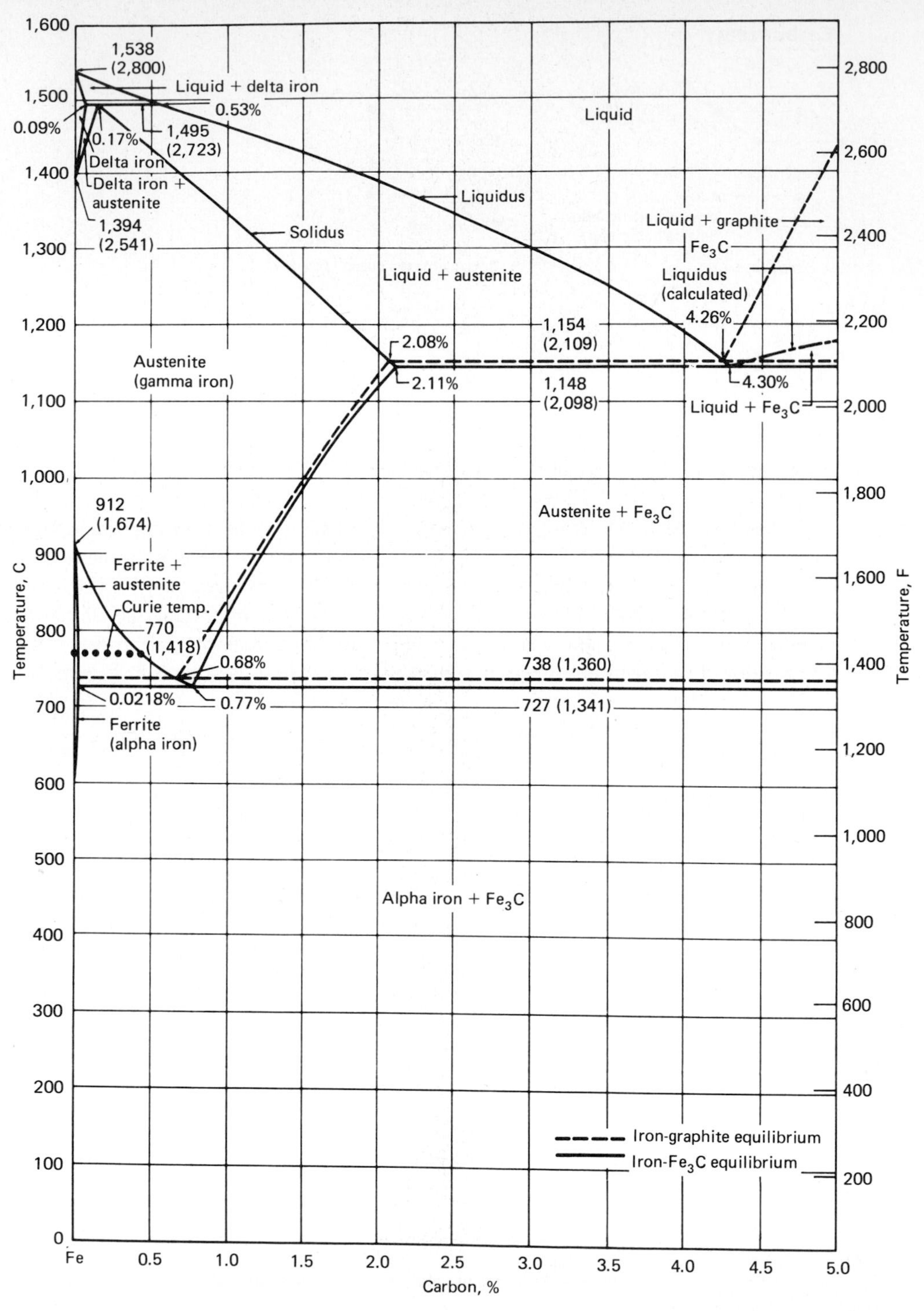

Figure 4–3 Iron-carbon equilibrium phase diagram (from *Metal Progress*, © ASM International, 1975).

a graph of the solid solubility of carbon in iron at various temperatures. The graph also shows that iron has different types of crystal structure at various temperatures. At room temperature, the crystal lattice has a BCC structure; what happens at elevated temperatures varies with the carbon content, but the important reaction is the ability of iron to transform to a FCC structure. This is called an allotropic transformation (a solid-state change of crystal structure) and it is the reason why steels can be quench hardened. The solubility of carbon in FCC iron is much greater than it is in BCC iron. If a steel is heated to above the critical transformation temperature (austenite region on the IC diagram), a lot of carbon will dissolve in the FCC iron, much more than could go into solution at room temperature. If a steel is quenched from this temperature, it will want to transform to its normal BCC structure. If quenching is rapid enough, the dissolved carbon is trapped in the lattice of the structure and the lattice distorts to a structure that is different from BCC or FCC; it forms martensite, which has a body-centered tetragonal structure. This means that one cube axis is elongated. The lattice is no longer an equal-sided cube. The strengthening effect of this transformation is dramatic. A steel with a tensile strength of 60 ksi (414 MPa) can have its strength increased to more than 200 ksi (1380 MPa). The hardness can increase from less than 20 HRC to 60 HRC. Solid solution strengthening usually results in strength increases that may be only 10 or 20 percent. Precipitation hardening and mechanical working can have strengthening effects that are more substantial than solid solution strengthening, but the strengthening effect of quench hardening of steels is by far the most useful hardening process. All the hardening processes that we have discussed are shown schematically in Figure 4–4.

Returning to our discussion of diffusion processes, the reason why carburizing was developed was that not all steels or irons respond to quench hardening. To form a 100 percent hard martensite structure in a plain carbon alloy (contains only carbon and iron), the carbon content must be at least 0.6 percent. The ancients who learned to carburize had only pure iron (wrought iron) filled with impurities such as slag. The iron contained very little carbon. They had to use a diffusion process to increase the carbon content to a sufficient level to allow quench hardening. The situation is still the same. The use of alloy additions has reduced the level of carbon that is required for 100 percent martensite, but the purpose of carburizing is to raise the carbon content at the surface of steel to a level that will allow quench hardening. With plain carbon steels, it is common practice to raise the carbon content of the surface to about 1 percent. With alloy steels, lower values may be adequate.

In summary, carburizing processes raise surface carbon contents to allow the surface to respond to quench hardening. All the available processes do this, but with different speeds and different ways of heating the part. The requirements for quench hardening ferrous metals are as follows:

1. Sufficient carbon content
2. Sufficient heating to allow transformation (from BCC to FCC)
3. Quenching rapid enough to prevent transformation to equilibrium structures

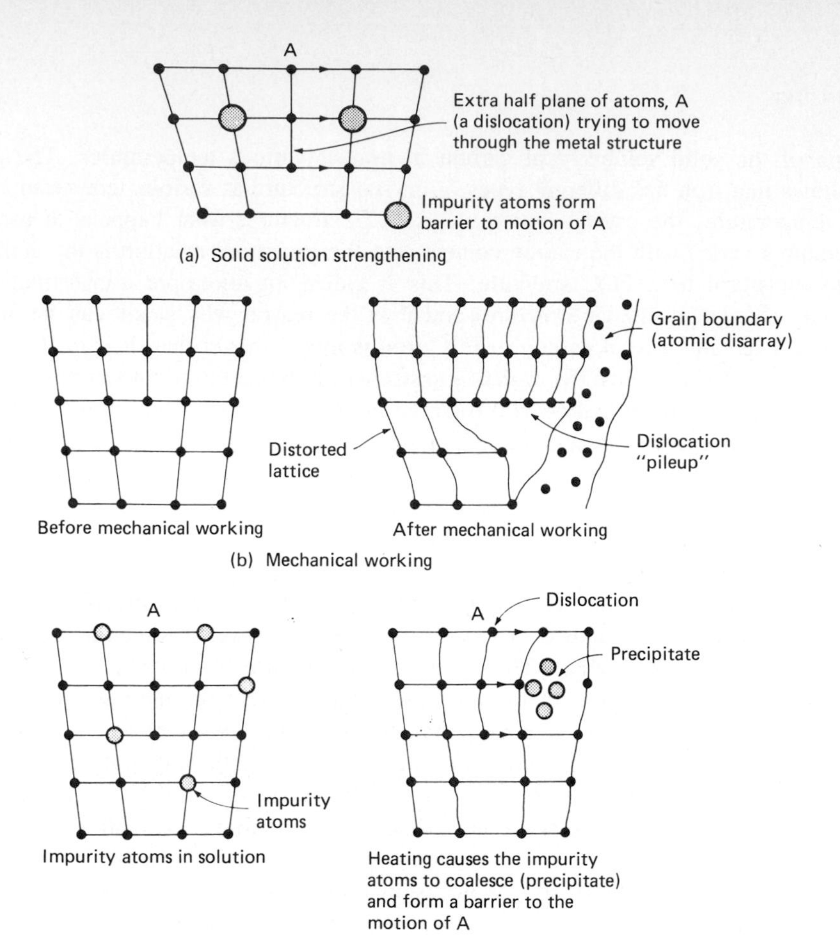

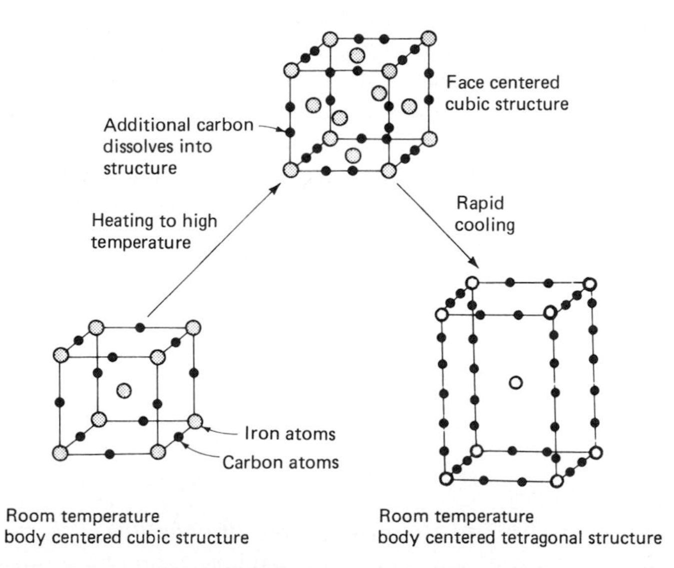

Figure 4–4 Strengthening mechanisms in metals (K. G. Budinski, *Engineering Materials: Properties and Selection*, 2nd ed. Reston, VA: Reston, 1983).

The carburizing process is intended to take care of the first requirement on steels that have insufficient carbon for direct hardening.

PACK CARBURIZING

Pack carburizing is the process of diffusing carbon into the surface of materials by heating them in a closed container filled with activated charcoal (Figure 4–5). This is the simplest and oldest carburizing process. It is not used for large volumes of parts but because it requires no equipment other than a furnace, a covered steel box, and the charcoal compound, it is still used where simplicity and minimal equipment requirements are factors.

The ''pack'' for pack hardening is usually a welded sheet metal or plate box with an open top and a flange for a sealing gasket. A cover is made to fit relatively tightly to the flange, and a gasket (usually asbestos) is placed on the flanged surface to minimize gas flow from the box. The box is filled with a carburizing compound, which is a granular charcoal (pea size) that is activated with chemicals such as barium carbonate that assist the formation of carbon monoxide, CO. The actual carburizing is done by CO gas and not by the charcoal per se. For diffusion to take place, the diffusing species must assume atomic form. For example, crystals of graphite will not diffuse into a steel; only carbon atoms will diffuse in.

The process works in this manner: the box is made of sufficient size to allow at least ½ in. (12.5 mm) of the carburizing compound around the part; the box is filled with the compound and the part is buried in the compound; the box is gasketed and the cover is put on; the box containing the part is put into any type of furnace that will allow uniform heating (usually a box furnace), and it is heated to the carburizing temperature for a prescribed length of time. The box can be slowly cooled from the carburizing temperature, or the box can be opened while still in the furnace, and the part is removed with tongs or lifting devices and quenched.

In the heated box the charcoal forms carbon dioxide, which in an excess of carbon (charcoal) converts to carbon monoxide, CO. The CO converts to atomic carbon at the part surface (and oxygen) and this carbon diffuses into the part. The box is sealed to reduce the amount of oxygen in the box, which keeps the charcoal from burning. The degree of carburization depends on the substrate (its carbon and alloy content), the carburizing temperature, and the time that the part is soaked at the carburizing temperature. The normal temperature range for carburizing is shown in Figure 4–6. Heat treaters usually have empirical data to tell them how long to leave the part in to achieve a desired degree of carbon diffusion.

The term commonly used for the entire field of surface-hardening processes is case hardening. The ''case'' referred to in this designation is the depth of the hardening below the surface. There are various ways to measure this depth, but a technique used in pack carburizing and some other carburizing processes is to place a notched pin of steel into the pack with the part to be carburized. A common pin size is ¼ in. (6 mm) diameter by 2 in. (50 mm) long. The notch is put in the center of the pin with a file.

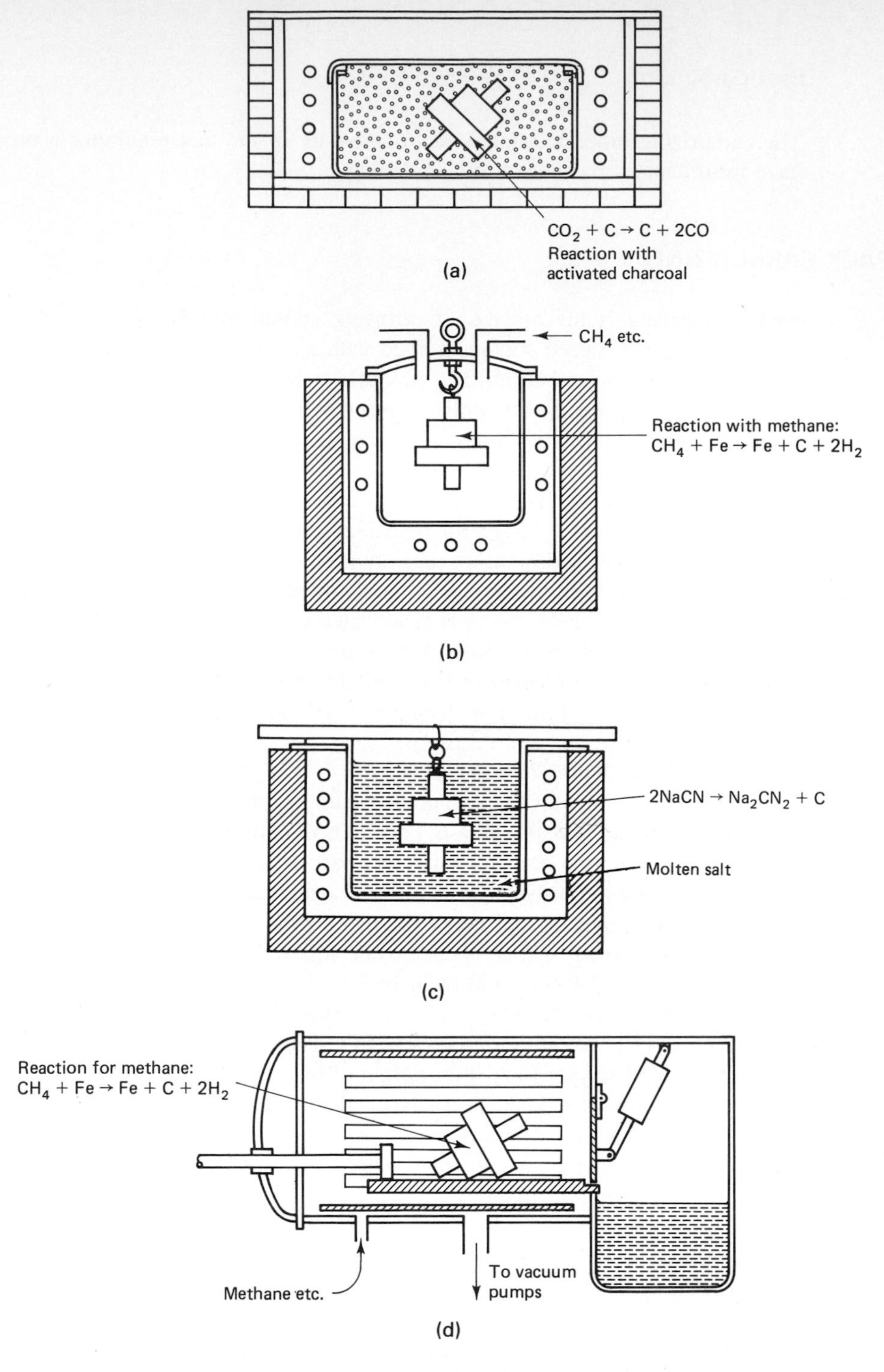

Figure 4–5 Schematic of various carburizing systems. (a) Pack carburizing. (b) Gas carburizing. (c) Salt carburizing. (d) Vacuum carburizing.

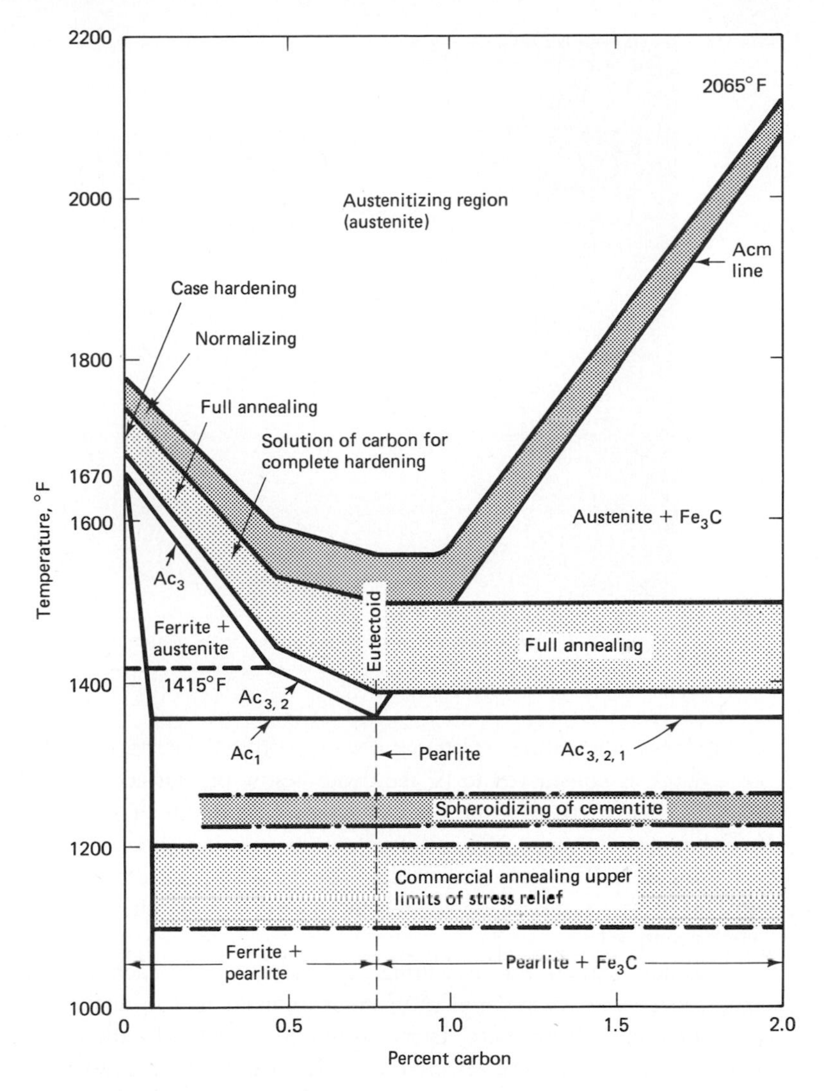

Figure 4–6 Temperature ranges for various heat treatments (G. M. Enos and W. E. Fontaine, *Elements of Heat Treatments*. New York: John Wiley & Sons, © 1963).

The pin is wired so that it can be removed from the pack while it is still in the furnace. It is quenched to harden the carburized surface and it is fractured. The depth of hardening from the surface is readily discernible in the fracture, and it can be measured with a loupe and an ordinary machinist's scale. The more sophisticated and more correct way to measure case depth is to use a microhardness test on a metallographic section taken from the carburized part. Figure 4–7 is a photomicrograph of a surface hardening showing

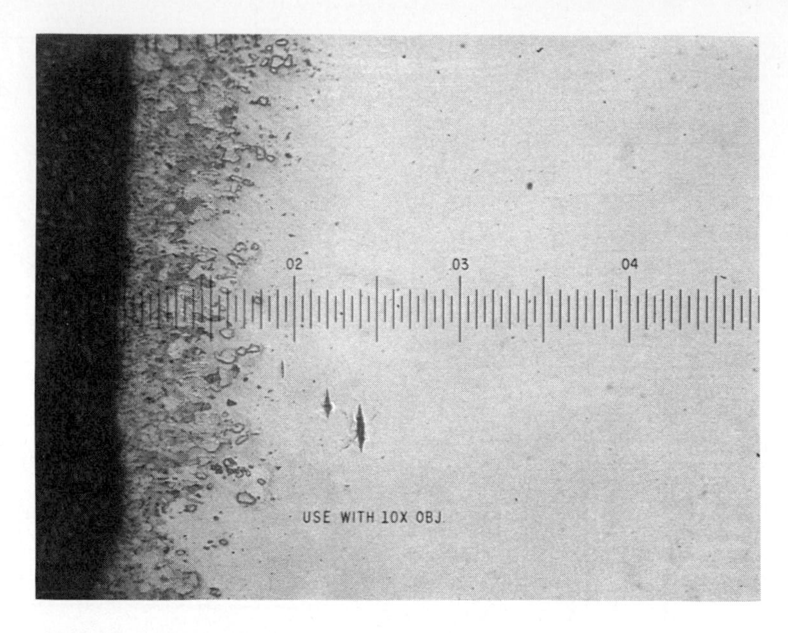

Figure 4–7 Microhardness indentations on the cross section of a borided steel.

microhardness indentations at various distances from the surface. The depth of hardening or case depth is considered to be the depth below the surface where the hardness falls below 50 HRC. This is called the effective case depth. The actual depth of carbon diffusion is not the same. The diffused carbon or other species does not stop abruptly. The hardness and carbon content decrease gradually until they are the same as the core. This is the true case depth, but the useful case depth is the depth of hard (over 50 HRC) material.

Another method for measuring case depth is to machine layers from the part surface (or a test bar) in 2-mil (50-μm) increments and to analyze the carbon content in the chips taken from these layers. A graph is made of carbon content versus depth below the surface, and the effective case depth is considered to be the depth at which the carbon content falls below 0.4 percent. There are additional ways to measure case depth. They apply to most surface-hardening processes. The most widely used technique is a metallographic section and microhardness survey. However, when specifying the depth of carburizing or other surface-hardening processes, it is recommended that the user be aware of the technique that the heat treater will use in measuring case depth. Ambiguity can be prevented by annotating the drawing specification to show exactly what is desired:

> Carburize, harden, and temper to a surface hardness of 60 HRC, case depth (hardness $>$ 50 HRC) shall be 0.04 in. (1 mm) minimum

Hardening of pack-carburized parts can be done by quenching from the pack or the parts can be reheated and quenched. Hardening from the pack presents some obvious problems. How do you remove a red-hot part from a sealed, red-hot steel box? Usually, the covers on pack carburizing boxes are only held by weights or mechanisms that can be actuated in the hot condition while still in the furnace. The cover can be moved to one side and the part can be pulled from the pack with tongs or by wires wrapped around the part. The parts are immediately quenched. Water or oil may be used depending on the alloy. The parts are then tempered. Tempering is usually minimal. Two hours at 300° or 350°F (150° or 175°C) is a common temperature. Case-hardened parts in general are tempered at low temperatures since the intent of all these processes is to achieve a very hard surface.

The alternative hardening technique is to slow cool the part in the pack to room temperature and reheat the part to the hardening temperature, followed by quenching and tempering. This procedure would be used if, for example, the part to be carburized was a 500-lb (227-kg) shaft that would not lend itself to easy removal from the pack. Many small parts would present a similar situation. There is another significant advantage to reheating for hardening; if the part is to be selectively hardened, it is possible to leave extra stock on areas that are to remain soft. The part can be carburized all over and the extra stock can be removed at this point (the case is removed), and when the part is reheated and hardened only the unmachined surfaces will harden. This procedure saves the expense of masking with copper plating or similar techniques. The disadvantage is that the reheating process drives the surface carbon deeper into the part, and the hardened case may be slightly lower in hardness (e.g., 56 HRC instead of 60 HRC) than if the direct quench is used.

In summary, pack carburizing is the least sophisticated carburizing process, but it remains a significant surface-treating process because for some users its advantages outweigh its disadvantages.

ADVANTAGES

1. Can be done in almost any type of furnace.
2. The equipment requirements are minimal (furnace, box, compound).
3. A wide variety of parts can be accommodated (as many as can be fitted and separated in a box, or as large as the box that will fit in the available furnace).
4. Very deep cases are obtainable (0.10 in.; 2.5 mm).
5. Requires lower operator skills than other processes.

DISADVANTAGES

1. Carburizing times are longer than for some of the other processes.
2. Not suitable for continuous production.
3. Labor intensive (pack loading, box maintenance, sealing, pack handling, etc.).
4. Unsuitable for thin, carefully controlled case depths.

GAS CARBURIZING

Gas carburizing is really the same as pack carburizing only the source of the carburizing gas is different and the furnace is pressurized with the carburizing gas. Any hydrocarbon gas can be used in gas carburizing (natural gas, acetylene, methane, manufactured gas, propane, and the like). The simplest system for gas carburizing is illustrated in Figure 4–5. A part can be put in a retort furnace and heated to the carburizing temperature with an inert gas such as argon flowing into the retort. At the carburizing temperature, the inert gas flow is discontinued and methane, CH_3, is introduced. The methane decomposes at the part surface to atomic (nascent) carbon and hydrogen, and the carbon diffuses into the part to produce the desired carburized case:

$$CH_4 + Fe \rightarrow Fe + C + 2H_2$$

The simple retort type of furnace shown in Figure 4–5 is seldom used for commercial carburizing operations because it would probably be even more labor intensive than pack carburizing, and it can result in risks of explosions. If there is any oxygen in a red-hot furnace when a combustible gas such as methane or propane is introduced, the ingredients of an explosive mixture are present. Almost all gas carburizing is done with a carrier gas into which a controlled amount of carburizing gas is introduced. Carrier gases are usually formed from natural gas, the same gas that homeowners buy from utility companies. The composition of natural gas varies with the source, but most natural gases are about 85 percent methane, with the remainder consisting of such things as ethane, carbon dioxide, nitrogen, and oxygen. Natural gas can be converted into a carburizing carrier gas or into a gas that is suitable as an inert heat-treating atmosphere by treating it in a device called a gas generator.

All metal surfaces should be protected from undesirable reactions with the prevailing atmosphere during heat-treating operations. Heating metals to elevated temperatures in air (over 1000°F) (538°C) can cause significant oxidation, decarburization, or other reactions that degrade surface properties. Gas-fired heat-treating furnaces can be adjusted so that the combustion products of the gas flames produce an atmosphere that protects the surface of metals from surface degradation, but the more prevalent technique is to use an endothermic gas atmosphere. The operation of an endothermic gas generator is illustrated in Figure 4–8. Natural gas is mixed with air in what is essentially a carburetor. This gas/air mixture is heated in a chamber that is fired by a separate gas supply or by some other external source of heat. The heated gas is passed through a catalyst bed that converts the input gas into a gas that can be inert, carburizing, or decarburizing. There are basically two types of gas generators for generating heat-treating atmospheres, endothermic and exothermic. As the name implies, with the endo generator, the heat to convert the gas is externally supplied; the exothermic gas generator partially combusts the input gas to develop the heat needed for catalytic conversion. Exo generators produce gases that tend to be decarburizing (they produce CO in the range of 1.5 to about 10 percent). This makes them less suitable for heat treating high-carbon steels and also less suitable for carburizing operations. Endo gas atmospheres typically have compositions that consist of about 20 percent carbon monoxide (CO), about 40 percent nitrogen,

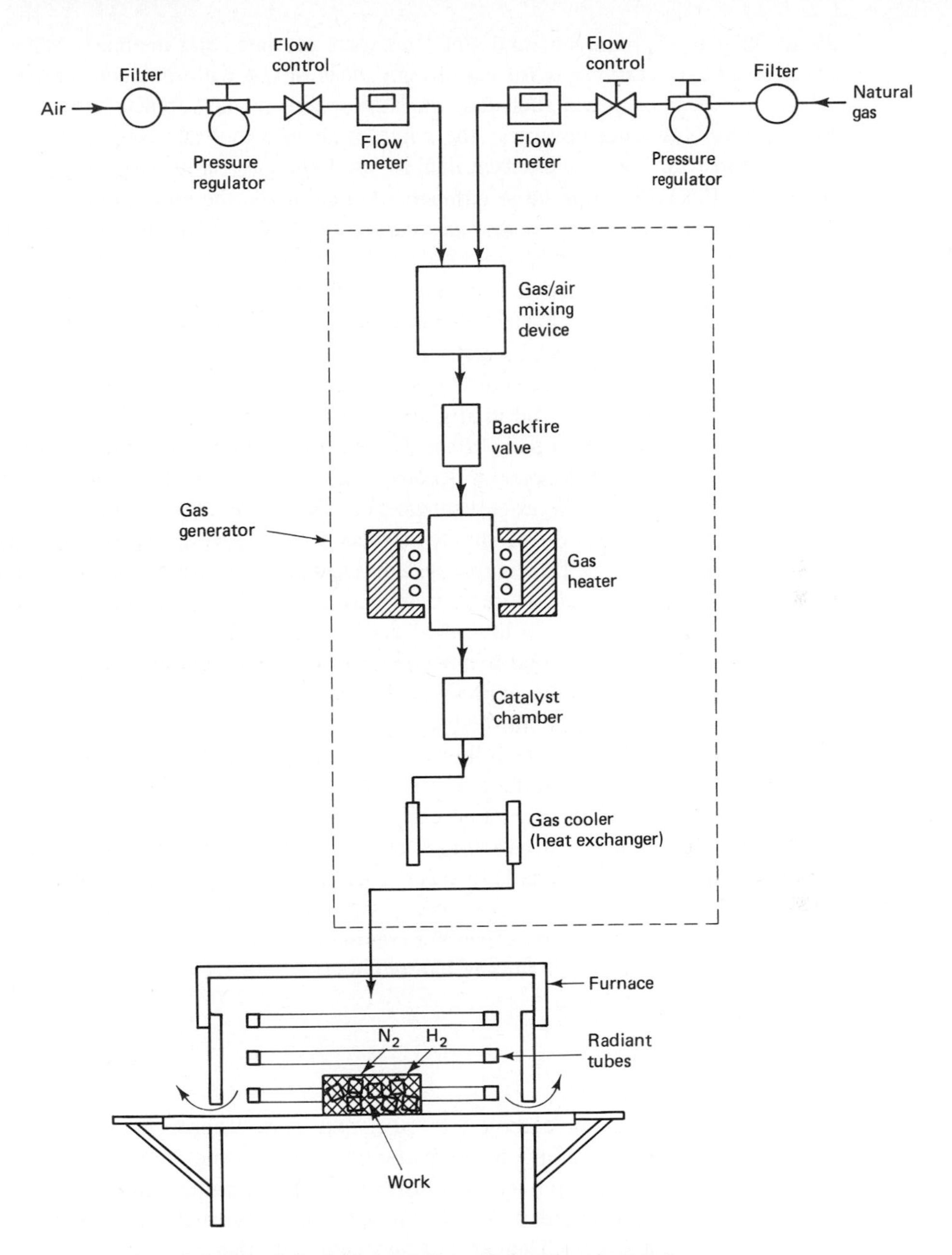

Figure 4–8 Schematic of an endothermic gas generator.

about 38 percent hydrogen, and small amounts of water and methane. When an endo gas is used as a carrier gas for carburizing, it is mixed with methane, propane, or one of the hydrocarbon gases such that the atmosphere in the furnace is endo gas with a hydrocarbon gas concentration in the range of about 3 to 10 percent.

Almost any type of furnace can be used for gas carburizing, but they must be leak tight. For carburizing large numbers of small parts, the work is loaded into baskets or fixtured in some form so that the carburizing gas can freely circulate around the parts. The parts are soaked for a prescribed length of time at the carburizing temperature and are then either slow cooled or quenched. Obviously, it would be most expeditious to quench right from the carburizing furnace. Some furnaces have integral quench chambers that allow this. Sometimes gas carburizing is done on conveyor hearth furnaces. Parts are placed on a wire mesh or similar conveyor; they enter a preheat zone in the furnace, then the carburizing zone, and finally are dropped into a quench bath. Controlled atmospheres are used in all furnace zones so that the parts receive the proper degree of carburization and do not oxidize or decarb in the preheat or quench zones of the furnace.

Gas carburizing's biggest advantage over pack carburizing is more accurate control of the composition and depth of the hardened case. Its biggest disadvantage is the equipment requirements. The control systems needed for gas control are more complicated than for pack carburizing. In fact, most gas carburizing is performed in continuous types of furnaces such as we described earlier. These types of furnaces require major capital expenditures. A second major concern with gas carburizing is safety. The gases used for carburizing can be explosive. The use of this process requires experienced personnel and reliable gas control systems.

The time required for gas carburizing is somewhat longer than for pack carburizing. It may take 8 hours to produce a case depth of 0.06 in. (1.5 mm) with the pack process; the same part and the same case depth may require a carburizing soak time of about 12 hours with gas carburizing. The user of a carburizing process does not need to be concerned with the details of gas carburizing, but he or she should be aware that gas carburizing is a relatively slow process that is capable of producing very accurate case depths. The process can be used for continuous carburizing systems; it is a candidate for high-volume production surface hardening.

LIQUID CARBURIZING

The process of diffusing carbon into steel and iron surfaces from molten salts that contain carbon-bearing compounds or carbon/graphite is called liquid or salt carburizing. The simplest type of equipment to perform this type of carburizing is shown in Figure 4–9. The system illustrated uses a metal pot that is externally heated with resistance techniques or by burners. The pot is filled with salts that melt and form a liquid that serves as the heat-transfer medium to heat the part to the carburizing temperature and also as the source of the diffusing species. The part to be carburized is immersed in the molten salt for a prescribed length of time to allow carbon diffusion from the salt.

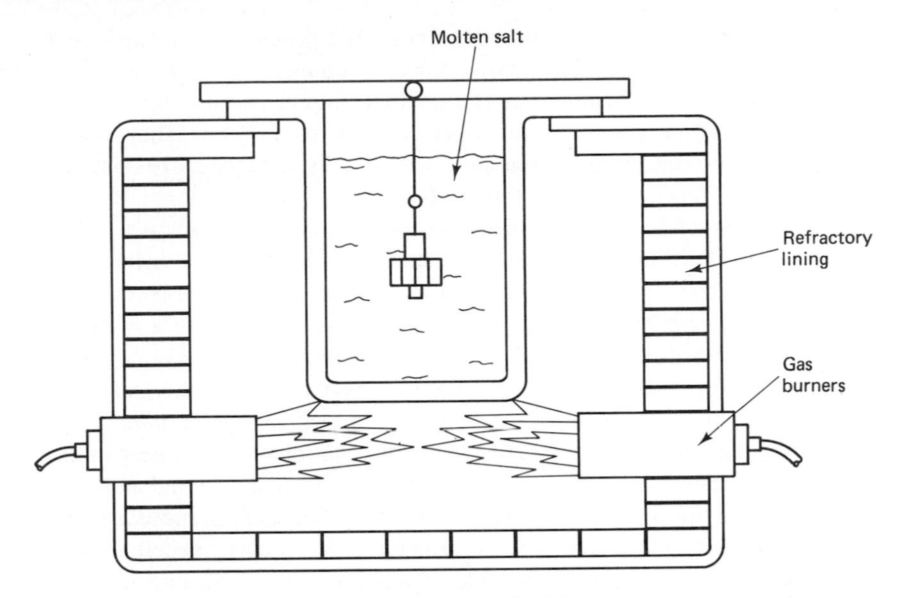

Figure 4–9 Schematic of a liquid carburizing system.

The part is then direct quenched or slow cooled and reheated for hardening of the case.

Many carburizing salts are proprietary in composition, but many have sodium cyanide, NaCN, as the active ingredient, the source of carbon as the diffusing species. These salts may become molten at a temperature of about 1000°F (538°C), and they can be used at temperatures as high as 1750°F (954°C). Two categories of salts are used in liquid carburizing, a low-temperature and a high-temperature salt. Low-temperature salts operate at about 1550° to 1650°F (845° to 900°C); high temperature salts operate at about 1650° to 1750°F (900° to 950°C). The compositions of the low- and high-temperature salts differ; low-temperature salts contain higher amounts of cyanide (10 to 20 percent), and high-temperature salts contain about 10 percent cyanide and barium chloride, which accelerates carbon diffusion. The nonactive ingredients in the baths are salts such as sodium chloride, potassium chloride, or compounds of alkaline earth metals. The mechanism of carburization may vary with the various bath compositions, but when cyanide salts are used, carbon for diffusion into the work can occur by the following reaction:

$$2NaCN \rightarrow Na_2CN_2 + C$$

The cyanide reaction with the metal to be treated can also result in the formation of carbon monoxide, which supplies carbon for diffusion, and nitrogen. Nitrogen usually becomes one of the diffusing species in CN baths; thus the case consists of carbon and nitrogen. Nitrogen in steels acts like carbon in solid solution strengthening, and it can also form hard iron/nitrogen compounds to make the case somewhat harder than a case obtained by the diffusion of only carbon. The presence of nitrogen in the case can be a disadvantage when parts are slow cooled, machined, and rehardened. The nitrogen makes the case more difficult to machine. Some liquid carburizing baths are cyanide free. Carbon/graphite particles are dispersed in a neutral salt. The carbon in the particles reacts to produce carburizing in a way that is thought to be similar to the way that charcoal carburizes in pack carburizing.

Low-temperature salt baths are usually used for producing thin cases of less than 0.030 in. (0.75 mm). The high-temperature bath carburizes faster and is used for heavy cases, up to 0.12 in. (3 mm). The lower-temperature bath is better for direct quenching. The lower temperature reduces distortion. Quenching from any heat-treating salt bath poses a number of problems. If a part is quenched into a low-temperature salt (500°F) (260°C), there are problems with salt compatability. If the salts are not compatible, there can be contamination or even violent reactions. If a part is air cooled, the salt will harden on the parts and create a removal problem. Quenching in oil creates probably the biggest problem. Salt gets stuck to the part and in blind holes and recesses. The reaction of the oil and salt can form a film that makes subsequent dissolution of the salts (in water) difficult. Water quenching is the best quench from the standpoint of salt removal. The salts are soluble in water and the water quench helps removal. From the user's standpoint, salt removal is not a significant factor. Heat treaters who use salt hardening and carburizing have developed techniques to effectively remove salt from most types of parts. The user needs to be aware that salt bath heat treatments present the risk that small blind holes may present salt-removal problems, and salts left on quenched parts can cause rusting or pitting on finished part surfaces.

In general, liquid carburizing is best suited for job-shop types of operations with few parts at a time. Salt bath carburizing can be automated with in-line salt baths for carburizing, reheating, and quenching, but such installations are capital intensive. The major advantage of liquid carburizing compared to the other techniques that we have discussed is speed. For example, a 0.060-in. (1.5-mm) case can be achieved at 1500°F (815°C) in a soak time of only about 20 min. The case is even slightly harder than with the other techniques because of nitrogen pickup from the CN (this does not occur in the CN-free baths).

The following are some other disadvantages of liquid heat-treating baths:

1. Cyanide salts are poisonous and can form poisonous gases; there is a disposal problem with used salts and rinse solutions and an operator handling problem.
2. Salt baths are a maintenance problem; pots and heating electrodes require periodic replacement.
3. Liquid carburizing is a batch process.
4. Salt baths require daily maintenance and composition checks.
5. Salt baths are difficult to shut down and to start up.

In spite of these considerations, liquid carburizing satisfies a need in the field of surface hardening; it is the fastest carburizing process, equipment requirements are reasonable, and for a job-shop type of operation it may be the best way to make money—more parts per unit time.

VACUUM CARBURIZING

The process of heating parts in a vacuum furnace and diffusing carbon into the surface of the parts by the action of a partial pressure of a cabonaceous gas introduced into the furnace is called vacuum carburizing. The mechanism of carburization is the same as gas carburizing, and the same gases can be used (natural gas, methane, propane, etc.). A vacuum carburizing system is shown schematically in Figure 4–10.

Vacuum furnaces are usually steel pressure vessels with water cooling capability on the outer vessel. The work chamber is a box or shell type of structure inside the vessel that contains the heating elements. The heating elements are usually resistance heated. The resistance elements do not touch the outer shell. Thus there is little heating of the shell since heat transmission from the elements by convection is minimal. Convection would be nil in a perfect vacuum. The work is heated by radiation from the elements. This is a very efficient system from the energy-consumption standpoint. The heat is almost entirely radiant, and molybdenum reflectors can be used to ensure that most of the radiant heat goes to the work. Heating is slower than in convection and salt furnaces, but more efficient.

Vacuum carburizing furnaces usually contain an integral quench chamber. The work can be manipulated to the quench section by a mechanism built within the furnace. To carburize, work is placed in the hot zone of the furnace in such a manner that the radiant heating will be as uniform as possible (large parts toward the elements, small

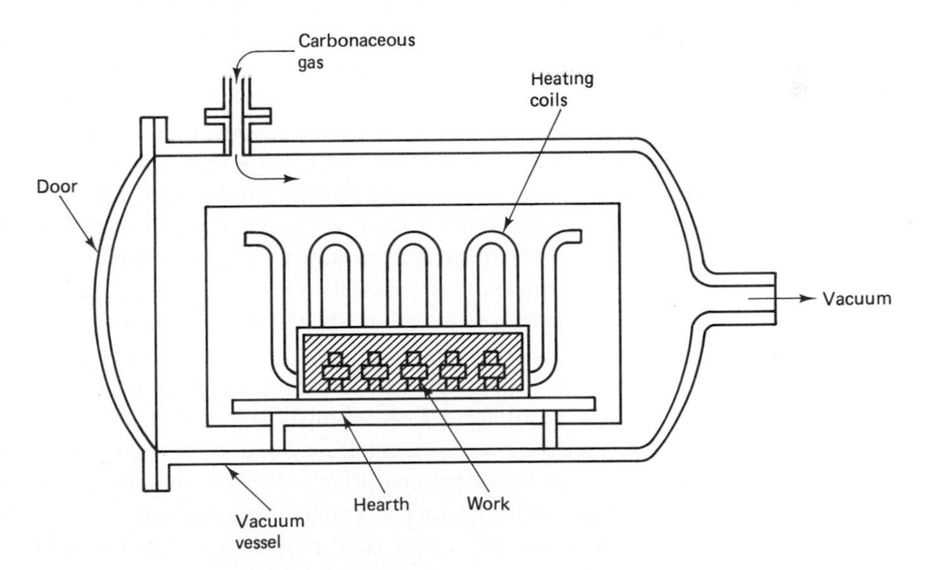

Figure 4–10 Schematic of vacuum carburizing equipment.

parts away from the elements). The furnace is sealed and evacuated, 10^{-4} torr (0.13 Pa) is usually adequate. The work is heated to the carburizing temperature. High temperatures are used to speed the process, 1800° to 2000°F (982° to 1093°C). When the parts are at temperature, the carburizing gas is introduced with a pressure of about 300 torr (3.9 Pa). The carburizing soak is commenced and the part may be soaked after the carburizing cycle to diffuse the carbon deeper into the part. The temperature of the parts is then lowered (to reduce distortion on quenching), and the parts are transferred into the quench tank in the quench zone.

The major advantage of vacuum carburizing over the other types of carburizing is freedom from scale and surface reactions. Parts stay clean and bright, and the case can be very uniform in composition and depth. It allows precise control of all the carburizing variables (temperature, atmosphere, soak, etc.). Times for vacuum carburizing are somewhat less than those for gas carburizing.

The disadvantages of vacuum carburizing are predominately related to equipment costs and throughput. Vacuum furnaces are very costly to buy and maintain. The work that can be put into the furnace (throughput) is a function of the size of the unit, and costs seem to go up exponentially with size. A furnace installation that can handle a 1000-lb (450-kg) load may cost $1 million. Vacuum carburizing can be done at commercial heat treaters without concern for the equipment costs. Thus this process can be a candidate for use whenever its advantages of clean surfaces and precise case control are service requirements.

NITRIDING

Nitriding is the process of diffusing nitrogen into the surface of ferrous metals to produce a hard case. Nitrogen is a small atom like carbon, and its effect on steels is similar to that of carbon, but with additional advantages and some disadvantages. It can make cases that are harder than carburized cases, but nitriding is slower and usually more complex than carburizing. As mentioned previously, nitrogen can solid solution strengthen steels, but the more important effect is the ability of nitrogen to form hard compounds with iron. These compounds can raise the surface hardness of some steels to as high as 70 HRC. This is a higher hardness than can be obtained in quench hardening even high-alloy steels. This very hard surface has utility in wear systems.

The mechanism of nitriding is generally known, but the specific reactions that occur in different steels and with different nitriding media are not always known. Nitrogen has partial solubility in iron. It can form a solid solution with ferrite at nitrogen contents up to about 6 percent. At about 6 percent nitrogen, a compound called gamma prime is formed with a composition of Fe_4N. At nitrogen contents greater than 8 percent, the equilibrium reaction product is epsilon compound, Fe_3N. Nitrided cases are stratified. The outermost surface can be all gamma prime; if this is the case, it is referred to as the white layer (it etches white in metallographic preparation). It is undesirable. It is very hard, but it is so brittle that it may spall in use. It is usually removed if present, and there are special nitriding processes aimed at reducing this layer or at making it less brittle.

The epsilon zone of the case is hardened by the formation of the Fe_3N compound, and below this layer there is some solid solution strengthening from the nitrogen in solid solution. Nitrogen will diffuse into any ferrous metal, but the reactions that we are discussing apply to ferritic structures. Conventional nitriding can be done on carbon steels, cast irons, alloy steels, and stainless steels, but the most significant surface hardening is obtained with a class of alloy steels of the Nitralloy type that contain about 1 percent aluminum:

C	Cr	Mo	Al
0.2/0.4	0.9/1.8	0.15/.45	0.85/1.2

When these steels are nitrided, the aluminum forms AlN compounds that can form discrete particles that strain the ferrite lattice and create dislocations that strengthen and harden. Other elements that enhance the ability of nitrogen to form hard compounds are chromium, molybdenum, vanadium, tungsten, and titanium. The Nitralloy types of materials form the hardest and deepest cases. Some alloy steels respond to nitriding, but the case depths are less and the surface hardness is lower (Figure 4–11). Because of the lower response of other steels to nitriding, it is not common to use conventional

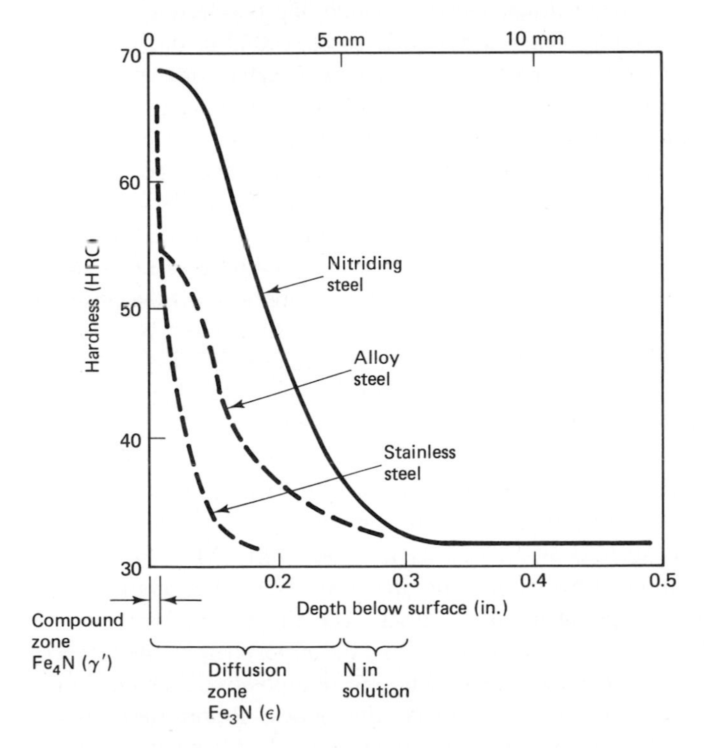

Figure 4–11 Nitride case profiles for various steels.

nitriding on carbon steel and cast irons, which typically contain no alloying elements to enhance nitride formation.

Nitriding is a subcritical hardening process. It is usually done in the temperature range of 900° to 1100°F (430° to 590°C). It does not involve a quench. Parts can be slow cooled from the nitriding temperature and they will be hard without subsequent heat treatments. When alloy steels are nitrided, they are hardened and tempered at a temperature that is at least 50°F above the nitriding temperature. This is done to promote a ferrite structure and to minimize distortion during the nitriding process.

There is no technical limit to the depth of case that is obtainable with nitriding. However, a case depth of 0.030 in. (0.75 mm) may take a soak at temperature of as long as 60 hours. The longer the nitriding time is the greater the tendency to form a white layer and the thicker the white layer becomes. For these reasons, nitride cases are usually shallower than carburized cases.

Another idiosyncracy of nitrided cases is a tendency to experience some surface growth and in turn some dimensional change. The formation of nitrogen compounds causes a volume expansion of the affected steel. This results in compressive stresses in the case that are often very desirable. These stresses improve fatigue resistance. Fatigue cracks start when a surface is subject to tensile stresses. If a surface contains residual compressive stresses, a fatigue crack will not start until the applied stress overcomes the negative stress (compressive) in the surface. There may be a tendency for distortion in nitriding if the part has a nonsymmetrical shape or if nitriding is selective. A rule of thumb used by some heat treaters is that heavy nitride cases, 0.020 to 0.030 in. (0.5 to 0.75 mm), cause a growth of about 0.0005 in. (12.5 μm) per surface on the diameters of round bars and a similar growth on inside diameters; length changes can be as much as 0.0005 in./in. (1 μm/mm) of length. Nitriding will occur in inside diameters if there is adequate exposure to the nitriding media. Liquid nitriding media go into most holes; gas media only go into holes if the gas flow is adequate. It is advisable to avoid sharp corners on nitrided parts. There is risk of chipping of the hard case.

We will now discuss the various processes that are used to nitride surfaces. Most of the precautions and use guidelines that we have discussed apply to all the nitriding processes.

GAS NITRIDING

A simple gas nitriding system is illustrated in Figure 4–12. Parts are put in a wire basket fastened to the top of a sealed retort. The work is heated to the nitriding temperature, with ammonia or an inert gas such as argon flowing into the retort. When the work is at temperature, nitriding is initiated by introducing ammonia, NH_3. The ammonia gas dissociates to nitrogen and hydrogen at the part surface. The nitrogen diffuses into the work in atomic form, and the hydrogen becomes a part of the atmosphere in the furnace. The liquid bubbler shown in the schematic is used to ensure a positive pressure in the furnace and to serve as a flow meter on the exhausting gases. Retort furnaces are usually resistance heated, but any form of heating could be used. After completion of

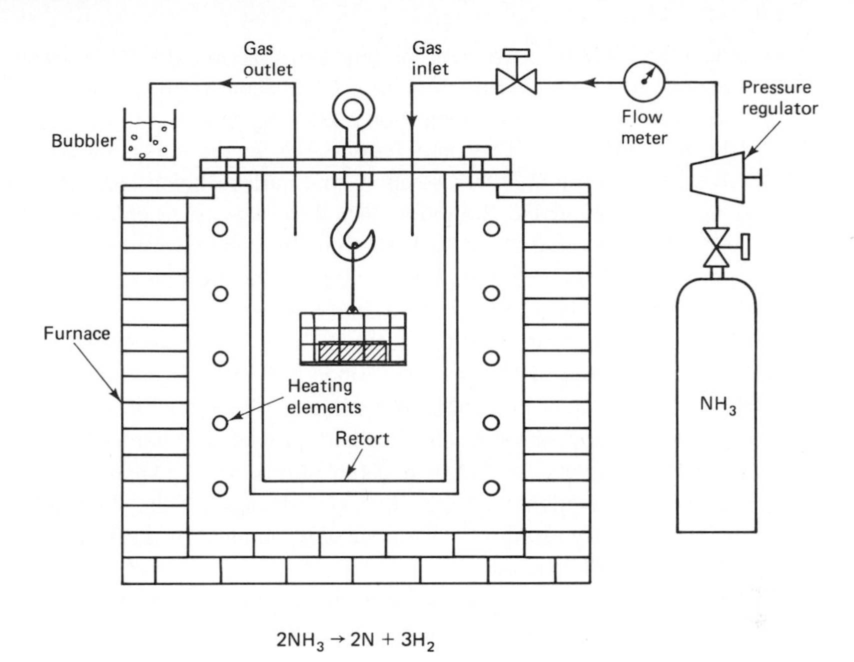

$$2NH_3 \rightarrow 2N + 3H_2$$

Figure 4–12 Schematic of gas nitriding.

the nitriding cycle, the entire retort may be removed and air cooled, or the furnace and retort may be allowed to cool to ambient temperature. Often cooling is done with argon or tank nitrogen flowing into the retort instead of ammonia. This ends the nitriding cycle. The parts are scale free and ready for use or for final grinding. Nitride cases are usually thin (less than 0.010 in.) (0.25 mm). Final grinding on nitrided surfaces is usually limited to a stock removal of 0.002 in. (50 μm) per surface. Heavy finish grinding will remove the case or the hardest part of the case.

Ammonia will dissociate into nascent nitrogen and hydrogen when it contacts steel heated to the temperature range of about 900° to 1100°F (480° to 590°C). To more precisely control the nitriding process, it is common practice to use external dissociators for the ammonia, in-line gas treatment devices that convert the ammonia to a desired degree of dissociation, rather than rely on the dissociation rate that occurs on the part surface. Some processes involve stepped temperature and dissociation cycles. The two-step or Floe process is used to minimize the formation of white layer.

Masking of parts for selective nitriding is done with techniques similar to those used for carburizing. Copper plating is the most widely used technique (0.5 to 1 mil; 12.5 to 25 μm), but there are a number of masking paints that work adequately if applied to recommended thicknesses. Nitrided cases are hard as they come out of the retort. It is not possible to remove them to machine off the nitride layer for selective nitriding as can be done with carburized parts.

Most stainless steels, 300 series, 400 series, and the precipitation hardening grades, can be gas nitrided. The chromium and some of the other alloying elements have a favorable effect on the formation of nitrogen compounds. With stainless steels, it is not necessary to have a ferritic structure. Case depth in most alloys is limited to about 0.007 in. (175 μm), but the precipition hardening grades can have case depths as deep as 0.012 in. (300 μm). The case hardness is in the range of 65 to 70 HRC. The problem with nitriding stainless steels is that surface finish can play an important role in making the process work. Smooth polished surfaces are troublesome. The most suitable surface for nitriding is a clean surface with fine random roughness (like that obtained with fine abrasive blasting). Proprietary processes are available from heat treating companies that are especially designed for stainless steel.

The final point that should concern the user of gas nitriding processes is the time and equipment details of the process. Since parts to be gas nitrided must be put in a sealed retort, there may be problems finding facilities that are suitable for large parts. For example, gas nitriding is an excellent process for hardening shafts that require high strength and a very hard wear-resistant surface. If the shaft is less than 4 ft (1.2 m) long, most heat treaters have retorts that are adequate. If the shaft is 7 ft long (2.1 m), it may be necessary to query a substantial number of heat treaters before one is found that can handle the job.

The time required for gas nitriding raises the cost. As is shown in Figure 4–13, if a heavy nitride case is required, the soak time in the furnace can be as long as 60 hours or even 100 hours. This ties up a furnace for a substantial amount of time, and the cost of this is reflected in the cost of the nitriding operation. In using gas nitriding, it is advisable to keep the case depth as thin as possible. Wherever possible, parts should be used with no finishing after nitriding. This can be done if the case depth is

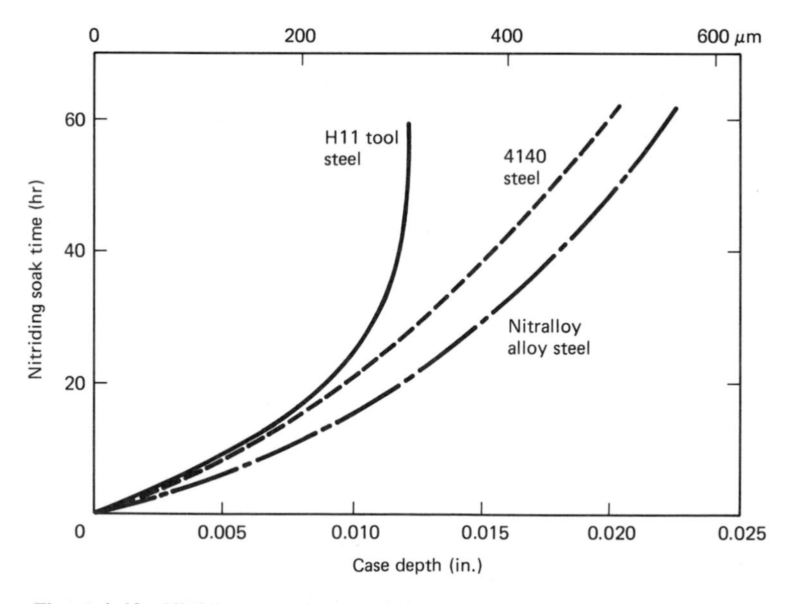

Figure 4–13 Nitriding times for gas nitriding of various steels at 975°F (523°C).

held to about 0.005 in. (125 μm). A nitriding specification that would allow this is as follows:

Material: Nitralloy 135M

Sequence:

1. Rough machine
2. Harden and temper to 28 to 32 HRC
3. Finish machine
4. Nitride all over 0.005 to 0.007 in. (125 to 175 μm) deep

A thin nitride will have a negligible white layer, and the size change due to growth on the nitrided surfaces will be only a few ten thousandths of an inch (5 to 10 μm).

SALT NITRIDING

Just as cyanide salts are used in liquid carburizing to serve as the sources of carbon for diffusion to carburize, these same types of salts can be used for a subcritical heat treatment to diffuse nitrogen in a surface. Salt nitriding is performed in baths that operate in the temperature range of about 950° to 1150°F (510° to 620°C). The baths consist of a mixture of sodium cyanide (NaCN), sodium cyanate (NaCNO), potassium cyanide (KCN), and some other salts. The active nitriding agent is the cyanate, CNO. The nitriding times, applicable materials, and the case depths obtainable are about the same as those for gas nitriding. The advantage of the salt process over the gas nitriding process is the easier handling of work. Parts do not need to be sealed in retorts; the handling costs are lower and work throughput can be greater.

Several proprietary salt nitriding process have evolved from the basic process. These processes are used for thin cases on a wide variety of steels. The salt composition and procedure is proprietary, but the process is similar to conventional salt nitriding. The baths run at about 1050°F (565°C); the time in the bath is usually 90 minutes and the parts are air cooled or water quenched. Typical case properties for the Tufftride® process are shown in Table 4–2. A unique feature of this process is the absence of white layer. The case consists of an epsilon Fe_3N compound layer that is usually less than 1 mil (25 μm) in depth and a diffusion layer that may be as deep as 0.030 in. (0.75 mm). This process is applicable to plain carbon steels. Normally, these steels do not respond to nitriding in conventional gas and salt nitriding processes. When carbon steels are nitrided in these baths, the compound layer is a fraction of a mil (25 μm) deep; it will have a hardness of about 55 HRC, and the hardness falls off very abruptly from the hard Fe_3N compound layer. Under a microscope, the diffusion zone shows the presence of needlelike compounds in the ferrite grains. These have been identified as Fe_4N gamma prime compounds. They do not contribute significantly to hardness, but their presence is thought to produce compressive stresses that improve the fatigue resistance of treated steels. The compound layer, Fe_3N, does not fall off abruptly in

TABLE 4–2. MATERIALS AND CASE DEPTHS FOR SHALLOW CASE SALT NITRIDING (TUFFTRIDING)

Material	Depth of case (compound) in inches (μm)	Typical hardness (HRC)
Carbon and low-alloy steels	0.0002–0.001 (5–25)	50–55
Tool steels	0.0001–0.0005 (2.5–12.5)	55–60
Corrosion- and heat-resisting steels	0.0002–0.001 (5–25)	58–65
Cast irons	0.0002–0.001 (5–25)	50–55

tool steels and steels that contain molybdenum, chromium, or other elements that enhance nitriding. The thin proprietary nitride cases are widely used to improve the wear properties of carbon steels and alloy steels that must be used at low hardnesses for toughness considerations.

The salt nitriding processes that we have just discussed present environmental problems in the disposal of the cyanide bath salts and in disposal of rinse solutions and their use is being phased out. A number of cyanide-free salt bath hardening processes have been developed to solve this problem. These baths produce cases similar to the proprietary salt processes (Tufftride, etc.), and the salt disposal problem is reduced by using salts of potassium cyanate (KCNO), sodium cyanate (NaCNO), potassium carbonate (KCO_3), and sodium carbonate ($NaCO_3$). The diffusing species comes from the cyanates. The noncyanide nitriding process is slightly different from the conventional salt processes. Parts are preheated to about 700°F (370°C) in air before going into the salt. This reduces distortion and allows the nitriding to proceed faster. After the soak in the salt, the parts are quenched in another salt that is maintained at about 750°F (400°C). This second salt removes some of the salt from the nitriding bath and it ends the nitriding cycle. Parts are then water quenched to room temperature and they are ready for use. The advantages of a noncyanide bath are obvious, but an offshoot of this process is a technique that involves a second treatment in an oxidizing salt bath. This latter treatment produces a black surface, and the parts have a compound layer that has enhanced corrosion resistance from the pickup of oxygen from the second salt treatment. The complete process involves a polish treatment between the nitride treatment and the second salt bath treatment. The temperature of the last oxidizing salt bath is about 750°F (400°C), and parts are water quenched from this temperature. The finished parts have shiny black surfaces. This latter treatment (QPQ®) reportedly significantly improves the corrosion resistance of the surface. It is widely known that nitriding imparts a modicum of atmospheric rust resistance to carbon steels. The double salt and polish process enhances this atmospheric corrosion resistance. In general, nitrided surfaces are not more resistant to chemical corrosion; in fact, nitriding of stainless steels reduces their chemical resistance.

ION NITRIDING

Ion nitriding is a combination of vacuum coating technology and heat treating. If a high voltage is established between two electrodes in a vacuum and a small amount of a gas is introduced, a plasma can be created between the electrodes. The plasma consists of ions of the gas that was introduced into the vacuum and electrons. The ions are accelerated by the potential between the electrodes and they strike the cathode. In ion nitriding, the work is made the cathode and the steel chamber of a vacuum furnace is the anode of a dc electrical system. The work is insulated from the chamber such that the only way that current can flow between the work and the chamber walls is through the ionized gas in the chamber (Figure 4–14).

Parts to be ion nitrided are cleaned of all carbonaceous material. They are fixtured on a work support that is insulated from all other parts of the system but is electrically connected to the power supply. Parts are spaced so that they do not touch, and the charge is stacked symmetrically. The chamber is pumped down to about 10^{-4} torr (0.01 Pa). The parts may then be cleaned by introduction of hydrogen or some other

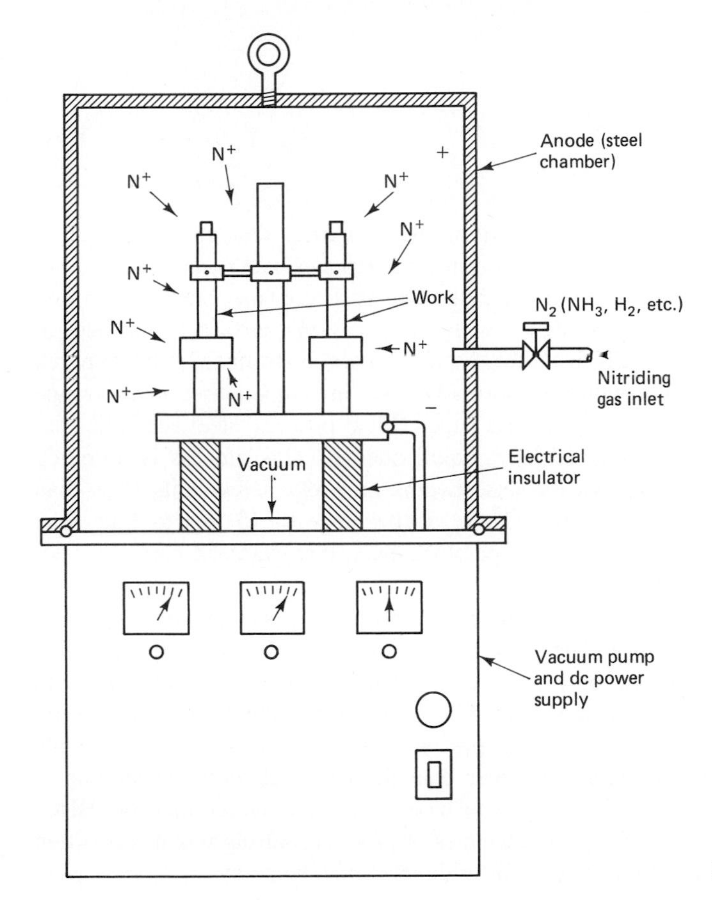

Figure 4–14 Schematic of ion nitriding.

gas into the chamber. The chamber pressure during cleaning may be 0.3 torr (40 Pa). The impinging ions remove oxides and other surface contaminants from the work. The nitriding process is commenced by feeding nitrogen gas into the chamber, and a potential of between 400 to 1000 V dc is established between the chamber and the work. A plasma glow will be established around the parts, and they will start to heat to the nitriding temperature, which is in the range of 650° to 1050°F (340° to 565°C). The glow is maintained during the nitriding cycle. The current density is about 1 mA/cm^2. The time to obtain a case can vary from about 15 min to about 30 hr. Treatment times are about one-half those required with gas nitriding, and there is no tendency to form a white layer. These are the major advantages of ion nitriding. The major disadvantage of this process is the cost of the equipment. Once again, if this process is obtained from a heat-treating firm, capital equipment cost is not a factor. The process provides excellent process control, which makes it suitable for use on critical parts. The case depths possible are similar to those for gas carburizing. If methane or some other carbonaceous gas is combined with the nitrogen when the glow discharge is established, the case will consist of both carbon and nitrogen.

CARBONITRIDING

Liquid Carbonitriding

The more usual term used for liquid carbonitriding is cyaniding. This process produces a very thin case of high hardness by immersion in a molten salt bath containing sodium cyanide, sodium carbonate, and sodium chloride. The case contains both carbon and nitrogen. The cyanide in the bath combines with oxygen at the surface of the bath to form sodium cyanate (NaCNO), which in turn produces carbon (from CO) and nitrogen for diffusion into the work. Cyaniding is normally done at temperatures in the range from 1400° to 1600°F (760° to 870°C). It is a supercritical process; steel substrates are austenitized, and hardening is produced by a liquid quench. This process is normally only done on low-carbon steels, and the case depths are only a few mils (less than 0.010 in.; 0.25 mm). Soak times in the salt are in the range of 15 min to 1 hr. The required quench and the high processing temperature mean that cyanided parts are very susceptible to distortion.

The salt baths used for cyaniding can be simple steel pots that are externally heated (similar to salt carburizing). There are even granular cyaniding compounds that can be used to produce a thin skin on parts. The surface to be hardened is heated with an oxy-fuel torch to red heat; the heated surface is immersed in the granular cyaniding compound and then quenched. This is a crude process, but it allows maintenance people to do case hardening with no equipment other than the oxy-fuel torch. Cyaniding is not a quality surface-hardening technique. The case is usually harder than 60 HRC, and it can produce good wear resistance, but the lack of process controls and the distortion tendencies limit its use to parts that require little dimensional accuracy.

Gas Carbonitriding

If a gas caburizing facility is modified so that ammonia is added to the carburizing gas, the resultant case will contain both carbon and nitrogen and this will be gas carbonitriding. There are advantages and disadvantages of this process over both carburizing and nitriding. A disadvantage compared to nitriding is that it is a supercritical process. The work must be austenitized and quenched, and this increases the tendency for distortion. An advantage over nitriding is that cycle times are much shorter (from 1 to about 10 hr). A disadvantage compared to carburizing is that case depths are shallower. Case depths are typically in the range from 0.003 to 0.030 in. (25 μm to 0.75 mm). Carburized case depths have no technical limit. Advantages compared to carburizing is that the cases can be harder and the process temperature is lower. Temperatures can be as low as 1400°F (760°C) and they seldom exceed 1600°F (870°C). These lower diffusion temperatures are due to the hardening effect of the nitrogen, and the advantage to the user is lower distortion. There are other advantages, such as freedom from salt disposal problems, but the lower hardening temperature and short cycle time are the principal reasons for use over competitive processes.

Carbonitriding can be applied to the same types of materials that are carburized and nitrided by other processes, but this process is most commonly applied to the low-carbon steels and carburizing grades of alloy steels. Carbon contents of applicable steels are usually less than 0.25 percent. Carbonitriding is produced in batch-type furnaces or in continuous furnaces, as illustrated in Figure 4–15. The source of the carburizing gas can be an endo or exo carrier gas with a carburizing gas addition; the source of the nitriding gas is ammonia.

In the continuous furnace in Figure 4–15, the preheat and quench sections would be supplied with a neutral generated gas, and the carbonitriding zone would be supplied

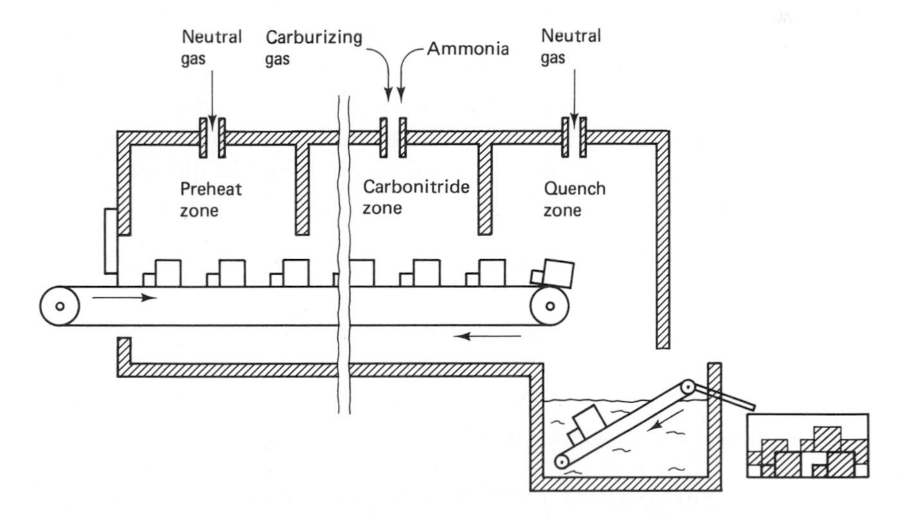

Figure 4–15 Carbonitriding in a conveyor hearth furnace.

with the neutral gas enriched with methane or the like and ammonia. The use of lower temperatures in carbonitriding favors the diffusion of nitrogen (carbon diffuses slowly at this temperature). Cases produced at low temperature may contain a compound layer (Fe_3N) as is found in nitrided cases. However, the use of a low temperature increases the cycle time; a temperature of about 1500°F (815°C) is more common. At this temperature, the case would probably contain almost equal concentrations of carbon and nitrogen. A typcial case on 1020 steel after 4 hr of treatment at 1550°F (840°C) and an oil quench would have a case hardness of about 60 HRC and a depth (greater than 50 HRC) of about 0.010 in. (0.25 mm). Carbonitrided parts require a liquid quench from the diffusion temperature. This is another disadvantage compared with gas nitriding. Liquid quenching always increases the risk of distortion. Carbonitrided cases tend to be harder than carburized cases because of their nitrogen content. This is another advantage of the process.

Carbonitrided parts can be tempered, but the nitrogen in the case makes the case resistant to tempering. For this reason, carbonitrided parts are often used as-quenched.

Carbonitriding possesses some of the advantages and disadvantages of both nitriding and carburizing, and it may appear that it is a compromise between these two processes; indeed it is.

Ferritic Nitrocarburizing

Ferritic nitrocarburizing is a subcritical heat-treating process to diffuse both carbon and nitrogen into the surface of steel from a gas atmosphere. The ferritic prefix is applied because the process is done in the temperature range of about 1050° to 1250°F (565° to 675°C). At these temperatures, the low-carbon steels normally treated with this process have a ferritic structure. The most commonly used process temperature is 1060°F (570°C). Alloy steels, cast irons, and some stainless steels can be treated, but the process is intended to produce a thin, hard skin of usually less than 0.001 in. (25 μm) on low-carbon steels. It is a way of imparting some wear resistance to formed sheet-metal parts, powder metal parts, small screw machined parts, small shafts, sprockets, sheaves, and the like. It is suitable for treating large quantities of parts in batch furnaces or in continuous furnaces. The furnace shown in Figure 4–15 for gas carbonitriding could be used; only the gases and soak temperatures would be different.

The ferritic carbonitriding process includes the following steps:

1. Degrease the parts (solvents are usually adequate, but smooth surfaces may require abrasive blasting).
2. Load the furnace and heat the parts to the process temperature (about 1050°F) (565°C).
3. Soak at temperature for the desired length of time (from 1 to 6 hr).
4. Immediately quench (usually oil).

The process gas is a mixture of a carburizing gas and ammonia. The carburizing gas can be the same as used in gas carburizing (endo carrier gas with an addition of methane or some other carbonaceous gas). The relative mixture of these two gases varies with the process, but they can be equal parts. The mechanism for the carburizing and the nitriding is the same as described previously for gas carburizing and gas nitriding. The carbon for diffusion comes from CO from the methane (or other gas), and the nitrogen comes from the ammonia (NH_3). There are very stringent controls on the gases to prevent explosive conditions.

The nature of the case is a compound layer that is about 0.001 in. thick (25 μm) and consists mostly of Fe_3N epsilon iron nitride, the same type of compound produced in other types of nitriding. Diffusion of carbon is sluggish at the low temperatures used in this process. The hard skin and the diffusion zone do not show a significant increase of carbon content at the surface, but the carburizing gas is still necessary to get the case. A 3-hr treatment on a 1010 steel will produce a compound layer that is about 0.0005 in. (12.5 μm) deep with a hardness of about 45 to 50 HRC. Nitrogen effects can be seen at depths of several thousandths of an inch from the surface. Nitrogen compound needles (Fe_4N) can be seen in the ferrite grains underlying the case. It is thought that the presence of the nitride compound in the diffusion zone under the case strengthens the substrate and improves the fatigue strength of treated parts.

As is the case with other nitride treatments, the nitrided surface has atmospheric corrosion resistance that is far superior to untreated material. Some ferritic nitrocarburizing processes employ subsequent oxidizing treatments and impregnations to further enhance corrosion resistance. This process has been successfully used to replace stainless steel parts used for resistance to atmospheric rusting, and the serviceability has been improved over the stainless parts because of the wear resistance afforded by the nitride case. This is a very competitive process for producing thin cases.

SPECIAL DIFFUSION PROCESSES

In Chapter 3, we discussed the metalliding process, which is an electrodeposition on a high-temperature salt electrolyte. This process is a special diffusion process when the coating is diffused into the substrate. This process is capable of diffusing many different elements into the surface of many different substrates. In this section, we will discuss some of the other diffusion processes that are used to change the surface and use characteristics of engineering materials. These processes are not widely used, but they can be used when other types of coating do not suffice.

Aluminizing

The process of coating materials with a thin layer of aluminum is called aluminizing. Metals are aluminum coated usually for one of two reasons: (1) atmospheric corrosion resistance, and (2) elevated oxidation and environmental resistance. Aluminum coating for atmospheric corrosion resistance (at ambient temperatures) is competitive with galvanized zinc coatings. When applied over steel, both of these materials are anodic to the steel, and scratched areas of the coating will be protected (the steel will be the cathode).

Aluminum is not used as widely as zinc for atmospheric corrosion, mostly because it is more expensive than zinc coatings, but there may be some conditions where its use is justified. Aluminum-coated steel for elevated temperature applications is much more cost effective. In this area, it competes with $10 to $20 per pound superalloys. There are very few metal alloys that can resist oxidizing conditions in the temperature range of 1600° to 2100°F (870° to 1150°C). Aluminized carbon steel can. It can be used for furnace parts, jet engine parts, flues, high-temperature fasteners, and other components that must resist oxidizing and sulfidizing conditions.

There are many different ways to aluminum coat steels and other metals: hot dip coating, rolled-in coatings, spray coating, pack cementation, and others. Hot dip and rolled clad coatings are usually applied as true "coatings." They are applied to the surface of metals, and they are buildups on the surface. They are suitable for the atmospheric corrosion types of applications, but for elevated temperature applications the diffused coatings are used. The melting temperature of aluminum is about 1200°F (650°C), but if aluminum-coated parts are heated to temperatures in excess of 1600°F (870°C), the aluminum does not melt off the surface; it diffuses in. Thus a substrate can be aluminum diffusion coated by cladding the part with a thin layer of aluminum and subsequently diffusing it in.

The process that is more germane to our discussion of carburizing and nitriding processes is pack cementation. Pack carburizing is pack cementation. Cementation simply means introduction of some species into a substrate by a diffusion process. Metals are aluminized by pack cementation by packing the parts in a metal container filled with a material that will provide a concentration gradient for diffusion of aluminum. Parts are cleaned usually by grit blasting; they are packed and spaced in the aluminizing media; the box or retort is sealed and the parts are heated for a desired length of time to allow the diffusion to proceed. The parts are slow-cooled and removed from the pack. The coating obtained can be from 1 to 40 mils (25 μm to 1 mm) thick. The outer surface is very rich in aluminum, and the diffusion zone has a decreasing content of aluminum. The aluminum in the diffusion zone is usually in solid solution since the solubility of aluminum in iron is quite good. It is this aluminum on the surface and in solution in the metal that imparts oxidation resistance.

The pack media is often proprietary, but one recipe is to pack the part in a mixture of aluminum powder, aluminum oxide powder, and a salt that will react with the aluminum at high temperature to produce an aluminum-rich gas that will bring the aluminum to the part surface so that diffusion can take place. The aluminum oxide is simply an inert spacer material to prevent agglomeration of the aluminum. A suitable salt would be aluminum chloride.

Aluminizing can be done on steels, nickel alloys, cobalt alloys, and copper. It is most often used on low-carbon steels. An aluminized low-carbon steel may outperform an expensive high alloy in resistance to some types of high-temperature oxidizing environments. If a design situation involves oxidizing or sulfidizing atmospheres from 800° to 2100°F (425° to 1150°C), it may be well to consider aluminized steel as a candidate material. Detailed corrosion data are available in the literature. Aluminized steels are used for combustion chambers on furnaces and in exhaust systems for internal combustion engines.

Siliconizing

Just as aluminum is diffused into steels, the same type of treatment can be done with silicon. There are pack and retort processes in which parts are subjected to gas atmospheres that react with the heated part surface to produce nascent silicon that diffuses into the substrate to be coated. One process involves tumbling parts in a retort filled with silicon carbide; when the work load reaches a temperature of about 1850°F (1010°C), silicon tetrachloride gas is introduced. This gas reacts with the part and the SiC particles to produce a concentration gradient of silicon on the part surface. The silicon diffuses into the part and forms a case as in pack carburizing. Case depths can be as thick as 0.040 in. (1 mm). The process proceeds at a rate of about 10 mils (0.25 mm) per hour. This process is normally done on low-carbon steels, and these steels develop a case that has a silicon content of about 13 percent. Siliconized steels have very good corrosion resistance to a number of oxidizing acids, and they can develop case hardnesses of about 50 HRC. Thus they have utility for applications involving corrosion and wear.

Chromizing

Chromium can be applied from a pack in the same manner as in siliconizing. Parts are packed in chromium powder and an inert filler such as aluminum oxide, and a salt is introduced that will go to a vapor phase at the processing temperature and serve as the carrier gas to bring chromium to the surface of the part. One such gas is ammonium iodide, which transforms to chromous iodide at the processing temperature of about 1800° to 2000°F (980° to 1090°C). A variation of this process without the pack is to put the parts to be treated in a retort and introduce the chromium rich gas with a carrier gas such as hydrogen. This latter process is called chemical vapor deposition (CVD), while the former is still pack cementation. The generic process of chemical vapor deposition can be used to apply many metals, ceramics, and other inorganic compounds. The species to be coated is put into gas form; this gas is introduced into the chamber containing the part to be coated, and the gas essentially condenses to produce a coating of the species that was in gas form. If the part is heated to a sufficient temperature, diffusion of the condensing species will occur and a diffusion coating is obtained. Diffused coatings are always preferred from the adhesion standpoint. The disadvantage is distortion from the high temperatures usually required for diffusion.

The retort process has been successfully used to make corrosion-resistant sheet steel from low-carbon steel. Coils of steel are loosened to allow gas flow through the convolutions of the coil. The entire coil is processed to produce a 1- to 2-mil (25- to 50-μm) diffused surface coating that can have a chromium concentration of 30 percent or higher. The core of the steel remains soft and ductile. This coil can be subsequently slit into smaller coils that can be used on punch presses and the like to make metal parts with essentially a stainless steel surface. One application of this type of material is for truck mufflers. The disadvantage of the process is the bare edge produced in slitting. Rust in service on the edges must be tolerated or eliminated by designs that cover this exposed edge.

If a high-carbon steel (e.g., 1080) is chromized, the diffused chromium can form chromium carbides and the 1- to 2-mil (25- to 50-μm) surface can have a hardness in excess of 60 HRC. Sheet material and parts processed to produce this type of surface are used for applications involving wear and corrosion.

Titanium Carbide

A variation of the CVD process for chromizing is to treat parts in a gas atmosphere of titanium tetrachloride, hydrogen, and a hydrocarbon gas. At a process temperature in the range of 1650° to 1850°F (900° to 1010°C), titanium and carbon will diffuse into the surface of the part and form a titanium carbide diffused case. The depth of the case is usually less than 0.5 mil (10 μm), but the surface is essentially TiC with a hardness of about 2200 HK, harder than any metal. This coating is most commonly applied to tool steels and hardenable stainless steel. Since the treatment is performed above the austenitizing temperature for these steels, the core must be hardened by quenching from the diffusion temperature or by reheating and quenching. This process produces an extremely wear resistant case, but the distortion that occurs in the process often limits its application. The case is too thin to allow grinding after treatment, and it is likely that the part will have a few thousandths of distortion. This process is excellent for parts that can tolerate a small amount of distortion. For example, razor blades treated with this process lasted in slitting operation ten times as long as untreated blades. A few thousandths of an inch (0.1 mm) of distortion in the razor blades did not affect their usability.

Boronizing

Boron behaves much like carbon in strengthening steel. It has about the same atomic size as carbon. It diffuses easily into steels and many other metals. Boronizing is the process of diffusing boron into metal surfaces to produce a hard surface layer that will enhance wear resistance. There are a number of techniques for boronizing. The metalliding process discussed in Chapter 3 is one technique. Chemical vapor deposition and pack cementation are other processes. The CVD process is performed by fixturing the parts to be coated in a retort that can be heated, and then a boron containing gas is introduced into the retort (Figure 4–16). The gases decompose on contacting the heated part, and atomic boron is available for diffusion into the substrate. The gases can be boron halides, such as BCl_3 and BBr_3, or the more complex boron hydrides, such as Ti $(BH_4)_3$. Diffusion of boron into the substrate can start at temperatures as low as 750°F (400°C). The pack process is similar to pack carburizing. The parts to be coated are packed in boron-containing compounds such as boron powder or ferroboron. Activators such as chlorine compounds and fluorine compounds are added to enhance the production of a boron-

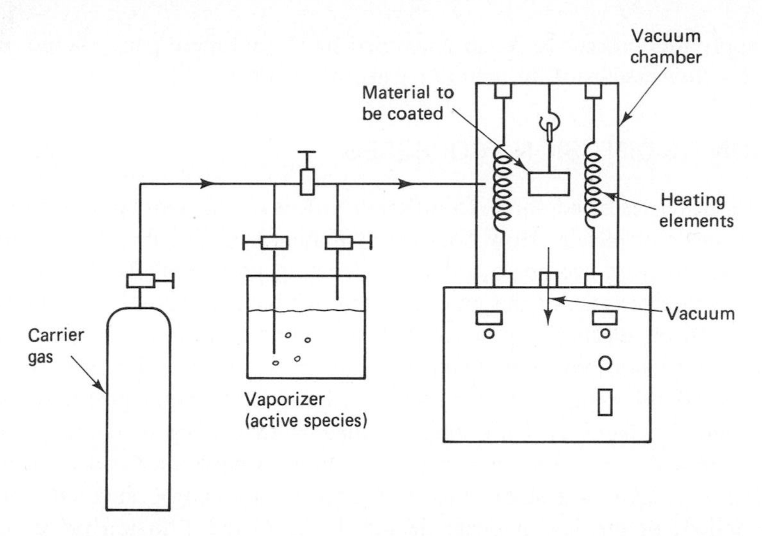

Figure 4–16 Schematic of a chemical vapor deposition (CVD) system.

rich gas at the part surface. The pack process can be performed at 1000°F (540°C) when processing hardened high speed tool steels that were previously quench hardened, or it can be done at other temperatures up to about 2000°F (1090°C). Higher temperatures enhance the diffusion rate and reduce the process time. Tool steels can be treated at their austenitizing temperature. They can then be quenched from the coating temperature to harden the substrate. The boron case does not need to be quenched to obtain high hardness.

The nature of a boronized case depends on the substrate that is treated. Usually, hard borides are formed. These compounds can have a hardness in the range from 1500 to 2000 HV. The hard compound layer can have a depth in the range from 0.0005 to 0.002 in. (12 to 50 μm). Many metals are suitable substrates for boronizing: titanium, nickel-base alloys, nitriding steels, cobalt-base alloys, tool steels, and cemented carbides. Some success has even been obtained with low-carbon steels, but the previously mentioned material systems are the preferred substrates.

Some boronizing processes are proprietary, and most users of this process send their parts to coating service companies. Boronizing fits into the spectrum of diffusion coatings in the category of thin coatings. It is most often applied over tool steels and other substrates that are already hardened by heat treatment. The thin boride compound surface further enhances wear resistance. For example, if a tool is made from a 60 HRC tool steel and it is still wearing at an unacceptable rate, a boron diffusion coating may be appropriate. The boride surface can have a hardness higher than obtainable with any steel alloy, and this very hard skin may provide a significant improvement in service life. This process is not widely used, and the risks in using boriding include distortion from the high process temperature and changes in surface finish due to chemical reactions in processing. Because the coating is thin, there is usually insufficient stock on the coating to take care of significant distortion by finish grinding. It is best to apply this process to finish machined parts, and these parts should be able to tolerate a few thousandths of an inch (75 μm) of distortion.

SELECTION OF DIFFUSION PROCESSES

We have described some 15 different processes that can be used to increase the surface hardness of metals. How does one go about selecting one process for use and how do these processes compare with other coating processes? Table 4–3 is a comparison of the characteristics of the processes that we have discussed. This table can be consulted to help in selecting a process, but let us first address the second part of the question, how diffusion processes compare with other coating processes. The answer to this is they do not compete if a buildup is desired. Diffusion processes are not intended to produce a buildup. They only compete with coating techniques on new applications where a decision must be made as to how to harden a surface for a wear application. If a seal area on a shaft must be hardened, it could be undercut and a coating can be applied, or the designer can decide if one of the diffusion processes may offer some advantage. Obvious advantages are that there is no need to prepare the surface for a coating process, and there may be a possibility of hardening the desired area with no subsequent machining after hardening. The latter is probably one of the most significant advantages of diffusion treatments. They can be applied in some instances where the size change in producing the hardening may be small enough that the parts can be hardened and used. Examples of this are parts such as small-diameter shafts, small sheet-metal levers and pawls, blanked gears, and most parts made on punch presses. These small wear components are often difficult to coat with other processes, but they are quite suitable for diffusion hardening processes.

Probably the first question to ask in considering the use of diffusion hardening processes is, How many parts? If the answer is many small parts, diffusion processes are probably logical candidates. If the answer is only a few parts, hardfacing and other coating processes may be cost-effective candidates. Diffusion processes can be continuous; conveyor hearth furnaces can be set up to process millions of parts. Even the batch diffusion processes can process multiple parts. In general, diffusion hardening processes are more cost effective than hardfacing for processing large numbers of parts. The cost of gas nitriding 100 thrust washers or some other simple part will be about the same as processing one. This is usually not the case with most wear coatings. Thus, if a hardening operation in question involves more than a few parts, diffusion hardening processes may be candidates.

Assuming that it appears that diffusion hardening is a viable process for a particular hardening operation, the next question to address is, Which process? A number of factors must be considered in answering this question. One of the biggest discriminators in diffusion hardening processes is the thickness of the hardened surface layer. There are really no theoretical thickness limits to diffusion processes. Diffusion processes are time dependent; if you have a compatible substrate, it will be possible to diffuse a species to any depth, even through a thickness. However, the time to do such a thing can be economically prohibitive. Figure 4–17 shows diffusion processes put into one of two categories, thin (a few mils) and thick. There is definitely a cost limit on thickness. In the wear community, it is an accepted axiom not to make a wear coating thicker than the wear that can be tolerated by the system. In keeping with this guideline, diffusion coatings are seldom made thicker than 0.030 in. (0.75 mm). Very few sliding systems can tolerate more than this amount of wear. If a system can, other hardening processes, such as welding hardfacing and local quench hardening, are used. In other words, our two categories could be labeled thin (a few mils [100 μm]) and thick (up to 30 mils; 0.75 mm).

Figure 4–17 can be used as a guide to the way that heat treaters view the various diffusion hardening processes. You can cyanide thicker than a few mils, but most people do not. The same is true for the other thin processes. The reason that people do not use these processes for the heavy surface-hardening jobs is cost effectiveness and other process constraints. Once again, they are not technical limits. Considering what wc have just discussed, select the process or processes that fulfill your thickness requirements.

Figure 4–18 addresses the substrate factor. Diffusion treatments can be applied to a number of alloy systems, but they are most often applied to steels. Most machine components are made from steels that fit into one of the four categories listed in Figure 4–18. This illustration also presents the processes that are most suitable for each steel category. Again, this chart is based on common practice; some heat treaters have successfully nitrided carbon steels, but most do not use this process for this class of steels. If an application requires a low-carbon steel because of cost or some other reason, then the processes listed for low-carbon steels should be made prime candidates. If an application can afford the cost of a nitriding steel, then nitriding is a hardening candidate. If the strength requirements of an application dictate thc through-hardened strength of a tool steel, the applicable processes will usually be the processes that can produce a surface that is harder than that of a hardened tool steel. Only a few processes are used on stainless steels, and it should be kept in mind that these diffusion treatments will

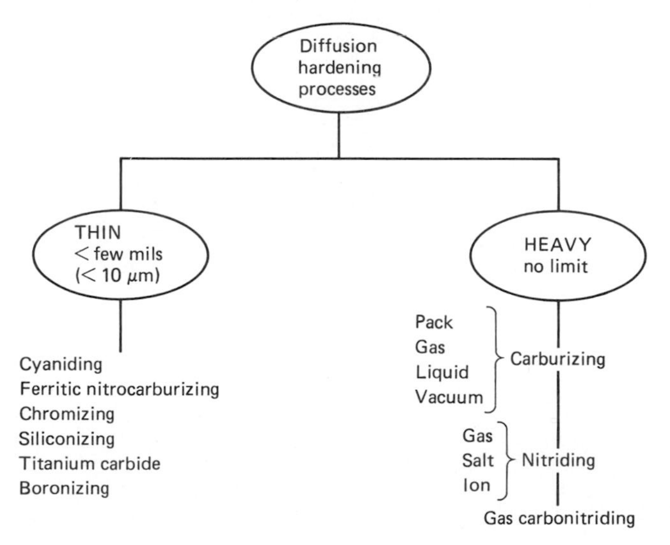

Figure 4–17 Categorization of diffusion processes by normal case depth.

TABLE 4–3 TYPICAL CHARACTERISTICS OF DIFFUSION TREATMENTS

Process	Nature of case	Process temperature in °F (°C)	Typical case depth in mils	Case hardness (HRC)	Typical base metals	Process characteristics
Carburizing						
Pack	Diffused carbon	1500–2000 (815–1090)	5–60 (125 μm–1.5 mm)	50–63*	Low C steels Low C alloy steels	Low equipment cost, difficult to accurately control case depth
Gas	Diffused carbon	1500–1800 (815–980)	3–60 (75 μm–1.5 mm)	50–63*	Low C steel Low C alloy steels	Good control of case depth, suitable for continuous operation, good gas controls required; can be dangerous
Liquid	Diffused carbon and possibly nitrogen	1500–1800 (815–980)	2–60 (50 μm–1.5 mm)	50–65*	Low C steels Low C alloy steels	Faster than pack and gas processes, can pose salt-disposal problem, salt baths require frequent maintenance
Vacuum	Diffused carbon	1500–2000 (815–1090)	3–60 (75 μm–1.5 mm)	50–63*	Low C steels Low C alloy steels	Excellent process control, bright parts, faster than gas carburize, high equipment costs
Nitriding						
Gas	Diffused nitrogen, Nitrogen compounds	900–1100 (480–590)	5–30 (125 μm–0.75 mm)	50–70	Alloy steels Nitriding steels Stainless steels	Nitriding steels give hardest case, quenching not required, low distortion, process is slow, usually a batch process
Salt	Diffused nitrogen, Nitrogen compounds	950–1050 (510–565)	0.1–30 (2.5 μm–0.75 mm)	50–70	Most ferrous metals including cast irons	Usually used for thin hard cases < 1 mil, no white layer, most are proprietary processes
Ion	Diffused nitrogen, Nitrogen compounds	650–1050 (340–565)	3–30 (75 μm–0.75 mm)	50–70	Alloy steels Nitriding steels Stainless steels	Faster than gas nitriding, no white layer, high equipment costs, close case control

Carbonitriding						
Gas	Diffused carbon and nitrogen	1400–1600 (760–870)	3–30 (75 μm–0.75 mm)	50–65*	Low C steels Low C alloy steels Stainless steel	Lower temperature than carburizing (less distortion), slightly harder case than carburizing; gas control is critical
Liquid (cyaniding)	Diffused carbon and nitrogen	1400–1600 (760–870)	0.1–5 (2.5–125 μm)	50–65*	Low C steels	Good for thin cases on noncritical parts, batch process, salt-disposal problems
Ferritic nitrocarburizing	Diffused carbon and nitrogen	1050–1250 (565–575)	0.1–1 (2.5 –25 μm)	40–60*	Low C steels	Low-distortion process for thin case on low-carbon steel, most processes are proprietary
Special Processes						
Aluminizing (pack)	Diffused aluminum	1600–1800 (870–980)	1–40 (25 μm–1 mm)	<20	Low C steels	A diffused coating for oxidation resistance at elevated temperatures
Siliconizing (CVD)	Diffused silicon	1700–1900 (925–1040)	1–40 (25 μm–1 mm)	30–50	Low C steels	For corrosion and wear resistance, atmosphere control is critical
Chromizing (CVD)	Diffused chromium	1800–2000 (980–1090)	1–2 (25—50 μm)	Low-carbon steel < 30 High-carbon steel, 50–60	High and low C steels	Chromized low-carbon steels yield a low-cost stainless steel; high carbon steels develop a hard, corrosion-resistant case
Titanium carbide	Diffused carbon and titanium, TiC compound	1650–1850 (900–1010)	0.1–0.5 (2.5 –12.5 μm)	>70*	Alloy steels tool steels	Produces a thin carbide (TiC) case for resistance to wear; high temperature may cause distortion
Boronizing	Diffused boron, boron compounds	750–2100 (400–1150)	0.5–2 (12.5–50 μm)	40–>70	Alloy steels Tool steels Co, Ni alloys	Produces a hard compound layer; mostly applied over hardened tool steels, high process temperature can cause distortion

* Requires quench from austenitizing temperature.

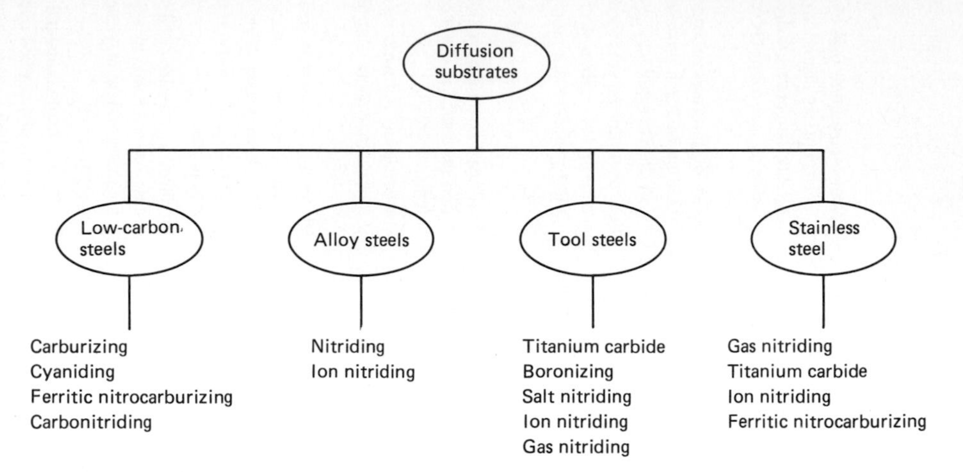

Figure 4–18 Types of steels used for various diffusion processes.

usually reduce their corrosion resistance. Thus, in this selection step, the thin or thick process is coupled with a substrate.

Figure 4–19 shows the hardness ranges that are common to the various diffusion hardening processes. Hardness can also be a selection criterion. If a tool steel is needed for an application, this steel could be capable of producing a surface hardness of 60 to 62 HRC. The only diffusion processes that may offer improved serviceability would be the diffusion processes that can produce hardnesses higher than that of the tool steel (nitriding, boronizing, and titanium carbide treatments).

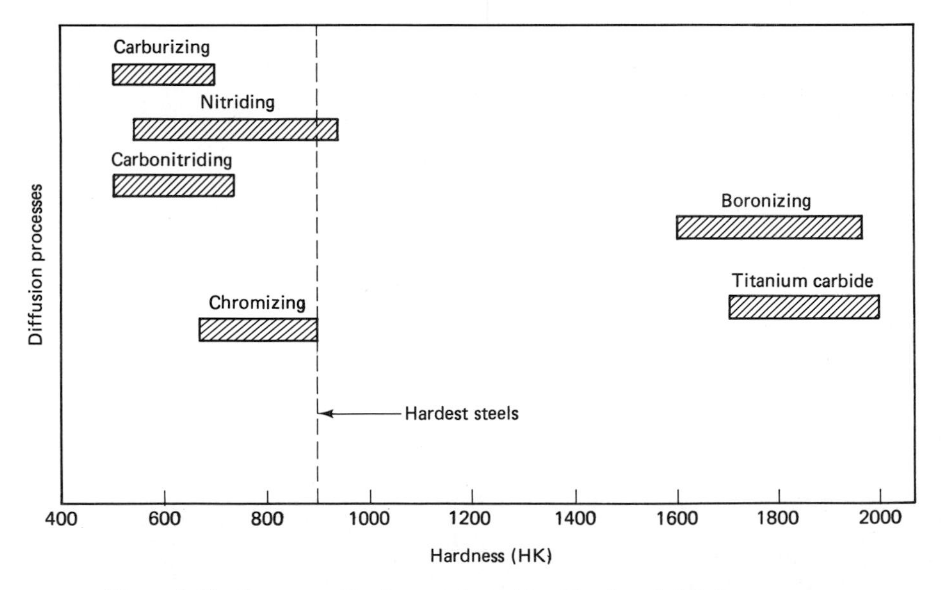

Figure 4–19 Spectrum of hardnesses obtainable with selected diffusion processes.

SUMMARY

In summary, diffusion treatments are not applicable to rebuilding parts, but they are candidates for producing hard wear-resistant surfaces on new parts. They are mostly used where the desired depth of wear surface is less than 0.030 in. (0.75 mm), and the specific process that is best for an application depends on mostly economic factors. Each process has a niche, something that it does better than others. Our discussion of these various processes should supply sufficient information to allow the choice of the right process for a particular application.

The final part of using a diffusion process is specification. A proper specification of a diffusion process should include the following elements:

1. Substrate (show exact material and condition of heat treat).
2. Depth of treatment (use a range).
3. Expected hardness of the surface (use a range and define case measurement technique).
4. Areas to be treated.
5. Applicable heat treatments (tempering, quench hardening, etc., for core properties).

Diffusion hardening processes are one of the oldest forms of hardening metals. They will continue to be candidates for most parts that need hardened surfaces.

REFERENCES

ASHBY M. F., and D. R. H. JONES. *Engineering Materials.* Elmsford, N.Y.: Pergamon Press, 1980.

BROOKS, C. R. *Heat Treatment of Ferrous Metals.* Washington, D.C.: Hemisphere Publishing Co., 1979.

DAWES, C., and D. F. TRANTER. "Process Modifications Widen Range of Nitrocarburizing," *Metal Progress,* Dec. 1983, pp. 17–22.

ENOS G. M., and W. F. FONTAINE. *Elements of Heat Treatment.* New York: John Wiley & Sons, 1953.

KOMATSU, N., and T. ARAI. "Development of Industrial Applications of Carbide Coating," *Research and Development in Japan,* 1981.

LESLIE, W. C. *The Physical Metallurgy of Steel.* Washington, D.C.: Hemisphere Publishing Co., 1981.

LINIAL, A. V., and H. E. HUNTERMAN. "Boronizing Process, A Tool for Decreasing Wear," *Wear of Materials—1979.* New York: American Society of Mechanical Engineers, 1979.

Metals Handbook, 9th ed., Vol. 5, *Surface Cleaning and Finishing.* Metals Park, Ohio: American Society for Metals, 1982.

OBRZUT, J. S. "Vacuum Heat Treating," *Iron Age,* March 7, 1983, pp. 24–29.

TONINI, G., L. CAPOROLI, and N. MERLINI. "Vacuum Heat Treating, Practical Guidelines," *Heat Treating,* Dec. 1962, pp. 28–32.

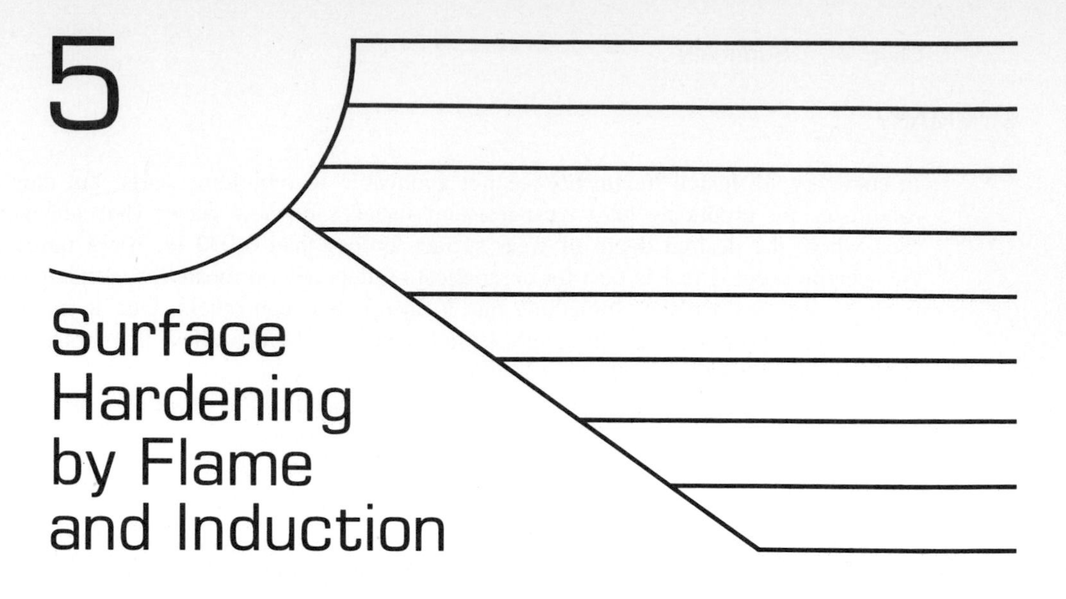

5 Surface Hardening by Flame and Induction

Artifacts dating to antiquity indicate that selective hardening of ferrous materials was practiced for centuries. When it was learned that certain ferrous materials would become hard when heated red hot and quenched, it was also learned that it is not necessary to heat and quench the entire shape. Metalsmiths learned that if they only harden the working surface of tools the remainder of the tool would remain ductile and less prone to breakage. This was the process of selective hardening, and this technique is still practiced. Surface hardening by flame and induction is a modification of these ancient techniques wherein only the surface (to some depth) is heated (austenitized) and quenched. Flame hardening is the process of applying heat from an oxy/fuel torch to a hardenable steel surface and subsequent quenching to produce a hardened area on the surface of a part. Induction surface hardening is the process of using induced electrical currents to produce local surface heating for hardening of ferrous materials.

Both processes have been used since the early part of this century, and they are still used in essentially their original form. Flame-hardening torches, fuel gases, and quenchants have been improved over the years, and induction-heating power supplies have become more efficient and sophisticated, but the concepts of both processes remain the same. Flame hardening accounts for an estimated 5 to 10 percent of the heat treating that is done in the United States. Induction hardening is more suited to mass-production techniques, and it is probably used on a similar fraction of mass-production parts that are hardened. In this chapter, we will discuss the details of these processes and how they compete with welding hardfacing and other surface-hardening processes.

FLAME HARDENING

Equipment

A basic flame-hardening system is illustrated in Figure 5–1. The basic elements of any flame-hardening system are as follows:

1. Oxy/fuel torches for heating the surface of the work
2. A system for rapid quenching of the heated work
3. A system to progressively heat the work (by moving the work or by manipulation of the torch)

Most flame-hardening systems use oxygen with acetylene as the fuel gas. Other gases, such as natural gas, propane, or synthetic gases, can be used. The combination of oxygen and acetylene produces a flame temperature of about 6000°F (3300°C). This temperature is adequate to heat the surface of any ferrous metal to its hardening temperature. The tip of the flame is the hottest portion, and surface heating is done by positioning the flame tip a slight distance (stand-off) from the surface. When the flame tip is imposed on the work, there is an ellipsoidal distribution of heat in the work centered about the flame tip. At the instant that the flame is imposed on the work, the temperature profile in the work will usually resemble the pattern illustrated in Figure 5–2a; after a second or so of elapsed time, the temperature profile will be something like that illustrated in Figure 5–2b. Melting has occurred at the point of impingement of the flame tip. The purpose of this process is to locally raise the temperature of hardenable steels to their

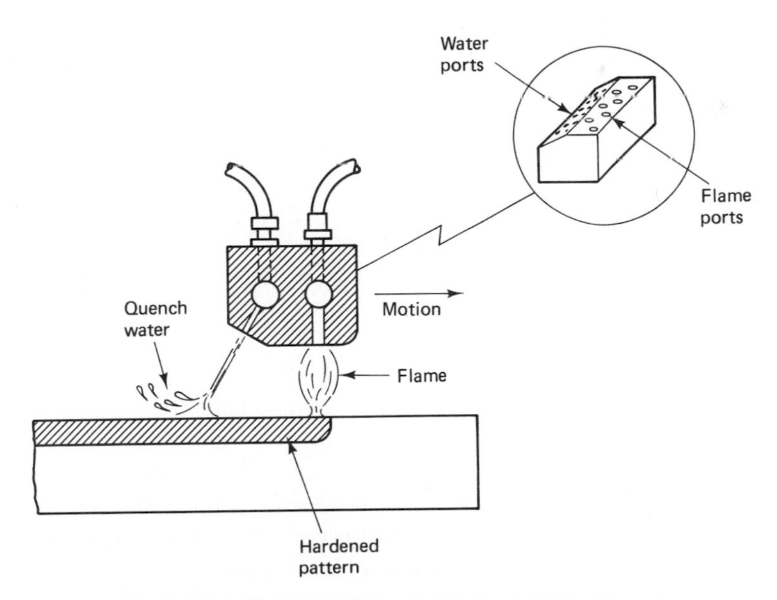

Figure 5–1 Schematic of flame-hardening torch for flat work.

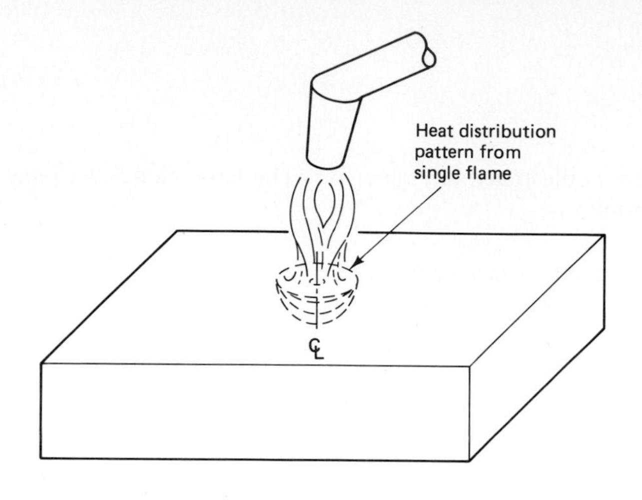

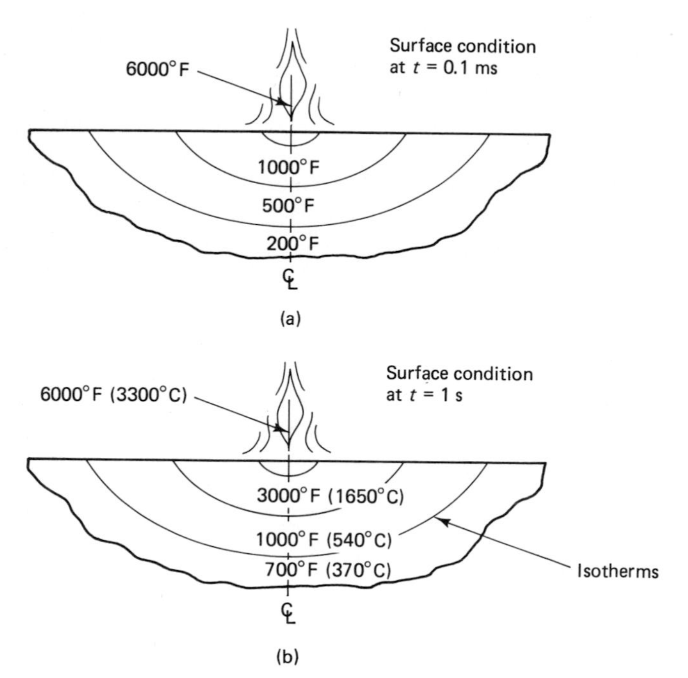

Figure 5–2 Work temperature profiles in flame hardening.

hardening temperature; melting is undesirable. This is why motion of the torch or work is a requirement of the process. Sometime between $t = 0$ and $t = 1$ in the example in Figure 5–2, the work surface for some depth below the surface will be in a suitable temperature range to raise most ferrous metals to their proper hardening temperature of about 1350° (730) to 1950°F (1065°C). The ''skill'' part of flame hardening is controlling the motion so that proper hardening temperatures are obtained. The heating time is usually so short that oxidation and diffusion processes such as decarburization do not have time to occur. The combustion gases can also be adjusted so that the hot gases are neutral or reducing in nature. They will prevent degradation of the work surface from the heating process. Flame-heated work will have a slight heat tint, but no surface deterioration or change in surface texture.

The zone to be hardened in the work is heated by convection and radiation from the hot flame gases on the surface, and conduction heats the work below the surface to the desired depth. The depth below the surface that achieves the proper austenitizing temperature will be the depth of hardening. This depth depends on the torch configuration, the time of heating, and the thermal diffusivity of the work. Most flame-hardening heads are made from copper alloys, and they contain internal cooling and multiple flame orifices. Overlapping flames are used to produce a uniform austenitizing temperature over the width to be hardened; the tip-to-work spacing is adjusted to produce the desired depth of austenitization. The head design, flame configuration, and quench nozzle configuration are the key to satisfactory hardening. The design of these heads is often considered proprietary by job shops that specialize in flame hardening.

There are basically three types of flame hardening heads:

1. Rectangular heads (Figure 5–1)
2. Toroidal heads (Figure 5–3)
3. Contoured heads to fit part shapes

The types of surface-hardening patterns that can be achieved are illustrated in Figure 5–4. The rectangular heads are used to harden areas on flat, horizontal, or vertical surfaces such as knife edges, machine ways, and the like. The toroidal heads are used in the horizontal or vertical position to progressively harden cylindrical parts, faces of rolls, bearing journals, insides of cylinders, and similar shapes. Special heads are used

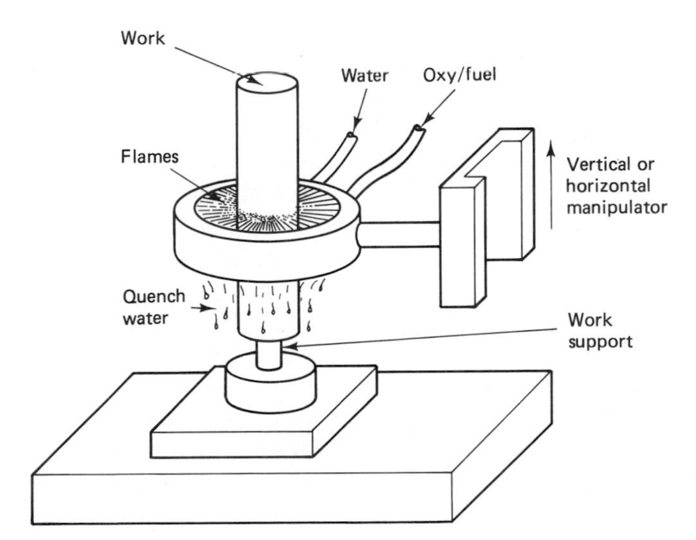

Figure 5–3 Schematic of flame hardening of cylindrical parts.

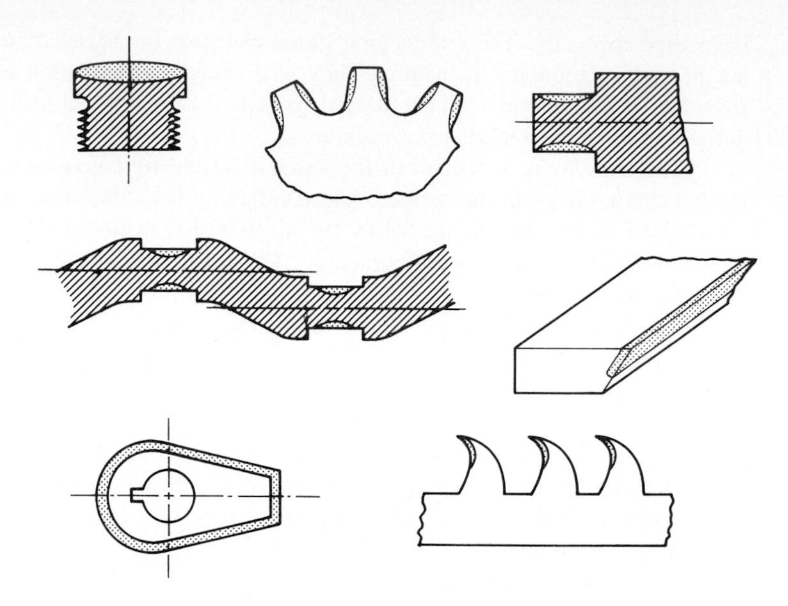

Figure 5–4 Typical flame hardening depth profiles.

to harden individual teeth on gears and sprockets, crankshaft throws, and similar parts that require that the flame conform to a shape.

The second important part of a flame-hardening system is the quench technique. The rectangular head illustrated in Figure 5–1 and the toroidal head illustrated in Figure 5–3 have integral quench systems. Coolant, usually water or water with an additive to slow the quench, is directed at the work at a low angle (15 to 30 degrees). In the hardening, the work is austenitized by the flame impingement; the head or work moves to produce the desired temperature, and the quenchant is directed out the trailing edge of the head to produce the cooling rate that is required to achieve a hardened martensite structure. There are many types of quench systems. For example, special machines are designed for flame hardening of gear teeth. The machine clamps the gear to minimize distortion; the gear is austenitized by flames positioned to achieve the proper hardening temperature and depth profile, and after the austenitizing cycle, the part is automatically lowered into a quench bath. Similar machines exist for other high-production parts, but one-of-a-kind parts are usually austenitized and quenched with rectangular or toroidal heads that can be easily adapted to produce desired hardening profiles.

Thus, the equipment for flame hardening of surfaces requires only a few basic components: the flame head or heads, a system for manipulating the head or work, and a quenching system. Tempering after hardening can be done in a batch-type furnace or with a suitable flame reheating system. These minimal equipment requirements mean that this process is not capital intensive. Flame-hardening heads can be as cheap as a few hundred dollars; the flame heads can be run off conventional oxyacetylene tanks, which are relatively low cost to buy and to refill, and the work manipulation can be done by a simple motorized torch manipulator or by modification to lathes or other machine tools. With a little ingenuity, this process can be used to surface harden areas on 25-ton (22,680-kg) parts. The work does not have to fit into a chamber or a furnace. There are no size limits and a 25-ton (22,680-kg) part can be flame hardened with a few hundred dollars' worth of equipment.

Material Requirements

A basic requirement of flame hardening is that the substrate be capable of quench hardening. In other words, the substrate must be a ferrous material with adequate carbon or carbon and alloy content to allow conversion to martensite when austenitized and quenched. With conventional furnace hardening procedures, carbon steel require at least 0.6 percent carbon to form 100 percent martensite on quenching. In flame hardening, the quench rate can be so rapid that steels with carbon as low as 0.35 percent may respond to form martensite (or at least bainite). Figure 5–5 shows the ranges of hardness that can be attained in flame hardening carbon steels. Alloying can increase the hardenability of steels so that it is possible to achieve a hardness of 55 HRC with carbon contents significantly lower than 0.6 percent.

Table 5–1 lists some commonly flame-hardened steels and cast irons and the hardness ranges that are commonly specified when these steels are flame hardened. As is shown in this list, this process applies primarily to carbon steels, a few alloy steels, cast irons with a pearlitic matrix, and some tool steels. As is the case with laser and electron beam hardening, the high alloys are not usually applied because of the sluggishness of these steels in austenitizing. The carbon in these alloys is often tied up in carbide phases. In furnace hardening, this carbon is put into solution in the matrix by soaking at the austenitizing temperature. When in solution, the carbon is available to provide matrix hardening. In flame hardening, it is possible to soak at the austenitizing temperature

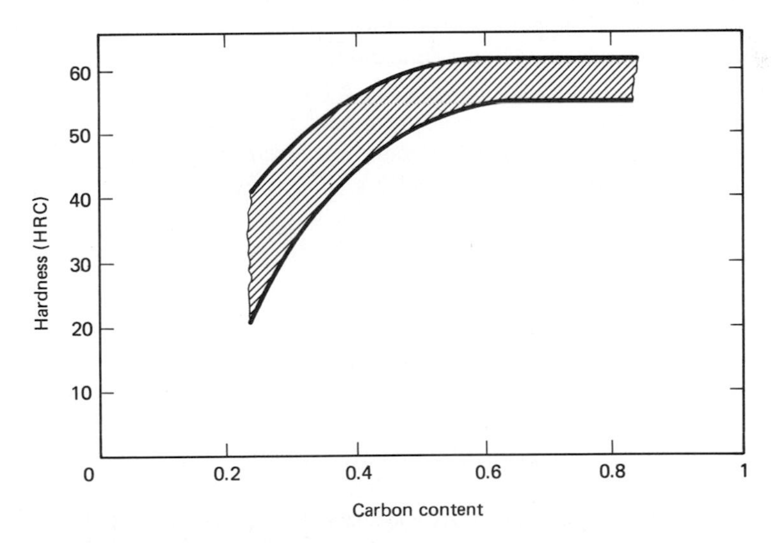

Figure 5–5 Hardnesses attainable in quench hardening carbon steels.

TABLE 5–1 MATERIALS THAT ARE COMMONLY FLAME AND INDUCTION HARDENED AND THEIR NORMAL HARDNESS RANGES

Carbon steels (HRC)		Alloy steels (HRC)		Tool steels (HRC)		Cast irons (HRC)	
1025–1030	(40–45)	3140	(50–60)	01	(58–60)	Meehanite GA	(55–62)
1035–1040	(45–50)	4140	(50–60)	02	(56–60)	Ductile, 80–60–03	(55–62)
1045	(52–55)	4340	(54–60)	S1	(50–55)	Gray	(45–55)
1050	(55–61)	6145	(54–62)	P20	(45–50)		
1145	(52–55)	52100	(58–62)				
1065	(60–63)						

to restore carbon to the matrix, but it is common practice to avoid these kinds of steels (high-alloy tool steels). Hardening is often not successful. A second problem with high-alloy steels is the propensity for quench cracking. High-alloy steels are intended to have deep hardening. When applied to flame hardening, these steels may harden deeper than anticipated, causing high stresses from the hardening size change; this can lead to quench cracking.

A rule of thumb used by some flame-hardening job shops is to never use higher hardenability than is needed to meet the hardness requirements of the intended service. For example, if a service hardness of 50 HRC is required, it is not necessary to use 4340 or one of the alloy steels that will quench out to much higher hardnesses. They would have to be tempered back to 50 HRC. If a hardness of 60 HRC is required, 1050 steel would be adequate. If the application required a core strength greater than can be achieved with 1050 steel (tensile strength of about 100 ksi or 689 MPa), a 4140 steel could be used. It would develop the same surface hardness of 59 to 61 HRC, but the core strength could be higher, about 150 ksi (1034 MPa).

An important point to remember is that flame hardening does not increase core strength or hardness. If an application, for example a shaft, requires a high strength and toughness with a hardened surface on bearing journals, it is recommended that the material be prehardened by conventional furnace-hardening techniques to the desired mechanical properties and then flame hardened in the areas that require high hardness. A typical sequence for making such a shaft if as follows:

1. Rough machine, leaving ⅛-in. (3 mm) stock allowance on all surfaces.
2. Harden and temper to 28 to 32 HRC.
3. Finish machine, leaving 0.005 to 0.008 in. (0.12 to 0.2 mm) grind allowance on surface to be flame hardened.
4. Flame harden indicated surfaces to a depth of 0.060 to 0.080 in. (1.5 to 2 mm) and temper to 59 to 60 HRC.
5. Finish grind flame-hardened surfaces.

This procedure would apply to alloy steels, such as 4140, 4340, and 6145, that can be through hardened to get a core hardness of 28 to 32 HRC. It would be difficult to do this on carbon steels such as 1050 or 1060. These steels have low hardenability and would probably require a water quench to through harden. This usually means unacceptable distortion. In addition, the poor hardenability of these steels precludes through hardening in sections that are thicker than about 1 in. (25 mm). Carbon steels are generally not suitable for applications that require high core strength.

A point that we have not discussed as yet is tempering after flame hardening. Sometimes tempering after flame hardening is neglected. This is not a recommended practice. Tempering reduces the brittleness of the surface-hardened area, and it also reduces the residual stresses in the hardened area to about one-half of what they would be without tempering. Flame-hardened parts should be tempered immediately after hardening, just like furnace-hardened parts. The tempering temperature should be at least the minimum tempering temperature that is recommended for a particular alloy. For example, if heat-treating data for 4140 steel show that 300°F (150°C) is the lowest tempering temperature that is used on this alloy, this would be the recommended tempering temperature to use after a flame-hardening operation.

Applications of Flame Hardening

The service characteristics of flame-hardened steels and cast irons, in the flame-hardened zone, are the same as the service characteristics of that alloy hardened by conventional techniques, with the exception of the residual stress pattern. If a flame-hardened zone is surrounded by unhardened material, the hardened zone will be in a state of compressive stress due to the volume expansion that occurs in the martensite conversion. There will be a residual tensile stress in the region where the hardened zone meets the unhardened part of the workpiece. The residual stress in the hardened zone is usually beneficial from the standpoint of wear and fatigue. Fatigue cracks do not start in material that is under compression, and this residual stress could help prevent surface fatigue or it could improve bending fatigue resistance. One precaution to be observed concerning residual stress in flame-hardened parts is that it is recommended that the transition from hardened to unhardened structure be located away from geometric stress concentrations. Figure 5–6 illustrates two techniques for doing this.

The depth of hardening that can be obtained in flame hardening depends on the type of material and the process parameters, but with normal techniques the depth of flame hardening is controlled within the limits of 0.05 to 0.25 in. (1.25 to 6.25 mm). It is very difficult to flame harden parts with a section thickness of less than about ⅛ in. (3.1 mm); they tend to through harden. On the other end of the scale, hardened depths of 0.25 in. (6.25 mm) or greater are possible on parts with heavy section thicknesses. To make it easier to obtain deep hardening it is usually advisable to use an alloy steel rather than a carbon steel. A depth of 0.25 in. (6.25 mm) is considered to be the upper limit for flame hardening, but there is no technical limit. When cases get this thick, it may be advisable to simply use a through-hardening process.

In summary, flame hardening is a process that is well suited to surface hardening of ferrous metals to substantial case depths. It can be done with minimal equipment requirements, but the process requires substantial operator experience, and parts are

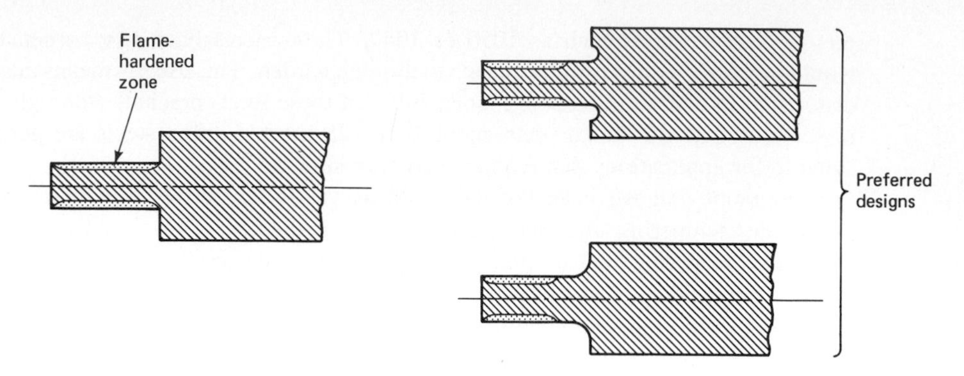

Figure 5–6 Locating flame-hardened zones away from geometric stress concentrations.

processed one at a time. If only one or a few hundred or a few thousands of parts are required, it may be advisable to send them to a heat treater that specializes in flame hardening. If hundreds of thousands or millions of parts are to be processed, it may warrant the setup of an in-house automatic or semiautomatic flame-hardening unit. Parts still need to be processed one at a time, but with suitable handling automation can be competitive with batch-hardening processes. One of the largest advantages of flame hardening is that it is easily adapted to very large parts. In this aspect it is competitive with welding hardfacing.

INDUCTION HARDENING

The basic mechanism of induction hardening goes back to Faraday's work with magnetism in the midnineteenth century. Faraday studied the rules of magnetism; he and others observed that conductors placed within a magnetic field have a current induced in them and this induction causes heating of the conductor. A simple transformer that has a primary coil magnetically coupled to a secondary coil, but not electrically connected, produces heating. This heating is due to two factors: (1) eddy currents, (2) and hysteresis. Eddy currents are power losses due to the resistance of a conductor to the current that is being induced in the conductor by a magnetic field (I^2R). The heating that occurs is proportional to the square of the induced current and the resistance of the conductor. *Hysteresis* in induction systems is the heating that occurs by reversals in the direction of the magnetizing force. Magnetic domains align with the magnetizing force; when the direction of the magnetizing force is reversed the domains realign and this cyclic realignment causes heating. Hysteresis heating only occurs in ferromagnetic materials up to the Curie point (the temperature at which ferromagnetic materials lose their ferromagnetism, about 1400°F [760°C] in steels). Heating by eddy currents is by far more important in induction heating. It applies to all conductors and induction hardening of steels requires heating to an austenitic structure where steels are nonferromagnetic.

Induction heating is essentially accomplished by making the work to be heated the secondary "coil" in an alternating current transformer. This concept is illustrated in Figure 5–7. When current flows in a wire there is a magnetic field around the wire with lines of flux oriented with respect to the direction of current flow (right hand rule). In the simple circuit illustrated in Figure 5–7, a coil is made from a conductor that is carrying alternating current, it could be the same 60 Hz current that is in every household. If a conductor (the work) is placed in the coil, current will be induced in the work. This current has no place to flow to since the work is not connected to anything. The current direction is reversing at the rate of 60 times a second and the net effect is heating of the conductor by eddy currents; if the conductor is ferromagnetic it also heats by hysteresis. This is a simple induction heating system.

To perform induction surface hardening, the conductor in the coil is the part that is to be heated. The work must be an electrical conductor for the system to work, but this is the only requirement. Heating occurs from the surface inward, and it is this phenomenon that makes induction heating suitable for surface hardening. The system can be set up so that only the surface heats to a controlled depth. Surface hardening is accomplished by controlling the heating parameters, time, current, and frequency such that the surface reaches the austenitizing temperature and the work is rapidly quenched by spray techniques or by releasing the work so that it falls into a liquid quench.

Induction-Heating Equipment

The simple system illustrated in Figure 5–7 can be used for induction heating, but present-day systems are much more sophisticated. There are basically three parts to an induction-heating system:

1. Source of alternating current
2. Heavy bus-bar type of conductor to bring the current to the work coil
3. Work coil or inductor

The 60-Hz induction system illustrated in Figure 5–7 causes deep heating because of the relatively low frequency of current reversal, and it would not be suitable for surface hardening. There has been continual change in the design of induction-heating equipment since the process was commercialized in the 1920s, but the heating concept

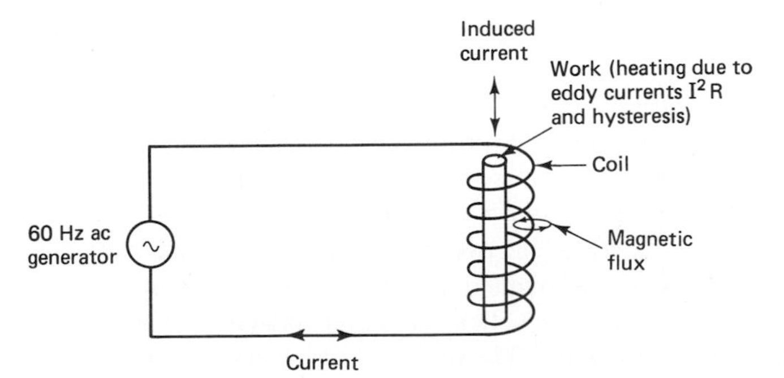

Figure 5–7 Simple induction heating system.

remains the same. The newest units use solid-state devices to control the heating parameters. Induction-heating units are sized by their power output. They range in size from 5 kW (18×10^6 J) to as large as 200,000 kW (72×10^{10} J). A 5-kW (18×10^6 J) unit may be used for heating a small electronic device; the 200,000-kW (72×10^{10} J) unit would be capable of melting a 100-ton (18×10^6 kg) heat of steel. A popular-size induction unit for surface hardening would be about 400 kW (144×10^7 J).

The parameters that are controlled in induction heating are as follows:

1. Coil current
2. Frequency
3. Coil inductance (turns)
4. Coil-to-work spacing
5. Heating time

Induction-heating units are between 50 and 90 percent efficient. Knowing this, a unit can be sized for a particular job by calculating the energy requirements of a particular heating assignment. The current output and frequency on modern induction-heating units are variable, and the induction units used for surface hardening usually use hundreds of amperes. This relatively large current flow requires large copper conductors to the inductor, and the inductor is almost always made from copper tubing or hollow copper shapes that are water cooled. The same surface heating that occurs in the work occurs in the inductor, and the water cooling keeps the inductor from melting.

The frequency of current reversal is a critical factor in an induction heating system. The depth of heating (d) that is achieved is a function of the resistivity of the work (r), the magnetic permeability (μ), and the frequency of the current reversals in the inductor (f):

$$d \sim \sqrt{r/\mu f}$$

From the practical standpoint, materials with low resistivity such as copper and aluminum are hard to heat only on the surface with induction. They tend to through heat, and d will be large. Materials with high resistivity such as graphite are easy to induction heat on the surface (small d). The depth of hardening is affected by the magnetic permeability of the material to be heated, but the effect is not as significant as the effect of frequency. The frequency of a modern induction-heating system can vary from 60 Hz to 50 MHz; thus the role of frequency is substantial. Most commercial induction heating units that are used for surface hardening operate in the frequency range of 200 to 400 kHz. Low-frequency induction-heating units are used for melting metals, heating of forgings, and similar deep heating tasks. The very high frequency units (over 10 MHz) are used for pulse hardening. This latter process produces extremely shallow heating (0.015 to 0.035 in.; 0.375 to 0.87 mm), which rivals electron beam and laser surface hardening. Figure 5–8 is an illustration of the effect of frequency on the depth of induction heating. Figure 5–9 is an illustration of a typical induction hardening system and the heating profile that is common in surface hardening. The desired heating profile for induction

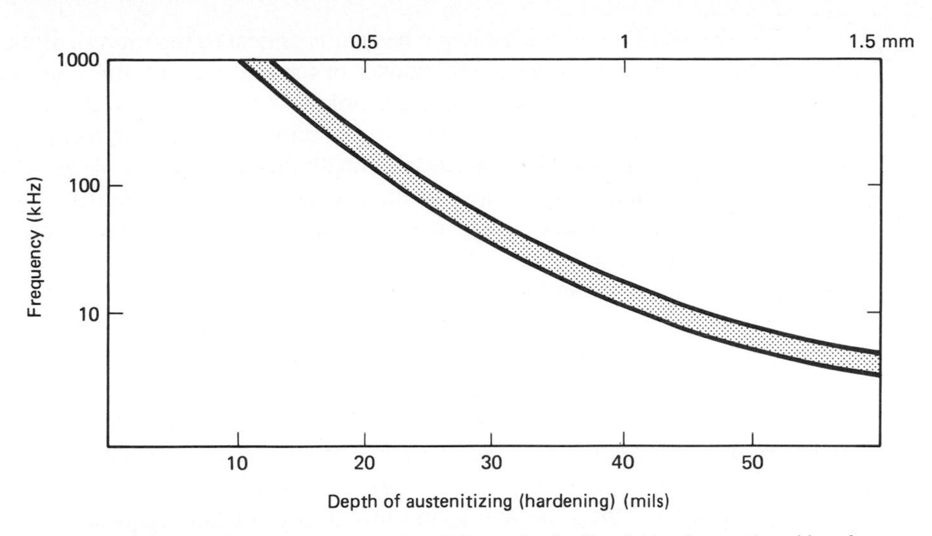

Figure 5–8 Effect of frequency on the minimum depth of hardening that can be achieved in induction hardening steels.

hardening is similar to the profile that is obtained in flame hardening. The shallowest depth of hardening that can be achieved in conventional induction hardening is about 0.020 in. (0.5 mm). The smallest diameter that will accept this 0.020-in. (0.5-mm) case depth is about 0.060 in. (1.5 mm) (with a high-frequency unit).

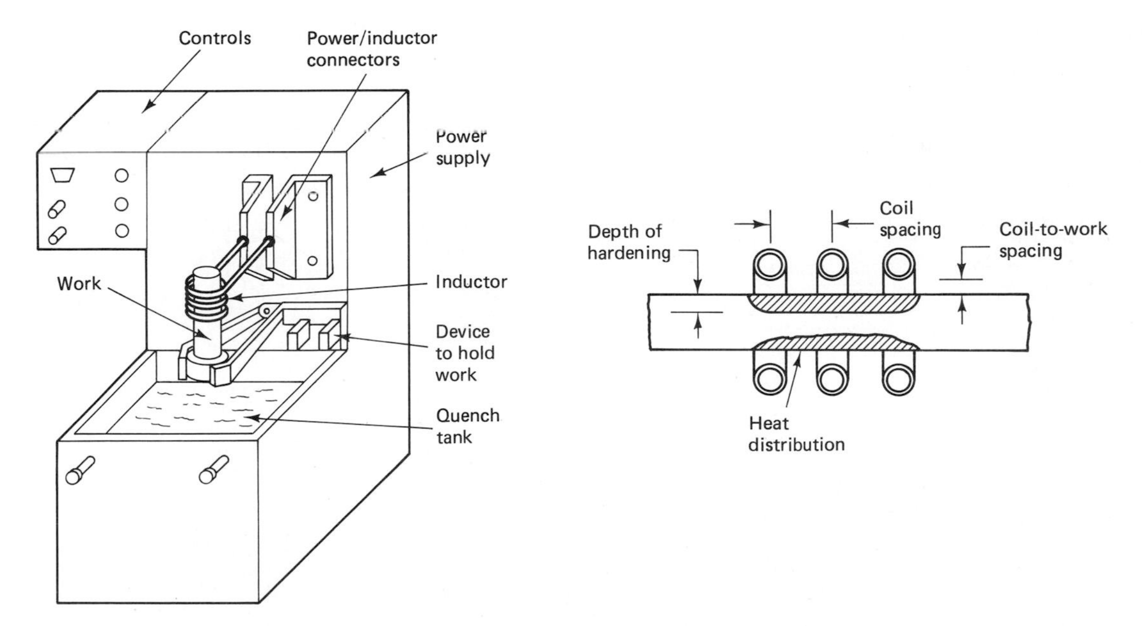

Figure 5–9 Typical induction heating system.

The design of a coil for induction heating is critical to the success of the operation. The spacing of the copper coils, the number of turns in the inductor coil, and the coil-to-work spacing determine the hardening profile that can be achieved with given power and frequency settings. Increasing the number of turns in the coil improves the uniformity of heating. The coil-to-work distance is normally in the range of ¼ to ½ in. (6.25 to 12.5 mm). The "skin effect" heating that occurs in induction heating also occurs in the coil; most of the current is carried by the surface. This is why inductors are usually made from tubing. The hollow center contributes little to the inductor impedance, and making the inductor from tubing provides a convenient means of water cooling. In addition, soft copper tubing is very easy to fabricate into coils that will fit a large variety of work. Some typical inductor configurations are illustrated in Figure 5–10.

The time required to heat the surface of a workpiece in induction heating is a function of the current available (kilowatts or joules) and the frequency of current reversal. Induction-hardening units are sized so that heating can be accomplished in about 0.1 to about 5 seconds. If times become longer than a few seconds, it is usually because the power available from the unit is too low or because the frequency available is too low. Slow hardening will allow conduction of heat to the core, and there will be a tendency to through harden.

To summarize, surface hardening by induction-heating requires a suitable induction power supply and a properly designed inductor. Conventional induction-hardening equipment does one piece at a time (usually in a few seconds), and the hardening profile is similar to that of flame hardening, but usually shallower. The equipment is more expensive (about $1000/kW) than flame-hardening equipment, but it is far less expensive than electron beam and laser surface-hardening equipment.

Applications

Induction heating is used for many applications other than surface hardening. We have already mentioned the use of this process for melting of metals. In this application, a

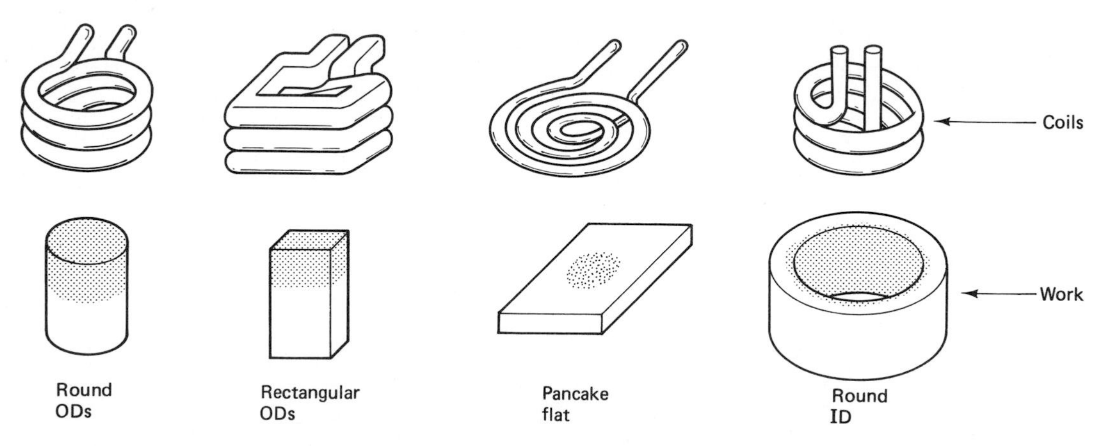

Figure 5–10 Typical inductor configurations for induction hardening.

nonconducting crucible is placed in the inductor coil, and the metal charged into the crucible is heated and melted. Induction heating is also used for heating for forging, brazing, and soldering, heating for interference fits, and for heat treatments other than hardening (annealing, tempering, etc.). The material requirements for these applications are simply that the material be an electrical conductor. The material requirements for the use of induction heating for surface hardening are essentially the same as those described in our discussion of flame hardening. Surface hardening by induction requires a ferrous material with enough carbon and alloy content to allow quench hardening. The part to be hardened must be placed in the inductor or suitably coupled with the inductor, and the material must be heated to the proper austenitizing temperature and then quenched.

The specific ferrous materials that are commonly applied to surface hardening by induction are the same as those shown in Table 5–1 for flame hardening. If the plain carbon steels like 1037 and 1045 produce lower than expected hardnesses, the boron-containing manganese alloy steels 15B37 or 15B45 can be used. Sometimes the H grades of alloy steels that have "guaranteed" hardenability are used instead of the standard grades of alloy steels.

The frequently used induction-hardening ferrous materials (Table 5–1) are normally water quenched. A spray ring can be placed below the inductor coil, and the part can be quenched by dropping it through the spray ring. The same procedure can be used with a water-filled quench tank under the coil. If water quenching produces cracking, the quench can be made less severe by adding about 5 percent emulsifiable oil to the water or by using proprietary quenchants, such as polyalkylene glycols in water.

Pulse hardening, which uses very high frequency induction and a capacitor discharge to produce shallower cases than conventional induction hardening, has the same material requirements as laser and electron beam. A typical pulse-hardening system is illustrated in Figure 5–11. The arrangement illustrated will produce a hardened stripe, and this system has the greatest advantage over conventional induction hardening when self-quenching is used. As in laser and EB hardening, heating of the surface of the substrate is so rapid that the bulk of the material stays at its original temperature. Conduction of heat from the treated zone can be fast enough to provide adequate quenching for hardening of many materials. This same type of thing is done in pulse hardening. The very high frequencies used (>1 MHz) produce rapid heating of the surface to a depth between about 0.004 in. (200 μm) and 0.035 in. (0.875 mm), and if the part has a mass that is at least ten times the mass of the induction-heated zone, the part will self-quench. Obviously, there are advantages to the use of this type of induction heating. The most suitable materials are the steels in Table 5–1, and high-alloy tool steels will be less responsive, as is the case with laser and EB surface hardening.

One final material consideration with regard to use of induction heating for surface hardening is electrical properties. Flame hardening only applies to ferrous materials, and they do not vary significantly in their ability to respond to heating by an oxy/fuel flame. Their thermal conductivities do not vary significantly. In induction hardening, the electrical resistivity and the magnetic properties of the material can produce significant differences in heating characteristics. This does not pose a problem, but it must be

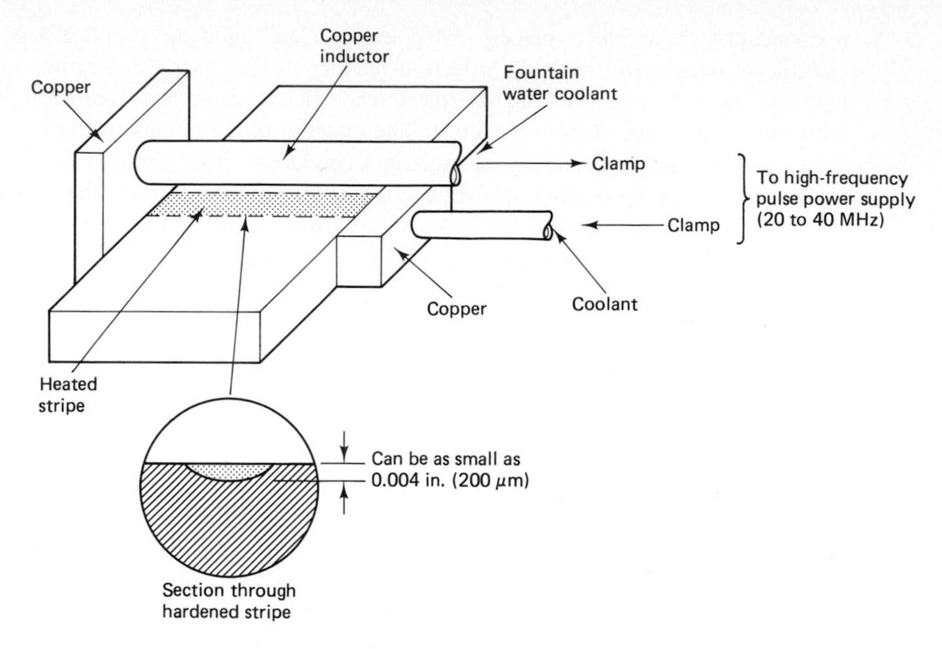

Figure 5–11 Schematic of induction pulse hardening.

recognized that different steel alloys may require differing induction-heating parameters. As shown in Figure 5–12, the electrical resistivity of, for example, a carbon steel can be significantly different from that of a cast iron. Each will require specific induction-heating procedures to achieve the same type of case hardening.

SUMMARY

In this chapter we have discussed two very old technologies that are still significant parts of the field of surface hardening. Both are somewhat underused because of lack of familiarity by many designers. These processes are competitive, but complementary. Each has advantages over the other, and each process has a niche where it is probably the best way to achieve a locally hardened surface. Compared to welding hardfacing and other wear surface treatments, both processes have the same disadvantages. They cannot be used to rebuild, they cannot be used to add material to a surface, and they cannot be used on any metal, only on hardenable ferrous substrates (if surface hardening is the goal).

Table 5–2 is an attempt to compare these processes to each other. In the category of equipment, induction hardening requires a capital investment that is probably at least ten times as great as that required for flame hardening. A do-it-yourself flame-hardening setup can be made for a few hundred dollars, while the cheapest induction-hardening system may cost $10,000. Induction heating is instantaneous and it is usually

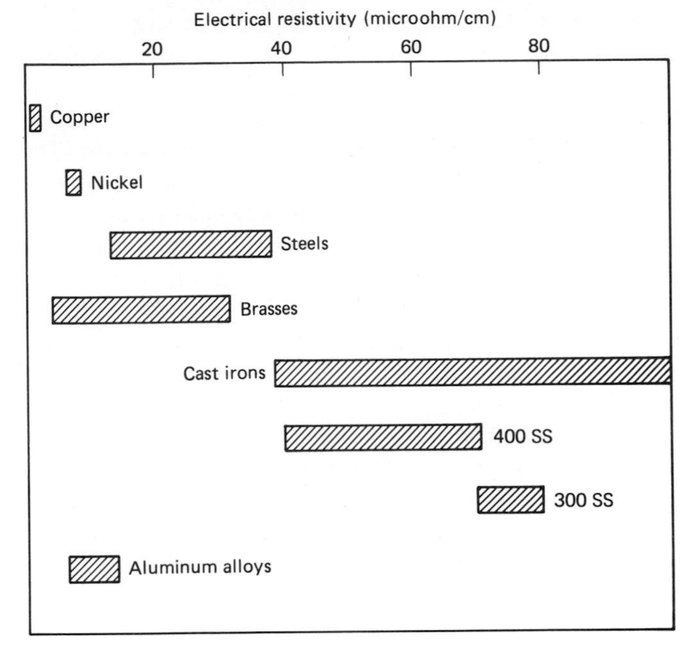

Figure 5–12 Electrical resistivities of some common metals.

TABLE 5–2 COMPARISON OF FLAME- AND INDUCTION-HARDENING PROCESSES

Characteristic	Flame	Induction
Equipment	Oxy/fuel torch, special head quench system	Power supply, inductor, quench system
Applicable material	Ferrous alloys, carbon steels, alloy steels, cast irons	Same
Speed of heating	Few seconds to few minutes	1 to 10 seconds
Depth of hardening	0.050 to 0.250 in. (1.2 to 6.2 mm)	0.015 to 0.060 in. (0.4 to 1.5 mm) (.004 for impulse)
Processing	One part at a time	Same
Part size	No limit	Must fit in coil
Tempering	Required	Same
Can be automated	Yes	Yes
Operator skill	Significant skill required	Little skill required after setup
Control of process	Attention required	Very precise
Operator comfort	Hot, eye protection required	Can be done in suit
Cost: Equipment	Low	High
Per piece	Best for large work	Best for small work

faster than flame hardening. The speed of heating depends on the part shape, but on almost any shape, induction heating is probably faster. Possibly the most significant difference between these two processes is the depth of hardening capability. Flame hardening, because it requires conduction of heat from the surface, is slower in heating the part, and it is not possible to austenitize a thin skin to produce a shallow case. The lower limit on case depth is about 0.050 in. (1.2 mm), while induction hardening with the pulse technique can produce thin cases on thin parts. The case can be as low as 0.004 in. (200 μm) with the pulse technique. On the other hand, induction hardening is not normally used for heavy case depths. It is capable of heating all the way through a part, but this is not normally done because with long heating times the part may oxidize since heating is done in ambient air. In flame hardening, the oxy/fuel flame can be adjusted to produce a reducing atmosphere around the work and oxidation does not occur. Slow heating for heavy cases is easy to do.

A common disadvantage of flame and induction hardening is that only one part can be done at a time. It is possible to have multiple heads for flame hardening and multiple induction units, but neither process can compete with the quantities of parts that can be surface hardened with batch surface-hardening processes like carburizing, nitriding, and the other diffusion processes. Thousands of parts can be done at one time.

Tempering after flame and induction hardening is often neglected, but this is not recommended. The service-property enhancement (better wear and fatigue resistance) justifies the cost. Tempering can be done with normal batch techniques in a conventional furnace. The extra cost can be minimal.

Both flame and induction hardening can be automated. Commercial machines are available, for example, to flame harden gears for automotive applications. Dial-type induction-hardening machines with a heating and quench station are commonly applied to mass-production parts. Both processes can be automated to the point where parts are simply loaded into a transfer type of machine and they come out surface hardened at the end of the machine. Automation of induction heating is easier than flame heating because of the attention required in maintaining proper flame conditions. Because of this factor, it can be said that there are some advantages to automating induction compared with flame heating. The induction-heating parameters remain very constant once the machine is set up. Similarly, less skill is required to make an induction-heating setup than to make a flame-hardening setup. Infrared thermometers can be used to accurately set induction-hardening austenitizing temperatures. These devices are more troublesome to use with flame hardening because the radiant heat from the flame compounds the difficulty of reading the part temperature. Flame hardening requires more skill and judgment on the part of the operator. Since only the work is heated in induction processes, there is less heating of the ambient area around an induction machine. They can even be a part of an assembly line filled with people. The same is not true for flame hardening. There is stray heat, and operators and people working in the area must wear eye protection to prevent eye damage from the bright oxy/fuel flame.

Where cost is a significant process-selection factor, flame hardening is much cheaper than induction if the equipment must be purchased. If heat-treating shops are to be used, induction heating will probably be cheaper for small parts, that is, parts that fit into your hand. When parts become massive, flame hardening is usually lower cost. Thus both processes have areas where each is the more cost effective.

From the standpoint of the overall field of surface hardening, induction and flame hardening are competitive processes. They cannot add material to a surface; in this regard, they do not compete with welding and coating processes. When they can be used, flame and induction hardening apply to the same types of materials, hardenable ferrous alloys and cast irons. Flame hardening usually offers the best cost/benefit ratio on massive parts or small quantities of parts of any size. Induction usually produces a better ratio on small parts and high-production parts. Both process are capable of producing the maximum recommended working hardness from the alloys that are applicable, and part serviceability will be comparable. They are current, viable processes for producing wear-resistant surfaces, and they should be considered for use when wear-resistant surfaces are required.

REFERENCES

CREAL, R. "High Intensity Induction: Alternate to Laser, Electron Beam," *Heat Treating,* Sept. 1982, pp. 21–28.

LEATHERMAN, A. F., and D. E. STUTZ. "Basic Induction Heating Principles," *Solid State Technology,* Oct. 1969, pp. 41–52.

OSBORN H., and others. *Induction Heating.* Metals Park, Ohio: American Society for Metals, 1946.

RUDD, W. C., and H. N. UDALL. "Selective Surface Hardening by High Frequency Reistance," *Heat Treating,* Dec. 1981, pp. 34–40.

TUDBURY, C. A. *Basics of Induction Heating.* New York: John Rider, Inc., 1960.

WALKER, E. "Beneficial Residual Stresses in Induction Hardening," *Metal Progress,* Sept. 1981, pp. 28–35.

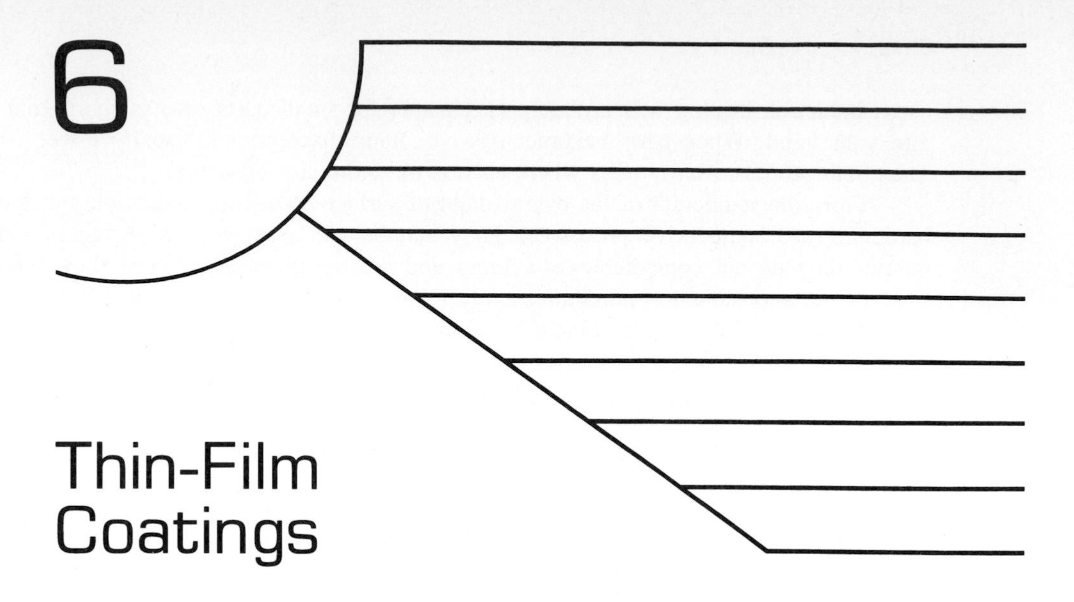

6

Thin-Film Coatings

We have previously discussed diffusion coatings, electrodeposited coatings, and chemical vapor deposited coatings. All these can produce what most people would consider to be thin coatings. In this chapter, the thin films that will be discussed are those that are applied by vacuum techniques. They may or may not fulfill the various opinions as to what is a "thin film." One prevailing definition of a thin film is a coating that is used for its surface properties. Coatings that are used for their bulk properties are not considered thin films. This is a reasonable definition, but this would make wear coatings bulk coatings, since they are predominantly used for their bulk properties (hardness, shear strength, etc.). Some people define a thin film as a coating with a thickness of less than 1 μm (40 μin.). This is still not a good definition in that many hard coatings that are used to enhance the wear properties of tools can be several micrometers thick.

Thin-film coatings are usually applied by the processes shown in Figure 6–1. They fall into two main categories: (1) coatings applied by physical vapor deposition (PVD), and (2) coatings applied by chemical vapor deposition (CVD). Physical vapor deposition processes, as the name implies, involve the formation of a coating on a substrate by physically depositing atoms, ions, or molecules of a coating species. Chemical vapor deposition involves the formation of a coating by the reaction of the coating substance with a substrate. The coating species can come from a gas or gases (as illustrated in Figure 4–14), or it can come from contact with a solid (as in pack cementation processes, Chapter 5). In this chapter, we will concentrate our discussion on coatings produced by physical vapor deposition, and our definition of thin-film coatings will be coatings applied with the PVD processes shown in Figure 6–1. The common denominator for these processes is that they are applied in vacuum equipment. Chemical vapor deposition processes can be done in vacuum equipment, but pressures lower than atmospheric

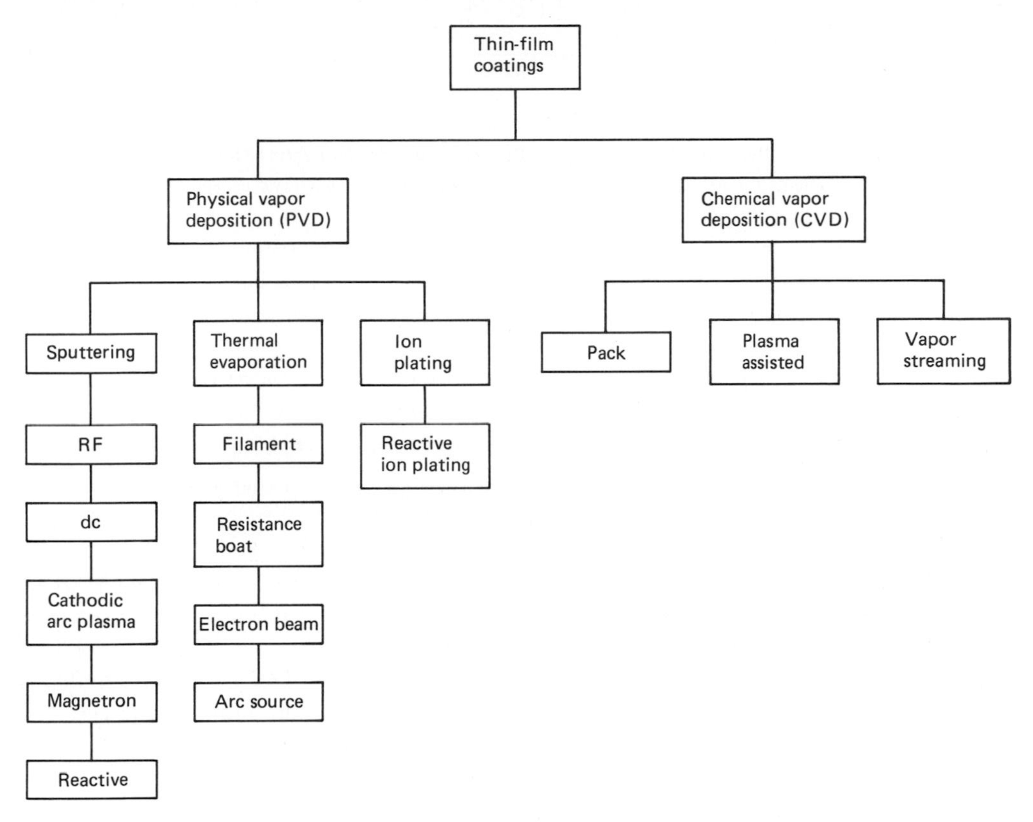

Figure 6–1 Spectrum of processes for thin-film coatings.

are not a requirement for many CVD processes, and they are not as widely used for wear-resistant surfaces as are the PVD coatings.

There are three main techniques for applying PVD coatings: thermal evaporation, sputtering, and ion plating. We will discuss these processes in detail, but, in brief, thermal evaporation uses heating of the material (by various techniques) until it forms a vapor and the vapor condenses on a substrate to form a coating; sputtering involves the electrical generation of a plasma between the coating species and the substrate; and ion plating is essentially a combination of these two processes.

There are many uses for the coatings produced by these processes, but the prevailing use categories are electronics, optics, decorative, and wear prevention. We will concentrate on the latter. Figure 6–2 is a list of some thin-film coatings and where they are used. The coatings used for wear applications are mostly hard compounds. Once again, we will emphasize these.

From the application standpoint, thin-film wear coatings can be used for the same types of applications that flash chromium electroplate is used for. Many wear systems cannot tolerate a wear depth of even a micrometer (40 μin.). Vacuum coating techniques have the potential of applying coatings that have higher hardnesses than any metal,

Coating	Application
Aℓ, Aℓ/Si, Aℓ/Si/Cu	Metallizing of integrated circuits
Ti/Pt/Au, Ti/Pd/Cu, W/Ti Cr/Au, Aℓ/Ni/Au	Contacting systems and backside metallization of semiconductor devices
Aℓ, Cr/Au, Ni/Cr/Au, Ta/Aℓ	Contacts and tracks for hybrid circuits
Cr/Cu, NiCr/Cu, Aℓ/Cu	Contacts for microwave circuits, capacitors, thermoprint heads
TaN, Ta/Aℓ, Ni/Cr	Resistive films
ITO	Contact layers for display elements
SiO_2, $A\ell_2O_3$, Ta_2O_5, SnO_2	Passivation, optics, wear parts, thermoprinting
$A\ell/SiO_2$, $A\ell/TiO_2$, $SiO_2/Cu/TiO_2$	Antireflective, IR reflection
TiN, WC	Wearing surfaces
Cd/Co/Cu, Ni/Fe, Nb_3Sn, Co	Superconducting films and films for magnetic discs
$Cu/Ti/Fe/TiO_2$	Dispersion strengthened films
Cr, Cu, Ni/Cr, Aℓ, alloys	Decorative coatings on plastics

Figure 6–2 Some thin-film coatings and their use (after Schiller et al.).

and they find use in these systems that cannot tolerate even microscopic wear losses. Many processes and coatings for producing wear-resistant thin coatings are still experimental, but the technology is becoming mature at a rapid pace. Eventually, thin-film coatings may become a part of the average designer's repertoire of wear coatings. In this chapter, we will try to present information on the coatings that are useful today and show how they fit into the spectrum of wear coatings.

THERMAL EVAPORATION

Mechanism

Thermal evaporation is the oldest and simplest process for applying thin coatings with vacuum techniques. In theory, it is no different than condensation of dew from moisture-laden air. In thermal evaporation, the material to be coated is heated until it becomes a vapor. The vapor cloud contains atoms or molecules of the evaporant, and it condenses as a film or coating on every surface that is exposed to this vapor cloud. The transformation of a solid to a vapor is accomplished by heating the material until the kinetic energy of the atoms or molecules in the material is such that some of the atoms will have sufficient kinetic energy to break free from the surface and form a vapor. If the material is in a container, the vapor will form an equilibrium pressure above the solid. The reaction stops. If additional atoms try to go to the vapor state, they collide with atoms already in the vapor state and essentially they go back to the solid or liquid. Molecules in the atmosphere above an evaporant tend to suppress this transformation to a gas phase. When a solid is heated in a vacuum, it is much easier for the solid to go to a vapor. The better the vacuum is (the lower the ambient pressure) the easier it is for the solid to go to a vapor. The same is true for temperature. The higher the temperature, the higher the kinetic energy of the atoms in a solid and the easier it is for them to go to a vapor state.

Figure 6–3 is a graph of the approximate pressure/temperature effects on the transformation of various metals from the solid to the vapor state. From the graph, at an ambient pressure of 10^{-3} torr (1.3 Pa), zinc will vaporize when it is heated to about 600°F (315°C). The zinc is not molten at this temperature. There are curves in the literature for most materials. These data can be used to determine the ease with which a material can be evaporated, and they also tell the user what kinds of pressure and temperature are needed to perform a desired vacuum coating operation. The general relationship that governs simple evaporation is

$$\frac{dN}{a\,dt} = c(2\pi mkT)^{-1/2(P^* - P)}$$

where N = atoms or molecules striking a substrate to be coated
a = area of the substrate
c = constant for a surface condition (largely a function of cleanliness)
m = molecular weight of the evaporant
k = Boltzmann's constant
T = absolute temperature
p^* = equilibrium vapor pressure of mercury at the surface of the substrate
p = chamber pressure

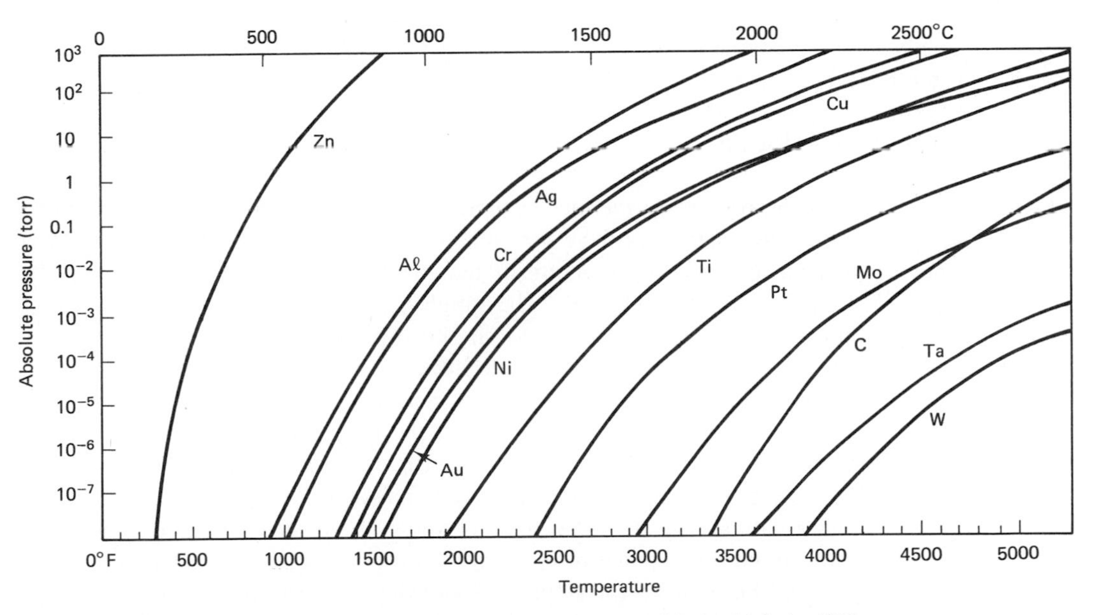

Figure 6–3 Approximate vapor pressures for various materials (multiply by 1333 to convert to pascals).

Thermal evaporation is performed in a vacuum chamber so that it is possible to obtain low enough pressures to allow easy evaporation. The material to be evaporated is heated also to assist evaporation. The pressures in the vacuum chambers are usually 10^{-4} torr (0.13 Pa) or lower. The evaporant is heated to whatever temperature is needed to obtain vaporization at the chamber pressure.

The mechanism of adhesion of evaporative coatings is usually similar to that of electroplating. If a coating is applied to an atomically clean surface, there will be a net attraction between the atoms in the coating and the atoms (or molecules) at the substrate surface. This is the familiar van der Waals bonding. There is usually no penetration of the coating into the substrate, but the substrate could be heated to obtain diffusion; this is seldom done. Evaporated coatings of metals on nonmetals are usually adhered by van der Waals or mechanical bonding or a combination thereof. If the evaporant and substrate are compatible, covalent types of bonds can be formed. If the evaporant is reactive with the substrate, chemical types of bonds can be obtained. The typical evaporative coating bond is mechanical in nature. Bond strengths are difficult to measure, but they do not compare with fusion bonds. The bond strength is mostly a factor of the cleanliness of the substrate.

The morphology of a thermally evaporated coating depends on the nature of the evaporant, the nature of the substrate, and the temperature of the substrate. Coatings initiate at nuclei of coating atoms. The nuclei grow together as evaporation proceeds, and, especially with metals, columnar grains are formed perpendicular to the substrate surface. As the temperature is raised, the coating can start to develop grain structure similar to those in wrought metals. At still higher temperatures, equiaxed grains with triple point boundaries are formed. Most metal coatings are crystalline in nature, but usually the grain diameters are not measured nor are they very important to the function of the coating.

Process Description

Physical vapor deposition by thermal evaporation requires a vacuum chamber with a pumping system adequate to achieve pressures of 10^{-6} torr (1.3×10^{-3} Pa) or lower, a heating system to induce evaporation of the species to be coated, and a system to hold or hold and manipulate the work. The simplest PVD system is illustrated in Figure 6–4. The containment vessel can be glass, as shown in our illustration, but the coating chamber can be steel or any material with sufficient strength to withstand implosion from atmospheric pressure when the vessel is evacuated. Porous surfaces tend to outgas and interfere with the vacuum. For this reason, the preferred vessel material is glass, smooth-finished stainless steel, or smooth electroplated steel (usually nickel plated). Work holders can be as simple as the ringstand shown in Figure 6–4, or they can be elaborate manipulators. The flux of evaporant varies in an ice-cream-cone shape from the center of a round pool source of evaporant. If a point source of evaporant is used, the coating would be uniform for a sphere about the point source. The evaporant flux is line of sight. If a perfectly uniform coating is desired on a flat substrate, it is necessary to manipulate the substrate during coating such that all areas on the substrate are exposed at the same radius from the source for equal time increments. It is seldom necessary to go to this extreme. If the substrate in the illustration in Figure 6–4 is held stationary during the coating operation, the coating thickness at the center of the plate will be about 20 percent thicker than the coating at the edges. This is acceptable for most coating applications. A more important consideration is the line-of-sight nature of the coating. It would definitely be necessary to manipulate a cylinder if the entire outside diameter of the cylinder had to be coated.

The vacuum system for a PVD coating operation can include any combination of pumps that will provide a pressure of 10^{-6} torr (1.3×10^{-3} Pa) or lower. The usual system has a mechanical pump and an oil-diffusion pump. Ionization gages and the like are used to monitor the vacuum. Bell-jar systems like the one illustrated in Figure 6–4 require coating parts in batches. The use of metal containment vessels allows attachments that can provide work load and coating chambers. Continuous coating systems are available where a coil of, for example, plastic can be unwound, passed over a coating stage and coated, and coiled up after coating. All operations are performed in the containment vessel. Pass-through coating systems have been developed for tin plating steel where coils of steel are continuously fed through a PVD coating chamber. Seals are maintained on both ends of the coating chamber. PVD coaters with bell-jar containment vessels are used for batch coating operations. Continuous coating systems require vacuum seals like those described on the tin plating steel mill operation. Batch coaters are by

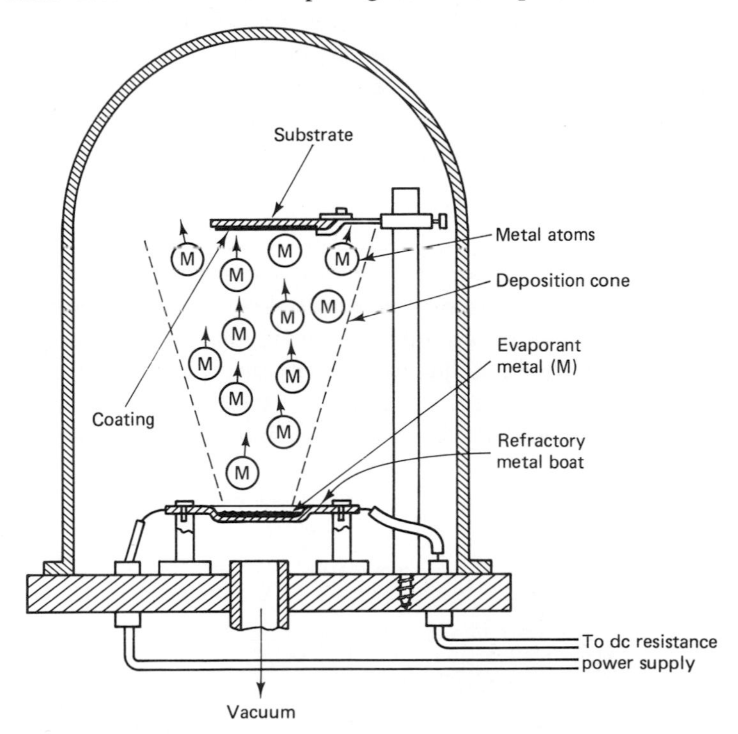

Figure 6–4 Schematic of thermal evaporation.

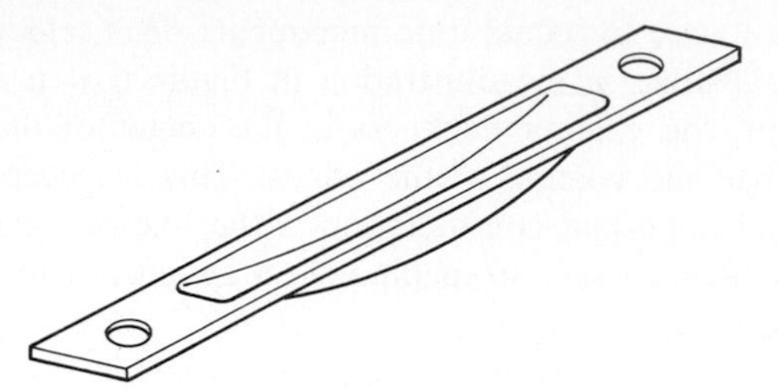

Figure 6–5 Refractory metal boat for use as a thermal evaporation crucible.

far the most prevalent systems. Chambers can be as large as an automobile. Fixtures and manipulators can be used to coat many parts in a single pump-down and coating sequence.

The most common way to heat evaporants in the batch-type coaters is to put the evaporant in a powder form into a refractory metal boat, as illustrated in Figure 6–5. The boat is resistance heated to melt the evaporant. For larger coating jobs, the boat can be continuously supplied with metal with a wire-feed device, like that used in GMAW welding, or by a powder feeder. Faster heating of the evaporant and melting of high-temperature metals can be done with an electron beam melter. As illustrated in Figure 6–6, the evaporant is placed in a ceramic or refractory metal crucible and is heated and melted by impingement of an electron beam. The source of the electrons is a filament within the gun. The electron stream is bent by magnets such that it curves and impinges on the evaporant in the crucible. The EB gun and crucible are water cooled. There are many details pertaining to melting to obtain the desired coating flux and flux distribution, but the user of the process does not need to specify these details. A typical PVD coating facility is shown in Figure 6–7.

Substrates

PVD coatings will coat any surface that is in the path of the flux from the evaporant source, but making the coating adhere to a substrate is another matter. Thermal evaporation coatings will deposit on any substrate: metal, plastic, ceramic, cermet, even on strange things such as leather. Adhesion to a substrate is dependent on the cleanliness of the substrate surface, the stress in the coating, the long-term stability of the coating in the use environment, and the chemical compatibility of the coating with the substrate.

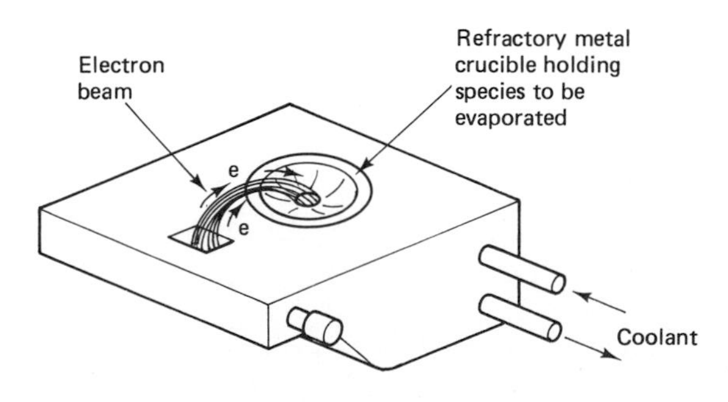

Figure 6–6 Electron-beam melter for thermal evaporation.

Figure 6–7 Typical vacuum coating system (can be used for thermal evaporation, sputtering, ion plating, and plasma-assisted CVD).

Surfaces to be PVD coated should be atomically clean for best adhesion. Metals should be free of nonadherent oxides. This can be accomplished by pickling or abrasive blasting. Smooth metal surfaces can be cleaned by vapor degreasing with solvents that do not leave a residue. Usually, several solvent systems are used, and the final solvent may be acetone or isopropyl alcohol. Many plastics pose coating problems by their very nature. Plasticized materials outgas in the vacuum, and the coating will essentially be deposited on the plasticizer, causing a dark coating with poor adhesion. It is preferred that these types of plastics be avoided. Sometimes plastics with minimal outgassing characteristics can be made coatable by sealing them with a lacquer that acts as a vapor barrier. The long-term stability of these types of systems should be tested before this approach is deemed acceptable for an application. Homopolymers with or without inorganic fillers are usually applicable to PVD coating without the use of sealers.

Ceramics and glasses can be solvent cleaned like metals. Glasses and smooth metals are often checked for cleanliness by a water break test. Water is misted on the surface to be coated. If the deposited water remains as a uniform fog on the surface, it is clean; if the fog coalesces into droplets, the surface is contaminated. Glasses sometimes require a precoat, but the coater will know when this is necessary. Sintered ceramics often contain porosity that can cause outgassing and coating adhesion problems. It may be necessary to vacuum and pressure impregnate these materials and similar materials such as powdered metals.

There are no standard tests for coating adhesion. Most coaters have developed their own tests. One of the most commonly used tests is to apply adhesive tape and forcibly strip the tape from the coated surface. Coating transfer to the tape is the measure of coating adhesion. Sometimes this test is made more severe by scratching the coating in a grid pattern before the tape is applied (ASTM B 347). Another simple test for adhesion is to abrade the coating with an eraser on a wooden pencil. A poor coating will be removed by the abrasion. Long-term adhesion and chemical effects between the coating and the substrate can be checked by repeating the adhesion tests after incubation in the intended service environment.

In summary, thermal evaporation coatings can be applied to any solid surface. Adhesion is best to surfaces that are atomically clean and do not react with the coating or outgas. Poor coating adhesion due to coating stresses usually does not become a factor unless the coatings become thick (>1 μm).

Coatings

Similar to the situation with substrates, any material that can be vaporized by heating in a vacuum can be evaporated coated. Unfortunately, many technical details interfere with what can theoretically be done. Pure metals can be melted and they will form a vapor in a vacuum. One significant problem in applying metals is reaction between the metals and the crucible or boat material. For example, a simple evaporant such as aluminum can react with a molybdenum evaporation boat to cause degradation of the boat. Tungsten or ceramic crucibles are often required. Refractory metals have melting points in excess of 3000°F (1650°C), and they are difficult to melt by simple resistance-heating techniques. An electron beam source may be needed to achieve sufficient energy for melting.

Alloys such as Inconel, brass, and the MCrAlY turbine coatings are constituted from elements that have different melting points. To obtain a coating with the desired stoichiometry, it may be necessary to alter the composition of the evaporant alloy so that the coating ends up with the desired composition. For example, if a 70 percent copper, 30 percent zinc brass coating is desired, it may be necessary to make the evaporant alloy 90 percent copper and 10 percent zinc.

Ceramics materials pose a similar problem with maintaining stoichiometry in the coating. In addition, many ceramics have such a high melting point that electron beam or other high-energy sources are needed to achieve evaporation. Some oxide coatings are obtained by depositing a metal and subsequently oxidizing the metal to form the desired oxide. For example, titanium can be deposited and oxidized to titanium oxide. Some ceramic materials such as silicon monoxide and magnesium fluoride evaporate and deposit with the proper stoichiometry. These coatings are used in optics. Silicon oxide is used as an abrasion-resistant coating and magnesium fluoride is used for antireflectance.

There are no theoretical limits to the materials that can be deposited by thermal evaporation, but, as we have tried to point out, many problems can arise from reactions of the evaporants with the heating supports, high melting temperatures, and maintaining the desired composition in the deposit. These problems have been studied and overcome for many systems. The following are some of the more common PVD coatings:

PURE METALS

Aluminum
Iron
Cobalt
Copper
Nickel
Cadmium
Silicon
Germanium
Tin

REFRACTORY METALS

Chromium
Tungsten
Molybdenum
Tantalum

PRECIOUS METALS

Platinum
Palladium
Rhodium
Gold
Silver

ALLOYS

Brass
Inconel
MCrAlYs

CERAMICS (METAL COMPOUNDS)

Silicon oxide
Tantalum oxide
Titanium oxide
Aluminum oxide
Magnesium fluoride

Many other coatings have been successfully applied, but the preceding coatings probably account for 90 percent of the coatings that are applied. They are applied for countless reasons. The metal compounds are usually applied for optics or electronic applications. The simplest and still most widely used application of a thermal evaporation coating is aluminum for reflectors. Most floodlamps and headlamps rely on this coating (on glass).

From the wear standpoint, only a few thermal evaporation coatings are used. We mentioned silicon oxide on plastics. These coatings have been shown to improve the scratch resistance of plastic lenses. Chromium is the only pure metal that has utility for improving abrasion resistance. It is used on photoetching masks to reduce scratching in handling. Some soft pure metals such as gold, tin, and silver may be used to reduce adhesive wear reactions in sliding systems, but their use for this application is not widespread. Essentially, none of the thermal evaporation coatings are used for wear systems that involve the scratching abrasion that one might encounter in handling hard particulates. They are usually only used on systems where microscopic attrition from a surface affects serviceability.

SPUTTER COATING

Sputter coating is a vacuum coating process that involves the use of ions from a gas-generated plasma to dislodge coating atoms or molecules from a target made from the material that is to become the coating. The plasma is established between the target and the substrate by the application of a dc potential or an alternating potential (RF) between the system electrodes (target and substrate). The dc process is used when the target and the substrate are both electrical conductors. The use of an RF power source allows a plasma to be generated between a nonconductor substrate (plastic or ceramic) and a target that may or may not be a conductor. An inert gas such as argon is introduced into the vacuum chamber to form the glow discharge plasma between the electrodes. The sputtering term refers to the process of dislodging atomic species from the surface of the target.

Figure 6–8 is an illustration of the simplest type of sputter coating system. The target can be a solid plate of the material to be coated, and the substrate can be a similar-size plate positioned parallel to the target. The target and the substrate are about the same size. The substrate to be coated can be attached to a heat sink for cooling, but the substrate must be connected electrically to the heat sink. The mechanism of coating can be complex because of the possible interactions that can take place in the glow region. As shown in Figure 6–9, the plasma region will contain target atoms or molecules, molecules and ions of the gas that is used to generate the plasma, and primary electrons from the target and secondary electrons from the substrate. With a dc potential between the target and the substrate, the positively charged gas ions accelerate out of the plasma toward the negative target material. Collisions of the plasma ions with the target cause target atoms to be dislodged from the surface. Their momentum carries them to the substrate to form the coating. The plasma also contains electrons that are emitted from the target during the ion-bombardment process. Electron bombardment of the substrate can cause secondary electrons to be emitted from the substrate. The substrate is in the plasma or glow region. Thus the substrate is also being bombarded

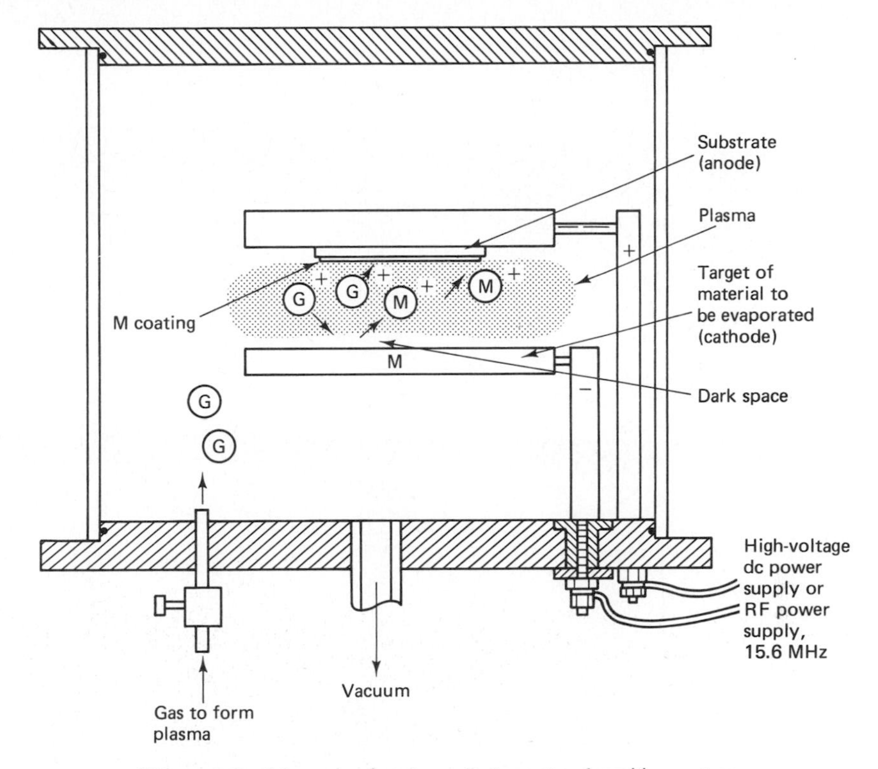

Figure 6–8 Schematic of a planar diode sputter deposition system.

by ions and primary electrons from the target. This makes the coating rate slow, but the effect of these particle collisions on the substrate is thought to enhance the coating adhesion. Because of collisions in the glow region, and because some of the coating can be removed by the action of the plasma on the substrate, the deposition rate is slow, possibly several hundred angstroms (10^{-10} m) per minute, while the rate with evaporation processes can be orders of magnitude higher.

A very useful modification of sputtering is the process of reactive sputtering. If a reactive gas is used in forming the plasma, it is possible to get reaction of the plasma gas with the target species to form a coating that is a compound of the target and the reactive gas. For example, if the target material is titanium and the plasma gas is ammonia (NH_3), it is possible to obtain a coating of titanium nitride on the substrate. Titanium nitride is a very hard compound that is commonly used as a wear coating on tools. Similar reactive coatings have been made from aluminum and oxygen to form aluminum oxide; titanium can be reacted with oxygen to form titanium oxide; the possibilities are numerous. Figure 6–10 shows a TiN coating on a molybdenum substrate.

In summary, sputtering differs from evaporation in that the mechanism of generation of the coating species is the action of ion bombardment from a plasma that is generated between the work and a target that is made from the species to be coated. It can be

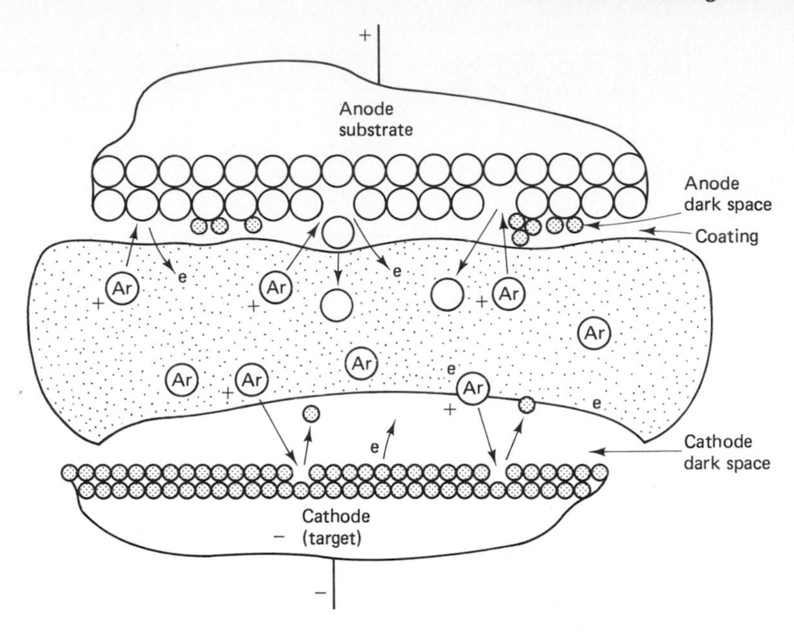

Figure 6–9 Surface reactions in sputter coating.

Figure 6–10 Plasma-assisted CVD coating of TiN on TZM alloy, 2 μm thick (600×).

used to coat compounds, and the substrate can be an electrical conductor or insulator. The use of reactive sputtering provides the potential for forming coatings that are reaction products of the plasma gas and the target material.

Process

Sputtering can be performed in the same vacuum chamber that is used for any coating with the addition of power supply feed-throughs and a plasma gas inlet. If the sputtering system is configured like the illustration in Figure 6–8, it is called a planar diode system. In small bell jar systems, the parallel electrodes are usually round disks with diameters in the range of 3 to 10 in. (7.6 to 25 cm). The spacing between the electrodes may be as small as 2 in. (5 cm) or as large as 5 in. (12.7 cm). If the target is a metal, a plate of that metal serves as the target and the cathode. The anode or work holder is usually a metal plate to which the substrates can be attached. They do not have to be conductors if the plate that they are attached to is metal. The power supply for dc diode sputtering usually has the capability of generating a potential of from 500 to 5000 V between the electrodes. If the target is a nonconductor, an RF power supply with a rating in the range of 500 to 5000 W is used. The target can be made from a plastic or ceramic. One limitation of sputter coating of nonmetals is obtaining a suitable plate of the substance to be coated, one that can take the process temperature and not degrade. For example, if a fluorocarbon coating is desired, it would be difficult to make a cathode plate from, for example, PTFE. What can be done is to put PTFE powder in a stainless steel tray. When the ions from the plasma bombard the tray, they will hit mostly the PTFE powder, which is spread uniformly over the tray. A similar process may be used for coating a lubricant such as MoS_2. Making cathode plates of the desired coating species often becomes the foremost problem in coating some inorganic and organic compounds.

Argon gas is the most common plasma gas for nonreactive sputtering. Gas pressures in the range of 0.1 to 10 Pa (0.013 to 0.13 torr) are used to establish the glow discharge. The gas pressure has an effect on the morphology of the coating. It is a parameter in achieving a desired coating. The vacuum system used for sputtering must be capable of pressures as low as 10^{-6} torr (1.3×10^{-3} Pa).

In the planar diode process, the target and work are positioned with the desired spacing in the chamber. They are electrically connected. The vacuum chamber is pumped down. The plasma gas is introduced, and the potential between the system electrodes is increased until a plasma glow is established. The coating process is now in operation, and the coating cycle is continued until the desired coating thickness is achieved.

Magnetron sputtering is a higher-efficiency modification of the process. A number of system configurations are in use, but essentially magnetron sputtering devices impose a magnetic field on the plasma to produce a desired electron motion in the plasma. Electron collisions with the plasma gas create ionization of the gas, and this ionized gas is needed to produce sputtering of the target material. Essentially, magnetron devices increase the efficiency of the process, and they are often used for production applications of sputtering. Planar magnetron sputtering devices have even been used to make continuous coatings on webs. Large magnetron devices have been used to produce solar reflection coatings on large glass panels for building purposes.

The process control parameters for sputtering are the cathode/anode area, the spacing of the electrodes, the plasma current, the potential between the electrodes, and the type and pressure of the gas used in forming the plasma. The composition of the gas in the chamber can be continuously monitored by a residual gas analyzer. The vacuum equipment requirements are generally the same as those used in thermal evaporation. The magnetron processes also require ancillary equipment to control the magnetic field on the plasma.

A recent development in sputter coating equipment is the cathodic arc deposition process. Cathodes carrying the species to be deposited can have an area of several square feet, and they contain permanent magnets to control secondary electron activity, as in magnetron sputtering. The unique aspect of this sputtering process is the addition of an arc to accelerate coating rates. After the plasma is established between the cathode and the work, an arc is generated on the surface of the cathode by a number of different devices, but the technique is not unlike striking an arc with a welding electrode. Once the arc is established, it is easily sustained and it "dithers" across the surface of the cathode at speeds as high as 100 m/s (20,000 ft/min). Each time the arc touches down on the coating species, microdroplets are produced that add to the fluence of the coating species. This is a higher-energy process than conventional sputtering, and the deposition rates can be as high as 10,000 Å/min. This type of equipment is commercially used for the production of optical mass memory disks, razor blade coatings, coatings on cutting tools, coatings on architectural glass, production of photovoltaic devices, and many other electronic devices. Coating job shops also use this type of coating equipment to produce TiN and similar wear-resistant thin-film coatings.

Substrates and Coatings

Sputter coatings can be deposited on any substrate that can withstand the temperatures that arise from the process. These can be substantial; temperatures are often in the range of 500° to 1000°F (260° to 540°C). Obviously, temperatures in this range can degrade a number of substrates. Plastic substrates may need to be fastened to a water-cooled anode plate. Magnetron sputtering devices can reduce substrate heating, and they are often used for coating heat-sensitive substrates such as plastics. From the standpoint of the substrate, process temperature is a prime selection factor for determining suitable substrates.

Cleaning of the surface is an important part of the thermal evaporation process. The impingement of plasma ions can be used to clean substrates that are to be sputter coated. Some sputter devices have a shutter than can be swung into position between the anode and cathode plate. A plasma can be established between this shutter plate and the substrate to sputter clean the surface prior to coating. After the cleaning operation, the shutter is positioned out of the plasma region, and sputtering can commence between the target and the substrate. Sputtering can even be used as a machining process. It is possible to mask a target and sputter off material in a desired pattern. The materials that can be sputter coated are (1) pure metals, (2) alloys, (3) inorganic compounds, and (4) some polymeric materials.

With reactive sputtering, it is possible to make coatings that are compounds of the plasma gas and the target material. The big advantage of sputter coating over thermal evaporation is that, if the system conditions are properly controlled, the target atoms or molecules will be sputtered from the target and deposited on the substrate intact. Alloy stoichiometry is maintained if the elements in the alloy are in solid solution. Special techniques may be needed for multiphase alloys.

In summary, sputter coating is a higher-energy process than thermal evaporation, and the deposition rates are somewhat lower (several hundred angstroms per minute) unless advanced processes such as cathodic arc are used. The bond of sputter coatings can be better than thermal evaporation coatings. Compounds can be applied if a suitable target of that material can be fabricated. Because of the higher energy involved in the process, substrate heating must be considered. From the application standpoint, sputtering is often used for depositing compounds and materials that are difficult to coat by thermal evaporation techniques.

ION PLATING

Ion plating is a vacuum coating process in which a portion of the coating species impinges on the substrate in ion form. Basically, the process is a hybrid of the thermal evaporation process and sputtering. Figure 6–11 shows a schematic of the simplest type of ion-plating system. The substrate holder is the cathode of a dc plasma power supply. The anode is a filament or resistance-heated boat, the type that is used in thermal evaporation. The vacuum system is pumped to 10^{-6} torr (1.3×10^{-3} Pa) pressure. Argon or another gas is introduced into the chamber until the pressure rises to 10^{-2} torr (13 Pa) or thereabouts and the high-voltage power supply is energized. A plasma is established between the filament and the substrate holder. Gas ions bombard the substrate, and surface atoms are sputtered off. This is the substrate cleaning portion of the process cycle. When cleaning is complete, the power supply to the resistance-heated filament is energized, and the filament material starts to evaporate and enter the plasma. Some evaporant atoms pass through the plasma in atomic form. Some atoms collide with electrons from the substrate and become ions. In ion form, they impinge on the substrate to become part of the coating. The ions pick up electrons from the substrate, and they return to atomic form when they become part of the coating. The evaporation rate is maintained at a rate that is higher than the rate at which atoms are sputtered from the substrate. Deposition of the evaporant in ion form, coupled with sputter cleaning of the surfaces, results in excellent coating adhesion.

Equipment

Ion plating can be done in any vacuum system that has the capability of thermal evaporation and sputter coating. The requirements are a power supply for producing evaporation

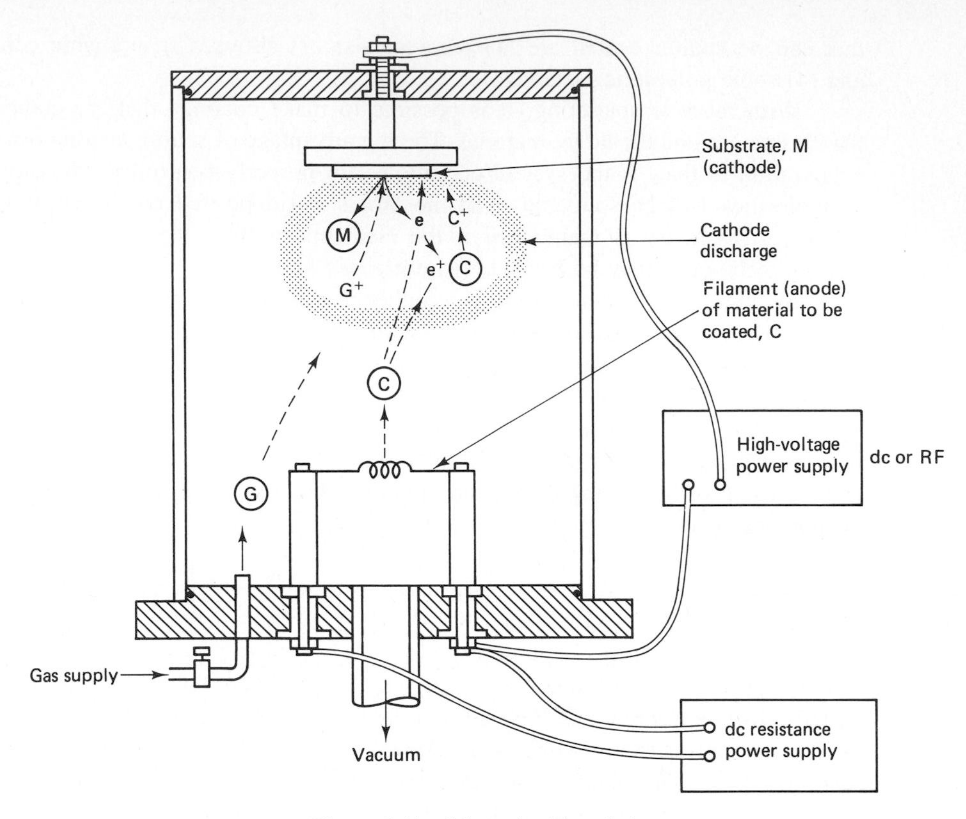

Figure 6–11 Schematic of ion plating.

and a high-voltage power supply for producing the plasma. To ion plate high-temperature materials, those with a melting point over 3000°F (1650°C), the evaporation is usually produced with an electron-beam melter. For ion plating of nonconductors, a RF power supply is used to produce the plasma. The plasma gas is usually argon, but a reactive gas can be used so that the coating species can react to form a coating that is the reaction product of the evaporant and the plasma gas.

The process-control parameters in this process are the same as for sputtering and for thermal evaporation. Since this process combines these processes, the user must be concerned with the evaporator electrical control, as well as the normal controls on the plasma system, gas pressure, plasma current, and so on. The fixturing of the substrate for ion plating is done as for thermal evaporation, with the exception that the substrate holder must be part of the plasma-generation system. The spacing between the evaporant and the substrate is larger than in sputter coating, and the glow discharge encompasses a larger volume of the chamber. Because of the collisions of the evaporant with ions and electrons in the plasma, the deposition of the coating species is better than line of sight. The coating ions impinge on the substrate in oblique angles, and an odd-shaped substrate such as a hemisphere can be uniformly coated over its entire exposed area.

Ion plating is successfully used for coating screw threads and similar shapes that would not coat evenly with thermal evaporation or sputtering.

The substrate requirements for ion plating are the same as for sputter coating. The substrate can be any metal or nonmetal, but nonconductors require the use of RF. The heating effects on the substrate are like those of sputter coating. Heat-sensitive materials such as plastics require cooling. Process temperatures can be in the range of 500° to 1000°F (260° to 540°C).

The materials that can be ion plated are those that can be evaporated. Metals are the easiest to evaporate, and they are the materials most commonly used in ion plating. Unlike sputtering, ion plating will not normally coat molecules of a compound unaltered. It is practical to assume that this process will work successfully to produce a metal coating, but other types of coatings may be difficult. The thermal evaporation rules for coating metal alloys apply. The stoichiometry of the alloy can be altered in the evaporation process, and an empirically determined composition may be required to get the desired stoichiometry in the coating. Coating rates for metals can be in the range of 3 to 5 μm (120 to 200 μin.) per minute.

In summary, ion plating is a process for obtaining very good adhesion between a substrate and a thin coating. It is a hybrid of thermal evaporation and sputtering, and the rules for both processes apply. Ion plating is mostly used for metallic coatings. It is even used to apply an initial metal coating on materials that are hard to electroplate, such as titanium, and conventional electroplating is performed over the ion-plated surface. Ion plating is one of the newer vacuum coating processes, and as equipment is developed to further enhance the capabilities of this process, it may become a candidate to replace conventional electroplating.

THIN FILMS FOR WEAR APPLICATIONS

We have discussed the three major processes that are used for depositing thin coatings by vacuum techniques: thermal evaporation, sputtering, and ion plating. Each process does something better than the other processes. These factors can serve as a basis for process selection. The following are some of the major features of each process:

ADVANTAGES OF THERMAL EVAPORATION

1. Lowest-cost equipment (from the standpoint of simplicity).
2. Lower substrate heating than high-energy processes.
3. Simpler process control.
4. Can coat nonconductors.
5. Continuous coating systems are commercially available.

DISADVANTAGES OF THERMAL EVAPORATION

1. Cannot deposit compounds (unaltered).
2. Deposits may have poor adhesion.

3. Line-of-sight deposition.
4. Deposition of alloys requires special evaporant compositions (to maintain stochiometry of deposit).

ADVANTAGES OF SPUTTERING

1. Atoms and molecules are transferred unaltered (can coat compounds and alloys).
2. Sputter cleaning of the substrate can enhance coating bond.
3. Still a line-of-sight process, but better throwing power than thermal evaporation.
4. Can deposit nonconductors on nonconductors and conductive materials.
5. Can create coatings that are reaction products of the sputter gas and the target species.

DISADVANTAGES OF SPUTTERING

1. Equipment is more complex than thermal evaporation.
2. Substrate heating can be substantial (cooling is often required).
3. Sputter targets of compounds and unusual materials are very expensive.
4. Process control is more difficult than in thermal evaporation.

ADVANTAGES OF ION PLATING

1. Best coating bond (compared with thermal evaporation and sputtering).
2. Best coverage (compared with thermal evaporation and sputtering).
3. High deposition rates (comparable to thermal evaporation and many times the rate of conventional sputtering).
4. Can create coatings that are reaction products of the evaporant and the plasma gas.

DISADVANTAGES OF ION PLATING

1. Complex process control.
2. Cannot deposit compounds unaltered.
3. Substrate heating can be substantial.
4. Difficult to coat nonconductors.

Because vacuum coating is expanding at such a great rate, there are probably equipment innovations in the offing that can alter any of the above statements, but in 1986 these statements apply to most commerical equipment.

To use the preceding process selection factors, it is necessary to list the most important criteria for an intended application. For example, if you want to apply PTFE on a metal, the only process that is normally used to do this is sputtering. If you wish to apply a compound such as titanium nitride, several processes can be used: sputtering with a TiN target, reactive sputtering with a titanium target and an ammonia gas (or other source of nitrogen) as the plasma species, or reactive ion plating (with a suitable evaporant and plasma gas). If you wish to aluminize plastic reflectors, thermal evaporation may be the most suitable process. Sometimes the desired coating or the nature of the substrate determines which process is the most suitable. The user must weigh the relative advantages of the processes and make a selection based on which process offers the most advantages.

Whatever the substrate, there is probably some vacuum coating process that can be made to work. Probably the biggest substrate consideration is substrate heating. There is no sense in putting a hard coating on a tool if the coating process will anneal or overtemper the substrate material. Similarly, plastic substrates can be destroyed by the coating process. In thermal evaporation coating, the only substrate heating that occurs is from radiant heat from the evaporant source. This can be low if the coating times are short. The high-energy processes, sputtering and ion plating, can produce substrate temperatures in excess of 1000°F (540°C) if adequate cooling is not used. In any case, the high-energy processes produce temperatures that are typically 500°F (260°C) or more. This can affect the use properties of many substrates. The best approach to substrate selection for the user of vacuum coating processes is to select the substrate that offers the desired use properties and then discuss the effect of candidate coating processes on the substrate and on coating adhesion.

We have already mentioned many of the coatings that can be applied by thermal evaporation, sputtering, and ion plating; many are technical coatings, coatings used for some physical property, but the coatings that have importance in tribological systems are relatively few. Table 6–1 is a tabulation of some the vacuum coatings that have been used to enhance the tribological properties of sliding systems. The soft metals such as gold, molybdenum, and silver can be applied to minimize fretting damage on electrical contacts or as a lubricant coating on otherwise unlubricated hardened metals

TABLE 6–1 THIN-FILM COATINGS FOR TRIBOLOGICAL APPLICATIONS

Thermal evaporation	Sputtering	Ion plating
Au	SiO	Cr
Ag	SiO_2	Mo
MCrAlY's	Cr	TiC
Cr	Mo	TiN
Mo	Au	Au
	TiC	Ag
	TiN	Si_3N_4
	Al_2O_3	
	WS_2	
	MoS_2	
	Si_3N_4	
	PTFE	
	TiB_2	

in sliding contact. The only hard metal that has utility is chromium. With a hardness that can be as high as 900 kg/mm^2, it can provide a modicum of abrasion resistance to substrates. The intercalative compounds such as MoS_2 and WS_2 have been successfully used to impart dry lubrication to rolling element bearings and to other sliding components that are used in the vacuum of outer space. PTFE sputter coatings have been used for release surfaces and for dry film lubrication on a wide variety of substrates.

The hard compounds are the most widely used vacuum coatings for tribological applications. Silicon oxide coatings are used for abrasion-resistant coatings on clear and translucent plastics. Titanium carbide, hafnium carbide, titanium–aluminum carbide, titanium nitride, and aluminum oxide are used to enhance the sliding wear and scratching abrasion of tool materials. All these coatings can have theoretical hardnesses above 2000 kg/mm^2. They are harder than the hardest metal. If they can be applied with good adhesion, they should improve the wear properties of most surfaces. They are usually applied in thicknesses of less than 3 μm. Heavier coatings often alter surface texture and dimensions.

One of the most significant factors limiting the use of TiN and similar wear-resistant thin-film coatings in sliding wear systems is the roughening that can occur during coating. The high-deposition-rate processes are particularly prone to the formation of surface nodules. These nodules are referred to in the trade as "macros" and they are similar to weld spatter. In cathodic arc sputtering, microdroplets are generated when the arc touches the cathode species. These droplets can be as large as 10 μm (400 μin.) in diameter. They are composed of the cathode material, titanium in the case of reactive sputtering of TiN. When they adhere to the work, they subsequently become coated with TiN and form nodules of abrasive that can abrade mating surfaces in sliding wear systems. The coated surface can become a piece of sandpaper. Figure 6–12 shows an example of this phenomenon. The formation of these nodules is usually of little

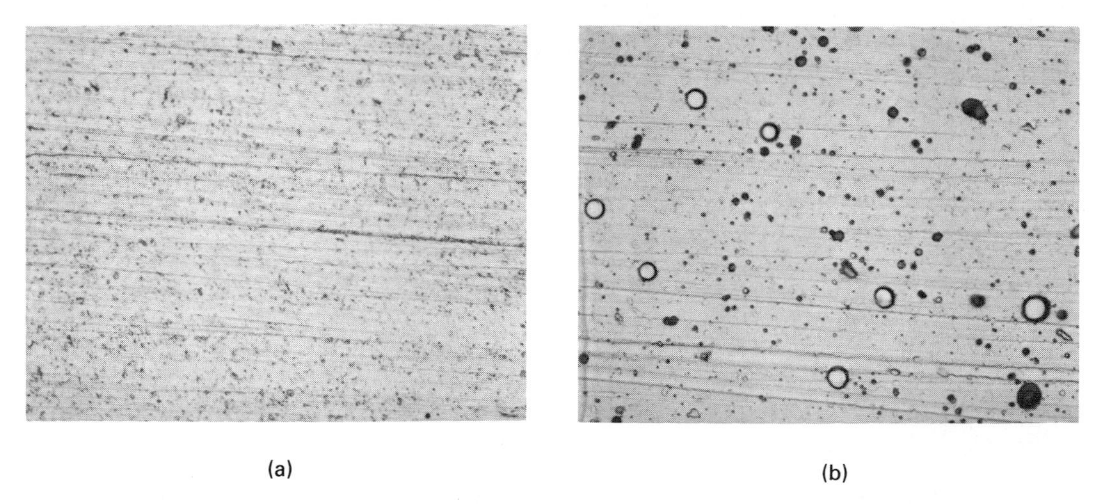

(a) (b)

Figure 6–12 Nodule formation from cathodic arc deposition of TiN. (a) Before coating. (b) After coating (800×).

consequence when these coatings are applied to drills and similar cutting tools, but for metal-to-metal and similar applications, the tendency for nodule formation should be made a process selection factor. There are ways to control nodule formation, and some processes are more prone than others. The potential user of these coatings should discuss the tendency to form these nodules with candidate coating companies.

Figure 6–13 presents some laboratory test data on the metal-to-metal wear characteristics of some stainless steels mated with a hardened steel with and without a cathodic-arc sputtered TiN coating on one member of the wear couple. The TiN coating was applied by the same equipment and all coated samples had the same thickness, 3 μm (120 μin.) In two instances the application of the coating on one member caused the system wear to increase. The coating only lowered system wear in one instance, when the coating was applied to a soft 316 stainless steel counterface. There were macros on the coatings of all three substrates and this probably is the explanation of the higher system wear rates on two of the couples, but the couple that showed an improvement also had macros. These data cannot be extrapolated to form global conclusions about the merits of thin hard coatings in unlubricated sliding systems. They do suggest that they may lower the wear resistance of some couples and that there may be a difference in the wear properties of the coating depending on the type of substrate to which they are applied. In a similar study applying TiN to nonferrous substrates it was determined that the TiN coatings increased system wear rates except when the coating was self-mated. There are many laboratories still studying the tribological characteristics of hard PVD coatings for metal-to-metal wear systems and it appears that the best course of

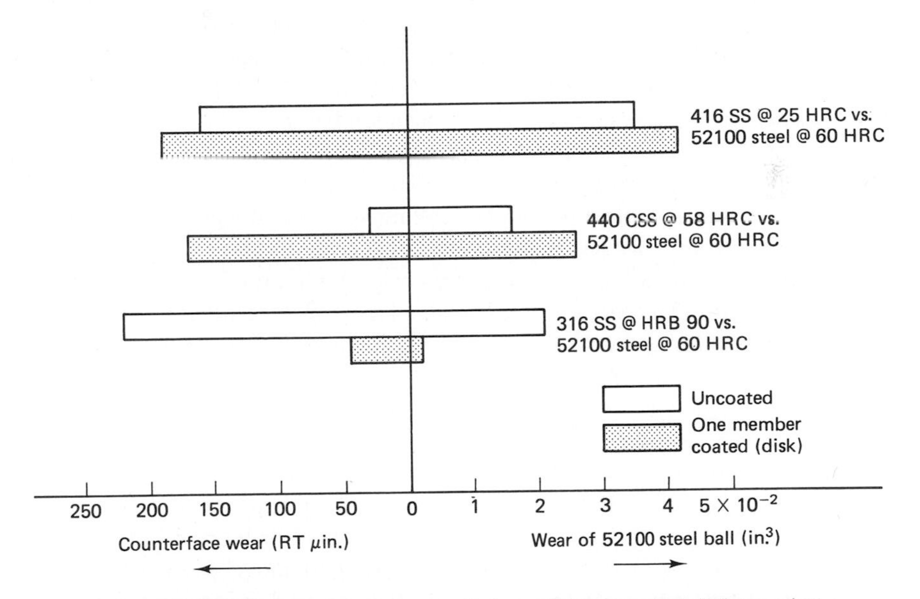

Figure 6–13 Metal-to-metal wear tests on 3 types of stainless steel sliding against a 52100 steel rider (ball) with and without one member (disk) coated with cathodic arc TiN (pin-on-disk test, 300 g normal force, 0.1 m/s velocity, 720 m sliding distance).

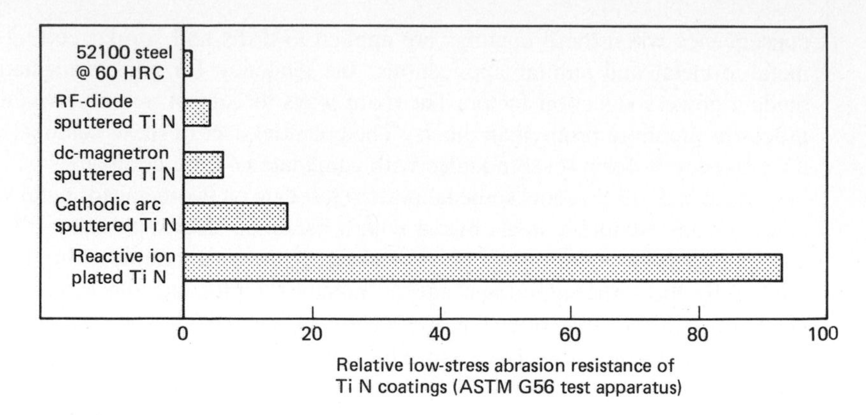

Figure 6–14 Effect of coating technique on the relative abrasion resistance of TiN on hardened steel applied by various processes (after Lee and Bayer).

action to take before using these coatings for these types of applications is to conduct compatibility tests under conditions that simulate the intended application.

Figure 6–14 presents some laboratory test data on the low-stress abrasion resistance of TiN coatings applied to a hardened steel substrate. These data clearly show that the TiN coating increased the abrasion resistance of the hardened steel, but there was a significant difference in the effectiveness of different application processes. The higher energy process, reactive ion plating showed more than an order of magnitude improvement over the RF diode sputtered coatings. Unrelated studies involving TiN coatings applied by various processes and by various coating houses showed significant differences in surface roughnesses, macro level and microhardness. From the user's standpoint these results suggest that application technique should be made a coating selection factor.

In service results from many sources suggest that thin PVD coatings on cemented carbide cutting tools allow higher cutting speeds and significantly longer tool lives over uncoated tools. In fact, it is estimated that over 50 percent of the commerical cemented carbide tool inserts that are sold are coated with a thin-hard coating with a thickness of about 3 μm (120 μin.). Commercial coatings include titanium carbide, titanium nitride, titanium-aluminum nitride, aluminum oxide, and graded composite coatings of the preceding. In machining metals, the role of these coatings is to reduce flank and edge wear. It is thought that these coatings prevent the chemical wear that comes from diffusion effects between the chips and the tool at the high temperatures that are encountered in high-speed machining operations. Thus, there are many case histories of success that suggest that thin-hard coatings are worth their extra cost on cemented carbide tools used for machining metals; the situation is not as clear cut on other tool-material substrates. Potential users of these coatings may want to move cautiously in applying them to high speed steels and other substrates that are used as cutting tools.

COATING SPECIFICATION

In considering the use of a thin-film coating for a particular application, the potential user must weigh the merits of the vacuum coating processes compared to other surface coating techniques. They are thinner than the bulk coatings of fusion welding processes. Some coatings, such as the MCrAlY's, have been successfully applied by thermal evaporation as thick as 0.5 mm (40 mils), but this is the exception. A commercial coater would have to develop a special process to do this. Vacuum coatings can be compared with thin electrodeposited coatings. Many electrodeposited coatings are used in thicknesses of 1 or 2 μm (40 to 80 μin.). They also compete with thin diffusion treatments, CVD coatings, and high-energy coatings such as ion implantation. The potential user of vacuum coatings must compare the relative merits of each process, the availability, and the economics. If it appears that a vacuum coating is the best approach to solving an existing or perceived wear problem, the user must first select a suitable substrate, then a coating, and then a process, and finally write a proper specification.

A proper vacuum coating specification should include coating material (including purity and stoichiometry if these are important), coating thickness (use a range not a single number), and areas to be coated (and allowable variation in thickness). In addition, the coater will need to know the sensitivity of the substrate to heating. What is the maximum temperature that the substrate can tolerate without affecting use properties?

SUMMARY

By our definition, thin-film coatings are those applied by vacuum processes and they usually have a thickness of less than 3 μm (120 μin.). They can be applied by various processes but most of the coatings that have utility in tribosystems are applied by either physical vapor deposition (PVD), or by chemical vapor deposition (CVD). The former coating technique is always done in vacuum equipment; CVD coatings can be applied by vacuum techniques or by the nonvacuum processes that we discussed in Chapter 4. In general, CVD coatings require high application temperature (1750° to 1920°F [950° to 1050°C]).

There are many PVD processes in use but the coatings that have the most utility in tribosystems are the hard carbides, nitrides, and similar coatings. Titanium nitride is the most widely available coating. These coatings are usually applied by sputtering or ion plating processes. There are at least six different variations of these basic processes and they differ in the energetics and mechanism of the deposition. The arc processes produce fast deposition rates; the magnetron processes use magnetic fields to direct the path of depositing species and the ion plating processes are intended to increase the energy of the impinging species. Wear tests on the same coatings applied by different processes suggest that the type of deposition process used may have an effect on coating serviceability. There may also be an effect on the nature of the substrate on coating serviceability. Most of the PVD processes involve a substrate temperature that is less than 1000°F (540°C); this makes PVD processes more suitable than CVD for applying coatings on heat sensitive substrates such as tool steels.

There are still factors to be learned about the use of these coatings for wear applications, but since the 1980s, these coatings have been widely applied to cemented carbide cutting tools and in most cases their use has significantly improved the serviceability of the coated tools. Users should have no qualms about using these coatings for tool inserts and similar cutting-tool applications. The use of thin-film coatings on other types of tribosystems may require some development work, but the indications for the future are that these coatings will be the high-performance coatings for the remainder of this century. They have the potential for replacing electroplatings, solid film lubricants, and possibly even liquid lubricants; they definitely should be part of every designer's repertoire of surface engineering processes.

REFERENCES

Bunshah, R. F., ed. *Deposition Technologies for Films and Coatings.* Park Ridge, N.J.: Noyes Publications, 1982.

Lee, E. J., and R. G. Bayer, ''Tribological Characteristics of Titanium Nitride Thin Coatings,'' *Metal Finishing,* July 1985, pp. 39–42.

Metals Handbook, Vol. 5, ''Surface Cleaning Finishing and Coating.'' Metals Park, Ohio: American Society for Metals, 1982.

Ramalingam, S. ''New Coating Technologies for Tribological Applications,'' in *Wear Control Handbook,* M. Peterson, ed. New York: American Society of Mechanical Engineers, 1982, pp. 385–409.

Schiller, S. and others. ''Advances in High Rate Sputtering with Magnetron Processing,'' *Metallurgical Coatings,* Vol. II. Lausanne: Elsevier Sequoia, 1979, pp. 455–467.

Zemel, J. N., ed. *Metallurgical Coatings.* Lausanne: Elsevier Sequoia, 1979.

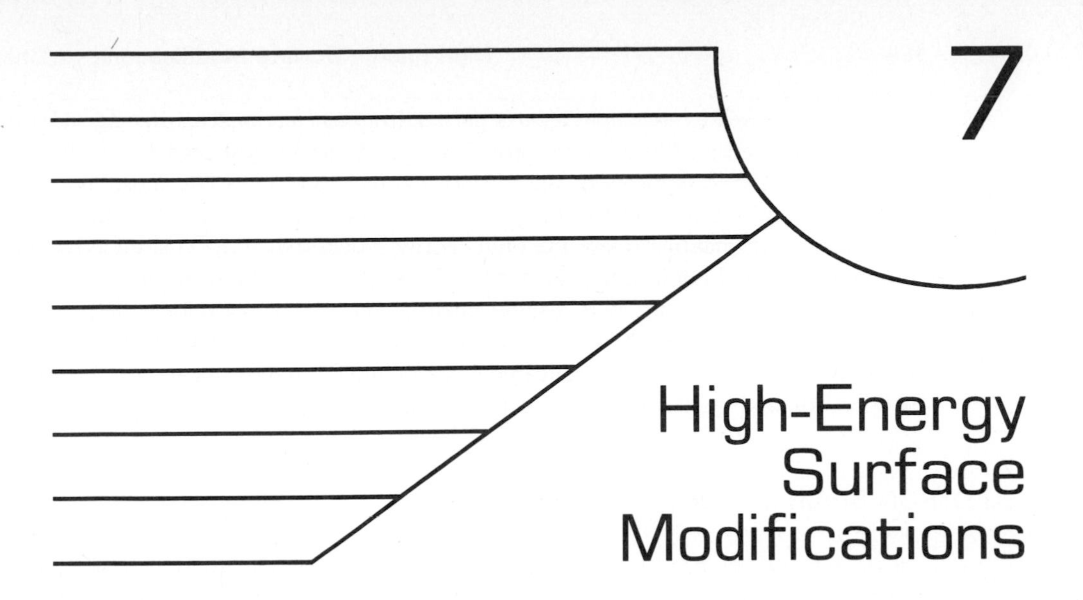

7 High-Energy Surface Modifications

In the next decade, it is predicted that traditional surface-hardening processes such as carburizing and nitriding will be replaced by more sophisticated processes, such as vacuum carburizing and ion nitriding. Concurrent with this change is the anticipated wide use of high-energy surface modifications for surface hardening:

Electron-beam hardening/glazing
Laser beam hardening/glazing
Ion implantation
Composite surfaces by laser and EB

The energy imparted to the surface with these processes can be as high as 20 million W/in.^{2}. As a comparison, the heating of a piece of steel in a convection furnace may require an energy density of less than 50 W/in.^{2}.

Electron beam and laser hardening use high-energy beams to austenitize the surface of hardenable steel, and self-quenching produces a surface hardening. Ion implantation involves the use of an impinging beam of ions to produce hardening or strengthening effects in the surface of suitable metals. Glazing is performed with lasers and electron beams. In this process, the surface is melted to produce structures that have different properties that can be achieved with conventional melting and solidification processes. Quench rates can be comparable to those used to produce amorphous metals. Composite surfaces can be made by using the beam melting processes to melt a surface layer, and a composite surface is made by introducing wear-resistant particles into the melted surface.

These processes are a significant departure from conventional hardfacing and surface-hardening processes, but they are probably one of the biggest competitors for "high-tech" applications, applications that do not involve wear rates like those that occur in the mineral-extraction industry. A valve seat, a knife edge, a cylinder liner, these are the types of applications where the high-energy processes can be cost effective and the more conventional hardfacing processes such as GMAW welding are not.

In this chapter, we will discuss each of these processes, how they work, the applicable materials, and how they can be used to produce wear-resistant surfaces. All these processes become more applicable with equipment improvements, but the principles of operation and guidelines for use remain relatively constant.

ELECTRON-BEAM SURFACE TREATMENTS

Electron-beam surface treatments are an outgrowth of electron-beam welding, and electron-beam welders historically were an outgrowth of the electron microscope. In the 1950s, there was considerable interest in developing microscopes that would produce higher magnification than could be produced by optical techniques. It was learned that these sought-after high magnifications could be obtained by impinging a beam of electrons on a very thin sample; the transmission of the electron beam through the sample could be used to record structural details by focusing the transmitted beam on photographic film or plates. Someplace in the application of this microscopic technique, it was learned that the focused beam of electrons could be used to do work. Thus in the 1950s electron-beam welders came into commercial reality.

There are a number of different types of EB welders, but they all have the basic elements that are shown in the welder schematic in Figure 7–1. The source of the electrons is a tungsten filament, not unlike those used in light bulbs. The filament is heated by a filament current power supply. Some distance from the filament is an anode plate at ground potential, and there is a very high potential between the filament and the anode plate. When the filament is heated, electrons are emitted and accelerated toward the anode plate by the high potential between the two. The anode plate has a hole in it, so rather than strike the anode plate, they pass through it. This assembly and its power supply are called the electron gun. Once the beam leaves the anode plate, it can be focused with magnetic lenses, which are electromagnets and permanent magnets. The permanent magnets are used to align the beam; the electromagnets are used to focus the beam and for beam deflection. Appropriate electrical inputs to the magnetic lens can produce beam oscillation or generation of circular patterns. The focused beam can be used to perform welding, cutting, or just heating of a workpiece. The electron gun must be under high vacuum (10^{-6} torr, 1.3×10^{-3} Pa) since gas molecules of any type will have a tendency to attenuate the electron beam. Electrons have very little mass, and collision with gas molecules will cause them to lose their velocity toward the work.

The work chamber for electron beam welders can be under high vacuum like the system illustrated in Figure 7–1, or the beam can exit the gun into a space that is open

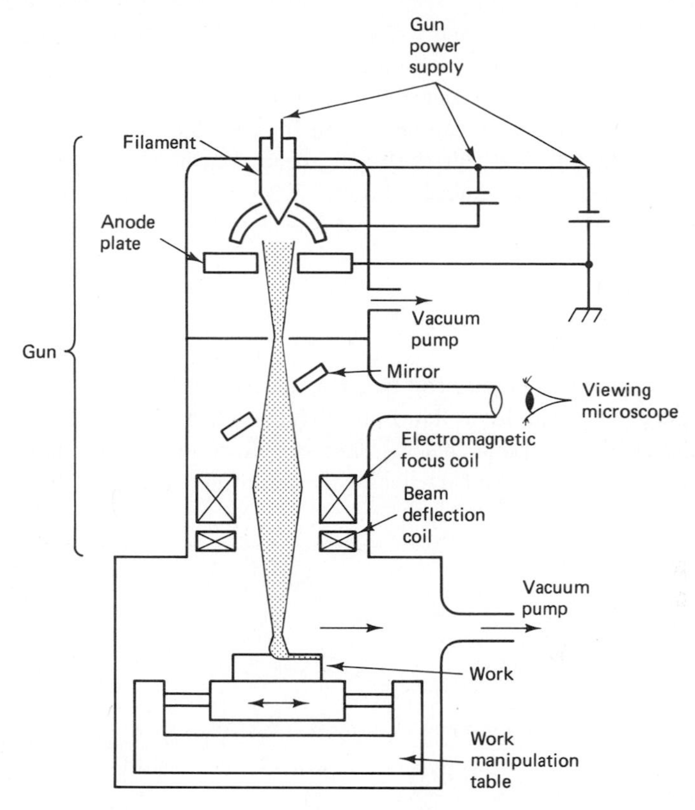

Figure 7–1 Schematic of an electron-beam system for welding, cutting, and surface treatment.

to ambient air but is subject to pumping from a mechanical pump. The systems like that in Figure 7–1 are called hard vacuum machines, and the other system is called a partial or soft vacuum machine. Hard vacuum machines require diffusion pumps. Partial vacuum machines only have diffusion pumps in the electron gun. In hard vacuum units, the work can be placed anywhere in the work chamber and the beam can be focused to any point that is in its line of sight. With a partial vacuum unit, the work must be very close (within several millimeters) to the exit of the beam from the gun. The workpiece heating produced by electron beams comes from the conversion of the kinetic energy of the stream of electrons to thermal energy when the electrons strike the work. The kinetic energy of the electrons in the partial vacuum units is far less than in the hard vacuum units (from gas molecule collisions), and the heating effect is greatly reduced. For example, a weld in a hard vacuum machine can have a depth to width ratio of 20. The depth to width ratio for partial vacuum machines may be only 1 or 2. The same ratio exists in heating for surface treatments. Hard vacuum machines are simply more efficient. A final detail about the nature of machines is manipulation of the work and the beam. Some welders have the gun mounted in the work chamber on a device that moves the gun in the x and y directions to make welds or to produce surface treatments. In the other technique used, the beam is stationary and the work moves under the beam on a servo-motor-controlled x/y positioner. In all types of machines, the movement of the gun or the work can be computer controlled so that any desired weld or heating path can be followed. Similarly, the beam can be made continuous or intermittent by analog or computer controls. Figure 7–2 shows a 3-kW (10.8×10^6j) hard vacuum EB welder.

ELECTRON-BEAM HARDENING

Electron-beam hardening can be performed on any EB unit that is capable of welding. This process is similar to flame and induction hardening with one big exception: heating of the steel is so rapid that it is not necessary to use the liquid quenches required by the other processes; self-quenching is used. The surface of hardenable steels is heated

Figure 7–2 Hard vacuum EB unit.

to the austenitizing temperature usually with a defocused beam (to prevent melting), and the mass of the workpiece conducts the heat away from the treated surface at a rate that is rapid enough to produce hardening. The material requirements for EB hardening are as follows:

1. Sufficient carbon content or carbon and alloy content to allow the formation of martensite.
2. Sufficient time at temperature to achieve solution of carbon in austenite.
3. Sufficient workpiece mass to allow the quenching requirements to be met (part thickness must be at least 10 times the depth of heating).

The first two requirements apply to quench hardening any steel by any process, but there are some factors that a user needs to know to successfully meet these requirements with EB hardening.

Materials

Steels are the most important basis metal that apply to EB hardening. A number of other metals form hardened structures by quench-induced allotropic transformations, some titanium alloys, some aluminum bronzes; but steels are the only materials that achieve hardnesses of 55 to 65 HRC. Which steels work? In conventional quench hardening, a carbon steel must contain at least 0.6 percent carbon to achieve formation of 100 percent martensite on quenching. The addition of alloying elements can reduce this carbon content requirement; but to achieve hardnesses of over 55 HRC, even high-alloy steels have carbon contents above 0.6 percent. The most suitable steels for EB hardening contain about this amount of carbon.

A second material requirement for EB hardening is that the carbon in the steel should be in such a form that it can quickly go into solution in austenite when the beam heating is applied. This factor is of equal importance with the carbon content requirement. For example, a ferritic cast iron may have a total carbon content of 3 percent, but it will not respond to EB hardening. Why is this? Ferritic cast iron has a matrix of ferrite, and graphite is dispersed within the ferrite in a three-dimensional pattern that resembles petals on a flower. The ferritic matrix will have a carbon content of less than about 0.2 percent. This is the only carbon that is available for the formation of martensite when EB heating is used. The heating is so rapid that there is insufficient time for additional carbon to diffuse from the graphite phase to assist formation of martensite. This is the most formidible problem with EB hardening; there is no soak time at temperature, and carbon that is tied up in carbides or graphite will not be available to produce matrix hardening. The ferritic cast iron mentioned in our example can be quench hardened with conventional techniques. In furnace hardening, the ferritic cast iron can be soaked for a half-hour or some adequate time so that carbon can diffuse from the graphite and go into solution in the austenite. It will now respond to quench hardening.

The most suitable materials for EB hardening are the same steels that are the most suitable for flame hardening:

1045 to 1080 carbon steels
Medium- to high-carbon alloy steels (4140, 4340, 8645, 52100, etc.)
Pearlitic matrix cast irons
W1, W2, O1, O2, L2, L6, S1, S2 tool steels

The common denominator for a steel that is well suited to EB hardening is a fine pearlitic matrix in the annealed condition. The high-alloy tool steels, steels with more than 5 percent total alloy, usually have working structures that consist of martensitic matrixes that contain hard carbides. These steels were developed to be more wear resistant than plain carbon steels through the action of the carbide phases. Unfortunately, these steels are usually manufactured with an annealed structure that consists of pearlite and spheroidized carbides. The spheroidization is done to make the steels more machinable. It does this, but it also depletes the matrix of carbon; much of the carbon that should be available for matrix hardening is tied up in alloy carbides. These steels may not respond to EB hardening because there is insufficient time to allow diffusion of carbon from carbide phases to the matrix. Common tool steels such as D2, A2, and M2 are very difficult to harden because of their carbide structure. Essentially, any steel that needs a soak at the austenitizing temperature will be difficult to harden without melting. Many high-alloy steels, tool steels, and 400 series stainless steels will become hard when the surface is melted, but surface melting is usually undesirable. If a surface-hardening process produces surface melting, a secondary finishing operation will be required. Obviously, EB hardening without surface melting is the preferred technique, and it is this technique that essentially only works with steels similar to those that we listed.

It has been learned that some of the EB hardening problems that are due to sluggish structures can sometimes be overcome by prehardening and tempering. For example, in an experiment, type 440C stainless did not respond to beam hardening. It was then hardened and tempered to its maximum hardness of 58 to 60 HRC and then it was beam hardened. The treated zone responded, and the hardness of the zone increased to 62 to 63 HRC. Experiments on 4150 steel showed that the highest hardness and deepest case depth could be obtained on steel that had been subjected to an isothermal quench and temper at 700°F (370°C). The steel was austenitized and quenched in salt at 700°F (370°C) and held there to produce a bainitic structure with a hardness of about 30 HRC. Pretreatments such as these destroy some of the economic advantages of EB hardening. This is an extra operation, and there can be distortion in the pretreatment processes. Prehardening to 25 to 30 HRC is the least objectionable of the pretreatments; the parts are still soft enough to machine after hardening, and distortion in the hardening process is not a factor. In any case, whenever possible it is advisable to use steels with known ability to harden with EB (previous list).

The preceding discussion concerns the first requirement for beam hardening, adequate carbon content (and suitable structure). The second material requirement for beam hardening, sufficient heating, is achieved by proper machine settings. The heating potential of the beam is measured by its energy density. There are ways to measure the energy density, but the common technique is to measure to beam diameter on the workpiece and assume that the beam's energy is evenly distributed over this area. It is not uniformly distributed; the energy of the impinging beam has a Gaussian distribution (because of varying electron velocities). But from the practical standpoint, a good approximation of the beam energy is simply the product of the filament current and the accelerating voltage. Thus the beam energy density can be approximated in units of kilowatts per unit of area. The spot sizes of electron beams can vary from about 0.010 in. (0.25 mm) in diameter to even rectangular shapes several millimeters on a side. The power of electron-beam units is determined as the product of the maximum available filament current and the maximum accelerating voltage. Accelerating voltages are usually in the range of 50 to 100 kV for "low voltage" welders and in the range of 100 to 200 kV for high-voltage welders. The smallest EB welders have a power of at least 3 kW; the biggest welder of the type used for heat treating may have a rating of 100 kW. The biggest EB power sources are used for melting ingots of refractory metals; these units are as large as 3000 kW.

Thus EB units are capable of producing energy densities that are adequate to austenitize any steel. The heating rate is a function of the beam power settings (filament current and accelerating voltage), the spot size, the speed of the work movement under the beam (or the beam movement over the work), and the thermal properties of the workpiece. Since it is possible to control the input of beam energy, it is possible to calculate all the conditions that are required to heat the surface of a given workpiece to the austenitizing temperature and to get proper quenching from the mass of the workpiece. This technique is done on a computer on advanced EB hardening units. The alternate is to use a setup piece. A beam setting and workpiece speed are selected on experience, and these hardening parameters are tried on the setup sample. If the part does not respond or if the desired width of hardening is not achieved, additional trials are performed on the setup piece. This latter approach is commonly used in job shops where only a few parts of a given geometry are required. From the user's standpoint, the most important concerns are that the part have adequate mass to allow self-quenching and that it have a geometry that is not prone to melting. Some parts that could cause problems in beam hardening are shown in Figure 7–3. Small projections, acute-angle cutting edges, very thin sections, and fine embossed characters are the types of things that can cause a problem; these geometries are prone to melting and, if they do not melt, there could be problems on quenching. Since the parts are in vacuum, convection cooling is negligible. The heated area of the part is quenched by conduction of heat to the bulk of the part. If the conduction path is small, as is the case of a small male projection on a part, it may not quench fast enough to obtain the desired transformation to hardened structure. A potential solution to lack of mass for quenching is to nest the part in a massive copper chill. This technique is successfully used, but a proper design for EB hardening should provide for adequate quenching by conduction to the bulk of the work.

In summary, the most suitable steels for EB hardening are steels with medium to

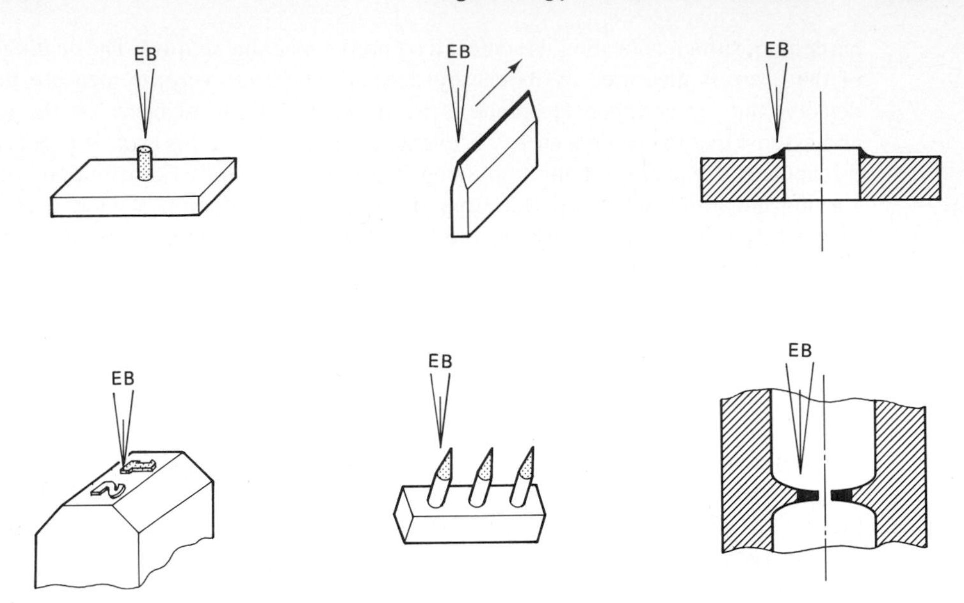

Figure 7–3 Parts that require special consideration for EB hardening.

high carbon content (0.4 to 1.2 percent) and low alloy content. The high-alloy steel may harden if the surface is melted, but this negates the advantage of using this process on finished parts. The other material consideration, sufficient mass to allow self-quenching, can be met on most parts, but chills can be used to overcome this problem.

Techniques

As mentioned previously, there are two basic types of electron beam guns, stationary and movable. When movable guns are used, the workpiece is fixtured and the gun is moved to produce heating for hardening. Stationary-gun machines require manipulation of the part under the beam. The area to be hardened on the workpiece must be in a line of sight with the beam. Gun movement or work manipulation can be accomplished by computer control to produce any desired pattern, and the beam can be pulsed or oscillated by standard controls on the machine or by computer control. Figure 7–4 shows some of the hardened-area configurations that are possible. The most common pattern is to harden stripes with a substantial spacing between stripes. Circular paths are useful on ends of punches and the like. These two patterns can be done manually (or by computer) on any machine. The dot pattern requires machine controls to pulse the beam and to move the work or the gun. Some sort of computer control is needed to follow an irregular shape, such as the edge of a die that blanks a kidney-shaped part. The beam oscillation patterns illustrated in Figure 7–4 can be obtained on any machine by programming in beam oscillation or pulsed beam deflection while the work is in motion with respect to the beam. The largest oscillation diameter is usually less than about five times the beam diameter.

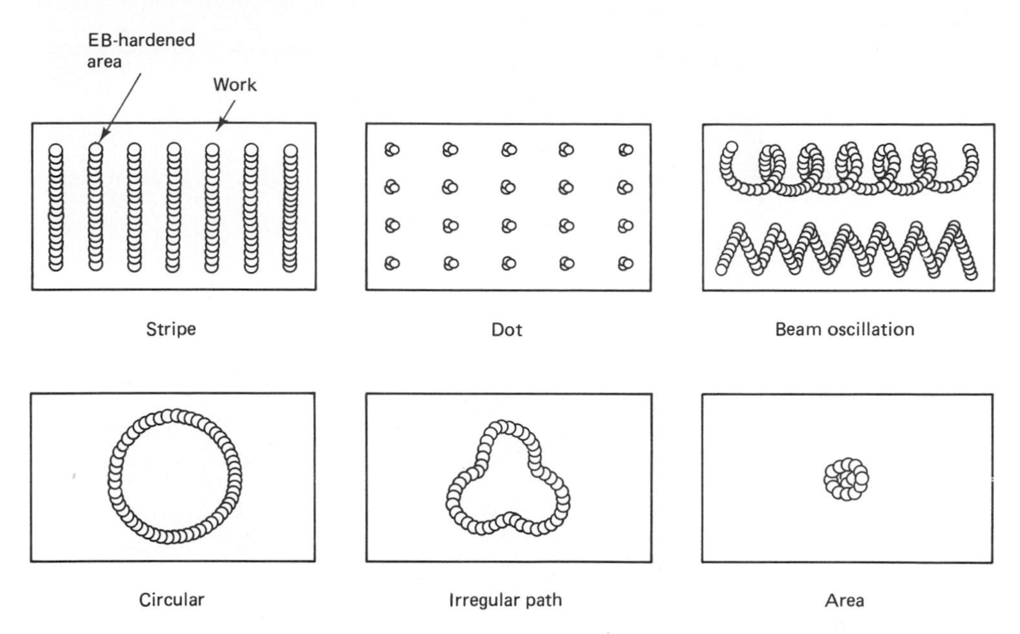

Figure 7–4 Possible EB hardening configurations.

Hardened areas are usually spaced to prevent tempering of previously hardened areas. As an example, if a stripe pattern 0.2 in. (5 mm) wide is used, the pitch of the stripes would probably be about 0.5 in. (12 mm). Since cooling of parts in hard vacuum EB units is only by conduction to contacting heat sinks, it is important to consider the effect of the hardening on the bulk temperature of the part. For example, if a rectangular piece 1/2 × 1 × 3 in. (12.5 × 25 × 75 mm) in size is to be hardened around its periphery to a width of 0.2 in. (5 mm), this amount of hardening on such a small part may cause the bulk temperature of the part to rise to, for example, 700°F (370°C) when the first edge is done. If this happens, the subsequently hardened edges may be low in hardness because of the lack of adequate quench. The structure of the areas hardened last will probably be bainite instead of martensite. This problem can be easily overcome by monitoring the bulk temperature of parts during the hardening operation and by using a heat sink or sequentially hardening limited areas on several parts (moving from one part to another so that they stay cool).

The more sophisticated EB hardening systems have computers programmed to provide a desired hardening profile. As mentioned previously, it is possible to calculate the thermal gradients produced in the work with a given beam power and travel speed. The critical aspects of beam hardening are in producing sufficient austenitizing temperature without surface melting. As shown in Figure 7–5, a properly beam hardened pattern will show no melting, only a heat line on both sides of the hardened zone. There are various theories to explain the visual appearance of an EB hardened zone. One reason why the hardened zone can be seen with the unaided eye is that the beam vaporizes surface contaminants. Another theory suggests that the hardened zone is visible because

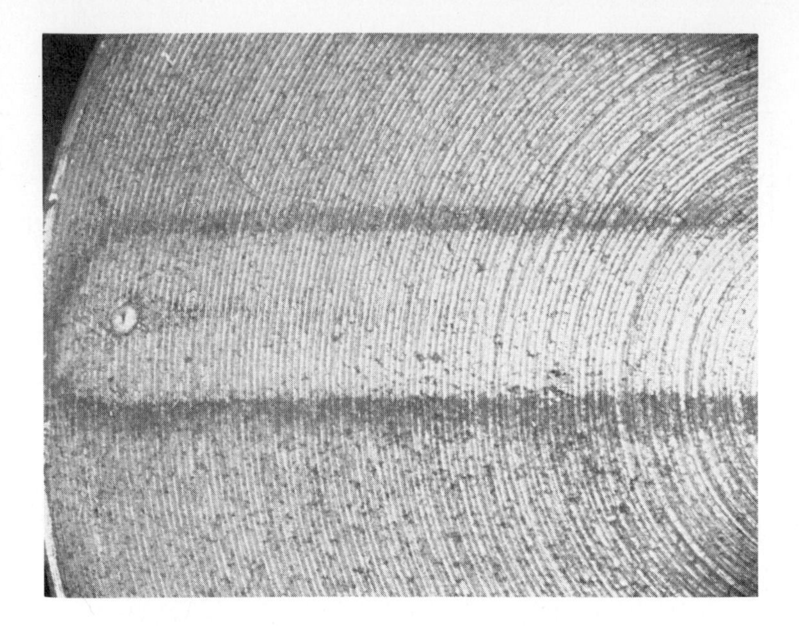

Figure 7–5 EB hardened stripe in 4140 steel (4× size).

the transformation to hardened structure causes a very slight increase in surface roughness because the lattice parameter of the martensitic structure in the hardened zone is larger than the lattice parameter of the unhardened material. This volume expansion causes a slight change in the surface roughness. Whatever the cause, the visual appearance difference of the hardened zone has no effect on the function.

Figure 7–5 shows a small pinpoint of melting where the beam was stopped on the work. This is the type of melting that is to be avoided. Starting the beam off the work and ending the beam off the work will minimize this condition. Figure 7–6 shows a typical profile of surface hardness for EB hardening. There is very little gradient in hardness across the width of the hardened zone, and the hardness drops rapidly to the core hardness. The transition zone from full hard to soft is very small. In the example shown in Figure 7–6, the width of the hardened zone is about 0.2 in. (5 mm), and the zone with a transition hardness is only about 0.020 in. (0.5 mm) wide. The cross section of a typical hardened stripe is shown in Figure 7–7. There is a varying depth of the hardened zone that corresponds with the gradient in electron velocity in the beam. Finely focused beams will produce a more triangular hardened profile and defocused beams produce shallower and wider profiles.

There is no technical limit to the size of the zone that can be EB hardened. The practical limit is the size of the welder. A hardened zone 0.25 in. (6 mm) wide with a depth of about 0.030 in. (0.75 mm) can be produced with a beam power of only 1 kW and EB units up to 20 kW are not uncommon. Hardened zones have been made with a width of 2 in. (50 mm), but it is more common practice to use hardened zone widths of about 0.250 in. (6 mm) with a depth of about 0.030 in. (0.75 mm). This hardened

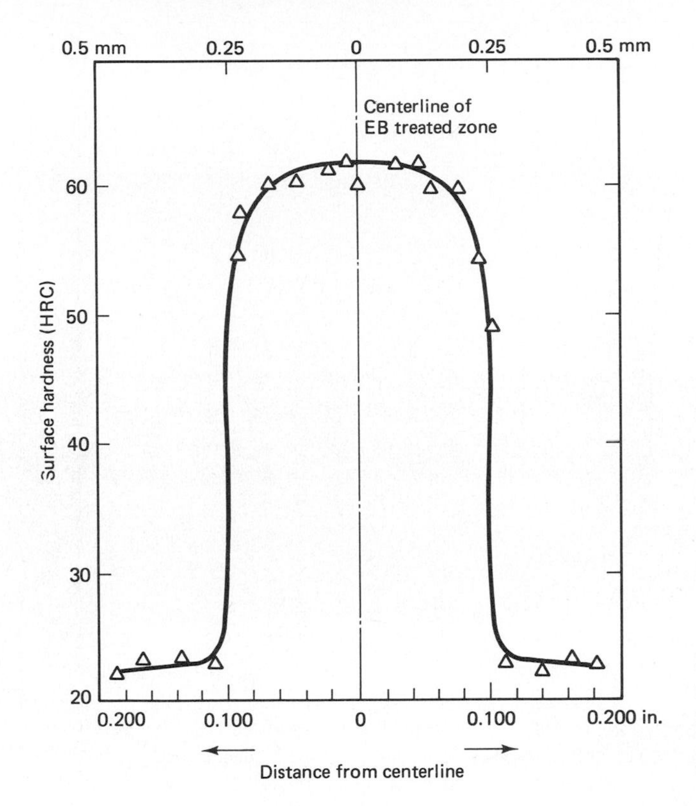

Figure 7–6 Surface hardness profile for stripe shown in Figure 7–5(4140 steel, 30 in./min (0.013 m/s) traverse speed. 50 kV, 10 mA, defocused beam, 1-in.-thick [25 mm] material).

zone width and depth is adequate for most tool edges. Large areas are usually hardened by using a raster pattern of stripes or spots with adequate spacing between zones to prevent unwanted tempering. In all EB hardening operations, it is advisable to temper the part after the hardening operation. Untempered martensite is so brittle that cutting edges and similar working surfaces can fail in fatigue without tempering. The usual procedure is to use the minimum temperature recommended for the specific steel that was hardened. A tempering temperature of 300°F (150°C) is adequate for the carbon steels and alloy steels that are most suitable for beam hardening.

The speed of EB hardening, like the width capacity, depends on the size of the unit, but most stripe hardening operations are performed at travel speeds of at least 30 in. (0.75 m) per minute. The larger size machines can achieve speeds of 120 in. (3 m) per minute. Hardening large areas with raster patterns of spots or stripes can be done at speeds such that areas of 10 in.^{2} (65 cm^2) can be covered with a pattern in a matter of a few seconds. Profiles can be followed at speeds up to 120 in. (3 m) per minute.

The equipment requirements for EB hardening, besides the EB unit, depend on the sophistication of the system. Hardening in a stripe pattern or in a circular pattern with beam deflection can be done on any EB unit. The part is moved under the beam or the gun is moved over the part for stripe patterns. Small-diameter circular patterns up to about 2 in. (50 mm) are done with the beam deflection controls. Sine wave

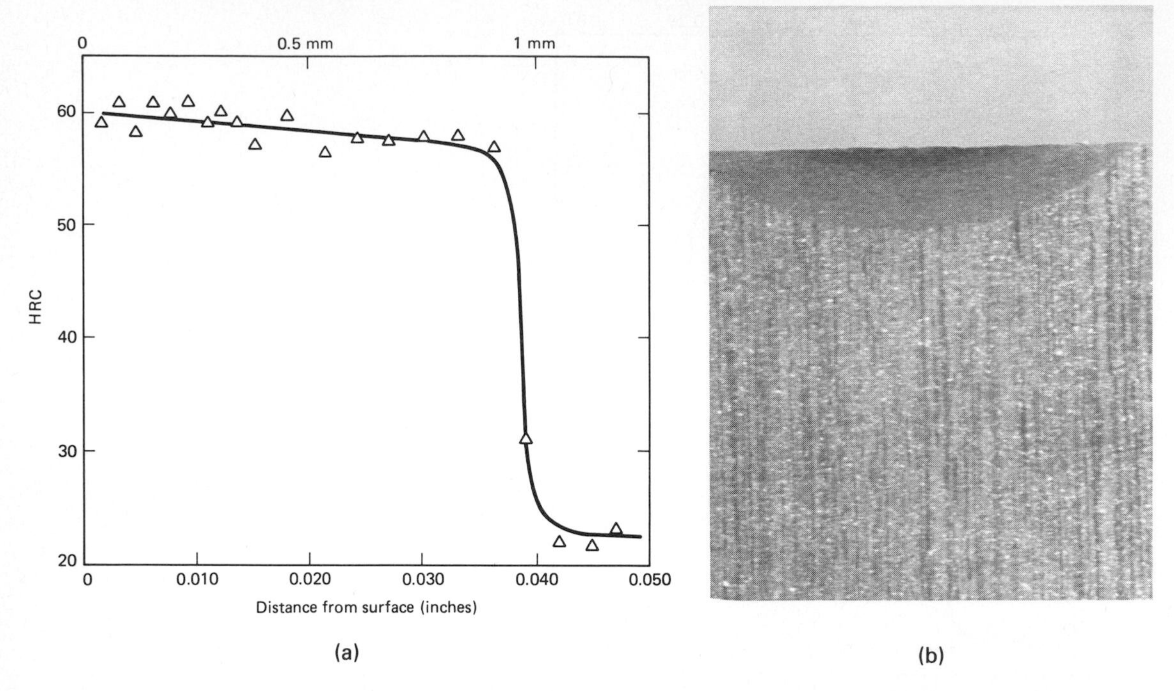

Figure 7–7 (a) Cross section of EB hardened stripe on O1 tool steel, and (b) the hardness profile from the part surface taken at the centerline of the hardened zone.

signals are put into the beam deflection magnets, and with the appropriate adjustment of the signal to the magnets, the beam can be made to continuously rotate in a circular path. Large-diameter circular paths are produced by putting the work in a rotary fixture with the beam stationary. Raster patterns and contours require NC or computer control of the beam and/or the work movement. These types of systems are usually developed for specific applications. This type of hardening system may not be available in a typical job shop that provides beam hardening services.

EB hardening in hard vacuum units obviously requires that the parts to be hardened fit into the chamber and that there be sufficient room in the chamber to move the part or the EB gun in the desired pattern. Out-of-vacuum units are not really out of vacuum. There will be shrouds around the work position, and a partial vacuum (10^{-2} torr; 13 Pa) is obtained in the work area by mechanical pumps. It is still necessary to minimize the amount of air that is present where the beam exits the unit. The maximum opening to the outside atmosphere when welding is progressing is usually narrow slots that are made as small as practical. Soft vacuum EB units can be set up to do production-type hardening at high rates. Parts to be hardened can be fixtured on conveyor nests that are cycled under the beam and hardened with a preprogrammed cycle. Batch tempering can be done after the EB hardening. Thus EB hardening can be competitive with other high-production processes such as flame or induction hardening. The depth of the hardened zone may tend to be shallower in the out-of-vacuum units compared with the hard vacuum units, but substantial hardened depths are attainable with both processes.

Specification

Electron beam hardening of wear surfaces is competitive with other selective hardening processes: induction, laser, and flame hardening. All these processes minimize the possibility of part distortion because it is not necessary to heat the entire part to the austenitizing temperature. EB hardening has advantages and disadvantages compared with each of these processes: it is more flexible than induction in its ability to follow contours and edges; it offers more precise control of the hardened zone than is possible with flame hardening; it does not have the surface reflectivity problems of laser hardening. But the disadvantage compared with these competitive processes is that parts to be hardened must be put into a vacuum chamber or at least a partial vacuum shroud. For hardening of tools, this may not be a significant factor; for hardening a hundred or thousands of parts, it may not be a significant disadvantage; if millions of parts are to selectively hardened, the vacuum requirement could become a factor that may make this process less cost effective than other processes.

If an in-house EB unit is available, this process definitely should be a candidate for almost all parts that may only need hardening in selective areas. If an in-house unit is unavailable, many job shops offer EB welding and hardening services, but to use EB hardening in any form it is necessary to specify its use properly.

The proper drawing notation for EB hardening is the same as that used for any selective hardening process:

> Electron beam harden area shown 0.20 to 0.030 in. (0.5 to 0.75 mm) deep; temper after hardening to a hardness of 60 HRC minimum.

The drawing should clearly show the length and width or pattern of the hardened areas. It is also important to specify where the EB hardening will fit into the manufacturing sequence. When EB hardening is used without surface melting, it is usually done as the second to last operation. The last operation would be tempering. If surface melting can be tolerated, the EB hardening and tempering could be followed by a grinding step to remove the positive bead that occurs with melting. This is risky since the hardened surface layer can be removed. The nonmelting technique is preferred.

In summary, EB surface hardening is competitive with welding hardfacing and many other surface-hardening processes, but it is most applicable to small tools and parts. It cannot be done in the field; it is impractical to use for applications that require substantial thicknesses of wear-resistant material (mineral-handling equipment and the like). At the present time, it is most effective on steels with low alloy content and matrixes with medium to high carbon contents (in solution). Thus there are process limitations; but where it fits, it is an excellent technique to create a very hard wear surface with low part distortion.

LASER HARDENING

Lasers have become commonplace in many fields, but their use in surface treatments is relatively limited. This is probably due to the high cost of the large lasers that are required for most metalworking operations, but this situation will change as techniques are developed to increase efficiency and reduce equipment costs. Laser hardening and hardfacing will become more important because they have unique advantages over competing processes.

The term laser is an acronym for light amplification by stimulated emitted radiation. There are three requirements for a laser: (1) the laser material, (2) a device to stimulate the laser material, and (3) the laser material must be retained in a cavity that will allow amplification by repeated reflection of the light. The early laser materials were usually ruby crystals (alumina/chromia); excitation of the crystal is obtained by pulsing of powerful flash lamps aimed at the ruby; the simplest laser cavity is obtained by applying reflective coatings on the ends of the ruby crystal. The exact mechanism of how a laser works is difficult to describe without getting into quantuum mechanics, but a simplified description is as follows: In neutral atoms, the electrons associated with an atom are in their ground state, a rest energy level; when the laser material is excited by high-intensity light, the electrons of some of the atoms will be excited from their ground state. They go to an unstable energy state; when the electrons from one of the excited atoms return to their ground state, they give off light energy in the form of photons. This photon emission can trigger a similar emission with the same energy and direction as the original photons; the photon emission is coherent, in phase. The flash lamps start this process, and reflection of the light (photons) in the laser material by the reflective ends on the crystal stimulates emission from other atoms. Eventually, there will be an energy inversion (more electrons in the excited state than in the ground state), and the crystal will lase through the end of the crystal that has a mirror that is only partially reflective. The coherent beam of photons that is emitted from the laser can be optically manipulated by lenses the same as any light beam. It is capable of being focused to perform work.

Laser materials can be solids, gases, or liquids. They must be transparent to the excitation energy (flash lamps). The active ingredient in the ruby laser is the chromium in the chromium oxide component of the ruby. Most solid-state lasers contain some ingredient that responds to the light excitation. Carbon dioxide is a common active ingredient in gas lasers, and liquid lasers usually use dyes that produce the lasing action. The wavelength of the light (electromagnetic radiation) produced by a laser is a function of the laser material, and the length of the laser cavity must be some function of the wavelength of the light to maintain coherence. The most common solid-state laser, neodymium doped YAG (yttrium/aluminum/garnet) produces light with a wavelength of 1.6 μm, and the most common gas laser material, CO_2, produces light with a wavelength of 10.6 μm in the infrared region. The ruby-based solid-state lasers usually have a power output of about 100 W, limiting their use to welding of small wires and similar applications. YAG lasers can be made as large as 2 kW, and they can be operated in pulsed or continuous mode. CO_2 lasers can be as large as 20 kW, with pulsed or continuous modes of operation. Beam diameters can be as small as 0.1 in. (2.5 mm) and as large as 2 in. (50 mm) (with very large lasers, >10 kW). Focused beams have a Gaussian energy distribution similar to that of electron beams, but other beam shapes such as an annulus are possible.

Lasers are not very efficient from the energy standpoint. The output energy may be only 10 percent of the input energy from flashlamps or arc sources. This is one of the reasons why lasers can be so expensive. There is no technical limit to the size of lasers. Research lasers used for atomic fusion experiments have energy outputs of 20 GW. Such devices cost many millions of dollars and complete buildings are needed to house them. Welding and heat-treating lasers generally cost in the range of $100,000 to $500,000. Laser systems for selective hardening of steels usually have a power output of at least 500 W.

Materials and Techniques

A schematic of a laser hardening system is shown in Figure 7–8. With a stationary beam, the work is attached to a table on an *xy* or rotary positioner and it is moved under the beam. The positioner movement can be controlled by NC equipment or by any suitable technique. Computers can be used to control both the positioner and the beam power. Lasers that are used for job-shop types of work (many different parts) usually have a coaxial microscopic viewing system like that used on EB welders. The

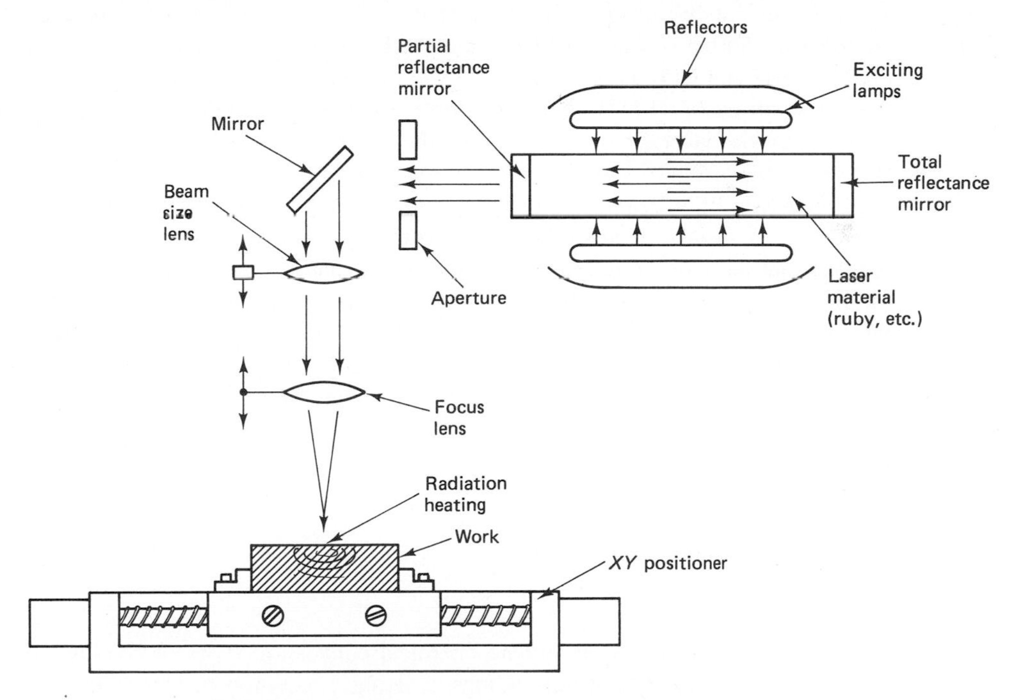

Figure 7–8 Schematic of surface hardening with a solid-state laser.

tracking of the system can be checked before the beam is turned on. Some electron beam units have movable electron guns. This is seldom done with lasers. If the work is to remain stationary during processing, beam deflection and manipulation are produced by moving the mirrors. For example, the inside diameter of a cylinder can be hardened by directing the beam down the centerline of the cylinder to a 45 degree rotating mirror inside the cylinder. This type of system would produce a hardened stripe on the ID of the cylinder. It is also possible to split the beam to produce, for example, four hardened spots in a single excitation of the laser.

Lasers do not require a controlled atmosphere. Attenuation of the beam power by gas molecules is nil. Ancillary gas shielding may be desired when heat treating to minimize oxidation. This can be done by directing a jet of inert gas at the surface to be hardened. Like any light beam that is focused by a lens, lasers diverge with the distance from the lens. The divergence is usually small, but this factor is dealt with by focusing on the surface to be heat treated. In welding, lasers are focused slightly below the work surface. The lens-to-work distance required in laser heat treating depends on the power of the laser, but it is common practice to keep this distance as small as possible. In most units, this distance is kept to several inches; however, on welding of hard-to-reach joints on large structures, the laser beam may have a focus point several feet from the final lens. Safety considerations, as well as the divergence, suggest that it is prudent to minimze the lens-to-work distance.

One of the most significant factors to be dealt with in using lasers to harden steels is surface reflectivity. The reflection of laser light from metal surfaces depends on the wavelength of the laser light, but with the most commonly used metalworking lasers, YAG and CO_2 as much as 90 percent of the incident laser light can be reflected from a shiny metal surface. This problem is usually not significant in welding, because as soon as the laser produces melting, the reflectivity of the surface decreases such that about 90 percent of the beam's energy is absorbed by the work. Melting proceeds by conduction of a beam energy by the molten keyhole that surrounds the beam. In hardening of metals, it is usually undesirable to allow surface melting (for the same reasons that were described in the section on EB hardening). To get a workpiece up to the hardening temperature without melting, it is necessary to make the laser beam heat the work by radiation. This is done by altering the surface of the work so that it absorbs rather than reflects the laser beam. A matte black surface is the optimum surface for the absorption of radiant energy. In laser hardening, this type of surface is achieved by a number of techniques:

1. Black chemical conversion coatings
2. Black electrodeposits
3. Carbon/graphite spray coatings
4. Heat tinting

The iron/manganese phosphate chemical conversion types of coatings are one of the most common laser coupling coatings. Coating is accomplished by a simple immersion and there is no residue after hardening. Electrodeposits and spray coatings can leave undesirable deposits. With some types of lasers, adequate emissivity may be obtained by using an abrasive-blasted surface or a surface that is dull from the surface texture from machining. In any case, the user of laser hardening must consider the reflection of the laser beam by the work surface. If coatings are required, it is the designer's responsibility to specify the type or types of coatings that are allowable for the function of the part.

The materials that are applicable to laser hardening are the same as those listed for electron-beam hardening. The carbon steels and low-alloy steels with about 0.5 to 1 percent carbon are the most likely materials to respond. The lack of soak time and the problem with dissolving carbides that were described when high-alloy tool steels are electron beam hardened are also present with laser hardening. Similarly, the hardening patterns, hardnesses, and hardening depth that can be obtained with EB hardening can be obtained with laser hardening. In addition, because a laser beam can be bent by a mirror, it is possible to obtain hardening patterns in areas that are blind to electron beams. This is a noticeable advantage over EB hardening.

Proper specification of laser hardening requires dimensions of the areas to be hardened, the depth of the hardening, and the expected tempered hardness. If black coupling agents are used, it may also be necessary to specify removal of the coupling agent. Thin conversion coatings and heat-tinted oxides are normally not removed as they have no adverse effects on most sliding systems.

Laser Hardfacing

The equipment problems associated with maintaining a vacuum in EB welding and hardening units prevent wide use of EB for adding filler metals for welding and surfacing. Since these problems do not exist with lasers, the use of lasers for "conventional hardfacing" is becoming more common. Hard, wear-resistant deposits can be made by fusing conventional hardfacing consumables with laser welders. Figure 7–9 shows some of the techniques that are employed in laser hardfacing. The work can be covered with a bed of hardfacing powder, and the beam is passed over the bed, melting the hardfacing powder and fusing it to the substrate. In another technique, an insert or solid overlay of the material to be applied is held in intimate contact with the substrate with a clamp or other technique, and the beam is used to fuse the material to the substrate.

There are problems with both of these techniques; with the preplaced bed of powder it is difficult to get a uniform deposit and the powder has a tendency to bead up and not fuse to the substrate; inserts are difficult to hold in intimate contact with the work. If the insert is not firm against the work, it may vaporize without fusing to the work. The only way that the substrate can be heated to the fusing temperature is by conduction of heat from the insert. Poor contact will prevent substrate fusing. The powder feed and wire feed techniques are the most highly developed approaches and they can be used to produce hardfacing deposits that are similar to those that can be obtained by PTAW or GTAW with less heat input. In the powder feed technique, the

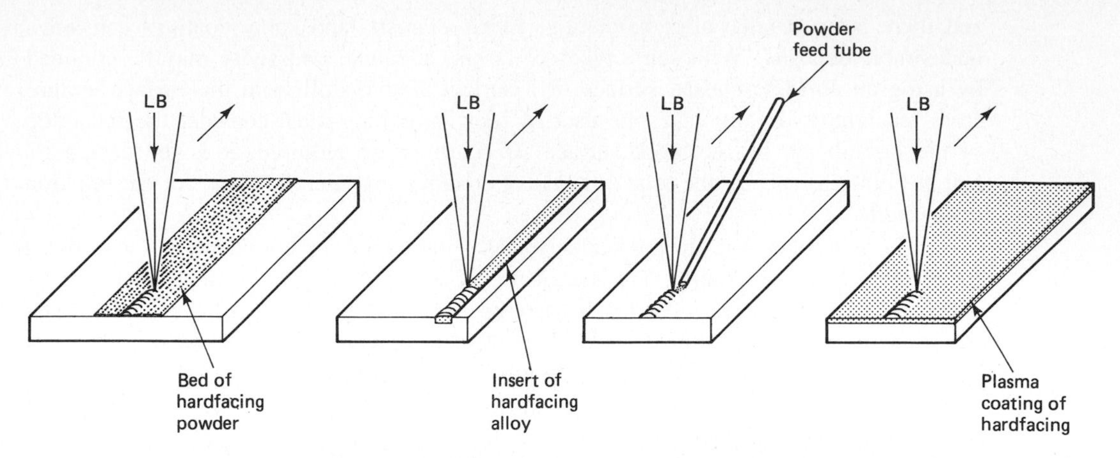

Figure 7–9 Techniques for application of hardfacings by laser fusion.

powder can be introduced into the molten pool by gravity or by using a powder jet propelled by an inert gas such as argon or helium.

The last type of hardfacing technique that is illustrated in Figure 7–9, fusing of plasma spray deposits, is less developed than the powder feed techniques, but it has significant potential. The application of the hardfacing alloy by plasma spray provides a more uniform deposit than the powder bed technique. Probably, the reason for the slow development of this technique is that the plasma deposits could be used as-deposited with the plasma torch. The major advantage of laser fusing is that the hard-facing will have a fusion bond to the substrate, instead of the mechanical bond produced by plasma without fusing.

A modification of these laser hardfacing techniques is surface alloying and embedding of ceramics and cermets to form composite surfaces. Instead of applying a wear-resisting alloy to the surface, it is possible to inject alloying elements into the molten pool produced by the laser to produce a new alloy on the surface. For example, the surface of a low-carbon steel can be made to have stainless steel properties on the surface by laser cladding with chromium by the preplaced bed technique or by injecting chromium powder by the powder feed process. Silicon can be injected into aluminum alloys to produce a high-silicon surface with good wear resistance. True composite surfaces are produced by introducing aluminum oxide, silicon carbide, and tungsten carbide particles into the laser melt pool. The surface to a depth of 30 or 40 mils (0.75 to 1 mm) can obtain a volume fraction of, for example, 50 percent aluminum oxide in an aluminum matrix (substrate). Many problems must be addressed in performing these techniques, but all the composite types discussed have been successfully demonstrated. They are not in widespread commercial use, but they are technically possible and can be used. At this time, much of the work in this area is concerned with measurement of the use properties that are obtained from these composite surfaces.

Laser Glazing

In the late 1970s, a significant amount of research was directed at rapid solidification of metals to alter properties. The product of this research was amorphous metals. It was determined that if cooling rates on solidification exceeded 10^6 degrees Celsius per second some metals would solidify without their normal crystalline structure. Various techniques were used to achieve these cooling rates, but one of the commonest techniques was to curtain coat a film of metal on a rotating water-cooled copper roll. Foils of nickel and some other metals were produced with amorphous structures. The amorphous structure enhanced mechanical properties and corrosion characteristics. Amorphous alloys are now commercially available for brazing fillers and the like.

An obvious extension of laser processing of metals is to use the rapid heating and quench rates to produce structure modifications and corresponding property changes. In laser glazing of metals, surface layers are melted and, with sufficient workpiece mass, the melted surface layer will solidify at quench rates that approach those used in making amorphous metals. It has been determined that rapid melting and solidification of many alloys can reduce segregation, alter hardenability, change the distribution of microconstituents, and produce a variety of other morphology alterations. In many cases, these changes improve the service characteristics. For example, laser-glazed and rehardened tool steels have shown improved wear characteristics over the same alloy at the same hardness without the laser treatment. The action of the laser treatment was to alter the distribution of the carbide phases that assist wear resistance. This process is still in the exploratory stage, but it has been successfully used to improve the service life of some alloys, and it has the potential for solving other special problems.

Summary

Lasers can be used to perform selective hardening, with hardening depths and material constraints similar to those for EB hardening. Laser hardening has some noteworthy advantages over EB: the parts do not have to be in vacuum, wider hardening profiles are possible, and there can be greater accessibility to hard to get at areas because of the flexibility of optical manipulation of light energy. The biggest disadvantage of lasers over EB in hardening is the necessity to use surface treatments to prevent beam reflection.

Conventional hardfacing with a wide range of consumables is also possible with laser welding equipment. Powder injection to the melt and wire feed devices makes laser welding competitive with GTAW and PTAW for surfacing of the same types of parts that are done with these other processes. In this mode of operation, lasers are only cost effective if they can apply hardfacings faster or with better quality than the competitive processes.

Laser surface modification is still limited in use, but it has potential for improving the properties of complex alloys by structure modification.

ION IMPLANTATION

Most semiconductors are made from single-crystal materials such as silicon and germanium. They are not semiconductors until they are doped with small amounts of other elements such as arsenic and antimony. There are a variety of ways of adding these active elements, but a technique that came into use in the early 1970s was ion implantation. This process allows precise control of the atomic dosage of these donor elements. Essentially, ion implantation is the process of adding elements to the surface by impingement of a stream of ions of the element to be added. The steps in the process are as follows:

1. The element to be implanted is converted into ion form by electron collisions in a plasma or some other technique.
2. The ions are focused by magnets into a stream.
3. The ions are accelerated by a voltage gradient such that a beam of high-velocity ions can be directed at a target (the part to be implanted).

In the early 1980s, ion implantation started to be applied essentially as a hardfacing tool. If ion implantation can be used to add elements to semiconductor substrates, why not use this technique to add elements to tool material surfaces to enhance their use characteristics. There are commercial treating companies and companies that sell equipment packages that will allow anyone with the cost of the equipment to become an ion implanter. This process has been tried by most progressive industries, but its application is in its infancy. Applications are still being investigated, but the process is here to stay. It competes with vacuum coatings and other thin surface-hardening techniques, and it warrants a discussion of how it can be used as a process to compete with the more traditional hardfacing processes.

Equipment

A rather old device that is capable of performing ion implantation is a linear accelerator of the type used in atomic physics. The lay term for these devices was ''atom smashers.'' They can be used to alter nuclei of atoms, to make isotopes, or to make elements that only exist for short times in these accelerators. A linear accelerator has an ion gun as the first component in the accelerator system. The remainder of the accelerator contains equipment to accelerate, focus, and refine the ion beam. A linear accelerator can implant ions of almost any element into a target. These types of devices are used for ion implantation, but they are not available for general industrial use. There are only a few of these devices in this country; they cost millions of dollars and are dedicated to research projects.

Ion implantation for industrial applications is performed in vacuum equipment designed specifically to do implantation. A schematic of an implantation device is shown in Figure 7–10. In this system, a gas of the element to be implanted is introduced in such a fashion that it passes through a stream of electrons that are being emitted by a filament toward a target plate. The atoms of the gas are ionized by collisions with the

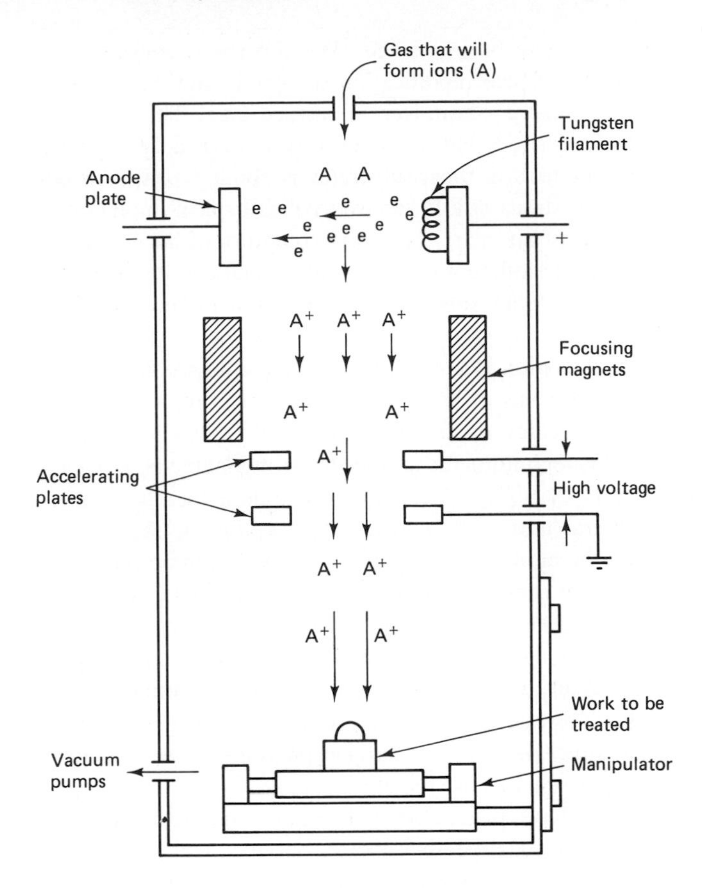

Figure 7–10 Schematic of ion implantation.

electron stream. For example, if argon is used as the implantation species, argon ions will be produced. These ions are then focused by magnets so that they form a beam of ions. The ions are now accelerated by essentially passing them through a column with a very large voltage drop from one end to the other. The beam of high-velocity ions exits into a work chamber where it strikes the work and produces implantation. The ion gun, ion accelerator, and the work are all in a vacuum chamber that can be pumped to pressures as low as 10^{-6} torr (1.3×10^{-3} Pa). Vacuum is required to prevent air molecules from attenuating the beam. During implantation, the gas that is to be ionized raises the pressure in the chamber, but it probably would not exceed 10^{-2} torr (13.3 Pa).

The parameters that are controlled in ion implantation are as follows:

1. Ion species
2. Ion flux (in atoms per unit area)
3. Treatment time
4. Impingement energy

Any element that can be ionized can be implanted. The fluence is controlled by the focusing of the ion beam, but for surface modification of metals and cermets, the flux is usually about 10^{17} atoms per square centimeter of surface area. The treatment time is determined by the penetration desired, but treatment times are usually in hours. The maximum depth of penetration of ions in most substrates is about 1 μm, with ion accelerating energies in the range of hundreds of kilo electron volts. Typical implantation depths are about 0.1 μm (4 millionths of an inch). The solid solubility of the implanted species in the substrate plays a role in the amount of implantation that can be achieved, but with direct implantation the limit of the substrate concentration of the implanted species is about 10 percent.

Recoil or ion beam mixing implantation is a technique developed to produce greater depth of penetration and to overcome one of the most formidable problems of ion implantation, obtaining the implant species in the form of a gas that is suitable for introduction into the implanter. In this technique the species to be implanted is deposited on the substrate as a thin film (usually by PVD). The coated substrate is then bombarded by a high-energy ion beam of neutral atoms such as argon, krypton, or xenon. The collisions of these neutral ions with the atoms of the coating cause them to be implanted into the substrate. With this technique, it is also possible to increase the atomic concentration of the implant species in the substrate to as high as 25 percent. Part of the reason for this higher concentration is the elimination of sputtering losses. In direct implantation, besides driving ions into the substrate, the ion collisions with the substrate surface cause sputtering of atoms from the substrate surface. As ions are implanted, the energy of the beam is also removing material. The recoil technique prevents sputtering of the substrate surface; only a portion of the coating will sputter.

The second advantage of the recoil process may be more significant than the improved atomic concentration. If it is desired to implant, for example, chromium, direct implantation would require that a chromium gas be generated and introduced into the ion gun. The details of obtaining gases of metals and many elements can be formidable. To do implantation of these materials by the recoil process, the implant species can be coated with any one of the well-developed vacuum coating techniques, sputtering, thermal evaporation, or ion plating. Thus implantation of metals such as chromium, titanium, and tantalum can be accomplished by simply applying these coatings to the work and driving them into the substrate surface with inert ions.

The flux of an ion beam can be measured by an electrical device called a Faraday cup that is put into the work chamber and bombarded with the ion beam. The implantation depth achieved in a particular implantation operation can be measured by a number of techniques, but in situ measurement can be done by Rutherford backscatter analysis (RBS). In this process, the surface to be analyzed is impinged with a high-velocity beam of helium ions (typically 2 MeV). These ions are elastically scattered from the nuclei of the substrate material, and the energy of the scattered ions is detected and analyzed. The energy of the scattered ions provides information on the mass of the implanted surface, which in turn can be used to deduce the degree of implantation. Since this process uses an ion beam, it can be done in the implantation unit by adding the capability of ionizing helium gas and by installing a detector unit to measure the energy of the recoil ions.

In general, ion implantation equipment is fairly complex. It is not portable and the parts to be treated must fit into a vacuum chamber. Implantation is a line-of-sight operation, and expensive fixturing may be required for work manipulation. The cost of a complete implantation system will probably be in excess of $500,000. Essentially, this process should not be purchased for use in an industry without documentation of the benefits over competitive surface-treatment operations. There are, however, adequate commercial ion implantation centers, so this process can be readily used by the average machine designer or materials engineer without fear of the capital equipment costs.

Metallurgical Considerations

Table 7–1 contains a list of some of the metallurgical reactions that can occur with ion implantation of metals. Since any element can be implanted in any substrate metal or alloy, the things that can happen to the surface are probably without limit. The reactions listed in Table 7–1 are known to have occurred with single-species implantation.

Going into solid solution is one obvious reaction. The ions become atoms by electron transfer from the substrate. Implanted species with small atomic diameters such as carbon, boron, and nitrogen go into interstitial solid solution; larger atoms such as chromium and tantalum go into substitutional solid solution. From the properties standpoint, increasing the alloy content of the implant layer can strengthen the surface; for example, yield strength, shear strength, and compressive strength can increase.

Transmission electron microscope (TEM) studies have confirmed that ion implantation can increase dislocation density. Since most metal deformation processes take place by motion of dislocations, increasing dislocation, densities by ion implantation has the net effect of making subsequent deformation processes more difficult; surface strengthening is the net effect. Similarly, implantation can produce other atomic defects, such as vacancies and lattice strains. These types of occurrences may also lead to strengthening. Conventional hardness testing techniques do not work on surface treatments that are less than 0.1 μm (4 μin.) in depth, but by using special techniques it has been determined

TABLE 7–1 ION IMPLANTATION SUBSTRATE EFFECTS

- Solid solution of implant element
- Production of dislocations
- Production of point defects
- Alteration of crystallinity
- Compound formation
- Alteration of the chemical nature of the surface
- Alteration of surface stresses
- Sputtering of surface atoms
- Heating

that implantation of certain species can increase hardness by measurable amounts. A treated surface can be 20 or 30 percent harder than unimplanted bulk material.

X-ray diffraction studies have shown that some implant species can cause amorphous structure in the implanted region of the surface. The effect of an amorphous structure on use properties may be good or bad depending on the substrate, but amorphous metals produced by some of the processes, such as laser glazing, are thought to have improved mechanical properties compared to the crystalline condition of the same metal.

The formation of compounds of the implant species and the host substrate have been detected in a number of implant systems. For example, extremely fine particles of iron nitride compounds have been detected in austenitic stainless steels that were implanted with a large dose of nitrogen. These compounds can improve wear properties, as they do in conventionally nitrided surfaces, or they can strengthen the surface by alteration of the state of stress in the surface.

It is also known that implantation of steels with elements such as chromium and tantalum can have an effect on the oxidation behavior of the implanted surface. Even implantation with ''neutral'' elements such as argon and helium are thought to have effects on the response of treated surface to chemicals and to processes where surface energy and the chemical nature of the surface are considerations such as in bonding and catalysis. Thus ion implantation can have an effect on surface reaction phenomenon, electrochemical reactions, bonding, friction, and adhesion.

The last two items listed in Table 7–1 are usually undesirable effects of ion implantation. Sputtering of surface atoms is usually not desired since the implanted surface is being removed as it is being treated. Recoil implantation was developed to minimize this effect. Heating of the substrate is undesirable for most implanted systems. One big advantage of ion implantation is that it does not add dimension to a surface. Finished parts can be treated. If the implantation process produces significant substrate heating, it can cause part distortion that will void the ''no size change'' advantage of the process. Most commercial implantation systems use heat sinks to cool small parts during implantation. There is no need to do this on massive parts; but, in general, cooling is used on all small parts, and there is an attempt to keep the part temperature below 300°F (150°C) during implantation.

In summary, there are many metallurgical aspects of ion implantation. The alteration of a substrate surface by ion implantation depends on the nature of the substrate and the implant species. Since there are countless combinations of implant species and substrates, the percentage of combinations that have been investigated to date is relatively small. Of the systems that have been extensively studied, there is no universal agreement as to the metallurgical reactions that occur. In other words, the mechanism of the surface modifications that occur is not fully understood. Since ion implantation started to be used for hardfacing types of applications, just about every metal system of commercial importance has been implanted. Cemented carbides and other cermets have been treated, and some work has been done with ceramics and polymers. There is no way to determine how many different implant species have been tried, but the species that are most commonly used are C, N, Ti, Ta, Mo, Y, B, P, S, and the precious metals.

From the user's standpoint, it may not be necessary to fully understand the metallurgical reactions that occur in ion implantation, but several of the possible reactions that we have discussed are probably occurring. Since this process is relatively inexpensive (unless you buy the equipment), most users simply treat parts with species recommended by the treatment facility and try the parts in service or evaluate laboratory samples treated with the recommended species. This is the most expeditious course for most users.

Applications

Now that we have discussed what can happen to a substrate surface in ion implantation, let us discuss the types of applications that have been successful. There are three main areas where this process is used:

1. Alteration of tribological properties
2. Alteration of chemical properties
3. Alteration of mechanical properties

Since the goal of this book is hardfacing techniques, we will emphasize the use of ion implantation to alter tribological properties (friction and wear). The applications where alteration of chemical properties are the major goal are primarily aimed at improving corrosion resistance. For example, chromium implantation of 52100 ball bearing steel is done to enhance corrosion resistance; phosphorus has been implanted in carbon steel to reduce atmospheric rusting. From the standpoint of mechanical properties, ion implantation is widely used to enhance fatigue resistance. It appears to be quite effective for this purpose in many metal systems.

In tribological applications of ion implantation, work has been done in modifying frictional characteristics and in trying to improve the resistance to all the various modes of wear that do not involve severe plastic deformation of the surface. For example, nobody seems interested in trying ion implantation to improve resistance to gouging abrasion, high-stress abrasion, and brinelling, but it has been used to reduce galling, fretting damage, polishing abrasion, surface fatigue, cavitation, liquid impingement erosion, and adhesive wear and to improve interactions with lubricants.

Table 7–2 is a list of some wear application results. As can be seen from this list, most wear applications of ion implantation are directed toward the types of tools that are used in many manufacturing operations: punches, dies, cutting devices, forming devices, and the like. The substrates for these applications are mostly tool steels, alloy steels, and cemented carbides. The implant species are usually the elements that can form compounds, such as titanium (TiC), and the small elements that assume interstitial sites in the substrate lattice and react with dislocations to impede motion and strengthen the surface (C, B, N).

Unfortunately, ion implantation as a process to compete with hardfacing is not mature enough to allow design guidelines on what species to put in what substrate. It cannot be said that nitrogen implantation will double the abrasion resistance of H13 tool steel. This application may have worked for somebody under some conditions,

TABLE 7–2 IMPROVEMENTS IN WEAR LIFETIME OBTAINED BY ION IMPLANTATION

Application	Material	Result
Paper slitters	1.6% Cr, 1% C steel	2× cutting lifetime
Synthetic rubber slitters	Co (6%), WC	12× cutting lifetime
Punches for acetate sheet	Cr-plated steel	Improved product
Taps for phenolic resin	M2 high-speed steel	Up to 5× lifetime increase
Thread-cutting dies	M2 high-speed steel	5× lifetime increase
Tool inserts	4% Ni, 1% Cr steel	Reduced tool corrosion
Forming tools	Carburized steel	Greatly reduced wear
Dies for copper rod	Co-cemented WC	5× throughput, improved surface finish
Deep-drawing dies	Co-cemented WC	Improved lifetime
Dies for steel wire	Co-cemented WC	3× lifetime increase
Injection molding nozzle, molds, screws, gate pads	Tool steel	4 to 6× lifetime increase
Swaging dies for steel	Co-cemented WC	2× lifetime increase
Fuel injectors and metering pump	Tool steel	100× less wear
Punch and die sets for sheet iron	Co-cemented WC	6× lifetime increase
Swaging dies, press tools for wheels	Co-cemented WC	Improved lifetime
Cam followers	Steel	Improved lifetime
Plastic cutting	Diamond tools	2 to 4× lifetime
Finishing rolls for Cu rod	H-13 steel	3× lifetime
Printed circuit board drills	Co/WC	3× lowered smearing, lower operating temperatures
Dental drills	Co/WC	2 to 3× (JKH) lifetime
N into thermally nitrided steel	17–4 PH alloy steel	Combination better than either process alone
Sheet steel slitter knives	Co/WC	3× lifetime
Plastic calibrator die	Tool steel	2× lifetime
Plastic profile die		2× lifetime
Beverage can scoring die	Tool steel	1.5 to 3× lifetime
Printed circuit drills	Co/WC	Significantly lowered operating temperatures
Hard chrome-plated wire guides		3× lifetime without significant wear
Rubber gasket cutting	Diamond tool	Significant improvement
Tinsel rolling	Chrome steel roller	Highly successful

Courtesy of Surface Alloy Corp.

but there is not enough known about the process to say that it will work for all applications of H13 steel. The user will have to perform his or her own tests and will probably have to rely on literature reports or treaters' recommendations to decide on a particular species.

To conclude this discussion, the statement that best describes ion implantation is that it is a process that can have a hardening and strengthening effect on a thin layer on the surface of almost any substrate. The depth of the ion implantation effect is usually less than 1 μm. If a wear or friction system might benefit from an alteration of the surface for a depth of only hundreds or thousands of angstroms, then ion implantation should be considered for use. It will compete with other thin-film coatings. Results to date indicate that it is a good candidate for wear systems where a very small amount of material removal will impair the use of a tool (cutting edges, die edges, etc.), and there are many systems of this type.

SUMMARY

The surface treatments that we discussed were described as high-energy processes, but a more descriptive title may be nontraditional surface treatments. These are the surface treatments that are not used for day-to-day problems. They are the processes to consider for special problems or for innovative solutions to ordinary problems. Electron-beam hardening is widely practiced by everyone who has an electron-beam-hardening facility; as laser hardening is practiced by everyone who has a laser facility. These processes work very well and they can be very cost effective. They are not developmental. Ion implantation is still developmental, but it is probably under investigation in every major industry for tribological applications.

Laser and electron-beam treatment of surfaces essentially replaces more distortion-prone processes such as quench hardening and conventional welding hardfacing. Ion implantation is best applied to zero-wear tribosystems, systems that have components that cannot tolerate wear depths of millions of an inch—tape heads, precision reference surfaces, precise cutting edges, and similar applications. All of these processes are growing in applicability and within the next decade they will probably not become as common as carburizing, but they will be used by most manufacturing firms.

REFERENCES

BACKISH, R. "Electron Beam Melting, Refining and Heat Treatment," *Industrial Heating,* Sept. 1985, pp. 26–30.

HITCHCOX, A. L. "Pinpoint Hardening with CO_2 Lasers," *Metal Progress,* April 1986, pp. 31–34.

IWATA, ATSUSHI. "Transformation Hardening by Electron Beam," *Bulletin of the Japan Society of Precision Engineering,* Vol. 18, No. 3, 1984, pp. 219–224.

JENKINS, J. E. "Some Factors That Influence Electron Beam Hardening," *Heat Treating,* Dec. 1981, pp. 28–32.

KUBEL, E. J. "Tailoring Surface Properties with Lasers," *Materials Engineering,* June 1985, pp. 41–44.

Metals Handbook, Vol. 6, Welding, Brazing, and Soldering. Metals Park, Ohio: American Society for Metals, 1983, pp. 802–804.

Oliver, W. C., and others. "The Wear Behavior of Nitrogen Implanted Metals," *Metallurgical Transactions,* Vol. 15A, 1984, pp. 2221–2229.

Preece, C. M., and J. K. Hirvonen, eds. *Ion Implantation Metallurgy*. Warren, Pa.: Metallurgical Society of AIME, 1980.

Saunders, R. J. "Laser Metalworking," *Metal Progress,* July 1984, pp. 45–51.

Savage, J. E. "Adding Wear Resistance via Ion Implantation," *Metal Progress,* Nov. 1984, pp. 41–44.

Shih, K. K. "Effect of Nitrogen Implantation on Impact Wear," *Wear,* Vol. 105, 1985, pp. 341–347.

Smidt, F. A. *Ion Implantation for Materials Processing*. Park Ridge, N.J.: Noyes Data Corp., 1983.

The Use of Ion Implantation for Materials Processing—NRL Report 5716. Washington, D.C.: Naval Research Laboratory, 1986.

Swartz, Melvin M., ed. *Source Book on Electron Beam Welding*. Metals Park, Ohio: American Society for Metals, 1981.

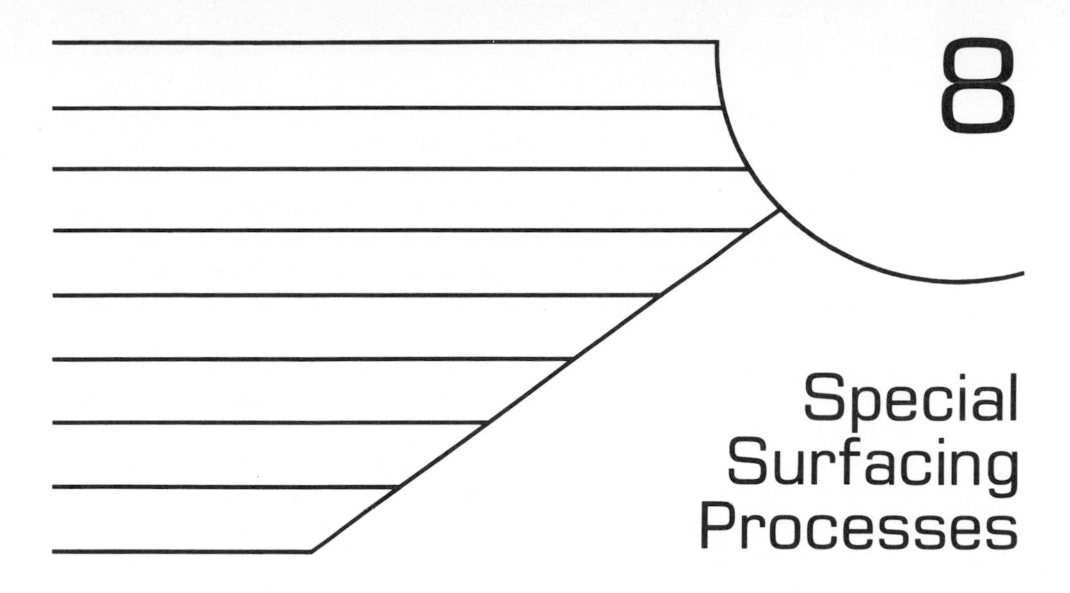

Special Surfacing Processes

We have discussed conventional surface-hardening processes and many of the coatings and surface treatments that are competitive with the welding processes. In this chapter, we will discuss processes that are likewise competitive with welding hardfacing, but that do not fit into a particular category. Some are proprietary and some only apply to specific types of applications; but all these processes can be used to enhance the wear properties of a substrate, and they play a role in surface engineering.

REBUILDING AND SURFACING CEMENTS

Anyone who works in the maintenance field has undoubtedly encountered a situation where a part is damaged by a gouge or a limited amount of scoring. One of the first things that comes to mind is to put an epoxy or autobody-type filler in the damage and put the part back in service. What we refer to as rebuilding cements are products that are designed to do just this. They are commercially available and they have different formulations, but most of these products are filled polymeric materials that are catalyzed, troweled on a surface, cured, and finished. They are intended as materials to rebuild worn parts, but there is no reason why they cannot be used on new installations as a wear-resistant coating or overlay. The types of wear that they resist depend on the composition of the cement, but in general these types of materials have utility in resisting scratching or low-stress abrasion and certain types of erosion such as slurry erosion, cavitation erosion, and liquid erosion. They resist scratching abrasion similarly to the way that concrete resists abrasion. Concrete is a combination of hard aggregate particles (stones), sand, and cement that binds the sand and aggregate together. When concrete wears in a highway, for example, the sand/cement phase wears down and the hard aggregate plays the major role in resisting wear. A casual inspection of any well-traveled concrete roadway will confirm this mechanism. Rebuilding cements are usually epoxies filled with hard materials, and they resist abrasion by transferring most of the scratching to the hard particles. They stand proud from the epoxy matrix.

In liquid erosion processes, if the polymer matrix is resistant to attack by the liquid, the corrosion component of the erosion is overcome. The hard particles in the polymer matrix resist the component of the erosion process that is caused by mechanical action. For example, rebuilding cements have been successfully used to repair eroded metal pump casings. Cavitation pits or eroded areas are filled in with the rebuilding cement, and the repair can make the pump better than the original. Many pump casings are made from gray cast iron. If a cast-iron pump is used for dirty water, material will be removed by corrosion from the water and by the mechanical action of hard particles in the water. A proper rebuilding cement will eliminate the corrosion from the water completely, and the hard particles in the cement will resist the material removal by the mechanical action of the entrained sand and the like that is invariably in many effluent systems.

Figure 8–1 is an illustration of a typical wear repair with a rebuilding cement. Figure 8–2 illustrates the nature of a typical rebuilding cement. Most rebuilding cements are thixotropic epoxies filled with various types of hard materials. The following is a list of some of the fillers used in commercially available rebuild cements:

1. Aluminum oxide (particles as small as several micrometers to spheres as large as several millimeters), see Figure 8–3

Figure 8–1 Cast-iron pump housing rebuilt with a trowel-on repair cement (center cavity light spots are remnants of previously used copper-based hardfacing).

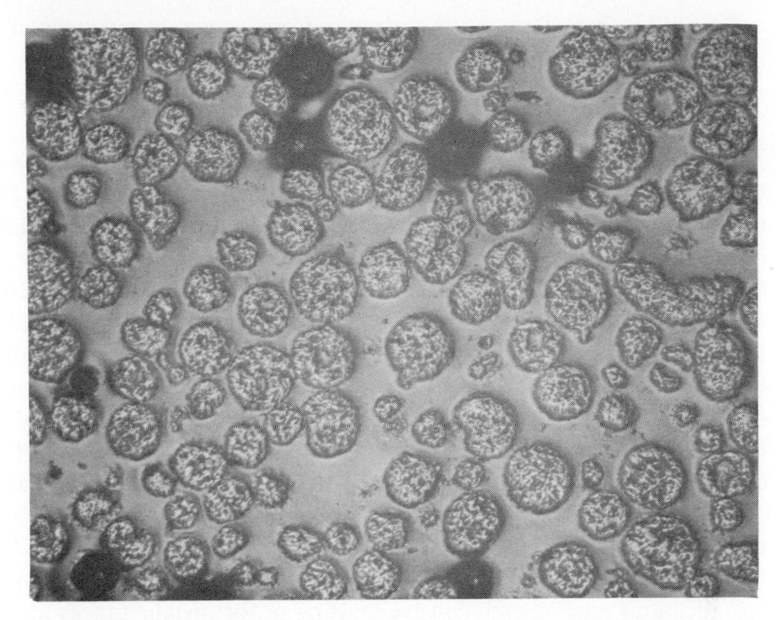

Figure 8–2 Epoxy rebuilding cement filled with aluminum oxide spheres (100×).

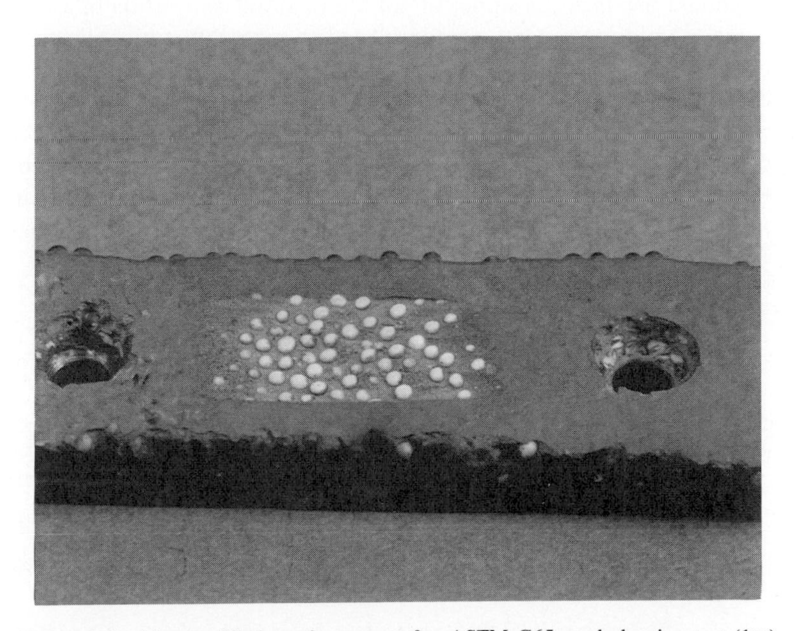

Figure 8–3 Alumina-filled repair cement after ASTM G65 sand abrasion test (1×).

2. Chilled iron (Particles in the range from several to hundreds of micrometers
3. Composite particles: ceramic-coated hard metal (several to hundreds of micrometers)
4. Glass beads (several to hundreds of micrometers)

All these fillers have hardnesses greater than many metals, and they can resist abrasion better than most metals. If the hard fillers are corrosion resistant, they will minimize the corrosion effects of erosion processes. The ability of the rebuilding cement to resist removal by mechanical action depends on the bond strength of the polymer matrix to the substrate and to the particles and on the nature of the particles (their size, distribution, and volume fraction). Cements with, for example, aluminum oxide spheres that are several millimeters in diameter will have relatively large regions of polymer exposed to abrasion. These types of cements are better suited to handling abrasion from large objects such as boxes and pieces of lumber. If the abradant is likely to be fine particles of sand in water, the cements with a very large volume fraction of particles and little exposed matrix are the most suitable. This concept is illustrated in Figure 8–4.

The application of repair cements is a key part of their use (Figure 8–5). Each manufacturer has detailed recommendations on application, and these instructions should be adhered to. The most suitable substrates are ferrous metals, and most preparation specifications recommend abrasive blasting of the metal surfaces to surface roughnesses in excess of 100 microinches R_a (2.5 μm). A factor to consider in the use of these cements is their ability to be applied to vertical surfaces and the thickness of the cement that can be applied. The more effective rebuilding cements can be applied to vertical surfaces in thicknesses of up to 1 in. (25 mm) without slumping or running. Another important use factor is sufficient pot life to allow complction of a job in one application. Good rebuilding cements are formulated to meet difficult application requirements, and ordinary thixotropic epoxies are not formulated to allow application like these rebuilding cements.

Since filled rebuilding cements are proprietary in nature, there are no generic specifications for the designation of these materials; but these products are advertised in maintenance journals and the like. If it appears that one of these products may fulfill a rebuilding or hardfacing need, consult several suppliers and obtain the details

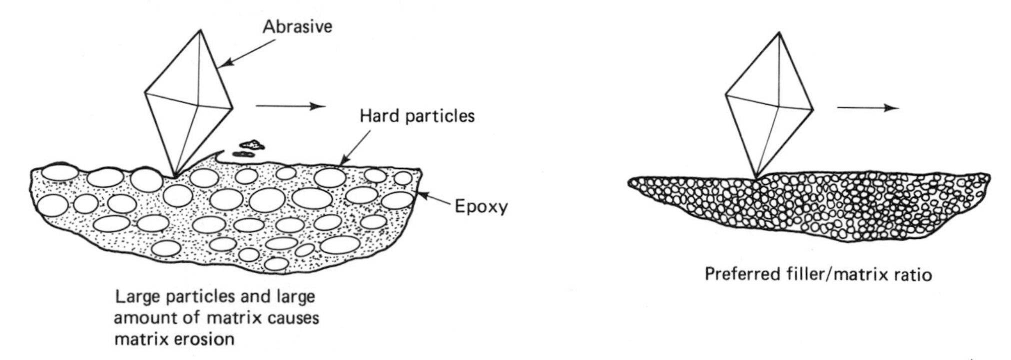

Figure 8–4 Abrasion of filled repair cements.

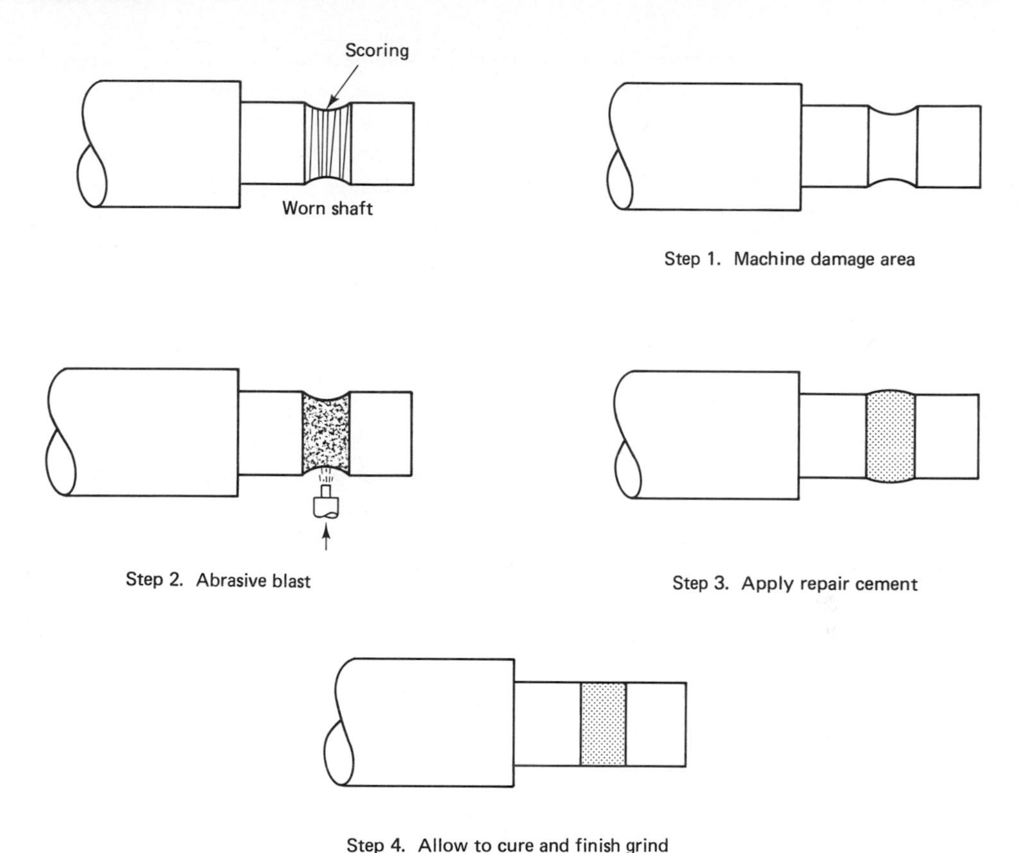

Figure 8–5 Steps in applying repair cements.

of the cements that they recommend for your application. A selection should be made based on applicability (thickness, pot life, slump characteristics, etc.), type and volume fraction of filler, wear characteristics, bond strength, and, where applicable, machinability. In general, these types of products are suitable for low-stress abrasion and some types of fluid erosion. They do an excellent job on many types of pump casings, pitted castings, propellers with cavitation damage, and severe gouges and grooves on parts that are difficult to weld repair. They have their place in solving wear problems, but the potential user must keep in mind that these cements have bond strengths only as good as epoxies; they are not for bulldozer bucket teeth.

WEAR TILES

For centuries "ceramic" tiles have been used to produce decorative effects in floors, as well as to prevent the wear of these floors. In buildings, these tiles can be marble, granite, or some naturally occurring mineral, or they can be made from earthenware, fired clays, or glazed and fired clays. This same concept is used to overlay machinery or materials-processing equipment to prevent wear and erosion (and in some cases corrosion). The only difference between the centuries old floor tiling and the application of wear tiles is the type of tile material and the application technique. Figure 8–6 illustrates some examples of industrial components that have been clad with wear tiles. The tiles come in different configurations and sizes. Industrial wear tiles are usually made from materials that are harder and more abrasion resistant than the tiles that are used for floors. Some of the more popular tile materials are cemented carbide, aluminum oxide, and basalt.

Cemented carbide tiles are usually quite small in size, about 1 in. (25 mm) square and ¼ in. (6 mm) thick. The type of carbide used is variable, but most have a composition of about 90 to 95 percent WC and the remainder is a cobalt binder. Aluminum oxide tiles can be almost any size, but they are usually about ¼ in. (6 mm) thick and their size is usually less than 6 in. (150 mm) square or hex. Basalt is a naturally occurring igneous rock that can be melted and cast into shapes while retaining a hardness between 8 and 9 on the Mohs scale. Basalt is composed of mixed oxides, about 48 percent silica, 16 percent aluminum oxide, 14 percent iron oxide, 12 percent calcium oxide, 8 percent magnesium oxide, with the remainder other oxides such as sodium oxide and titanium dioxide. Basalt materials melt at about 2280°F (1250°C), and basalt wear tiles are made by casting to desired shapes. Tiles can be sand or permanent-mold cast, and centrifugal casting can be used to produce a monolithic lining in pipes and similar shapes that can be rotated for casting on their inside diameter. Basalt tiles are brittle

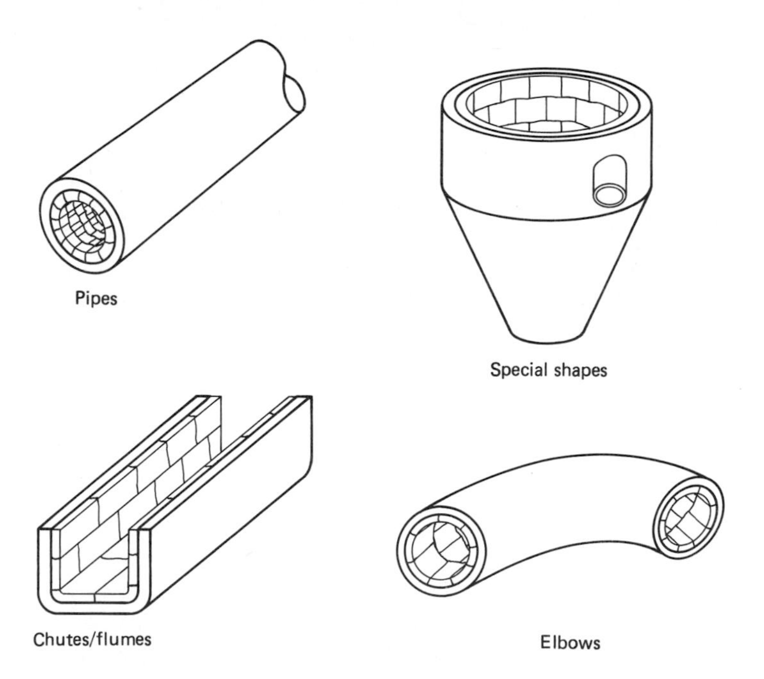

Figure 8–6 Industrial applications of wear tiles.

like most ceramic types of materials, and they are usually cast in shapes that are about 1 in. (25 mm) thick and less than 100 in.^{2} (0.06 m^2) in area. Aluminum oxide tiles are made in the same way that most ceramic shapes are made: the aluminum oxide is mixed in powder form with a clay-type binder and it is compacted by isostatic techniques to a shape and sintered.

Wear tiles are adhered to substrates by cements. If the temperatures and chemical conditions are suitable, the carbide and aluminum oxide tiles can be put in place with epoxies. Basalt tiles are usually applied with mortars. The cement joints between tiles need to be as narrow as practicable since the cements do not have the erosion resistance of the tiles. The small ceramic and aluminum oxide tiles are sometimes available prebonded and spaced on 1 ft^2 (0.1 m^2) papers to ease installation on large areas.

There are other tile systems than those mentioned. White iron wear plates can be plug welded onto surfaces in a manner similar to tiles; ultrahigh molecular weight polyethylene (UHMWPE) lined pipes are available, and UHMWPE sheets can be treated to allow adhesive bonding to surfaces like the hard tiles. Elastomers can be used in a similar fashion. Other hard-tile materials are cubic boron nitride and silicon carbide.

The relative low-stress abrasion resistance of the hard surfacing tiles to relatively soft abrasives such as sand is approximately proportional to their hardness:

Silicon carbide: 2600 HK
Aluminum oxide: 2100 HK
Cemented carbide: 1800 HK
Basalt: 1500 to 1700 HK
Mild steel: 200 HK

Testing may be required to determine if a candidate tile is resistant to a particular abrasive or corrodent, but in general all the preceding outwear mild steel (1020) by a factor of 5 to 10 under conditions of low-stress scratching abrasion. A moderate-hardness elastomer (60 Shore A durometer) and the UHMWPE can compete with some of the hard tiles in abrasion resistance in some systems. They compete quite well in water slurries containing fine abrasives and in, for example, coal chutes that carry pea-size particles of coal. These are situations where the stress imposed on the elastomer or polymer by the abrasive is very low. The hard tiles are preferred when the stresses on the abrasive are sufficient to cause crushing of the abrasive. The cemented hard wear tiles are suitable for the types of abrasive-handling devices shown in Figure 8–6. They have an advantage over weld hardfacing in that large areas can be covered usually with low-skill labor, and there is no heating of the substrate. The major disadvantage compared to welding hardfacing is the higher cost of the overlay and the lead time required to obtain these materials. Basalt and UHMWPE- or rubber-lined pipe are off-the-shelf items, but cast shapes and tiling of odd shapes can involve substantial lead time (to obtain materials and applicators). Thus welding hardfacing has an advantage on odd shapes. Welders and consumables are readily available. The cost of wear tiles and lined pipes depends on the material and the shapes, but of the hard materials, cemented carbides may cost several hundred dollars per square foot (0.1 m^2) simply based on the cost of raw materials (WC). Aluminum oxide and silicon carbide shapes are similarly dear, not as raw materials but as sintered shapes. Basalt is competitive in price with things like furnace bricking. Plastic, basalt, cement, and elastomer lined pipes are far cheaper than the ceramic/cermet tiles. These pipes may only cost several times the cost of unlined pipe.

In summary, wear tiles are available to solve wear problems that are difficult to solve with monolithic materials, hardfacing, and conventional surface treatments. The use of lined pipes is often easily justified; the hard tiles are more expensive and they may only be justified for severe wear problems that could not be solved by lower-cost techniques.

ELECTROSPARK DEPOSITION COATINGS

In the electrospark coating process, an electrode of the material to be coated is brought in contact with the surface to be coated and arcing occurs between the electrode and the work, causing some of the electrode material to be deposited on the substrate. There is instantaneous melting of the electrode and the substrate at the point of contact. Figure 8–7 is a schematic of this process. In the most common type of equipment, the electrode is cemented tungsten carbide; the power supply produces variable direct current up to about 50 A and the voltage is less than 25 V. There may be additional control for polarity and fine tuning of the sparking and deposition rate. Repetitive arcing or sparking is accomplished by putting the electrode in a vibrating holder similar to vibrating tools that are used for engraving identification on metal objects. Each time that the electrode strikes the work, current flows to the work, causing a spark and heating of the electrode tip and substrate, and some of the electrode is deposited on the substrate. Deposition rates are controlled primarily by the current and the amplitude of vibration of the electrode. Newer spark-deposition devices use switching in the power supply to

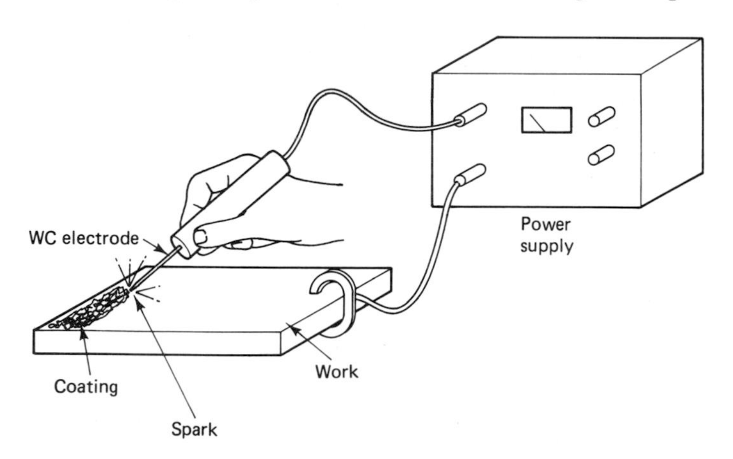

Figure 8–7 Schematic of spark discharge coating.

Figure 8–8 Cross section of a spark-discharge coating (100×).

achieve repetitive arcing instead of the vibrating electrode. The surface to be coated must be traced, like shading an image in a coloring book. This traversing of the surface can be mechanized, but hand-held techniques are more common. Spark-deposition power supplies and consumables are commercially available from several manufacturers, and the coating process can be obtained from job shops.

Figures 8–8 and 8–9 show typical examples of the appearance of a carbide coating produced with a vibrating type of device. The coating will have approximately the same composition as the stylus from which it was deposited. Coating thicknesses up to

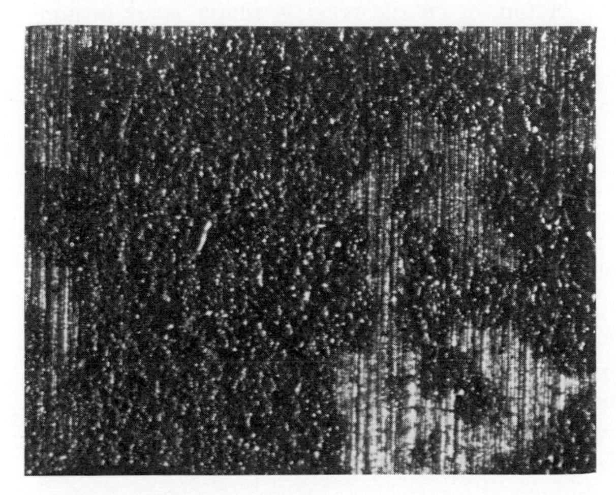

Figure 8–9 Surface of a spark-discharge carbon steel (20×).

5 mils (125 μm) have been reported, but with the vibrating type of applicator the coating is somewhat discontinuous and porous, with a thickness in the range of 0.0003 to 0.001 in. (7.5 to 25 μm). The wear resistance of these coatings depends on the continuity, thickness, and nature of the coating. The stylus for spark-deposition coating can be any metal or cermet, but cemented carbide is almost exclusively used. If the coating is continuous and 100 percent cemented carbide, it should have the wear resistance of cemented carbide. Thus this is the most logical material to deposit. The effectiveness of these coatings in enhancing the wear properties of a substrate depends on the intended service. Some laboratory tests have shown that these coatings applied on a soft steel or gray cast iron improve metal-to-metal and abrasion resistance somewhat, but the coatings do not perform like solid cemented carbide. On the other hand, these coatings have been found to increase the service life of paper-cutting dies and similar tools that do not require a keen die or knife edge. A major limitation of this process is the roughness and porosity of the deposit. The coatings are too thin to grind after coating, and many tools cannot tolerate a surface roughness of about 200 microinches (4 μm CLA), which is typical with the vibrating devices. If an application requires a hard, rough surface, these coatings may offer some benefits, but it is still advisable to apply these coatings over a 50 to 60 HRC substrate if the properties of the carbide are to be utilized. Thin, hard coatings on soft substrates often wear by a mode of deformation of the soft substrate under the coating.

In summary, electrospark deposition coatings are not substitutes for bulk electroplates or hardfacings, but these coatings can improve the abrasion resistance of some metal surfaces if applied properly. The equipment is low cost, from less than $200 to several thousand dollars depending on capabilities; it is portable and easy to use, and coatings can be applied to most metal substrates. There are job shops that apply these coatings on a commercial basis. The user must keep in mind that the coatings are used as deposited and they are very rough. There are some places where such a coating is desirable. For example, they have been successfully used to build up the inside diameter of blanking die holes. The die holes were worn oversize and slugs were pulling through the die hole instead of falling out the bottom of the die hole. A simple spark coating around the die hole restored the tight punch-to-die fit, and the dies could be used for millions of additional parts. Thus there are places where these coatings work, and they are candidates to compete with some of the other coatings that we have discussed.

FUSED CARBIDE CLOTH

A unique method of hardfacing is the application of cemented carbide on metal substrates by fusion of a carbide/binder cloth. This process is proprietary, but coatings with this technique are available on a job-shop basis. The cloth that is the unique part of this coating process is composed of tungsten carbide or other carbides and nickel, cobalt, or nickel/chromium-type binder particles bonded together with fluorocarbon fibrils. The cloth has sufficient strength to be handled, cut with ordinary scissors, and draped over the surface to be coated. Techniques are available for holding the cloth to round bars, on cam surfaces, and even inside bores. After applying the cloth to a substrate, the part and cloth are put into a controlled atmosphere furnace and heated to the brazing temperature of the binder. The organic binder is removed in the heating process, and at the brazing temperature of 1650° to 2100°F (900° to 1150°C), the binder melts and fuses the carbide particles together and the carbide coating to the substrate. The cloth and a coated steel block are illustrated in Figure 8–10.

Various types of carbides and binders are available and the carbide particle size can vary from about 2 μm (80 μin.) to about 500 μm (0.02 in.) in diameter. This coating can be applied to any metal substrate that can tolerate the brazing temperature (ferrous materials, nickel-base alloys, etc.). The cloth shrinks significantly in the brazing operation, and this must be allowed for in placement and size of the cloth. After brazing, the coating can have a thickness in the range of 5 to 60 mils (0.12 to 1.5 mm), and the fused coating can have a carbide concentration of up to 75 percent by weight. The coating is metallurgically bonded to the surface with the brazing alloy, and it can be ground and processed as if it were a conventional cemented carbide. Figure 8–11 is an illustration of some typical shapes that can be coated with this process.

The cost of this coating process depends on the usual factors: coating area, number of parts, and part complexity. With large numbers of parts, this process probably competes with surface-hardening processes such as carburizing and nitriding. It may not be lower cost for these types of parts, but from the standpoint of what is accomplished by this coating, there are few competitors. This process can apply a 0.060-in. (1.5 mm) thick carbide coating over an area of 50 or 100 in.^{2} (300 to 600 cm^2) in a simple furnace fusing process. If an application could benefit from this type of surface, there is good

Figure 8–10 Carbide brazing cloth and a block overlayed with this cloth (1 mm thick).

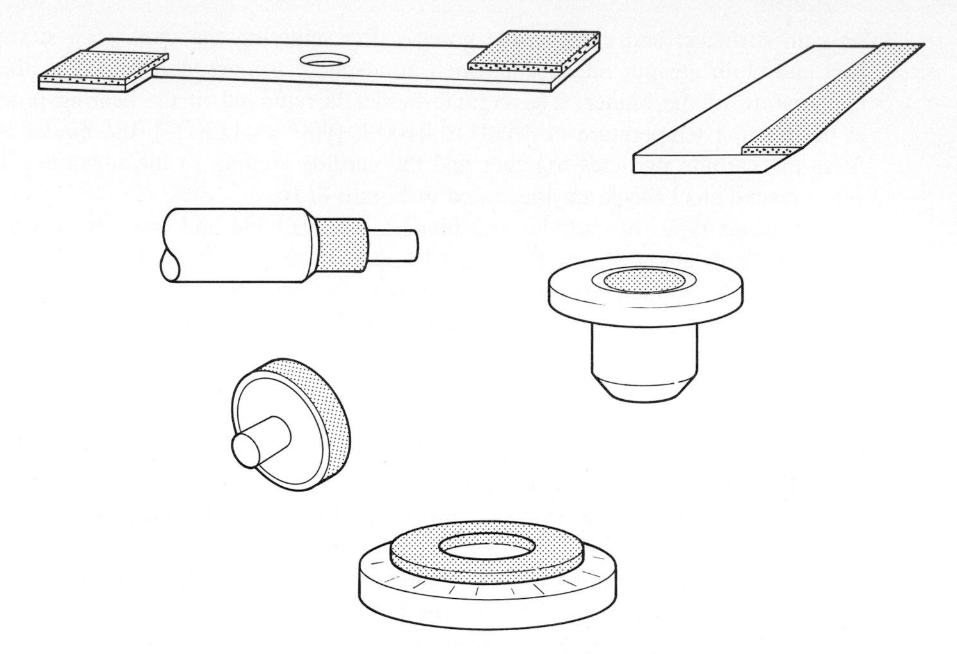

Figure 8–11 Typical parts that can be hardfaced by carbide brazing cloth.

justification for using this process. In fact, this process has been successfully used to overlay the inside of bores as small as ½ in. to a depth of several inches (12 to 150 mm). There may not be another surface coating process that can do this. Thus the advantage of this process is that it can be used to apply relatively thick WC/CO coatings to shapes that would be difficult or uneconomical to coat to these thicknesses with other techniques.

There are some obvious disadvantages in using brazed carbide cloth; it cannot be done in house, and it requires heating the substrate to high temperatures that can cause distortion and soften previously hardened parts. These limitations may preclude its use for some parts, but it is a useful process and has a definite role in the spectrum of wear-resistant surface coatings.

THERMAL/CHEMICAL DEPOSITED CERAMIC COATINGS

This process is used to produce ceramic coatings that are competitive with the ceramics and cermets as applied by thermal spray processes. It is proprietary in nature, but it is basically a chromium oxide type of ceramic that is deposited by repeated application of a liquid slurry and thermal treatments. The coating produced is aluminum oxide particles (submicrometer) in a matrix of chromium oxide. The coating constituents in slurry form are brushed or sprayed on the surface of the substrate to be coated. They are dried and baked in a furnace or, for on-site coatings, in a kiln built around the part to be coated. The heat treatments convert the chromium compounds in the slurry to chromium oxide. The coating can be as thick as 0.005 in. (125 μm), but the usual thickness is 0.002 to 0.003 in. (50 to 75 μm). Different thicknesses are obtained by varying the number of spray/bake cycles.

These coatings can be used for the same types of applications that apply to thermal spray coatings: rolls, seal faces, spray nozzles, and the like. The bond strength of these coatings is similar to that of a good thermal spray coating. The hardness of these coatings can be as high as 1800 HV, and the porosity can be as low as several percent. The fact that these coatings are applied like paints, by brushing or spraying, allows their use on internal diameters and on complex shapes. Ferrous metals are suitable substrates. The ability to coat small internal diameters as small as 0.030 in. (0.75 mm) gives this process an advantage over thermal spray coatings. Thermal spray torches can be used to coat ceramics on the inside of diameters as small as 1 in. (50 mm); diameters that the torch cannot fit into can only be coated to depth equal to the internal diameter dimension.

In summary, thermal/chemical deposited ceramic coatings look like chromium oxide plasma coatings after they are applied, and their use properties are similar. They have the advantage over thermal spray coatings in that they can be applied to the inside of small holes and to ''unsprayable'' geometries. They can be obtained from a number of licensees around the United States.

CENTRIFUGAL-CAST WEAR COATINGS

A significant application of hardfacing is coating of flights on extruder screws. These screws are used to extrude plastics and other substances. They are subject to abrasion from inorganic fillers in the material that is being extruded, and they are subject to metal-to-metal wear when the screw rotates in the extruder barrel without hydrodynamic support. The usual counterface for hardfaced extruder screws is a barrel lined with a hardfacing type of material that is applied by centrifugal casting. Extruder barrels can have internal diameters as large as 16 in. (40 cm) and their length can be as much as 40 ft (12 m). Coating the inside diameter of these barrels is outside the present limits of fusion welding technology. The technique that is most used is centrifugal casting. In this process, the bore of the barrel is charged with a wear-resistant alloy in pellet form; the ends of the barrel are capped, and the barrel is heated to the melting point of the alloy while rotating in a furnace or induction coil. The alloy charge melts and, when cooled, the barrel is lined with a hard layer with a thickness of hardfacing of up to 0.10 in. (2.5 mm). The cast lining is finished to the required dimension by honing techniques. These coatings are used on other components besides extruder barrels; they can be applied to any cylindrical part that can be rotated and take the process temperatures. Some other parts that can be coated with this technique are pump cylinders, compounder barrels, valve bores, slurry piping, and the like.

Most steels are suitable substrates for centrifugal-cast cylinder linings. Extrusion barrels are subjected to pressures as high as 20,000 psi (137.9 MPa), and for this reason the usual substrate material is 4140 steel or a similar alloy steel. Several main classes of lining alloys are available:

1. White irons (alloy irons with a hardness of about 60 HRC).
2. Nickel/chromium/boron alloys (these materials are similar to the nickel-base hardfacing consumables; hardness in the range of 35 to 60 HRC).
3. Cobalt-base alloys (similar to the cobalt-base hardfacing consumables; hardness in the range of 45 to 60 HRC).
4. Composite materials (nickel or similar matrixes containing up to 50 percent carbides; see Figure 8–12).

All these alloys form a metallurgical bond to the substrate. The bond to the substrate is enhanced by boron additions; the boron makes these materials self-fluxing. The white irons have abrasion resistance similar to the iron/chromium hardfacing consumables, and the cobalt- and nickel-base alloys have wear properties similar to the hardfacing alloys that have comparable composition and hardness. The composite alloys are somewhat different than the composite welding consumables in that carbide additions must be designed so that they do not concentrate at the lining substrate interface during centrifugal casting. Since tungsten carbides are much heavier than the matrix metals, the centrifugal force of the casting alloys can make the carbides concentrate at the lining/substrate interface instead of at the ID of the cylinder, where they are needed for wear resistance. This problem is sometimes overcome by using mixed carbides; titanium carbides are

Figure 8–12 Cross section of a composite cylinder lining showing carbide size and distribution (25×).

lighter than the matrix metal and will not migrate to the lining interface. Some lining suppliers use mixtures of WC, VC, and TiC to get a controlled distribution of carbide throughout the lining. It is generally agreed that the composite linings will have at least twice the abrasion resistance as the noncarbide lining alloys.

Centrifugal-cast cylinder linings are useful materials for extrusion equipment and any other equipment that require wear resistance on the inside diameter of a cylinder. They usually have better wear and corrosion resistance than nitrided or carburized cylinders, and with a thickness in the range of 0.040 to 0.10 in. (1 to 2.5 mm), they can tolerate more wear than many of the diffusion hardened surfaces. These linings are available from a number of suppliers in the United States and in Europe, and lined cylinders can, in some cases, even be purchased off the shelf. Thus centrifugal-cast cylinder linings are definite candidates for cylindrical parts that are subject to wear.

WEAR SLEEVES

A favorite technique of maintenance personnel for repair of worn shafts is to sleeve the damaged area. In fact, this is often done in preference to hardfacing. This preference often arises from the fact that most machine repair people operate out of a machine shop where they have the facilities to make a sleeve, but may not have immediate access to a welder who is familiar with hardfacing or the machine repair person is not familiar with hardfacing. In any case, the use of sleeves competes with welding hardfacing, thermal spray buildups, and plating.

There is nothing wrong in concept with sleeving worn shafts, except that most repair mechanics sleeve worn shafts with low-carbon steels or other machinable materials. If a seal or bushing wore a shaft in the first place, it follows that the sleeve-repaired shaft will wear in the same fashion if the sleeve is not made from a hardenable material. The second problem that exists in using sleeves is putting the sleeve on with the proper interference. There is a rule of thumb used by some repair mechanics for interference fits: 0.001 in./inch (1 mm/m) of diameter. This sometimes works, but the proper use of sleeves requires calculations of the proper interference that determines the stress tending to split the sleeve, the stress tending to deform the shaft or tube on which the sleeve is shrunk, and the clamping pressure of the sleeve on the shaft. Figure 8–13 presents some simple formulas for making these calculations. Parts should never be sleeved without going through these calculations. The allowable stress on the sleeve and the shaft should be the yield strength with some safety factor applied (possibly 2).

These calculations will solve the problem of proper interference fit, but the problem still remains: if the sleeve is not more wear resistant than the original shaft material, it will wear out again. Case-hardened or through-hardened materials can be used for the sleeve, but this presents the problem of size change during hardening. To get the proper interference between the shaft and sleeve after hardening, a costly grinding operation may be required. One way to bypass this problem is to use purchased wear sleeves that are available in a wide variety of sizes for use as wear sleeves. Some vendors of these sleeves make them from cast hard alloys such as cobalt/chromium or nickel/

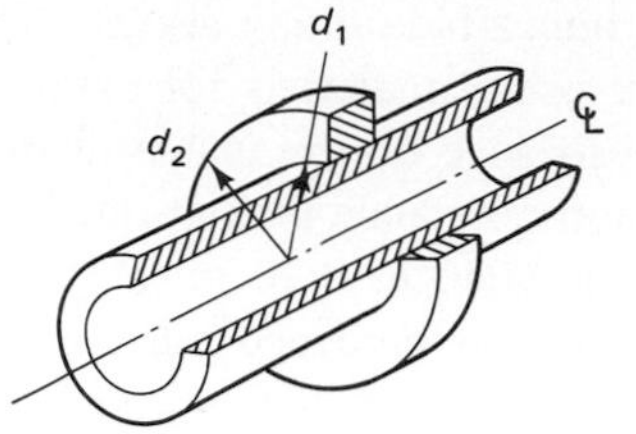

d_1

d_2

℄

Stress at d_1. $\sigma_1 = \frac{E\epsilon}{d_1}\left(\frac{d_2^2 + d_1^2}{2d_2^2}\right)$

Stress at d_2. $\sigma_2 = \frac{E\epsilon d_1}{d_2^2}$

Clamping pressure at d_1. $P' = \frac{E\epsilon}{d_1}\left(\frac{d_2^2 - d_1^2}{2d_2^2}\right)$

where E = modulus of elasticity (assuming shaft and sleeve have the same modulus)

ϵ = diametrical interference

Figure 8–13 Formulas for calculation of interference fits on wear sleeves.

chromium/boron. Some manufacturers of wear sleeves chromium plate about 1 mil (25 μm) thick the sleeve outside diameter for wear resistance. The use of these types of sleeves may prevent repetition of the original wear failure. One manufacturer of wear sleeves makes the wall thickness less than 0.030 in. (0.75 mm) so that these sleeves can be applied on shafts worn by compliant seals without machining of the shaft or sleeve. Compliant seals such as lip seals can usually tolerate the increase in shaft diameter of less than 0.060 in. (1.5 mm) caused by the wear sleeve.

The effectiveness of chromium-plated or hard metal wear sleeves in preventing recurring failures on shafts subject to local wear depends on the application, but a thick, fully hardened wear sleeve will be effective for many types of wear; the chromium-coated sleeves will probably be quite effective for preventing lip seal wear. Shafts that are worn from inorganic fillers in rope-type packing may require through-hardened sleeves or sleeves with a substantial thickness of case hardening or chromium. In any case, the use of sleeves to repair worn shafts competes with welding hardfacing, and the use of sleeves should be weighed using some of the considerations that we have mentioned.

WEAR PLATES

If large areas need resistance to abrasion or metal-to-metal wear, a wide variety of plates are available made from wear-resistant metals or with wear-resistant surfaces. Figure 8–14 is an illustration of some common configurations. Cast wear plates are available made from cobalt/chromium wear-resistant alloys and from white-iron types of alloys. These plates usually come with countersunk holes for mounting with mechanical fasteners. For very large surfaces, plates are available that are hardfaced by submerged arc or bulk welding techniques. These often come with holes that allow fastening with plug welding. Hardfaced sectors are also available for lining of pipes. These can be hardfaced with any alloy, but a significant application for these types of plates is for coal-fired boilers where elevated temperatures require hardfacings with hot hardness. Pipes lined with these plates are used to convey coal to the boilers. They must resist solid-particle erosion at high temperature. They are usually fitted such they can be held in place with simple tack welding or by their geometry.

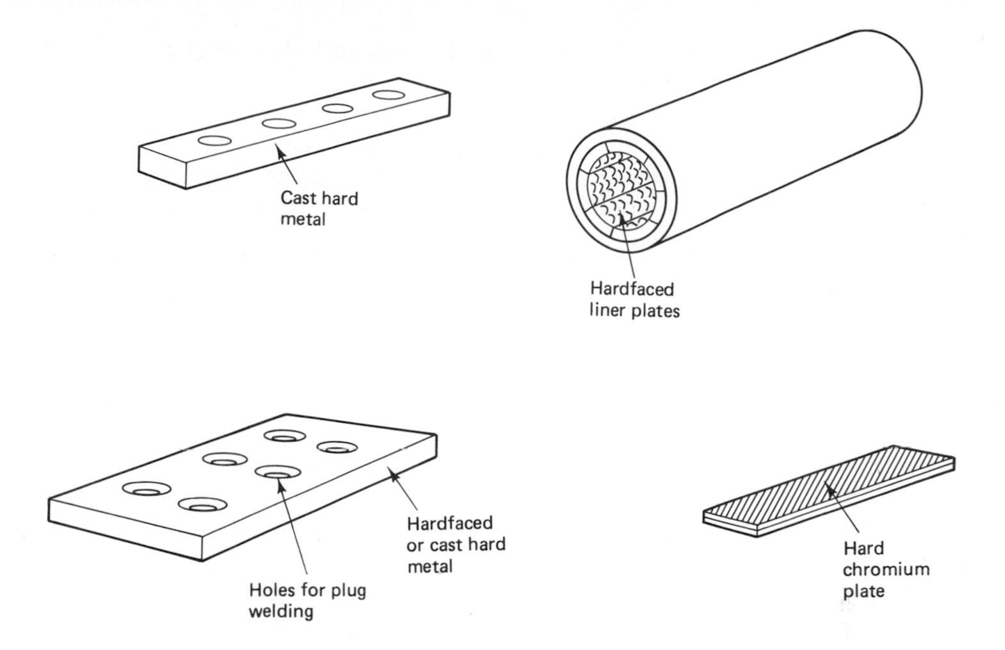

Figure 8–14 Available wear plate configurations.

Chromium-plated sheet metal is available for wear-resistant surfaces on machine ways and the like. These wear "plates" can be epoxy bonded to the surfaces to be protected.

Finally, there are wear-resistant plates that have acceptable weldability. These can be applied to surfaces by fillet welding or any other form of welding. Weldable wear plate materials include wrought Stellites®, manganese steels, and high-strength, low-alloy steels.

Two grades of Stellite are available in sheet and plate form, alloy 6B and alloy 6K. The former has a hardness of about 43 HRC; the latter has a hardness as high as 48 HRC. Manganese steels are available in several compositions in wear plates and in cast shapes that can be welded to earth-moving equipment and mining equipment. The high-strength, low-alloy plates are available in thicknesses from ¼ to several inches (6 to 100 mm). Most of these types of materials have compositions that are adjusted to achieve a particular strength and hardness. Carbon contents are kept as low as possible so that weldability is maintained. These plates are prehardened at the mill, and the hardness can range from about 36 HRC to as high as the low forties on the C scale. These types of materials are more abuse resistant than they are wear resistant. Their hardness and low carbon content make them less suitable for scratching abrasion applications than hardfacing alloys. Their high strength (yield strengths can be as high as 150 ksi or 1034 MPa) makes them suitable for applications where they will be pounded but not subject to extensive scratching abrasion. For example, these types of steels are used for bodies on off-road trucks. These truck bodies have to resist deformation from falling rocks. They lose material due to scratching abrasion, but this is a lesser factor than resisting deformation.

The role of wear plates in the spectrum for wear-resistant surfaces is that they should be considered when very large areas have to be protected from wear. It is usually too costly to hardface truck bodies and the inside of coal hoppers by conventional welding techniques. Cladding with wear plates can be the cost-effective way to protect this type of equipment.

SUMMARY

In this chapter, we covered a wide range of processes and products for wear reduction. Others methods could be put into this category, but the ones discussed here have been in use for many years, and it is assumed that this is testimony that they work. Unfortunately, it was not possible to obtain quantitative wear data comparing these products with hardfacings and other surface treatments. For this reason, it is recommended that potential users take steps to confirm the applicability of these products to a particular wear system. Keep in mind that a wear plate or wear sleeve does not resist all forms of wear. Each product discussed is really directed at reducing a particular form of wear; some may resist scratching abrasion; some will resist gouging; some will resist liquid erosion. In general, these products should be considered when hardfacing, plating, and hardening heat treatments are inappropriate.

REFERENCES

Abresist Handbook. Urbana, Ind.: Abresist Corp., 1973.

The Bimetallic Cylinder, An Introduction for Plastic Processors, Bulletin BC 721. New Brunswick, N.J.: Xaloy Corp., 1972.

Formaflex Carbide Coatings. Cleveland, Ohio: Imperial Clevite Corp., 1973.

K-ramic Ceramic. Colorado Springs, Colo.: Kaman Sciences, Inc., 1980.

A Plastic Processors Guide to Bimetal Barrels and Screws (Wales, Wis.: Bimex Corp., 1984.

SHELDON, G. L., and R. N. JOHNSON. Electrospark Deposition, in *Wear of Materials 1985*. New York: ASME, 1985, pp. 388–396.

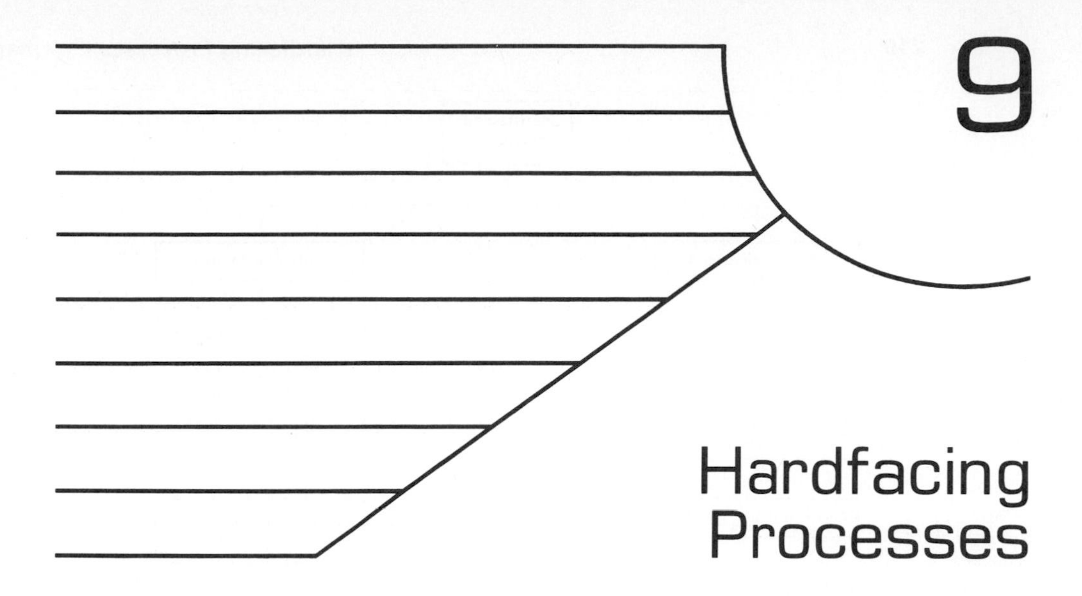

9 Hardfacing Processes

Hardfacing alloys can be applied with any welding process, but the most important commercially available processes are shown in Figure 9–1. This illustration categorizes hardfacing processes into fusion and nonfusion groups for an important reason. Fusion processes have a metallurgical bond, and the others do not. From the user's standpoint, this means that alloys that are fused have a bond strength at least equivalent to the strength of the weaker of the fused metals. The nonfusion processes produce a deposit that is held on with the same kinds of forces that hold on electrodeposits such as chromium plating.

A qualifying statement is necessary regarding fusion deposits; the strong metallurgical bond can only be achieved if the hardfacing and the substrate have the ability to form a solid solution. Weldability considerations cannot be violated. As an example, a fusion deposit of any iron-base hardfacing on titanium will crack and come off. This is because titanium forms a brittle intermetallic compound when it is fused with all but a few other metals. This will be discussed later in more detail, but the user should keep in mind that fusion processes cannot be used to apply any hardfacing to any substrate.

Since nonfusion processes do not require the alloying of the facing and the substrate at the bond line, they can be used to apply almost any alloy on almost any substrate that will stand the temperature of the spray. Ferrous metals can be sprayed on aluminum. Ceramics can be sprayed on copper substrates. Thermal spray coatings have been successfully applied to thermosetting plastics. This is the biggest advantage of the nonfusion processes. The sacrifice in bond strength can be substantial. A good bond of a thermal spray coating would be 10,000 psi (69 MPa), a strength equivalent to a very good epoxy. These mechanical bonds are not suitable for applications that involve operating stress levels that would be high enough to cause deformation of the nonfusion deposit.

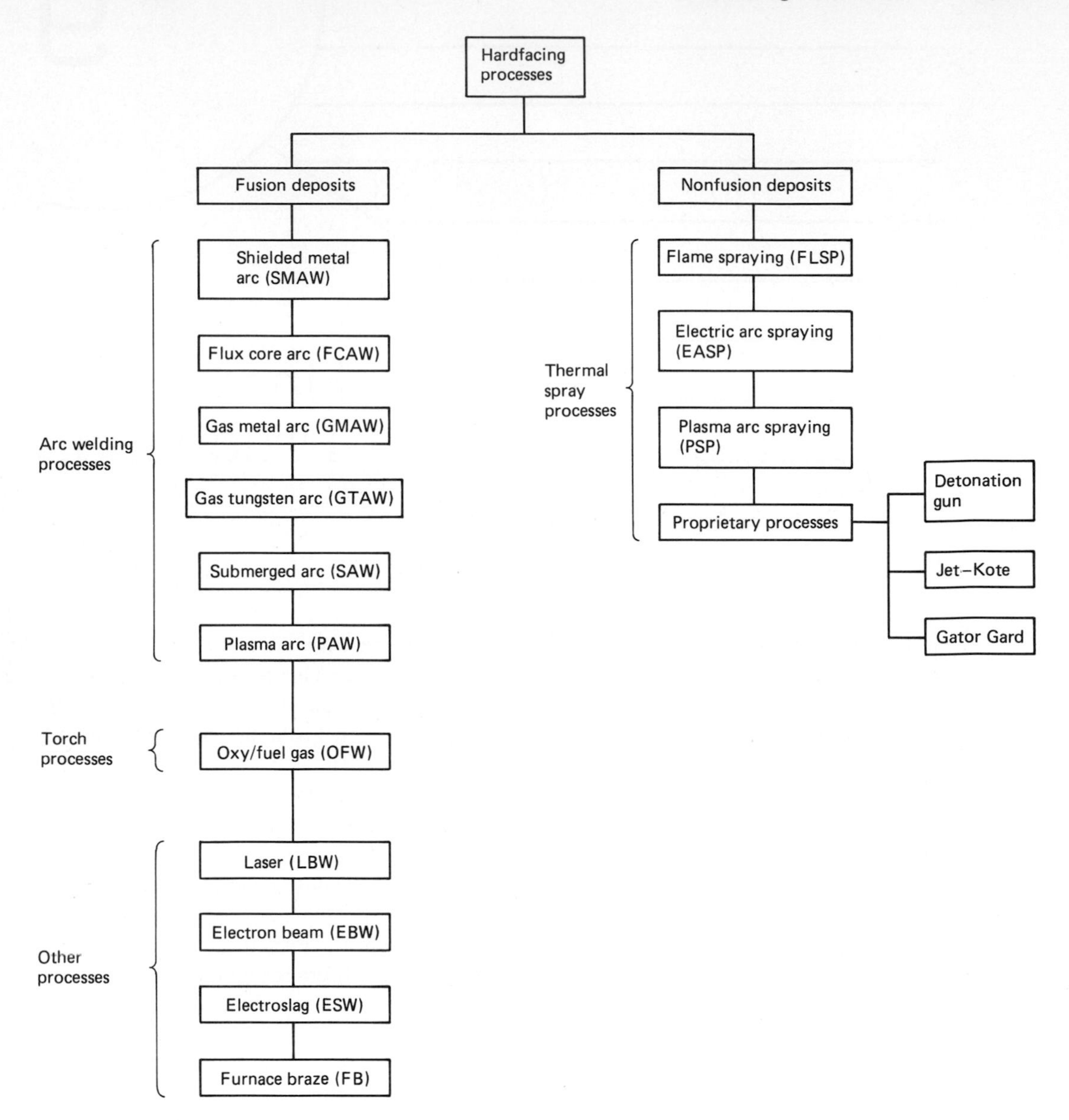

Figure 9–1 Spectrum of commercially available hardfacing processes.

Figure 9–2 illustrates the bond line condition that exists with the various hardfacing processes. Arc welding processes usually produce significant alloying of the surfacing and the substrate. Since the first layer is contaminated or diluted with alloy from the substrate, it is customary to apply at least two layers of hardfacing. Only the second layer will have the intended wear properties of the surfacing. Oxy/fuel surfacing processes

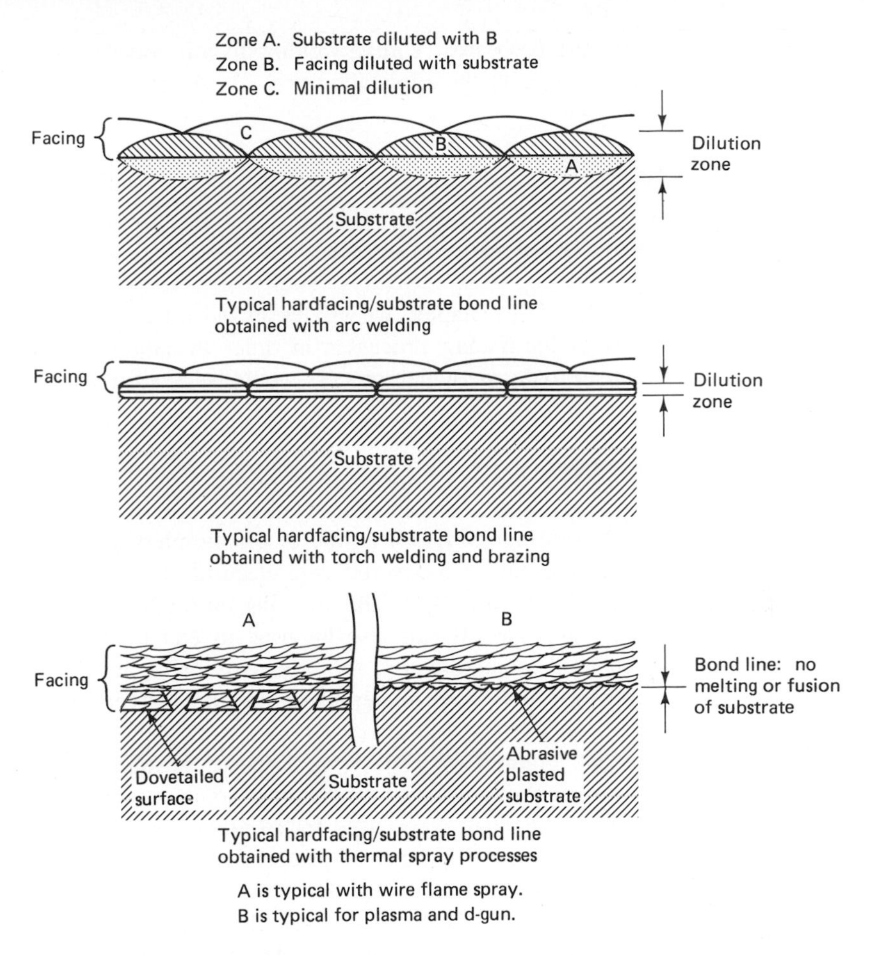

Figure 9–2 Facing/substrate bonding with various hardfacing processes.

produce less melting and thus less dilution of the surfacing with the substrate. One-layer deposits are possible with careful application by the welder. Some hardfacing alloys are even designed to be used as one-layer deposits. The illustration of a nonfusion bond line shows a lamellar type of deposit. The thermal spray hardfacing processes all involve generation of molten droplets of the material to be deposited. These droplets ''splat cool'' when they hit the surface, and the deposit contains possible oxides from reaction with the propelling gases, as well as possible porosity from the lack of flow between ''splats.'' Bond strength can usually be enhanced by roughening or grooving the substrate. Abrasive blasting of the substrate is the usual preparation for plasma spray processes. A grooving and dovetail preparation is common in wire-sprayed round shapes.

With this discussion of the bonding differences between fusion and nonfusion processes behind us, we will move on to the main agenda of the chapter, the description of available hardfacing processes. Unlike welding to join metals, the use of hardfacing requires that the user specify the application process. A proper hardfacing specification includes the following:

1. Application process
2. Alloy
3. Thickness and the areas to be surfaced

In the remainder of this chapter, we will briefly describe the differences, advantages, and disadvantages of hardfacing processes in order to supply enough information to allow a potential user to decide which process would be best for an intended application.

SHIELDED METAL ARC WELDING

Shielded metal arc welding (SMAW) is one of the simplest arc welding processes. A weld is achieved by melting of a covered wire electrode in an arc formed between the work and the electrode (Figure 9–3). The covering on the electrode decomposes in the arc to form a shielding gas. It can also be used to add alloy and to promote weld cleanliness. When used for hardfacing, the process remains the same. The electrode is simply made from an alloy, core material, or coating that produces a wear-resistant deposit.

The equipment required for SMAW is a source of high current at low voltage. Most SMAW power supplies are motor generators or transformers/rectifiers with a current control. The process is manual; when used for surfacing, the welder covers the area to be hardfaced with the required number of passes to produce the specified thickness of deposit. There is no thickness limit except that some alloys will show excessive cracking when applied in over two layers. In such cases, the welder usually "butters" the area to be hardfaced with enough layers of lower alloy buildup material so that only a few passes of the hardfacing are required to meet the specification. This buildup process is often used with the other hardfacing processes we will be discussing.

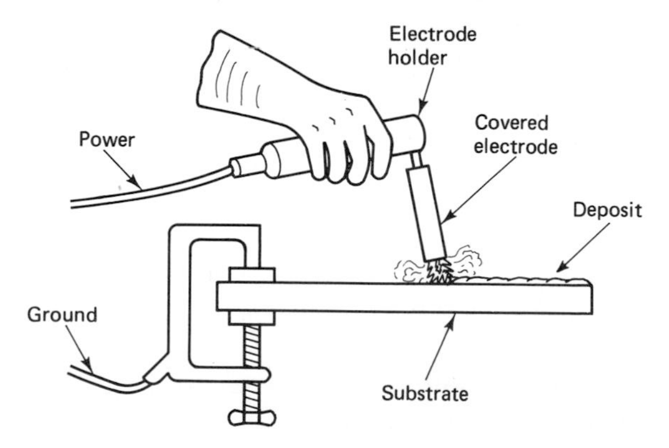

Figure 9–3 Shielded metal arc welding (SMAW).

One big advantage of SMAW is that the equipment is readily available in all welding shops. Another is that hardfacing consumables can be purchased in small quantities (5 lb; 2.2 kg), and it is not much trouble to keep a supply of hardfacing alloys on hand. Another advantage of SMAW is that deposits of many alloys can be applied out of position. As we will see when we discuss some of the other processes, some require flat position welding. SMAW is applicable to ferrous hardfacing alloys, as well as to cobalt, nickel, and composite tubular electrodes. The biggest disadvantage of the process is that the deposition rate is low; it is up to the ability of the welder. Typical deposition rates are less than 5 lb (2.2 kg) per hour. The SMAW process is most suitable for small deposits or for field welding where the portability of the equipment is important.

GAS TUNGSTEN ARC WELDING

Gas tungsten arc welding (GTAW) is significantly different from most of the other hardfacing arc welding processes in that the arc that causes melting of filler metal is between the work and a nonconsumable electrode. The torch contains a tungsten electrode that does not melt in the arc. Filler metal is applied by manually dipping a bare rod into the arc area or by some ancillary device that introduces bare wire into the arc region. The process in manual mode is illustrated in Figure 9–4. The arc area is gas shielded by a gas flow from the torch.

GTAW welding equipment is widely available in weld shops. It requires a power supply not unlike that required by SMAW, in addition to shielding gas equipment that includes the tank of gas, a pressure regulator, and a gas flow control. In manual operation, hardfacing alloys are obtained in bare rods, and they are melted and flowed on the substrate to form the deposit. In automatic operation, this process usually requires custom alloy feed devices and work manipulation devices.

Almost any metallic surfacing can be applied with GTAW. The main disadvantage of the process is that it is slower than almost all other processes when used manually.

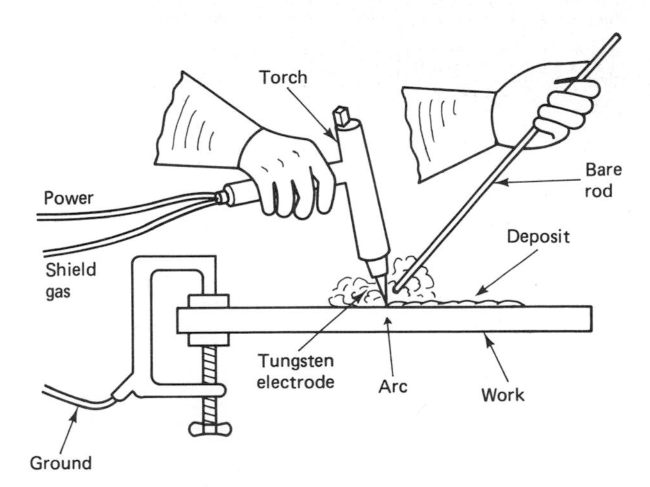

Figure 9–4 Gas tungsten arc welding (GTAW).

In automatic operation, the deposition rate will depend on the setup, but it is not considered to be a process for large deposits and is usually most suitable for shop welding.

GTAW is best suited to surfacing of small parts such as tools and dies. It produces a very neat deposit, an advantage on finish-machined parts (die repair and the like). It is a likely candidate for automatic equipment designed to do production surfacing on small parts.

GAS METAL ARC WELDING

Gas metal arc welding (GMAW) uses a continuous filler wire fed through a gun. The arc for fusion of the filler metal is established between the wire and the work. Welding can continue until the coil of wire is exhausted. Shielding gas to protect the molten metal is fed through an annulus around the exiting wire in the tip of the torch (Figure 9–5). Deposition of the filler metal can be accomplished by several mechanisms: the wire can short against the work or the wire will melt off without touching the work. The former process is called short-circuiting transfer or short arc, and the latter is called globular transfer. A third variation is called spray transfer. From the hardfacing standpoint, the transfer mechanism can affect dilution and bead profile, but the occasional user of hardfacing can rely on the welder's judgment on the choice of transfer options. Short-circuit transfer is the preferred option for out-of-position welding.

Gas metal arc welding equipment is moderate in cost; a power supply, wire feed, and gas control are required. The GMAW gun can be hand held, but in automatic operation it is usually manipulated, as is the work. Multiple wire feed can be used for very large surfacing jobs. A recent variation of this basic process uses filler metal in

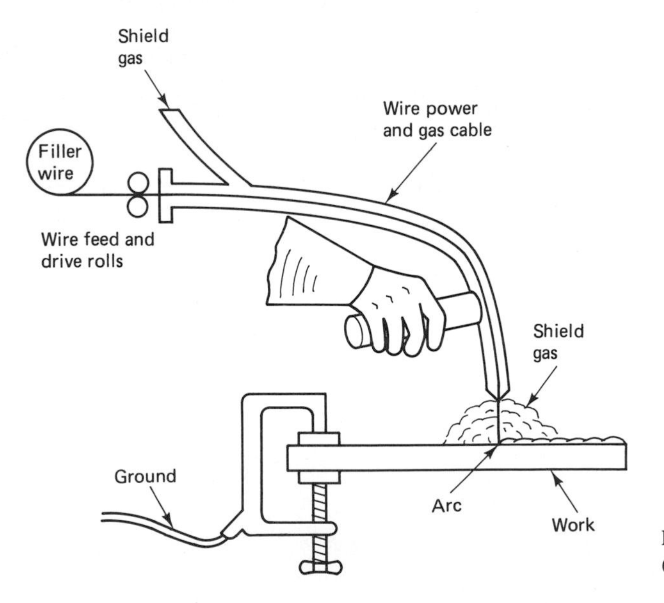

Figure 9–5 Gas metal arc welding (GMAW).

the form of metal strips that may be several inches wide to achieve extremely high deposition rates. In this form, the process can be used, for example, to clad the inside of a very large tank or to clad the surface of a very large roll. Many alloys are available in spooled wire form for hardfacing with GMAW (ferrous, nickel base, copper base, cobalt base, etc.). Some hardfacing consumables are applied with this process with the shielding gas turned off (open arc).

GMAW is a suitable process for field welding of hardfacings on large equiment, and it is also suitable for large shop welding jobs. Since the equipment is bulky, it is not usually used for small jobs or for hard-to-get-at areas. Its limitation on small jobs is also due to the high costs of filler metal when it must be purchased as a 25- or 50-lb (10- or 20-kg) spool. If only a few inches of hardfacing are required, the purchase of a 25-lb (10-kg) coil of wire can be a significant extra expense.

FLUX CORED ARC WELDING

Flux cored arc welding (FCAW) (see Figure 9–6) has only been in wide use since about 1975. Its distinguishing characteristic is the use of filler metal that is made by rolling thin sheet metal into a tube. The tube is filled with a flux that contains materials that volatilize to provide protective shielding for the weld pool (self-shielding). Metal powders can also be added to the core material to combine in the weld pool to create an alloy weld deposit. In operation, the coiled tubular wire is fed through a gun, and an electric arc is established between the wire and the substrate. Gas shielding may also be used to supplement the shielding effect of the flux.

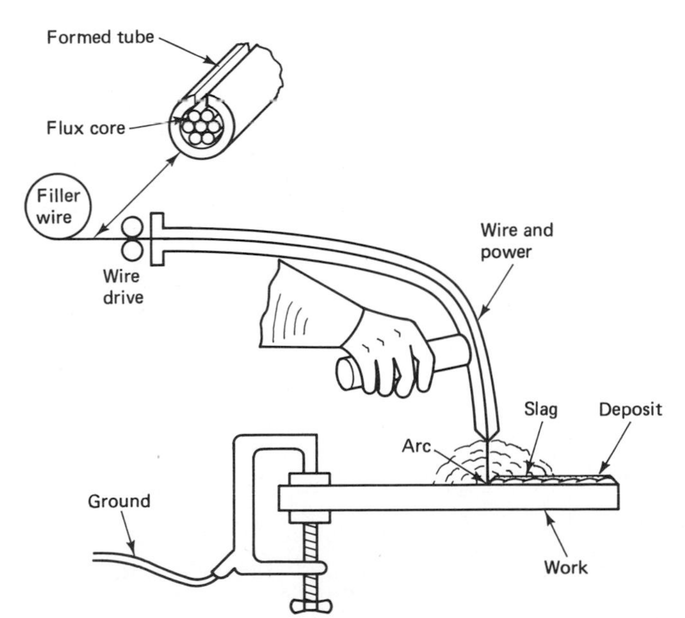

Figure 9–6 Flux cored arc welding (FCAW).

FCAW can be used semiautomatically, as well as in automatic modes. The equipment requirements are a power supply, wire feeding equipment, and shield gas equipment (if used). In conventional welding, the flux core process is used to provide deeper penetration on heavy material than is available from the forerunner of this process, gas metal arc welding. When used for surfacing, it has an advantage over other processes in that deposition rates can be very high. The biggest disadvantage of this process for hardfacing is that many hardfacing alloys are unavailable in the form of flux cored tubing. In 1986, most FCAW filler wires were ferrous based.

FCAW welding is a less important candidate process for the occasional hardfacing user. It is best suited for very large shop-type surfacing jobs on ferrous systems.

SUBMERGED ARC WELDING

Submerged arc welding (SAW) is a high deposition rate process using a consumable wire and a flux applied by ancillary equipment. The arc is between the work and one or more spool-fed filler wires. During welding, the arc is covered with a heavy blanket of a granular flux; this is the origin of the "submerged" term. The flux is fed from a hopper that follows the welding head. Sometimes additives in the flux intentionally become part of the alloy deposit.

Submerged arc welding (Figure 9–7) requires equipment similar to GMAW, with the exception that the shield gas equipment is replaced with the flux hopper. The flux is usually gravity fed, with the work in the flat position. Out-of-position welding requires dams to retain the flux. The bulk of the equipment makes this process most suitable for automatic shop welding. SAW typically produces very sound, wide overlays at deposition rates that can easily reach 50 lb (25 kg) per hour. There is usually no need to remove slag by chipping; the fluxes are usually formulated so that the slag lifts off

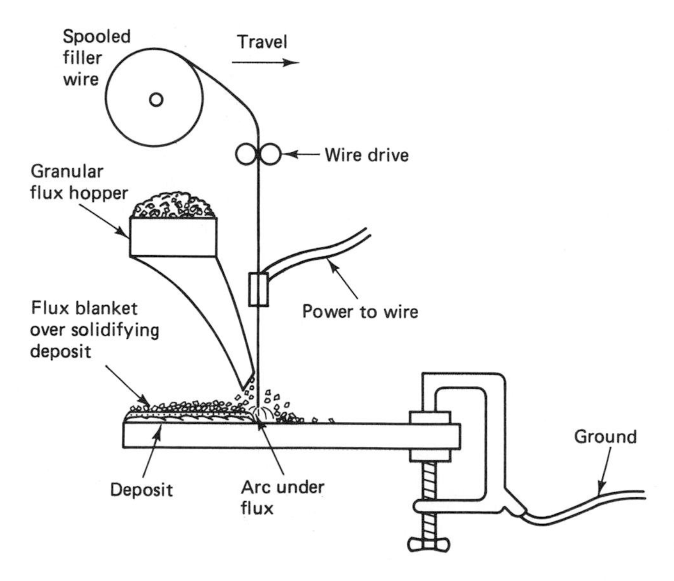

Figure 9–7 Submerged arc welding (SAW).

by itself. The surfacing alloys applicable to this process are any of those that are available in spooled wire. However, it is most used on ferrous systems.

This process is best suited for heavy surfacing of large tanks, plates, rails, and the like—anything that can be positioned for flat position welding.

PLASMA ARC WELDING

A plasma is essentially an ionized gas. In plasma arc welding (PAW), an inert gas is flowed through a constricted electric arc in the welding torch to form a plasma. The plasma ''flame'' is used to melt filler metals for welding or hardfacing. The plasma exiting the torch can have a temperature in the range of 30 to 50,000° Kelvin (50° to 90,000°F). Thus a plasma welding torch can have extraordinary penetrating capability. If the plasma flame is obtained by an arc between the plasma gun and the work, the process is called transferred arc plasma (PTA). If the arc is maintained within the welding torch between a tungsten electrode and a water-cooled nozzle, the process is called nontransferred arc plasma.

In concept, the PAW process is a modification of the GTAW process. The arc is between a nonconsumable tungsten electrode and the work or between the electrode and a copper annulus within the welding torch. The formation of a plasma by passing inert gas through a constricted electric arc creates a condition that produces a much hotter arc or flame than can be achieved in GTAW. Unfortunately, plasma welding equipment can be quite complicated and expensive. The basic equipment required is a power supply for the arc, a plasma gas supply and controls, shielding gas and controls, a water-cooling system for the torch, and controls to coordinate all these systems (Figure 9–8). In its simplest form, small plasma arc welders are available for manual application

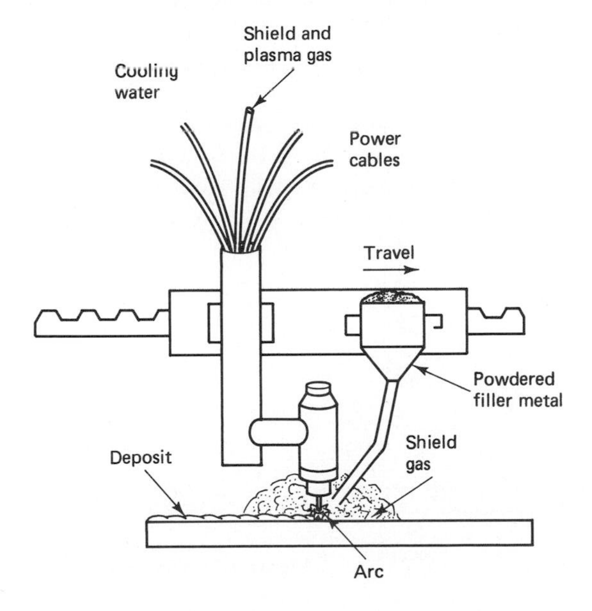

Figure 9–8 Plasma arc welding (PAW).

of hardfacings to tools and the like. More sophisticated welding machines are usually set up for automatic shop surfacing of production parts. Filler metal addition is accomplished by spool feeding of bare wire or powder into the arc zone. The transferred or nontransferred arc option can be used as desired to obtain different degrees of dilution.

The advantage of plasma arc welding for surfacing is speed that is usually better than automatic GTAW with smooth, accurate weld profiles. The major disadvantage of this process is the high equipment costs. In general, this process is best suited for special production hardfacing of parts in automatic equipment. PAW surfaced parts usually require minimal machining.

OXYACETYLENE WELDING

Oxyacetylene welding (OAW) (Figure 9–9) is the oldest and simplest hardfacing process. The hardfacing deposit is made by simply heating the substrate with the torch so that the surface is glassy, on the verge of melting. The flame is then directed at the bare filler rod to get the hardfacing to melt. The droplets are heated until they wet the substrate and form a continuous deposit.

The only equipment required for oxyacetylene welding is a torch, tanks of fuel gases, pressure regulators for the gas, and some bare filler metal. Some hardfacing alloys require flux similar to brazing flux. This process is most widely used for cobalt, nickel, and copper base hardfacings. There are other fuel gases other than acetylene for torch brazing and the like, but in general it is recommended that only acetylene be used for hardfacing.

Oxyacetylene welding is probably the slowest hardfacing process, but is also the best process for minimum alloy dilution. Thus OAW hardfacing is still used for critical applications where dilution cannot be tolerated. Because of the slowness of the deposition rate (2 to 5 lb or 0.9 to 2.2 kg per hour), OAW is usually used only for small jobs or for field welding where a torch is the only equipment available.

FURNACE FUSING

Some hardfacing alloys are available as a paste or a metal cloth that can be applied to a surface and furnace fused to form a hardfacing deposit. All these materials are proprietary

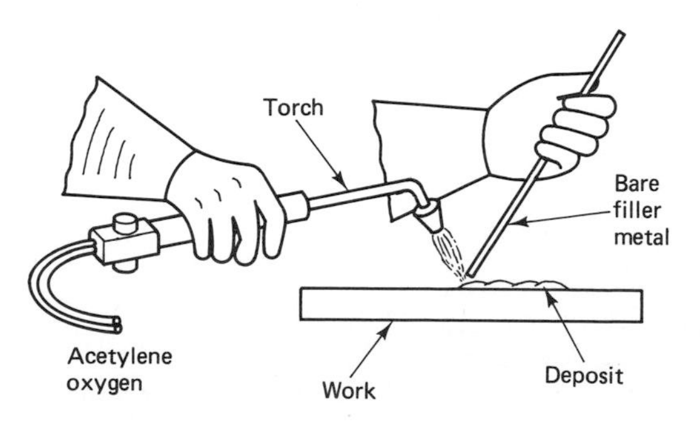

Figure 9–9 Oxyacetylene welding (OAW).

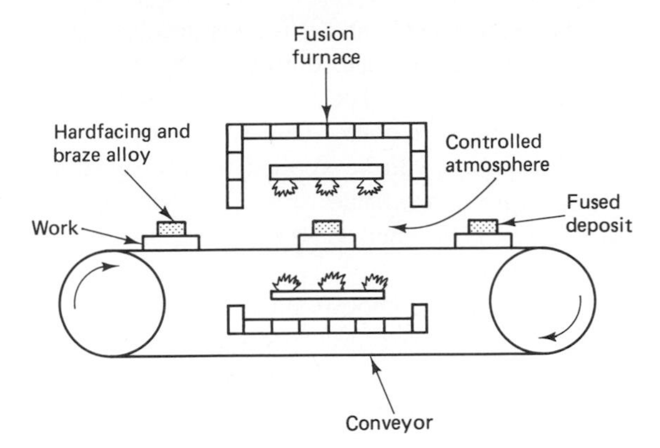

Figure 9–10 Furnace fusing of hardfacings and brazed-on surfacings.

in nature, but they are readily available. A typical application is illustrated in Figure 9.10. The surfacing material is simply applied to the substrate and fused in a furnace at sufficient temperature to cause melting. Most of these types of materials have fusing temperatures in the range of 1600° to 2100°F (870° to 1150°C). These surfacings are usually composites (e.g., hard compounds such as tungsten carbide that are held in a low melting binder such as a brazing alloy). The brazing alloy forms the matrix for the hardfacing and it also produces the bond to the substrate. These coatings can be as thick as 0.08 in. (2 mm) and they can be applied to a variety of substrates, but the normal substrate is a ferrous material.

Furnace-fused coatings are usually used for production-type parts. It can be costly to develop the necessary furnace controls, part fixtures, and surface preparation procedures if only a few parts are required.

THERMAL SPRAY PROCESSES

In our introductory remarks in this chapter, hardfacing processes were broken into two major categories, fusion and nonfusion. We put many processes like flame spraying and plasma spraying into the nonfusion category because most deposits applied by these processes are not melted on the substrate to allow coalescence between the hardfacing and the substrate—a metallurgical bond. This is not quite correct in that hardfacing alloys can be applied by a nonfusion spray process and subsequently melted to form a fusion deposit on the substrate. This secondary fusion operation is most commonly done with a simple oxyacetylene torch, but GTAW, plasma, or other heat sources could be used.

Thermal spraying is the generic category of welding processes that includes all the processes that apply a consumable in the form of a spray of finely divided molten or semimolten droplets to produce a coating. The melting of the consumable can be accomplished in a number of ways, and the consumable can be introduced into the welding torch or heat source as powder, rod, or wire. Thermal spray consumables can be metals, ceramics, cermets, and even plastics. Any material can be sprayed as long as it melts or becomes plastic in the heating cycle and if it does not degrade in heating. Because the deposit does not fuse with the substrate during the spraying operation, it is possible to ignore metallurgical compatibility. The coating does not have to form a solid solution with the substrate to achieve a bond. This is an extremely significant feature of thermal spray processes; coatings can be applied to any substrate that will not degrade from the heat of impinging consumable droplets. Substrate temperatures seldom exceed a few hundred degrees Fahrenheit (150°C). Hard metals or ceramics can even be applied to thermosetting plastics.

The spraying action can be produced by combustion gases or by using compressed gas or air to achieve atomization. There are two basic ways of achieving the heat for the melting of the consumable material in the spraying operation, gas combustion or an electric arc. Figure 9–11 shows how the commonly used thermal spray processes fit into these two categories. When the deposit is to be left unfused, the bond to the substrate is assisted by roughening the surface to achieve mechanical locking action. Figure 9–12 is an illustration of some of the techniques employed. Grooving is no longer necessary for most thermal spray processes, but grooving or rough turning are options suitable for cylindrical parts. Only two "joints" are employed in thermal spray hardfacing processes; either the coating is applied in an undercut or on the substrate surface. Undercutting is the preferred technique because it minimizes the chance of chipping the coating on the edges in use or handling. The depth of the undercut determines the finished deposit thickness. Repair jobs may require undercuts as deep as 0.050 in. (1.25 mm) to remove scoring, but on new work the deposit thickness and undercut depth are usually in the range of 0.003 to 0.010 in. (50 to 250 μm), unless the part can tolerate more wear than this in service. Thermal spray specialists often employ bond coats of special materials that assist the mechanical bond of the hardfacing deposit. Where these bond coatings can be tolerated, the undercutting is made deeper to allow for the thickness of the bond coat, which may be only 0.002 to 0.005 in. (50 to 125 μm). If a coating is applied to a substrate without undercutting, the surface still requires roughening (usually abrasive blasting) even if a bond coating is used.

In summary, thermal spray processs are the welding processes that apply a consum-

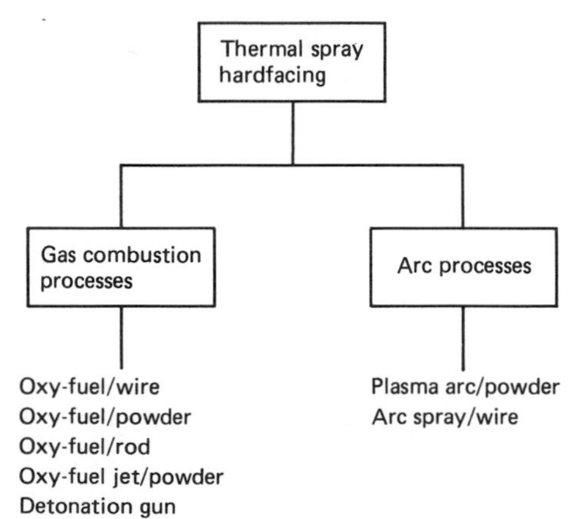

Figure 9–11 Thermal spray processes.

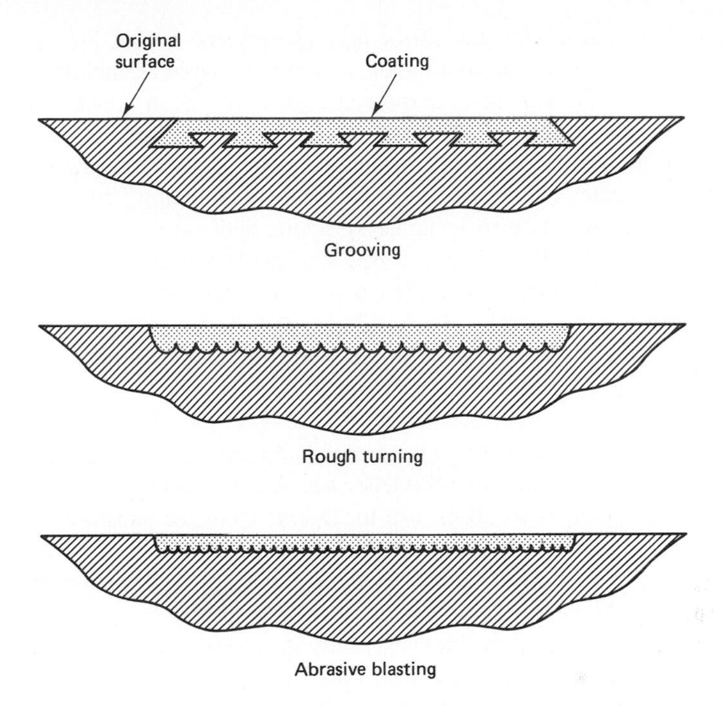

Figure 9–12 Surface preparation techniques for thermal spray coating.

able in an atomized form to form a deposit on the surface of a substrate or in a machined undercut. Almost any material can be deposited on almost any substrate. There are about six basic processes for applying thermal spray deposits, and many different pieces of equipment are available in each process category. In the remainder of this chapter, we will describe the most widely used thermal spray processes and types of equipment.

Flame Spraying

Wire Processes As shown in Figure 9–11, five processes fall into our category of gas combusion processes. The accepted term for all these processes is flame spraying (FLSP), but there are some problems with this term in that some special processes such as detonation gun and jet spraying are very much different in nature, and heating of the consumable is not done by the action of a conventional "flame." In the detonation gun process, the consumable powder is heated by ignition and detonation of an explosive gas mixture. In the jet process, the heat source is combustion of fuel gases within a chamber with an action that is not unlike the action of burning propellents in a rocket engine. Thus, in Figure 9–11 we opted to classify these processes as gas combustion processes; all these processes involve combustion of fuel gas in some manner.

All thermal spray processes are offshoots from the "metallizing" process that was invented in 1910. In the original process, wire or powder was introduced in a controlled manner into the tip of an oxyacetylene torch. The consumable melted into droplets that formed a coating on a substrate without melting of the substrate. This original process evolved over the years to a process in which a wire from a reel is fed by drive rollers into an oxyacetylene flame, and the melted droplets are atomized by air jets in the torch to produce an atomized metal spray that forms a coating on impingement with a substrate. This process is still called ''metallizing'' by some equipment suppliers, but a more correct term is flame spraying with an oxy/fuel wire gun; the process is shown schematically in Figure 9–13. The torch is lit like a conventional oxy/fuel torch; the flame is adjusted to specified conditions and, when a trigger is pressed, the wire feeds into the flame and compressed gas, usually air, causes the melting wire to atomize into fine droplets that impinge on the substrate and splat cool to form a coating. A roll spraying operation is shown in Figure 9–14.

This process has been used for at least the last 40 years with the equipment described in Figure 9–13. The process requires a wire feed gun, compressed air, a tank of oxygen, a tank of acetylene, and a device to pay out the wire to the gun. The gun is held about 4 to 10 in. (10 to 25 cm) from the substrate to be coated. The torch produces a round or elliptical pattern with a diameter of about 3 or 4 in. (7.5 to 10 cm) at normal spray distances. If the torch were held stationary during spraying, the deposit would have a gradient in thickness from the center to the edges where the coating stops and has no thickness, only intermittent nodules called ''overspray.''

The wire feed guns are a bit bulky because of the wire drive mechanism and the gas hoses, but these guns are frequently hand held and hand manipulated. When the job permits, the torch is put on a manipulator. A lathe works well for cylindical parts, and rack and pinion torch manipulators can be used for flat surfaces. Present technology also allows the use of robots for torch or work manipulation.

The materials that can be sprayed with the wire flame spray process are any

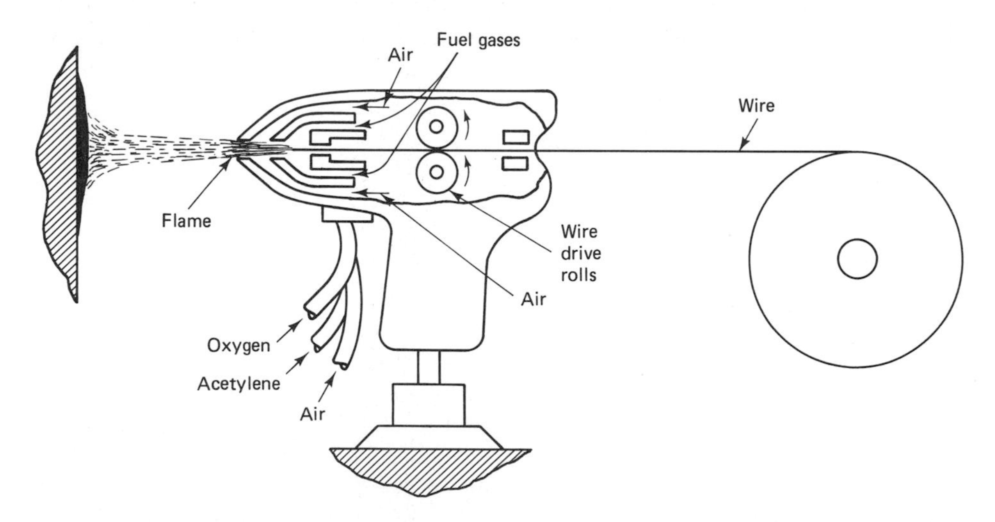

Figure 9–13 Thermal spraying, gas combustion/wire gun (wire flame spraying).

Figure 9–14 Wire spraying worn journals on a roll.

material that can be made into flexible wire that will melt in the oxyactylene flame. Many consumables are available from many suppliers, but the commonly sprayed materials are zinc, aluminum, machinable steels, hard steels, 300 series stainless steel, bronzes, and molybdenum. The stainless steels, aluminum, and zinc are primarily used for corrosion protection. Aluminum and zinc are commonly sprayed on water tanks, bridges, and similar structures. The bronzes and hard steels are used for wear protection, and the soft steels and molybdenum are used as rebuilding material for repair jobs.

The surface preparation for this process is at least abrasive blasting, but the grooving process shown in Figure 9–15 was used for many years for heavy deposits for rebuilding.

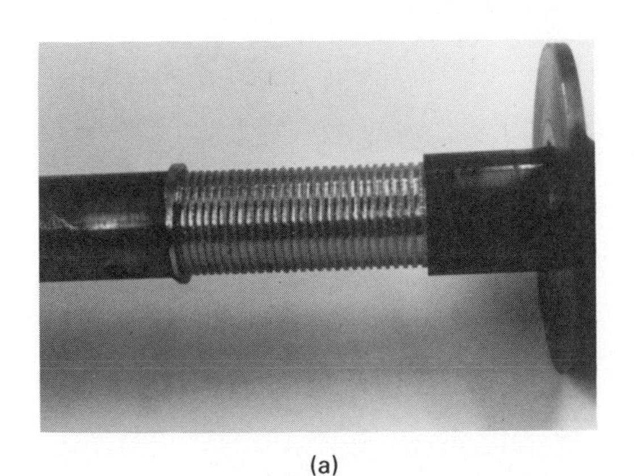

(a)

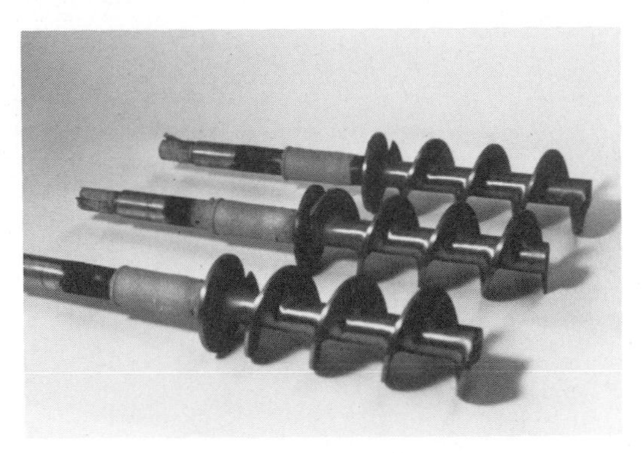

(b)

Figure 9–15 (a) Groove preparation for wire spraying; (b) parts after spraying.

The use of bond coats has reduced the need for grooving in many applications. There is no thickness limit to this process, and deposits as thick as 0.25 in. (6 mm) have been made. A common thickness for wire spray deposits is 0.030 to 0.050 in. (0.75 to 1.25 mm) for wear applications and for rebuilding. For corrosion applications, deposits can be as thin as 0.001 in. (25 μm). The deposition rates for wire spraying vary with the consumable and the equipment, but rates can be as high as 1000 ft^2 (93 m^2) per hour per mil (25 μm) of coating.

The wire spray process is most widely used for coating metals such as aluminum and zinc on carbon steels for rust protection and for heavy rebuilding tasks. It is quite economical for these types of applications, and the equipment costs are reasonable. Soft steels and hard steels are the most commonly used materials for wear applications. The soft steels are used to restore dimensions; they are machinable. The hard steels are used where the deposit is a wear surface; they are usually finished by grinding. Wire spray deposits have significant porosity, and their bond strength is inferior to plasma and other high-energy thermal spray processes. Thus this process is not used for sophisticated applications, those that require high-alloy or ceramic deposits, or for applications that are not subject to rigorous service conditions.

Powder Spray Processes The process of introducing powder consumables into a conventional oxyacetylene torch to achieve a spray deposit is still used. Powder flame spraying can be done with oxyacetylene torches that are modified in design to allow powder introduction into the fuel gas stream by simple gravity feed and siphon action, as shown in Figure 9–16. Because there is no high-pressure air to assist atomization of the consumable powder, the deposition rates are usually slower than that of the wire process, but the process will produce a coating with serviceability that is adequate for some applications. The porosity is even greater than that of the wire spray process and the bond strength may be lower than that of a wire spray deposit, but this simple torch can spray a much wider variety of materials; materials that cannot be easily made into wire that will feed through the wire feed torch. These torches often sell for less than $500, and this low price allows their use in shops that may infrequently need to apply a nonfusion hardfacing. In fact, there are self-fluxing hardfacing powders that can be

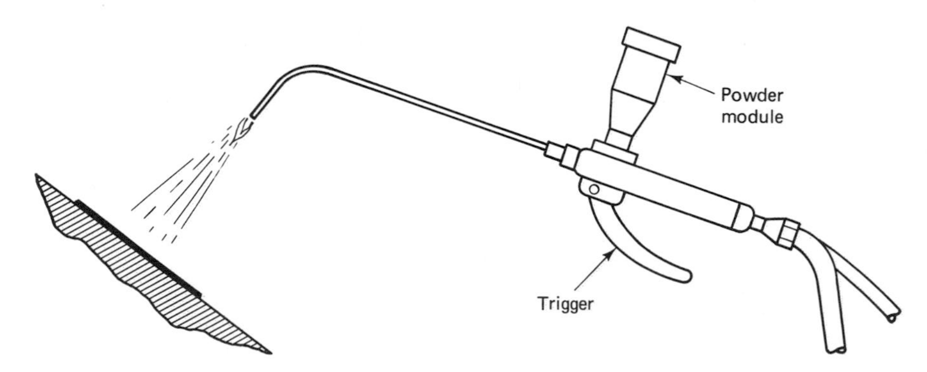

Figure 9–16 Flame spraying with a simple oxy/fuel torch.

applied with these simple torches and subsequently fused to the substrate. Thus these torches can do fusion as well as nonfusion deposits. The biggest disadvantages of these simple powder torches are slow deposition rate and low coating bond strength. Both are lower than can be achieved with more sophisticated equipment.

One of the more sophisticated powder spray torches is illustrated in Figures 9–17 and 9–18. These torches use compressed air to increase the degree of atomization and the deposition rate and to improve bond strength (through increased particle velocity). The torch is lit like a conventional oxy/fuel torch, and the flame is adjusted to prescribed conditions with the compressed air on. Spraying of the deposit commences when the trigger action starts the flow of powder into the flame. The mechanism of forming the deposit is the same as it is for all thermal spraying processes. The consumable powder is melted or made plastic in the flame, and the atomized consumable splat cools on the substrate to form the deposit.

This type of torch is often mechanized. For cylindrical surfaces the torch is mounted in the tool holder on a lathe, and track mounting of the torch is used for coating flat surfaces. Manual manipulation can be used to coat irregular surfaces. Like wire spray units, these torches can also be manipulated by robots.

There are many more consumables available for the powder thermal spray guns. Available consumables include bond coats, carbides, high-alloy steels, stainless steels, cobalt-base alloys, and even ceramics. The air-assisted powder spray guns produce faster deposition rates and better coating bond than the simpler torches that do not use air, but they are somewhat more complicated (require more maintenance) and they cost more.

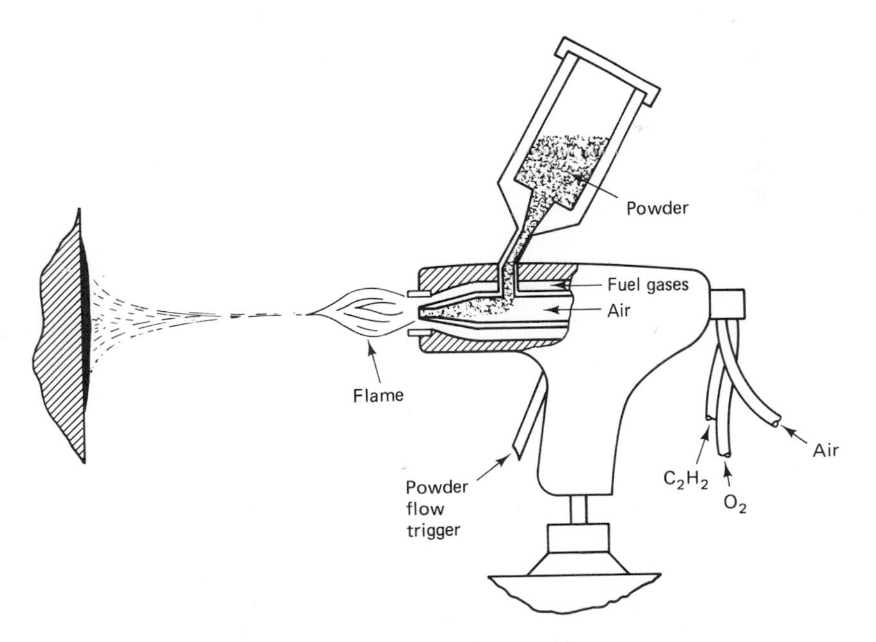

Figure 9–17 Thermal spraying with a gas combustion/powder gun.

Figure 9–18 Oxy/fuel powder spray gun.

Rod Consumables Ceramic powders can be sprayed with the equipment that we just described, but it is not usually done. The flame temperatures in conventional oxy/fuel torches are usually only about 5000°F (2760°C). Many ceramics that have utility as wear coatings have melting points that are at least 3500°F (1920°C). Some materials such as zirconium oxide require temperatures of almost 5000°F (2760°C). Conventional oxy/fuel flames simply do not have enough heat to produce a good ceramic coating. One oxy/fuel torch that is designed to spray ceramics uses a solid rod of ceramic consumables with air to assist in atomization. This type of torch is shown schematically in Figure 9–19.

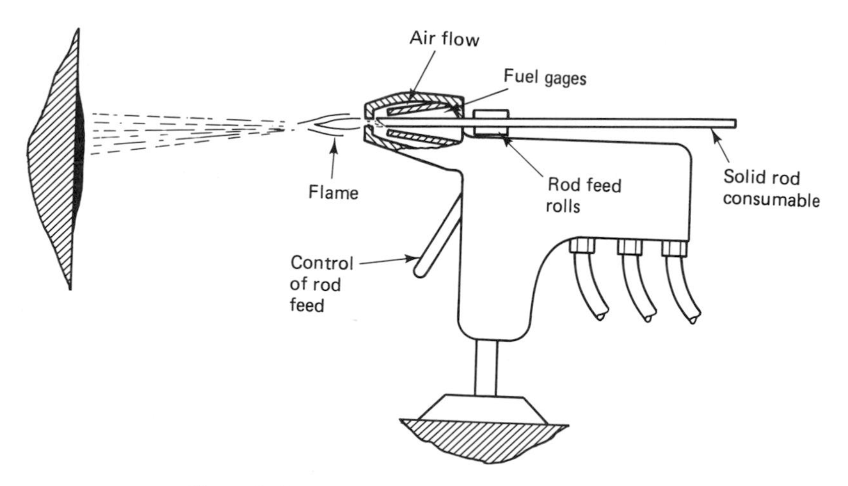

Figure 9–19 Thermal spraying with a rod feed gun.

Rod consumables are available in aluminum oxide, chromium oxide, zirconia, and ceramic mixtures. Spraying is done in the same manner as for the other flame spray processes, and it is claimed that the ceramic becomes fully molten and the atomized consumable droplets achieve an impact velocity of 550 ft/s (2.8 m/s). This process fills a gap in the capability of the flame spray processes. The wire guns are not used because ceramics are not presently available in the form of flexible wire; powder guns are not too widely used for spraying of ceramics because coatings are usually too friable to give good service life; this process is only used to spray ceramics. Thus there are flame spray guns that spray consumables that range from soft metals to ceramics and carbides.

Detonation Gun Coating The detonation gun or d-gun process is a proprietary process that also involves gas combustion for coating deposition, but it does so in quite a different fashion than the flame spray torches that were previously described. This process is at least 30 years old, and it remains proprietary mainly because of the many details involved in producing the proper parameters for successful coating application. The basic torch is illustrated in Figure 9–20. Powder is fed into the gun with a proprietary powder feeder under a small gas pressure. Valves are open to introduce a prescribed mixture of acetylene and oxygen into the gun's combustion chamber, and the explosive gas mixture is detonated with a spark from a spark plug. The temperature of the detonated fuel is about 7000°F (3870°C), a temperature sufficient to melt most materials. The detonation produces a particle velocity of up to 2400 ft/s (730 m/s). Detonations are repeated four to eight times per second, and nitrogen gas is used to flush the detonated gases from the combustion chamber after each detonation. The powder consumables have a particle size of about 60 μm (0.0024) and the spray pattern per detonation is about 25 mm (1 in.) in diameter. Each detonation produces a coating thickness of a few micrometers, and the gun or work is manipulated to get an area coverage (overlapping splats). Typically, deposit thicknesses are kept in the range of 3 to 5 mils (75 to 125 μm), but this is not the limit of the process, simply an economical coating thickness.

As one might expect from the process description, the gun produces substantial noise. Spraying is usually done in a soundproof room with 18-in. (45-cm) thick concrete

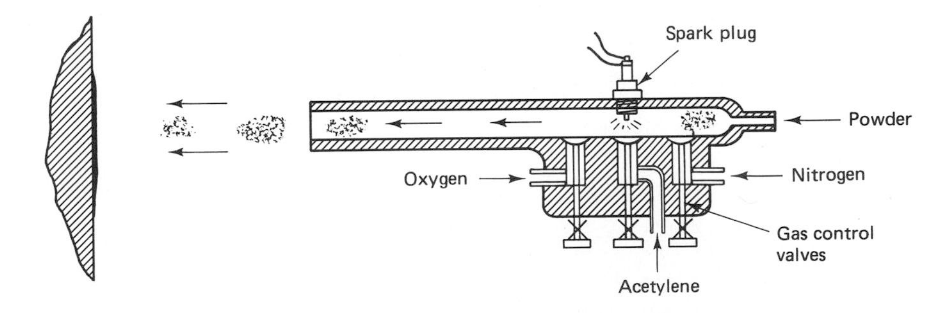

Figure 9–20 Thermal spray coating with the detonation gun process.

walls. The operator is outside the room. Obviously, this means that the coating process must be mechanized. Many techniques are used, including robot devices. The complete spray system requires a soundproof room, fuel gases and controls, purge gas and controls, a sophisticated powder feed mechanism, electronic controls for ignition, and a manipulation system.

Detonation gun coatings can often produce deposit densities that are as high as 98 percent of theoretical density. Almost any material can be sprayed, but this process is most widely used for spraying the "high-tech" coatings, carbides, ceramics, and complex composites. Bond strengths are typically above 10,000 psi (69 MPa), and coatings can be finished to very low surface roughness. D-gun coatings are often considered the premier thermal spray coatings.

Combustion Jet Spraying This thermal spraying process was first offered on a commercial basis in 1981. It is intended to be competitive with the d-gun process in coating quality, but the spraying equipment can be purchased; d-gun equipment is not sold and spraying must be done at one of the 20 or so centers that have the equipment. The heat source and the carrier for consumable droplets is a continuous gas combustion jet emanating from a combustion chamber within the torch. The torch is shown schematically in Figure 9–21. The torch is an offshoot of "rocket torches" that are used for cutting of granite rocks. These torches work like a rocket; the propellent is burned in confinement, and the exiting gases provide the thrust to lift the rocket. Rocks that contain water fracture (cut) when the intense heat from the rocket jet changes the state of the moisture in the rocks. In the combustion jet thermal spray torch, oxygen and a fuel gas, such as hydrogen, propylene, or other hydrocarbon gas, are ignited by a pilot flame in the combustion chamber of the torch, which is at right angles to the nozzle of the torch. The consumable to be sprayed is introduced into the center of the jet stream from a powder feeder using a carrier gas that is compatible with the fuel gas mixture. The combustion gas pressure may be in the range of 60 to 90 psi (400 to 600 kPa).

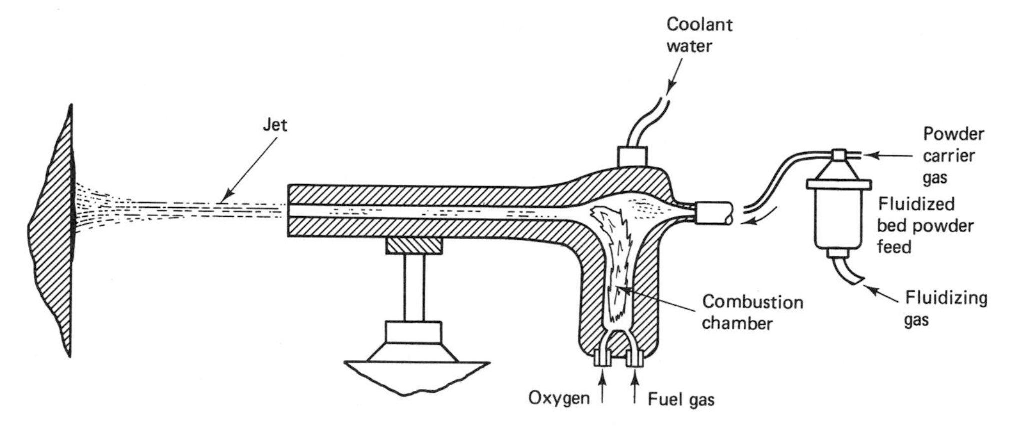

Figure 9–21 Thermal spraying with fuel jet (Jet-Kote) process. Jet-Kote is a registered trademark of Stoody/Deloro Stellite Inc.

The flame temperature at the point of powder introduction is about 5400°F (2980°C). The exiting gas combustion jet can have a velocity as high as 4500 ft/s (1370 m/s), faster than that of d-gun, and about four times the velocity of sound in air. Particles are melted or made plastic as they are in other gas combustion thermal spray processes, and the bond strength of the coating is a product of the particle velocity and temperature. Deposit densities are reportedly better than 90 percent of theoretical density. The spray pattern is about 1 in. (25 mm) in diameter, and the coating deposition rate can be about 10 lb (45 kg) per hour. The high combustion gas pressures mean that gas usage becomes a significant cost of the process. Oxygen flow rates can be 1000 ft^3 (28 m^3) per hour. The oxygen and fuel gas costs may be about $50/hour.

Since this is a relatively new process, not as many consumables have been tested for suitability for jet spraying. Tungsten carbide/cobalt cermet powders are the most popular consumables for wear applications, but the process has been successfully used for cobalt-base hardfacings, ceramics, stainless steels, and other corrosion-resistant materials. Bond strengths are comparable to those of d-gun and plasma spray deposits, usually above 10,000 psi (69 MPa).

This process is known by the trade name of Jet-Kote and by the generic term "high velocity combustion spraying." This process produces coating with better density and bond strength than other thermal spray processes and reportedly comparable to d-gun. The advantage over the d-gun process is that the equipment can be purchased, and it has an advantage over plasma spraying processes in that the equipment is lower cost. The first jet spraying torches sold for only a few thousand dollars, and a plasma installation may cost in the range of $50,000 to $80,000. In 1986, jet spraying equipment cost about $30,000. The biggest disadvantage of this process compared with the mature thermal spray processes is the lack of applicable consumables. Only a handful of materials have been successfully sprayed, and there are literally hundreds of consumables available for well-developed thermal spray processes like plasma. A second disadvantage is the safety requirements of confining a rocket-type combustion reaction within the torch. The torch is water cooled, but safe handling of the fuel gases is a process concern. The first jet torches used hydrogen for a fuel gas, which caused safety concerns. The jet spraying process appears to be accepted to the point that it is now a permanent part of thermal spraying, and future developments will undoubtedly increase the applicability of the process.

Arc Processes

Wire/Arc Spraying As illustrated in Figure 9–11, two thermal spraying processes use an electric arc as the heat source for obtaining the molten droplets that form the thermal spray deposit: arc spraying and plasma arc spraying. The wire/arc spraying process is identified by the AWS acronym EASP, electric arc spraying. This process has a history similar to that of oxy/fuel/wire spraying. It has been in use for many years for the same types of applications as the combustion gas wire spraying process, but because it uses two consumable wires instead of one it produces much higher deposition rates. The wire/arc spraying torch is illustrated in Figure 9–22. The consumable material

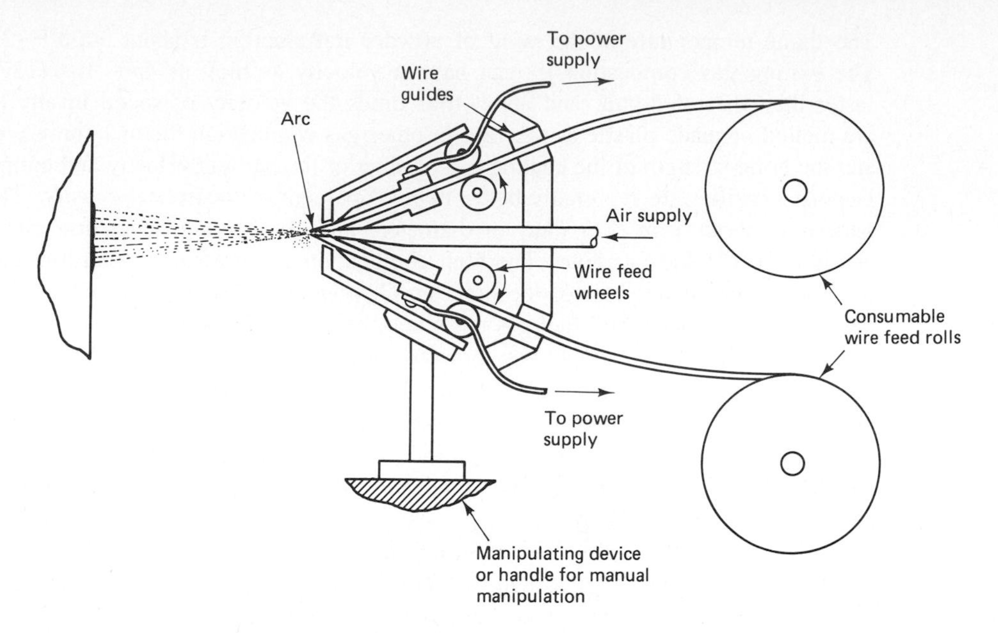

Figure 9–22 Thermal spraying with arc spraying process.

is introduced into the torch as two metal wires from reels. The wires are fed by motor-driven feed rolls, and at the tip of the torch these wires are inclined on an angle such that they meet each other. The wires are electrically insulated from each other, and each wire is connected to a welding power supply. When the torch is energized for spraying, the two wires are driven into contact and an arc is established that melts the wire. An air jet in the torch atomizes the molten metal into droplets that are sprayed at the substrate to form the coating. The wires can be as large as 1/16 in. (1.5 mm) in diameter, and the arc current can be as high as 3000 A. The wires are vertical in some torches and horizontal in others. Square wire can be used to increase the deposition rate over the rate that is obtainable with round wires. The gun can be hand held and manipulated, or it can be put on a lathe or some other manipulation device.

The nature of the arc spray deposits is similar to the oxy/fuel deposits. The material is splat cooled onto a roughened substrate, and the deposit contains substantial porosity and oxide inclusions from oxidation of the consumable in the atomization air. The available consumables are essentially the same as those that are available for oxy/fuel spraying; but because of the high deposition rates, this process is very commonly used for spraying soft metals for corrosion protection. Deposition rates can be as high as 100 lb (45 kg) per hour, and very large structures such as bridges have been sprayed with aluminum or zinc for atmospheric corrosion protection. This process is not widely used for hardfacing, usually only when a large surface area must be covered.

The forté of this process is the high deposition rates; the disadvantage over the oxy/fuel wire spray process is the more complicated and more expensive equipment. Both processes are designed to do the same job. This does it faster. Like the other simple thermal spray processes, the deposit quality is not as good as the deposit obtained with plasma, d-gun, jet spraying, and the other high-velocity thermal spray processes.

Plasma Arc Spraying (PSP) Plasma arc spraying (PSP) is the workhorse of thermal spray processes. It is the process of applying a material by melting and atomizing it in powder form in a plasma that is obtained by passing a gas through an electric arc between nonconsumable electrodes contained within the torch. The equipment required for plasma spraying is shown in Figures 9–23 and 9–24. The electrodes for establishing the arc that produces the plasma are usually tungsten, and the gun is water cooled to carry away the portion of the arc heat that is absorbed by the torch. The arc power supply also contains the controls for the water cooling and the flow of the gas that is used to establish the plasma. A plasma is essentially an ionized gas that also contains electrons, ionized gas atoms, and even some molecules of the plasma gas. The gas used to form the plasma is a process variable, but gases such as hydrogen and nitrogen can be used. The plasma that is formed by the gas going through the constricted arc in the torch can have a temperature in excess of 50,000°F (28,000°C), well above the flame temperatures attainable in the gas combustion processes. The material to be sprayed is introduced in powder form into the exiting plasma gas stream either in the torch nozzle or just outside the nozzle. The powder size is usually in the range of about 30 to 100 μm (0.001 to 0.004 in.) in diameter. The powder is fed by a powder feeder that usually consists of a hopper pressurized with an inert gas, a vibrator to keep the powder from clumping, and a gear pump to meter the powder into the carrier gas stream. The velocity of the plasma/droplet stream that exits the nozzle is usually subsonic, but high-velocity plasma torches are also available. The velocity of the spray is usually increased by constricting the torch nozzle. Reportedly, this torch modification can increase the particle velocity to mach 3.

A number of parameters must be controlled to obtain a good spray deposit. These include nozzle-to-work distance, powder size and type, point of powder introduction, arc current and voltage, type of plasma gas, particle carrier gas type, and powder fluence. Argon and helium are popular plasma gases, and if these are properly used, the spray deposit will contain very little oxide. This is a process advantage over other thermal spray processes that use air as the gas to produce the droplets that comprise the consumable spray. The porosity of PSP deposits is usually in the range of 5 to 15 percent. The porosity in d-gun deposits is usually less than 5 percent. Porosity of gas combustion process thermal sprays can be as high as 50 percent.

The size of the spray pattern depends on the spray distance and the nozzle configuration, but a spray pattern of about 1 in. (25 mm) in diameter is common. The intense heat of the plasma does not produce substantial substrate heating. Most of the heat to the substrate comes from the heat from the impinging consumable droplets and substrate temperatures usually do not get over 300°F (150°C). When substrate heating becomes a consideration, on for example thin-wall tubes, auxiliary jets of inert gas are used for substrate cooling.

Hundreds of powder consumables are available from at least a score of suppliers; they fall into about four different categories: metals, ceramics (primarily oxides and

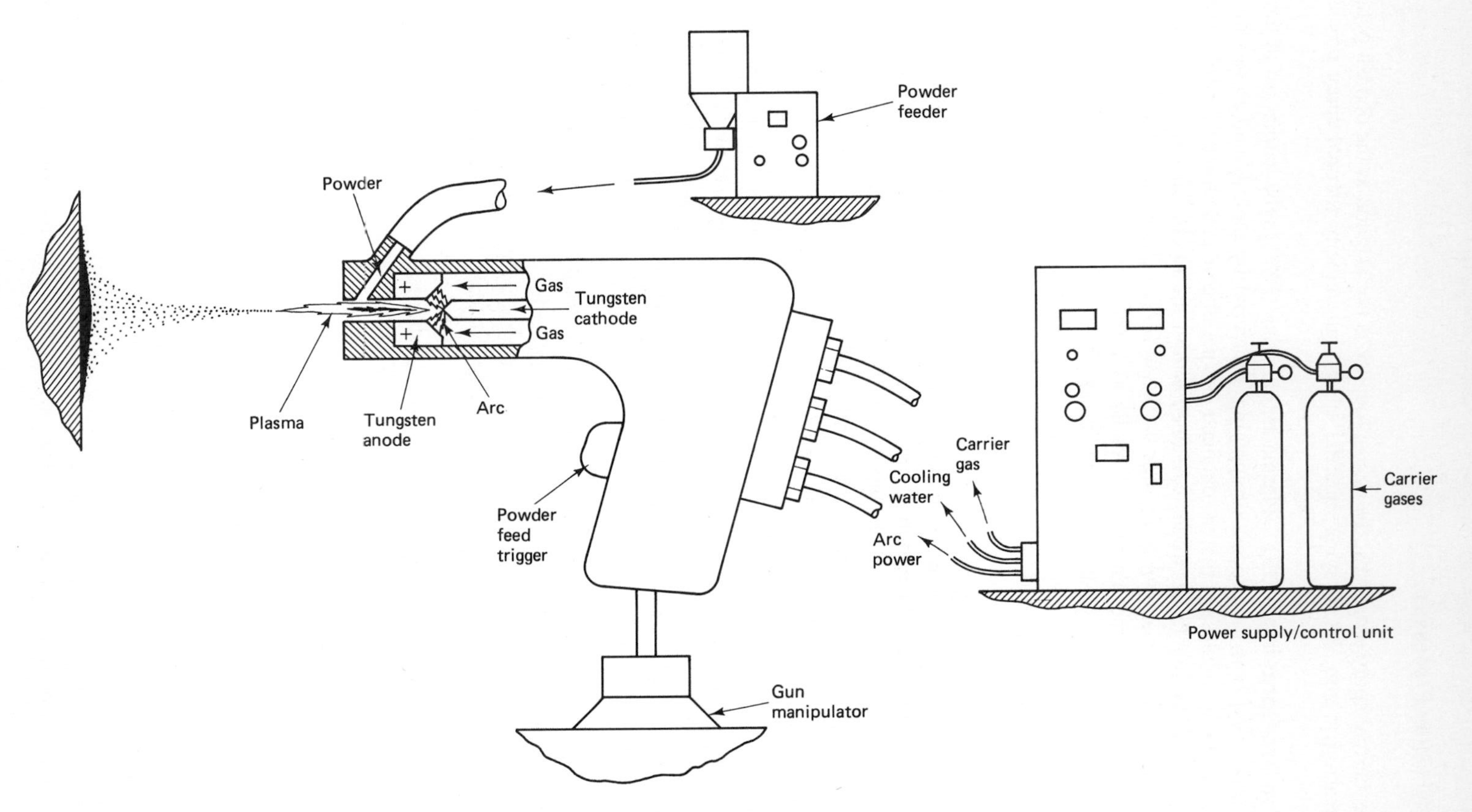

Figure 9–23 Thermal spraying with plasma arc process.

Figure 9–24 Plasma arc spraying of a flat surface.

carbides), cermets, and composites. Metals vary from soft metals such as aluminum and zinc for corrosion applications to cobalt-base hardfacing materials for wear applications. The most popular ceramic coatings are aluminum oxide and chromium oxide or mixtures of chromia and silica. These are used mostly for wear applications. Ceramics such as yttria-stabilized zirconia, magnesium zirconate, and calcia-stabilized zirconia are used for thermal barrier coatings on engine components and the like. Aluminum oxide and magnesia/alumina are often used for electrical insulation applications. The most popular cermet consumable used in plasma spraying is tungsten carbide/cobalt. This material is the thermal spray counterpart of the familiar cemented carbides that are used in cutting tools and in a wide variety of wear components. The composite consumables are usually used for special applications, and they include such things as metal/graphite powders and metal/molybdenum disulfide powders.

The particle size of the powders is a process variable. The properties of the deposit such as hardness and porosity can be affected by the particle size and its flowability in the gun feeder systems. The mechanical, physical, and chemical properties of a thermal spray deposit may be different than the properties of the same material in the bulk. This is due to factors such as oxide inclusions, porosity, changes in morphology during spraying, and changes due to the splat cooling effect of the powders impacting the substrate. Plasma spraying has been so extensively used in critical components that there is a significant data base on the properties of many deposits. Coating suppliers can be asked to furnish deposit properties if this is important to an application. In general, the differences between spray and bulk properties are inconsequential.

For example, the properties of a PSP deposit of aluminum oxide will differ from sintered aluminum oxide in such things as hardness and electrical properties because of the deposit porosity, but the differences are not substantial. PSP aluminum oxide is

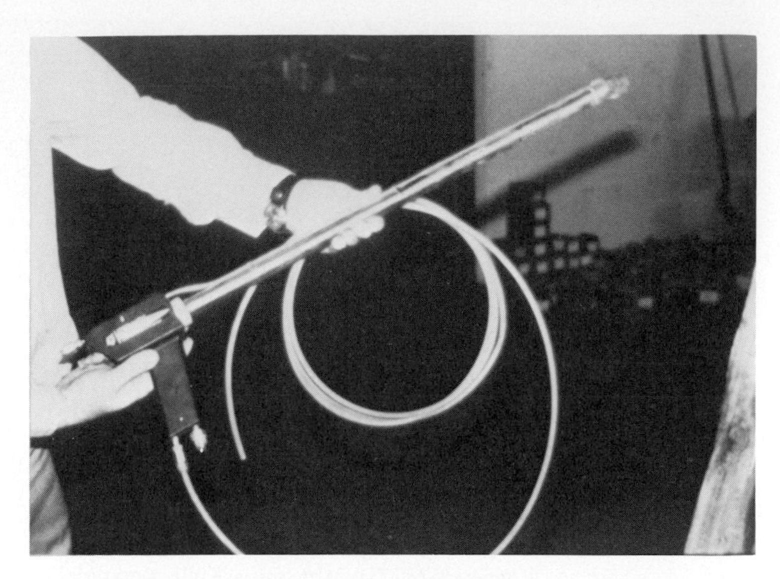

Figure 9–25 Plasma spray torch for small internal diameters.

still very abrasion resistant and it is still a good dielectric. The porosity in PSP deposits may affect its ability to protect surfaces from corrosion. The porosity in thermal spray deposits may be interconnected, and thus these deposits cannot serve as corrosion barriers unless they are sealed. Sealing is common and it can be done by pressure impregnation of furfural alcohols, epoxies, and fluorocarbons. For corrosion applications, it is always a good practice, with or without sealants, to use a coating substrate that is resistant to the anticipated environment. This will prevent corrosion under a deposit if the corrodent penetrates to the substrate.

The biggest disadvantage of plasma spraying over other thermal spray processes is the cost of the equipment. It is probably the most expensive of the processes that can be purchased (about $80,000). It is a line-of-sight process, as are all the thermal spray processes. The equipment is complicated and often bulky, but there are special torches for spraying the inside diameters of rolls and the like. Torches are available that will allow coating of the ID of holes as small as 1 in. (25 mm) in diameter (Figure 9–25), and the coating can be as much as 30 in. (75 cm) deep into a bore. The main advantage of plasma spraying over other thermal spray processes is coating quality. This advantage is followed closely by the wide availability of consumables. As mentioned in our introductory remarks, plasma spraying is the workhorse of the thermal spray industry, and it will probably remain in this role for some time to come.

APPLICATION OF THERMAL SPRAY PROCESSES

We have described the major thermal spray processes, what the equipment looks like, and how the deposits vary. We will discuss thermal spray consumables in a subsequent chapter, but a few process considerations should be mentioned before concluding this discussion of thermal spray processes: suitable substrates, coating evaluation, and application guidelines.

Substrates

Thermal spray coating can be applied to any substrate that is not degraded by the heat from the thermal spray torch and the heat from the impinging particles. This includes most metals and plastics that have a continuous-use temperature of about 300°F (150°C). Generally, most thermosetting plastics are compatible. One extremely important substrate factor should be kept in mind when designing for thermal spray coatings. It is necessary to roughen the substrate to get good coating bond. This roughening is usually done with abrasive blasting with abrasives such as aluminum oxide (often 80 grit). With a soft steel or other soft metal substrate (less than 200 HB), this will produce a surface roughness greater than 100 microinches (2.5 μm R_a), which is adequate for coating adhesion. If a 60 HRC tool steel is abrasive blasted with the same abrasive, the roughness that can be obtained on the surface may be less than 50 microinches (1.25 μm (R_a). This affects bond strength to the point that many thermal spray jobbers refuse to spray hardened steels. Bond coats can be used to enhance the bond of thermal spray deposits on hard materials, but the bond coats are typically soft metals and the coating can lose some durability simply because it has been applied to a soft metal. What to do? Try to avoid thermal spray coatings on substrates with a hardness greater than 50 HRC. If there is no alternate to spraying on a hard substrate, discuss the application with the spraying vendor and use his judgment on an appropriate surface preparation and bond coat.

Another factor that pertains to the nature of the substrate is section thickness. Thin tubing and fragile parts are prone to distortion during the abrasive blasting operation and from the heat of spraying. For example, thin wall tubes (for rolls) can take on a barrel shape from the heat of spraying. The barrel shape gets "locked in" by the spray deposit, and on finish grinding the coating can penetrate in the center of the roll. Problems like this can be solved by ancillary cooling during spraying. The distortion problem from grit blasting can be lessened by using low-pressure blasting techniques. The major point to be made is that fragile parts may need special precautions. If in doubt, discuss the part with the coating contractor.

Coating Evaluation

The bond strengths of thermal spray deposits are most commonly determined by tensile shear tests, similar to those used for platings and other coatings. The ends of two strips of the substrate under study are sprayed with the desired consumable. The coated ends are epoxied together, and the uncoated ends of the adhered assembly are put into a tensile tester and pulled to failure. A good coating will produce an adhesive failure; the epoxy will fail and the coating will remain intact. The bond strength is recorded as "greater than ×××," where the ×'s are the tensile shear strength of the epoxy that was used in the test. The strongest epoxies have a tensile shear strength in the range of 12 to 20,000 psi (82 to 138 MPa). If the thermal spray deposit comes off before the epoxy fails, the apparent stress at the point of failure is recorded as the coating bond strength. Another version of this type of test is performed by epoxying a small dumbbell-shaped device to a flat surface that has been sprayed with a particular consumable. A special fixture pulls the dumbbell off the coated surface, loading the coating in tension; the pulling device is calibrated to yield the load at failure. There are other destructive tests, but most of these types of tests are used to develop consumables or to solve some particular application problem. The normal user of thermal sprayed coatings is only interested in a test that will not destroy the part (nondestructive evaluation). There are some simple and some not so simple NDE tests.

The simplest nondestructive thermal spray coating test is visual inspection. The things to look for are porosity, cracks, adhesion to the edges of undercuts, impurities in the coating, blisters, and simply spots that do not clean up in finishing. Porosity in thermal spray coatings is to be expected, but the porosity should be uniform and consistent with what can be achieved from the particular process that was used. Figure 9–26 is an example of acceptable porosity in a plasma sprayed ceramic coating. Notice that the coating is quite smooth between the pores. The overall coating may have a finish that looks almost polished to the unaided eye, but when measured for roughness with a profilometer, it may show a roughness that is higher than expected. This is indigenous to thermal spray coatings that contain porosity. Even the very best d-gun coatings may only produce a roughness of 20 microinches (0.5 μm R_a) because of the influence of porosity. Metal coatings can achieve a roughness that may be as low as 5 microinches (0.12 μm R_a), but the best finish obtainable on a ceramic coating is probably only 20 microinches (0.5 μm R_a) with normal spraying and finishing techniques.

A simple test that can be performed to investigate porosity and the integrity of a

Figure 9–26 Porosity in a plasma hardfacing deposit (aluminum oxide) (100×).

coating bond is to apply a vinyl tape such as duct tape to the surface and strip it from the surface in a peeling fashion. If powder particles are deposited on the tape, the coating integrity may be questioned. Another very simple test is to tap the coating with the plastic or wood handle of a screwdriver. This takes some practice, but if there is a lack of bond under the coating, it will produce a thud instead of a solid sound when wrapped. Another part of visual inspection is to look for cracks or lack of bond at the end of a deposit. This will require a loupe, but problems at the end of a deposit are often easy to uncover with this technique. This same type of inspection can often uncover contaminants in the coating from foreign materials. Sometimes powder hoppers are not thoroughly cleaned between spray jobs, and it is possible to have a ceramic deposit that contains some obvious metallic particles. These probably came from a contaminated powder feed systems, and they can cause service problems if the coating is to be used in an environment that will chemically attack the metal contaminants.

The more sophisticated NDE techniques for thermal spray coating include such things as ultrasonics, thermal imaging, acoustic emission, and laser holography. None of these are widely used by spraying vendors for routine spray jobs. In general, most of these techniques are in the research mode, with possibly the exceptions of ultrasonics and thermal imaging. A very dense metal deposit will respond to ultrasonic techniques, but the ceramics and other materials with 10 to 15 percent porosity may cause a problem. If the deposit is a dense metal, the coating can be traced with the ultrasound transducer, and if the coating is tightly adhered, the bond line will be invisible to the sound wave. If there is a void, the sound will be reflected and the extent of the void can be mapped.

The thermal imaging process is based on the theory that a void under a coating will alter the thermal diffusivity of the coating material. Heat will not flow through the coating at a void as it will if there are no voids. The substrate is heated such that the path of heat flow is from the substrate to the coating. Infrared imaging is used to observe the heat-flow pattern on the coating surface. A void will show a cool area. There is a laser heating method that heats the coating surface, and the heat flow from the surface inward is measured, but the basic principle is the same. As mentioned previously, the sophisticated techniques are not as developed as one would hope; on most jobs, a visual inspection and judicious inspection with a loupe is adequate.

Application Guidelines

We will discuss the details on how to properly specify thermal spray coatings on engineering drawings in a subsequent chapter; but to summarize this discussion on hardfacing welding processes, we will mention some factors that should be kept in mind when considering the use of thermal spray coatings for producing wear-resistant surfaces. We have described a score of thermal spray processes, but it may not be clear when to use one or the other. In many ways, they all do the same thing. This is true, but the different processes do it with different speeds and facility, and they produce different coating qualities. Figure 9–27 provides the author's opinion on where to use the various processes. Another hardfacing user may have another opinion, and a coating service may have another opinion. In terms of coating quality, most users would agree that

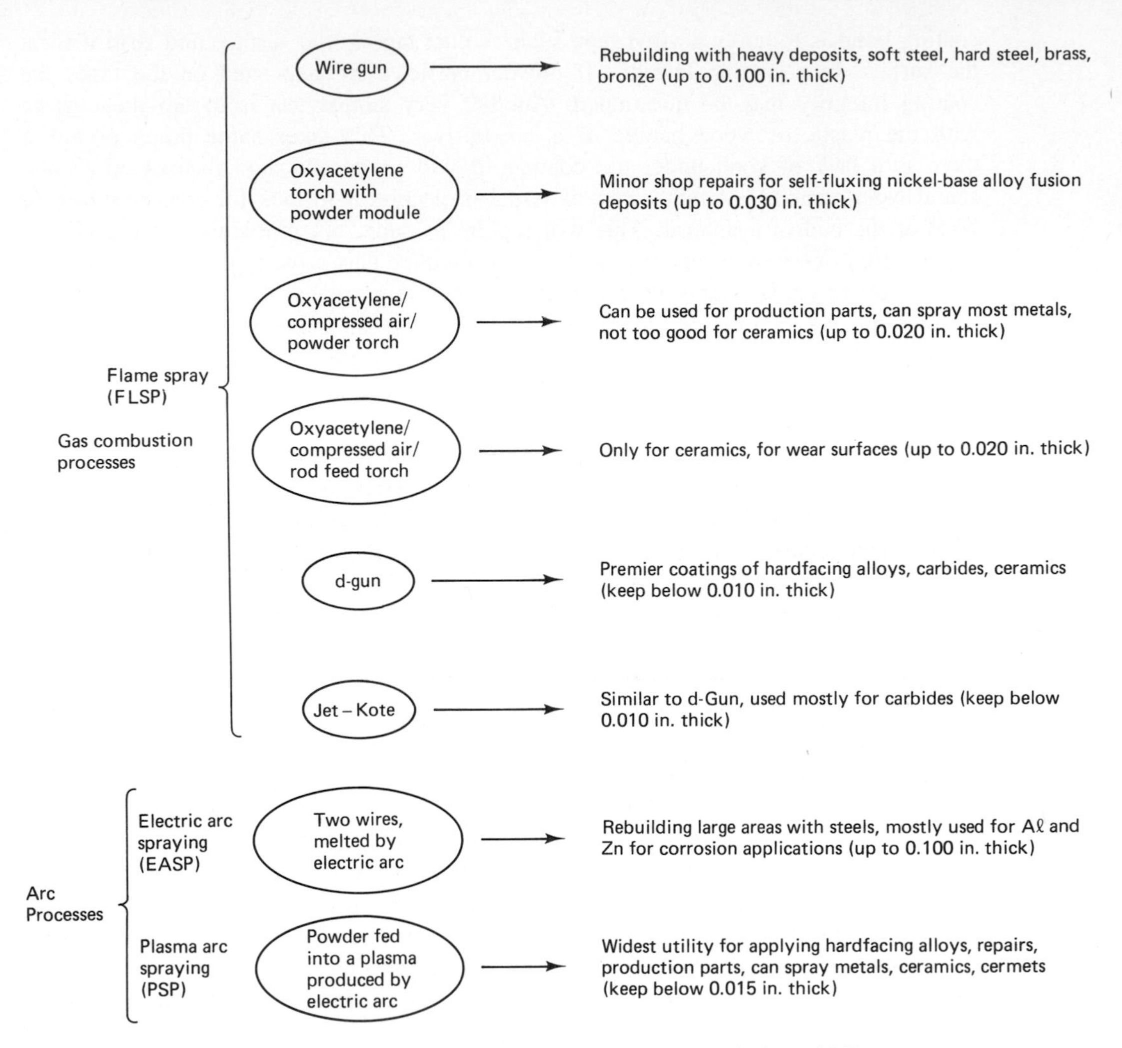

Figure 9–27 Thermal spray processes: application guidelines.

the d-gun, Jet-Kote types of processes produce the densest coating with the best bond. Next in quality are plasma sprays, air-assisted flame sprays, and the as-sprayed coating from simple oxyacetylene torches with a powder module, which produce the coatings with the highest porosity and the lowest bond strength. The low-cost torch, on the other hand, excels for applying coatings that can be sprayed and subsequently fused. This torch is also the one that can be afforded by almost any shop. Thus there are advantages and disadvantages to each process, but all have their place and all are capable of producing a wear-resistant surface that will meet many application requirements.

Figure 9–28 illustrates some things that should be done and some things that should not be done in applying thermal spray coatings. A major application rule is to

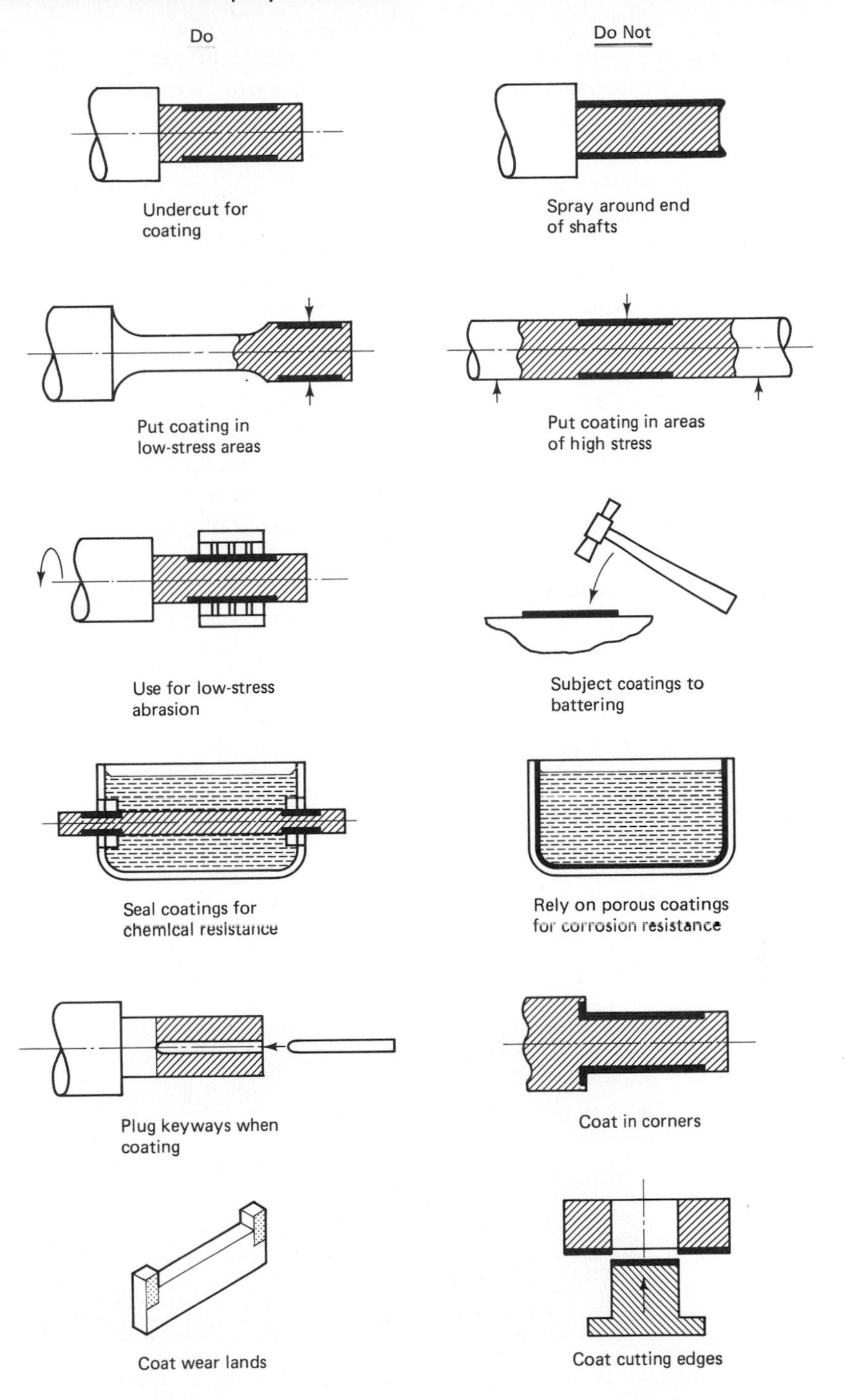

Figure 9–28 Some do's and don'ts for thermal spray coatings.

avoid spraying to the ends of parts such that the coating can be subject to chipping in handling. The recommended approach is to apply the coating in an undercut leaving a small metal land to prevent edge chipping. If undercuts are used, remember that these undercuts are stress concentrations. Radius the ends and put the undercuts on bosses whenever possible.

Because thermal spray coatings adhere by a mechanical bond, they should not be used where they will experience impact, battering, and the like in service. These coatings are stripped by abrasive blasting; any impingement of this type in service will cause bond failure. If the coatings are to be used for corrosion resistance, remember that they are porous; apply sealants and put them on a substrate that will resist attack. Avoid bond coats that can be attacked in the corrodent.

All these spraying processes work best when the gun is normal to the substrate surface. The bond degrades as the angle of spray impingement decreases. Ideally, spraying should not be done at an angle to the surface of less than 60 degrees. For this reason, it is not advisable to spray into keyways and on faces that are not normal to the torch nozzle. It should also be kept in mind that masking of parts can be expensive, and the deposit should be designed to minimize masking. Masking is done with metal shields, lead tape, waxes, and a variety of other techniques. All thermal spray processes produce overspray, a loosely adhered deposit adjacent to the intended deposit. If this overspray is unacceptable on a part, then specify masking. Be aware that these things increase cost.

The last "do not" in Figure 9–27 concerns the use of thermal spray coatings on things like knife edges and die edges. Because of the coating's mechanical bond and its porosity, it is extremely difficult to make a cutting edge from a thermal spray deposit. Thermal spray coatings a few mils (75 μm) thick invariably chip in grinding keen 90 degree or lower-angle edges. Thin carbide deposits of a few micrometers are often used in the as-sprayed condition on knife edges, but these knife edges are not keen and the coating roughness is used to produce a sawing effect. Some applications can tolerate this type of edge; in fact, many times these sprayed knife edges excel in service.

To end this discussion, Figure 9–29 shows two shafts that are coated with a

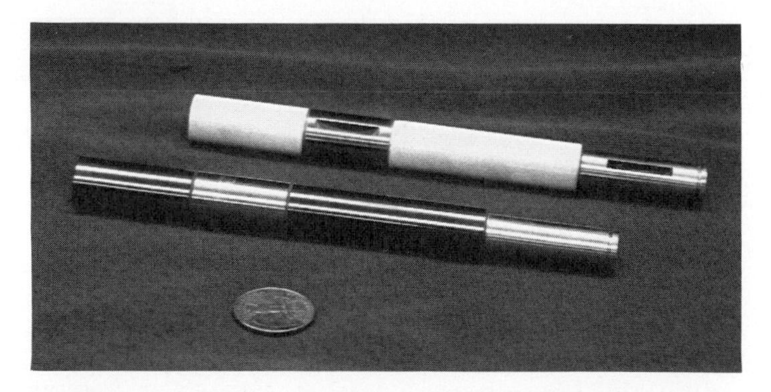

Figure 9–29 D-gun coatings on stainless steel shafts (white coating is aluminum oxide; black is chromium oxide).

thermal spray process. In this application, thermal spray coating competed with a number of other coating processes and solid materials and it won. This is the significance of the thermal spray processes; they are an essential part of hardfacing and they can do things that cannot be done by other processes.

SUMMARY

This chapter reviewed the welding processes that are used to apply hardfacings, one of the under-utilized surface engineering processes. Most designers and engineers are familiar with welding techniques, but as applied to hardfacing these processes have areas of applicability that are different than in using welding processes for joining. To use hardfacing for surface improvement the designer must specify the type of process that is to be used. This is often not the case in using welding for joining; it is usually left to the welder. The process details presented are intended to provide the basis for a proper process selection.

The fusion processes are for applications that can tolerate the heat of melting the surface of the substrate; the nonfusion processes are for applications where this degree of heating cannot be tolerated. The fusion processes are the obvious candidates for the heavy-duty wear processes, those where wear losses may be measured in millimeters. The thermal spray processes are more suited to wear processes where allowable wear losses may be measured in micrometers. We have discussed the advantages and disadvantages of all types of hardfacing processes. These factors should be recalled when deciding which surface engineering process to use to increase the serviceability of a particular part. Process selection is an important part of the surface engineering process.

REFERENCES

IRVING, ROBERT R., "Jet Kote," *Iron Age*. June 18, 1984.

LONGO, FRANK L., ed. *Thermal Spray Coatings*. Metals Park, Ohio: American Society for Metals, 1985.

MERINGOLO, V., ed. *Thermal Spray Coatings*. Atlanta: Tappi Press, 1983.

TAKEUCHI, S. "A Newly Developed Arc Spraying Device," *Surface Engineering International*, Vol. 1, No. 1, 1986.

TUCKER, R. C. "Plasma and Detonation Gun Deposition Technique," in *Deposition Technologies for Films and Coatings*, R. Bunshah, ed. Park Ridge, N.J.: Noyes Publications, 1982.

The Welding Handbook, Vol. 2, 7th ed. Welding Processes—Gas Welding and Cutting, Brazing and Soldering. Miami: American Welding Society, 1980.

The Welding Handbook, Vol. 3, 7th ed. Welding Processes—Resistance and Solid State Welding and Other Joining Processes. Miami: American Welding Society, 1980.

Welding Terms and Definitions, AWS A3.0–80. Miami: American Welding Society, 1980.

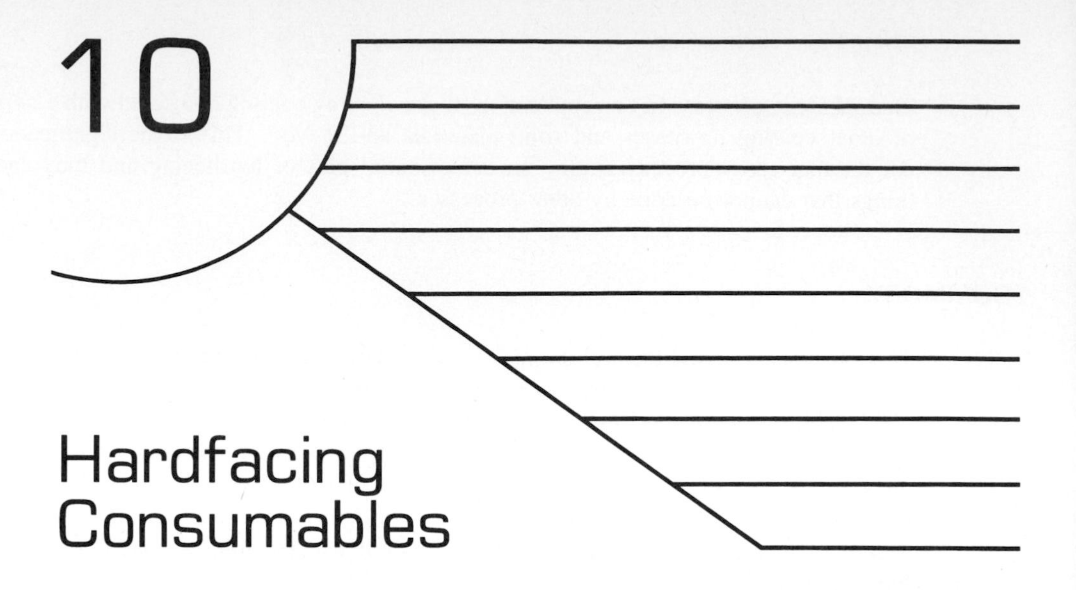

10 Hardfacing Consumables

A major factor that limits the application of hardfacing by average engineering and maintenance personnel is the lack of knowledge about what can and what should be applied to a substrate to solve a particular problem. This chapter will address the former: what can be applied.

The term consumable is used to describe materials in rod, electrode, wire, or powder form for deposition by welding or an allied process. The material is "consumed" by melting as it is deposited on the surface; hence the term consumable.

Hardfacing consumables fall basically into just two categories: (1) materials that are applied by fusion to a substrate and (2) those that are applied by thermal spray processes and do not require fusion to obtain a substrate bond. As shown in Figure 10–1, the nonfusion materials are powders, wires, or solid rods of the type used in some flame spray torches. A wide variety of product forms are available for fusion deposits. A very simple guideline to remember regarding hardfacing consumables is that the high-hardness ceramics cannot be applied by fusion processes because they have high melting points and poor compatibility with most substrates from the standpoint of solid solubility. Ceramics are usually compounds of a metal and a nonmetal. The ceramics that are important in hardfacing are usually carbides, compounds of carbon and a metal, and oxides, compounds usually formed by the combination of a metal and oxygen. Ceramics can be applied by thermal spray processes, and they are often used as hard phases in metal matrixes. In the latter form, they are called cermets, a ceramic type of material in a metal matrix. Composite consumables are cermets. They have hard ceramic-type compounds embedded in a metal matrix.

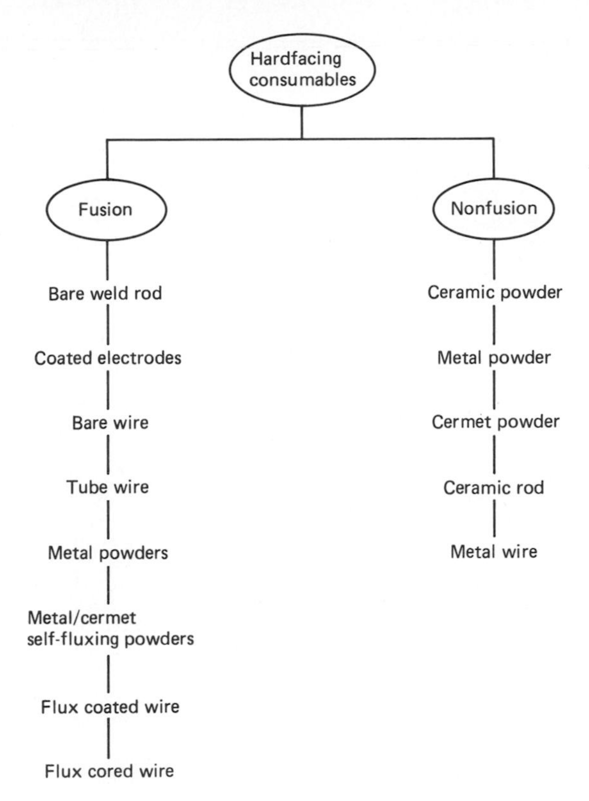

Figure 10–1 Categories of hardfacing consumables.

How does a user become familiar with the hundreds of specific hardfacing consumables that fit into the two categories and dozens of product forms? The answer is not as complicated as one might think at first. There may be hundreds of specific hardfacing materials available, but the fusible metal alloys fit into about ten alloy systems, and the ceramics and cermets that are in wide use fall into even fewer material systems. The basic systems are illustrated in Figure 10–2.

To make things even more manageable, the American Welding Society has established a hardfacing consumable identification system that allows alloy identification without resorting to the use of trade names. In the remainder of this chapter, we will discuss the AWS designation system and specification for solid surfacing welding rods and electrodes (ANSI/AWS A5.13–80), the specification for composite surfacing rods and electrodes (AWS A5.21–70), and generic specifications for the ceramic and cermet types of hardfacing consumables. The goal of this chapter is to provide an understanding of the available consumables, their chemistry, their general properties, the available forms, and how they should be identified on drawings and specifications. These alloys will be married to application needs in Chapter 11. This chapter is to show what is available.

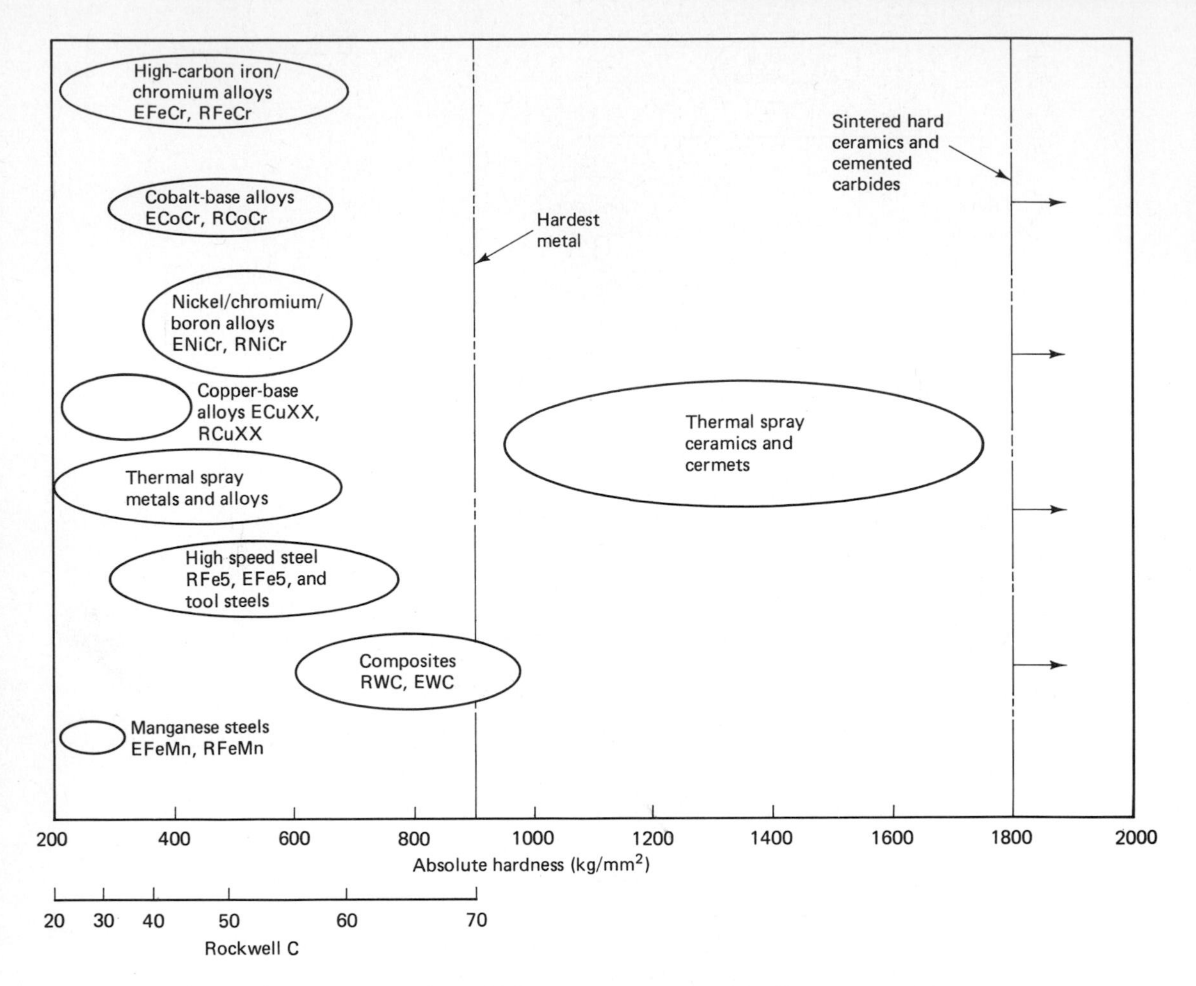

Figure 10–2 Hardness spectrum of hardfacing consumables.

HARDFACING METALLURGY

Before getting into a discussion of all the different hardfacing systems, we will discuss some of the materials principles that are the basis of these systems. As shown in Figure 10–3, hardfacing materials are used for many different applications, everything from rock crushing to repair of the cutting edges on small milling cutters. The thermal spray coatings are used for everything from turbine rotors to tiny electronic parts. The common denominator for these applications is resistance to some forms of wear. We have stressed many times throughout this text that parts do not wear; they wear by one or more modes of wear. Using the same principles, a particular hardfacing material cannot solve

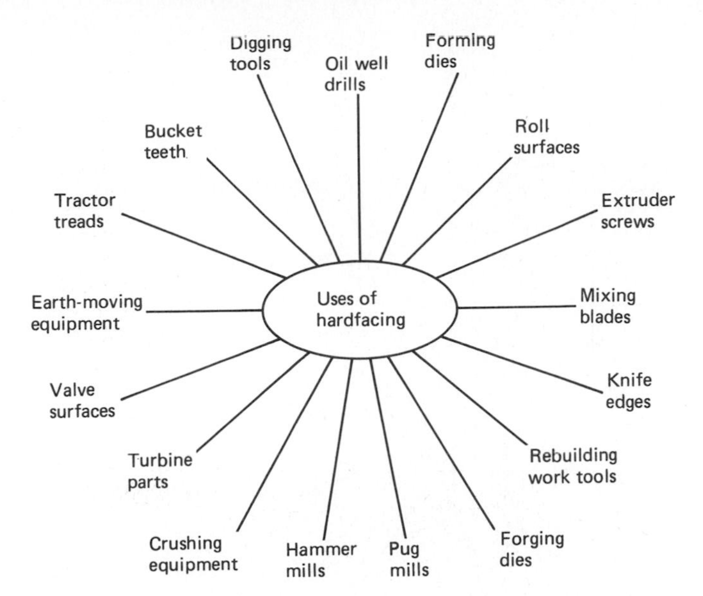

Figure 10–3 Uses of hardfacing.

all types of wear problems, only some types of wear. Thus there are quite a number of hardfacing material systems; each system has a somewhat different area of applicability, but there are some basic metallurgical and materials engineering concepts that were used to develop the hardfacing systems that we will subsequently describe.

What material properties make a material wear resistant? Figure 10–4 lists some material properties that are known to affect wear processes. This illustration shows that wear processes usually depend on a combination of material properties. For example, hardness by itself does not control wear. Diamonds used for cutting of glass plates in the manufacture of photographic plates wear quite rapidly, even though they are our hardest material; material is removed from the diamond's cutting edges by adhesion of microscopic amounts of the diamond on the shapes being cut. On the other extreme, soft gold platings are very effective in preventing fretting damage even with contact stresses that are well above the compressive strength of the gold; the gold acts as a lubricant, and the noble nature of gold prevents the formation of oxide products that are known to accelerate fretting damage. Thus many material properties affect wear, and hardfacing alloy systems have been formulated to have a combination of some of the material properties that are known to affect particular types of wear. Hardfacing alloys do not work simply because they are hard, or simply because they have high strength. They work because they have a combination of properties that can address a particular form of wear.

The best place to start in discussing the general materials science of hardfacing alloys is to review the solid materials that are used in all industries to resist wear. Figure 10–5 shows the relative use of engineering materials for tools, devices that

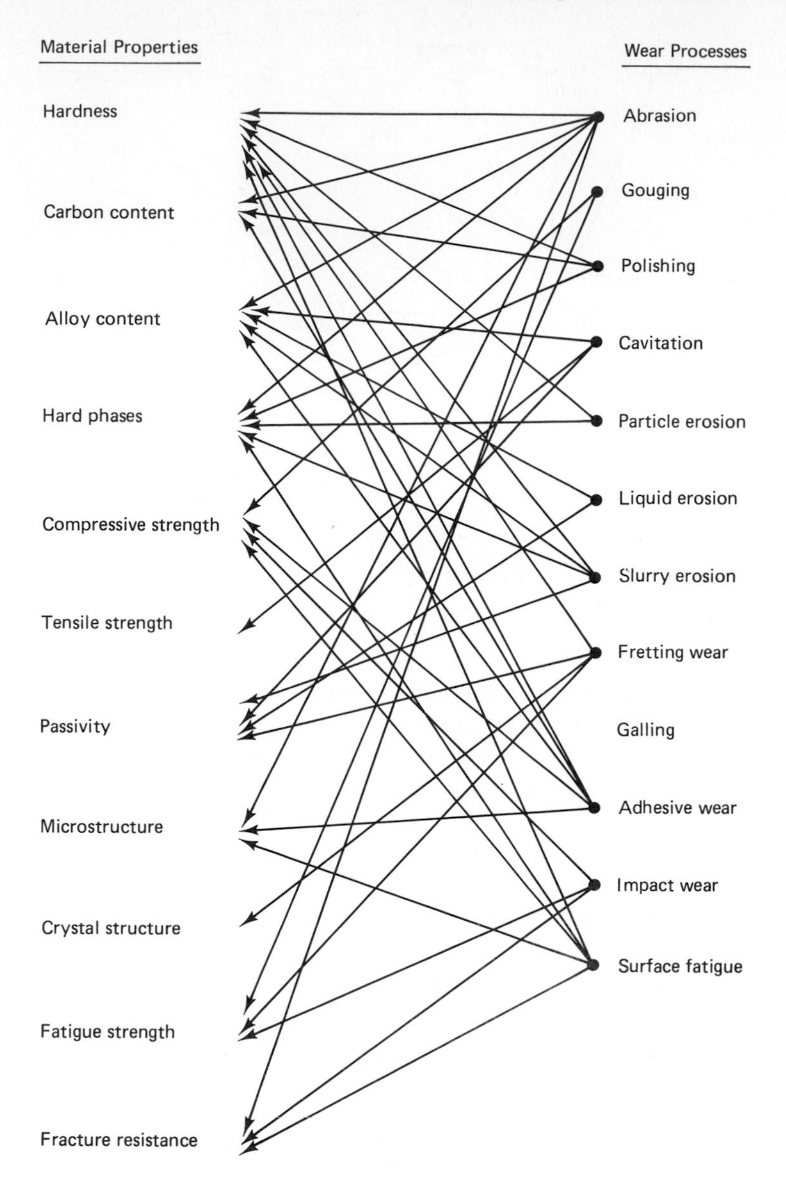

Figure 10–4 Role of material properties in solving wear problems.

form, shape, or work other materials. The most important tool materials are tool steels, followed by alloy steels, carbon steels, cemented carbide, stainless steels, nonferrous metals, and then ceramics. These material systems are really the basis of hardfacing systems, and we will preface our discussion of hardfacing consumables by discussing the material factors in these systems that relate to wear; they are essentially the same factors that are the basis of hardfacing systems.

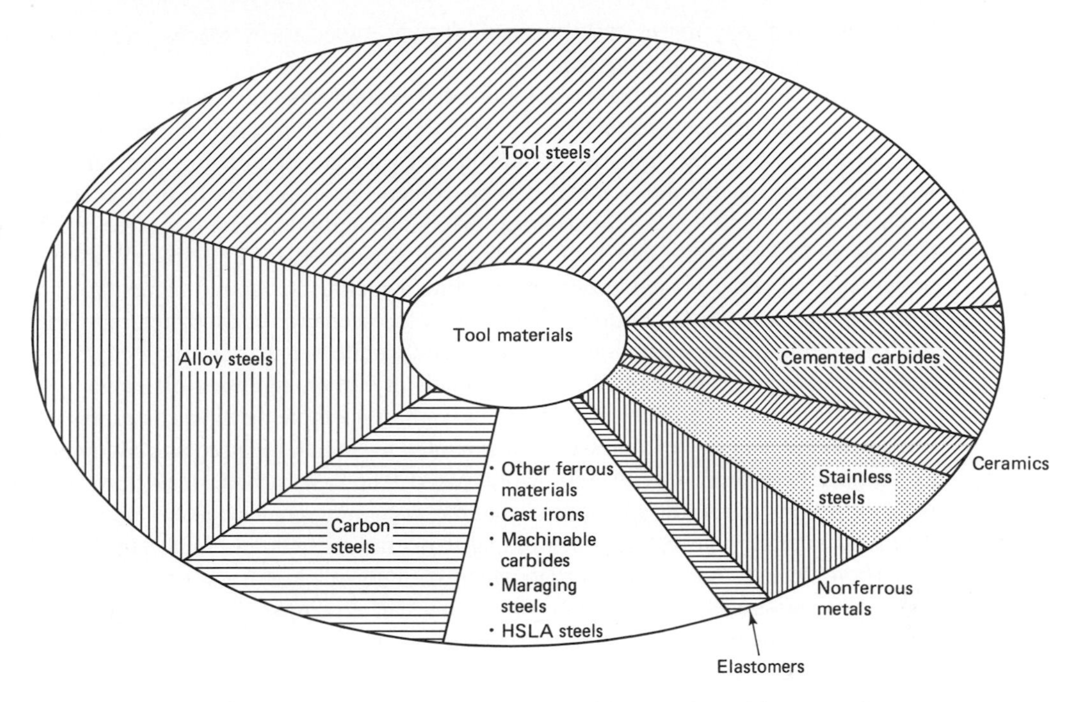

Figure 10–5 Spectrum of tool materials showing estimated frequency of use.

Carbon Steels

Carbon steels are steels with up to 2 percent carbon and only residual amounts of other elements, with the exception of aluminum or silicon, which is added for deoxidation, and manganese that is varied to increase strength. These steels cannot be direct hardened (by heating and quenching) unless the carbon content is above about 0.4 percent. To obtain fully hardened structure, it is generally felt that the carbon content should be about 0.6 percent. The hardenability of carbon steels is very low, and it is common practice to use low-carbon steels for parts that will never be quench hardened or for parts that will be carburized or subject to other diffusion treatments for hardening. Medium-carbon steels, with carbon in the range of about 0.4 to 0.8 percent, can be direct hardened, but section thickness should be kept below about 1 in. (25 mm) to assure response to quenching. Because of their low hardenability, these steels require rapid quenching. If they are not cooled from their hardening temperature within seconds, the hardened structure, martensite, will not completely form. The rapid quenching that can be obtained with selective hardening techniques such as flame and induction hardening gets around this problem, and medium-carbon steels are often selective hardened when they are used for wear applications. High-carbon steels are essentially steels with carbon contents higher than 0.8 percent, but from the practical standpoint the highest carbon content normally used is about 1 percent. These steels still require rapid quenching for hardening, and they are used for such things as low-cost saw blades, steel strapping, doctor blades, knives, and the like.

The composition of the commonly used carbon steels is shown in Table 10–1. This table identifies these steels with the AISI (American Iron and Steel Institute) designation system, where the first two letters identify the major alloying elements and the last two digits specify the nominal carbon content in hundredths of a percent. Carbon steels starting with a "11" alloy designation contain sulfur to make them free machining. Carbon and alloy steels can have a range of structures: ferrite, "pure" iron with a body-centered cubic crystal structure; pearlite, lamellar ferrite and cementite (an iron/carbon compound, Fe_3C); austenite, iron with a face-centered cubic structure; martensite, the hard, body-centered tetragonal structure; and bainite, a metastable aggregate of ferrite and cementite and carbides that are compounds of carbon and other elements.

From the standpoint of hardfacing, low-carbon steels are the most common substrates for hardfacing, and low-carbon steel consumables are sometimes used as buildup materials to support hardfacing on severely worn parts. The low-carbon steels have excellent weldability, but the sulfur additions to the free machining grades make them unweldable; they tend to crack on cooling from welding temperatures. Another important relationship of carbon steels to hardfacing is that, when medium- or high-carbon steels are used as hardfacing substrates, the rapid quenching that can accompany arc welding (from mass effects) can cause quenching cracks. Preheating of the substrate before hardfacing reduces the quenching effects and cracking tendencies. Carbon steels with a carbon content in excess of 0.3 percent should be preheated (and usually postheated) for arc welding processes. The carbon content is not a factor with nonfusion thermal spray processes.

It is well known that the hardness of a steel plays a role in its wear resistance. The Archard type of equation that was discussed in Chapter 2 shows wear resistance to be inversely proportional to hardness:

$$W = k\frac{PL}{H}$$

where W = wear volume
k = constant for the tribo system
P = applied load
L = sliding distance
H = hardness of the softer member of the mating couple

For abrasive wear systems, this equation can have a multiplier of 3 on the hardness term; some investigators apply an exponent such as 3/2 on the hardness term. The exact role of hardness is still debated, but there is no debate that hardness does relate to wear. Carbon steels can have hardnesses that range from below 80 HRB to as high as 62 HRC (Figure 10–6). The high hardnesses will increase the resistance of these steels to many wear processes. However, hardness is not the only factor that determines the wear resistance of these steels and other steels. Microstructure affects wear in ways that are not quite as obvious as hardness.

Steels and high alloys are not homogeneous materials; they contain microconstituents that essentially make them composite materials. Concrete is composed of aggregate

TABLE 10–1 NOMINAL COMPOSITIONS OF SOME AISI/SAE (a) CARBON STEELS AND (b) ALLOY STEELS

AISI-SAE No.	Nominal composition wt. %								
	C	Mn	S	Si	Ni	Cr	Mo	V	P
Carbon steels, carburizing grades									
1008	0.1 max	0.4	0.040 max	0.22					0.35 max
1018	0.18	0.75	0.040 max	0.22					0.35 max
1020	0.20	0.45	0.040 max	0.22					0.35 max
Free machining									
1117	0.17	1.15	0.11	0.22					0.35 max
1118	0.17	1.45	0.11	0.22					0.35 max
Carbon steels, direct hardening									
1040	0.40	0.65	0.040 max	0.22					0.35 max
1050	0.51	0.65	0.040 max	0.22					0.35 max
1060	0.60	0.55	0.040 max	0.22					0.35 max
1080	0.82	0.75	0.040 max	0.22					0.35 max
1095	0.97	0.40	0.040 max	0.22					0.35 max
Free machining									
1137	0.35	1.50	0.11	0.22					0.35 max
1141	0.41	1.50	0.11	0.22					0.35 max
1144	0.44	1.50	0.11	0.22					0.35 max

(a)

AISI-SAE No.	Nominal composition wt. %								
	C	Mn	S	Si	Ni	Cr	Mo	V	P
Alloy steels, carburizing grades									
4320	0.21	0.55	0.040 max	0.22	1.70	0.50	0.25		0.35 max
4620	0.21	0.55	0.040 max	0.22	1.70		0.25		0.35 max
4820	0.21	0.60	0.040 max	0.22	3.50		0.25		0.35 max
8620	0.21	0.80	0.040 max	0.22	0.65	0.50	0.22		0.35 max
9310	0.11	0.55	0.040 max	0.22	3.25	1.20	0.11		0.35 max
Alloy steels, direct hardening grades									
1340	0.41	1.75	0.040 max	0.22					0.35 max
4047	0.47	0.80	0.040 max	0.22			0.25		0.35 max
4130	0.31	0.50	0.040 max	0.22		0.95	0.20		0.35 max
4140	0.40	0.88	0.040 max	0.22		0.95	0.20		0.35 max
4150	0.50	0.88	0.040 max	0.22		0.95	0.20		0.35 max
4340	0.40	0.70	0.040 max	0.22	1.83	0.80	0.25		0.35 max
5140	0.40	0.80	0.040 max	0.22		0.80			0.35 max
5150	0.50	0.80	0.040 max	0.22		0.80			0.35 max
5160	0.60	0.88	0.040 max	0.22		0.80			0.35 max
6150	0.50	0.80	0.040 max	0.22		0.95		0.15	0.35 max
8630	0.30	0.80	0.040 max	0.22	0.65	0.50		0.20	0.35 max
8655	0.55	0.88	0.040 max	0.22	0.55	0.50		0.20	0.35 max
8740	0.40	0.88	0.040 max	0.22	0.65	0.50		0.25	0.35 max
9250	0.60	0.88	0.040 max	2.00					0.35 max

(b)

Source: Budinski, K., *Engineering Materials: Properties and Selection,* 2nd ed., Reston, 1983.

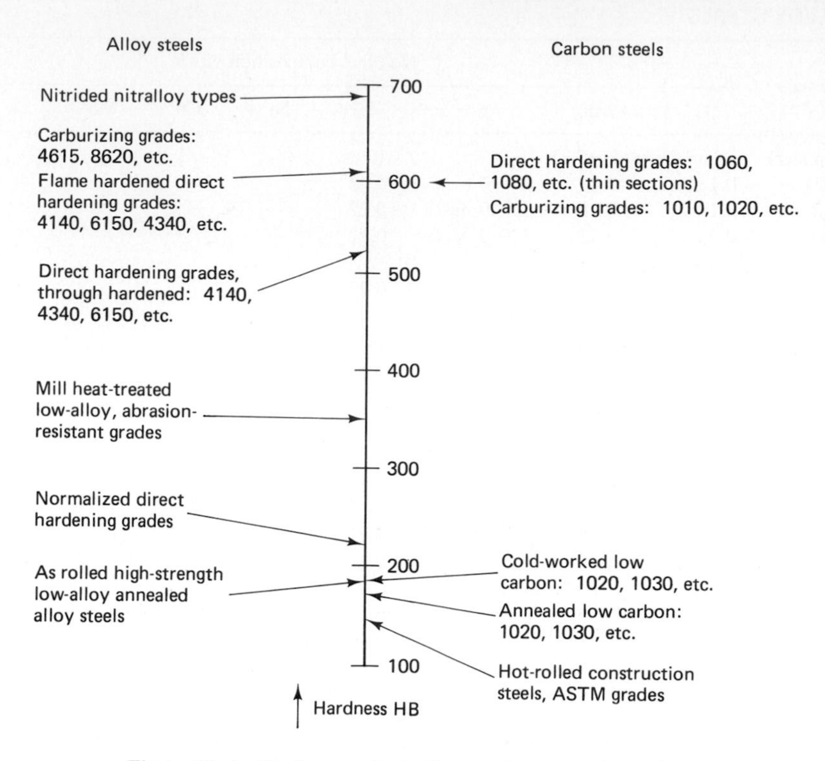

Figure 10–6 Hardnesses obtainable in carbon and alloy steels.

(stones) bonded by cement. When concrete pavement wears, the stones start to stand proud from the surface. They resist tire abrasion better than the cement matrix. The same sort of thing happens in steels and many hardfacing alloys. Table 10–2 lists the microconstituents that can occur in steels.

In nonhardened steels, ferrite is the matrix and the hard microconstituents that can be held in the matrix are carbides, pearlite, and free cementite. These microconstituents are harder than the ferrite matrix, and when they are present they will have an effect on the response of a steel to most wear processes. In the quench-hardened condition, the matrix will be martensite, and the harder phase can be metal or alloy carbides. Figure 10–7 shows how steels with the same hardness can have different wear characteris-

TABLE 10–2 MICROCONSTITUENTS OF STEELS

Steel Phase	Microhardness
Ferrite	70–200 HV
Pearlite	250–320
Cementite	840–1100
Martensite	500–1010
M_7C_3 carbides	1200–1600

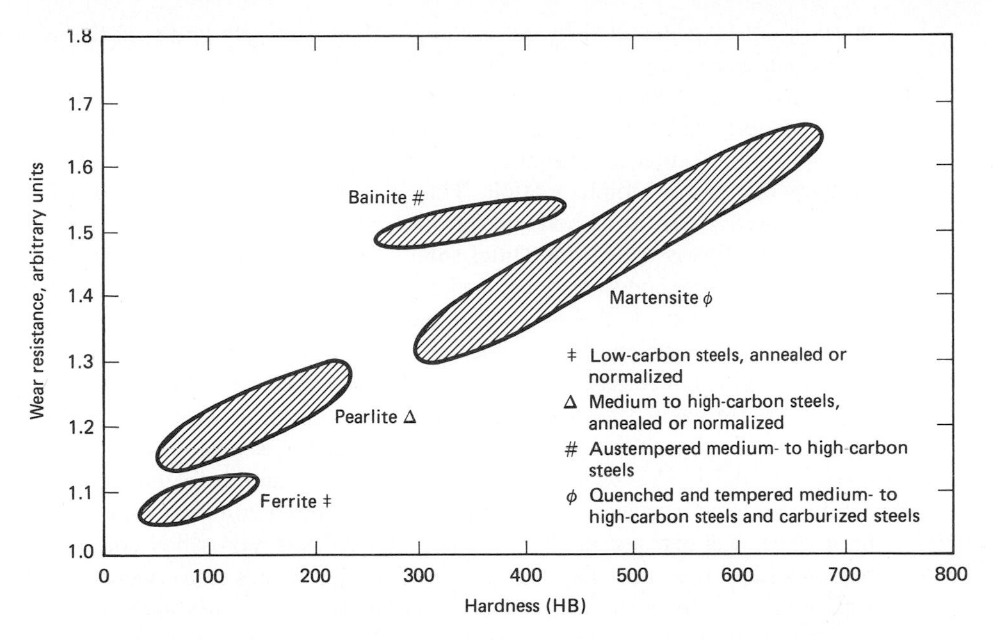

Figure 10–7 Effect of structure and hardness on abrasion resistance (after Zum Gahr).

tics depending on their microstructure. This is a very important concept because it is used in most hardfacing alloys. These alloys can have the same hardness as a carbon or alloy steel, but they will have significantly different wear characteristics because they have a microstructure that contains microconstituents that alter wear properties. In addition, some hardfacing alloys have matrixes that produce wear characteristics that cannot be expected from a simple steel with the same hardness. For example, the wear characteristics of a steel that contains 10 percent carbides in a pearlite matrix will have lower wear resistance than a steel with 10 percent carbides in a martensite matrix.

In summary, carbon steels are not widely used for hardfacing consumables, but they are common substrates for hardfacing, and the principles of what can be done with them in heat treating and the principles of their microstructural possibilities are the basis for many hardfacing alloys.

Alloy Steels

There are three principal ways to strengthen and harden a steel: (1) cold work, (2) alloying, (3) heat treatment. Cold working and alloying work in all metal systems, and hardening by heat treatment only works on some metals. The plastic deformation produced by cold working metals strengthens by a mechanism of dislocation production and multiplication. Dislocations are crystal defects characterized by a line or row of atoms that is displaced from its normal position in the crystal lattice. The disordered crystal structure produced by plastic deformation simply makes additional deformation more difficult; atoms cannot move as easy in a distorted atomic structure as they can in a structure that has not been distorted by deformation. This, in simple terms, is the mechanism of strengthening by cold work.

Alloying works in a similar way; the presence of alloy atoms in a lattice of host atoms inhibits dislocation motion, which in turn strengthens a metal. Large atoms in an iron matrix, such as chromium or nickel, tend to substitute for host atoms and produce barriers to atomic motion. This is called substitutional solid solution. Small alloying atoms such as carbon and nitrogen take up lattice positions between the larger host atoms. This is called interstitial solid solution. Alloy steels contain alloy atoms of various types and concentrations, and they are strengthened by solid solution effects; but more important is the effect that these alloying elements have on the ability of carbon steels to harden. By mechanisms that are more complex than we care to get into, the addition of certain elements to steels increases their ability to form hard structures when they are quench hardened. Hardening temperatures can be altered, and the time constraints for quenching are reduced. For example, a heat treater has only about 1 second to cool 1340 steel from its hardening temperature to the black heat range to get a fully hardened surface. The heat treater has about 1 minute to do the same quench on a 4340 steel that has the same carbon content, but more alloy content; it has better hardenability. If a steel has high hardenability, it is easier to quench harden; it will harden deeper and it may achieve a higher hardness. If a very large amount of alloy is added to a steel, it may also form hard second phases that add wear resistance to the steel.

Some common AISI alloy steel compositions are shown in Table 10–1. The important alloying elements that are found in alloy steels include chromium, nickel, manganese, molybdenum, and vanadium. All these elements increase the hardenability of steels, but to different degrees. Molybdenum is the most effective of this group and nickel is the least effective. Silicon is used to increase hardenability and to impart toughness; vanadium increases hardenability, refines grains, and promotes the formation of hard carbides.

As shown in Table 10–1, alloy steels usually only have fractional percentages of these alloying elements. In these small amounts, the formation of alloy-rich, hard second phases is minimal. Figure 10–8 shows a typical alloy steel microstructure in the hardened condition (completely martensite, no second phases). As a class of materials, alloy steels are widely used for structural applications: high-strength power transmission shafts, cams, gears, levers, frames, and the like. For wear applications, they are hardened to the usual range of 55 to 60 HRC, often by selective hardening. Some grades of alloy steel are not normally used in the direct hardened condition (4615, 8620, 9310, etc.). These grades do not have sufficient carbon content to form high hardness on quenching. They are usually carburized for a hard surface, and the role of the alloy addition is to make it easier to get the case hard and to produce a stronger core than can be obtained with carbon steels.

Many hardfacing materials are simply alloy steels. For example, a 4130 steel consumable is available for rebuilding the popular medium-carbon alloy steels, 4140 steel, and 4340 steel. Rods and electrodes with less than 5 percent total alloy are used as general-purpose hardfacings that can have a deposit hardness of anywhere from about

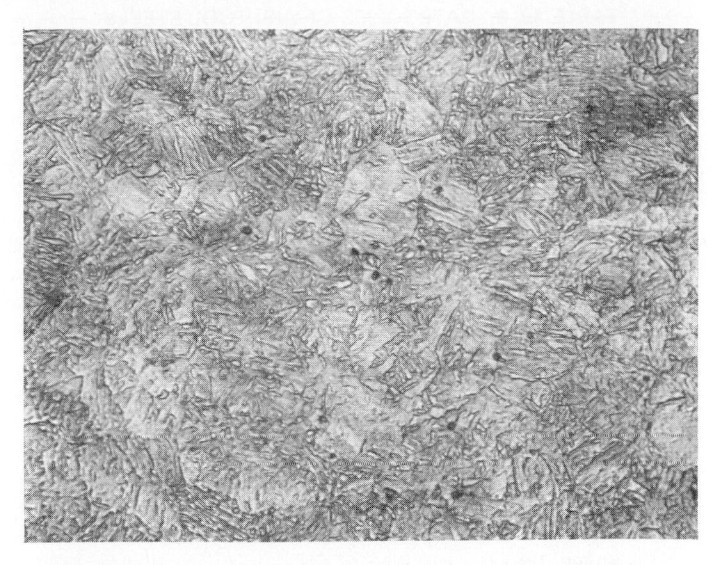

Figure 10–8 Structure of hardened and tempered 4140 steel (750×).

40 to 55 HRC. Appendix III lists many alloy steel consumables under the heading of iron-base consumables. They are available as bare rod for GTAW welding, as coated electrodes, and as GMAW wire, and some thermal spray consumables are essentially similar to the AISI alloy steels in composition. The low-alloy content makes these types of consumables low cost. Their use characteristics parallel those of the AISI alloy steels. The medium-carbon alloy steels offer good strength and abuse resistance without the brittleness that characterizes high alloys and tool steels. Fusion welds with these materials are usually crack-free. Low-alloy hardfacing consumables have significant utility for applications that cannot tolerate the brittleness of the high alloys or the softness of low-carbon buildup consumables.

Tool Steels

Figure 10–5 shows that tool steels are the most widely used materials for tool materials. The handbook definition of a tool steel is simply a carbon or alloy steel that is capable of being hardened such that it can be used to work or shape other materials. The carbon steels and alloy steels that we just described fit this definition; they can be and are used for tool applications. Tool steels are different from carbon and alloy steels even though the definition could include them. To metallurgists, the term tool steel means a class of steels that are capable of being hardened and that are made to quality requirements that make them more acceptable than carbon and alloy steels for severe service. They have better cleanliness; their hardenability is often guaranteed; they receive extra inspection at mills for defects; they are often given special heat treatments at the mill (spheroidization) to allow uniform hardening. Thus these steels are different from

TABLE 10–3 CATEGORIES OF AISI TOOL STEELS

Class	Class symbol	Discriminating factor
Water hardening tool steels	W	Carbon is the major alloying element
Cold work tool steels	O	Oil hardening
	A	Medium alloy air hardening
	D	High carbon, high chromium
Shock resisting tool	S	Medium carbon, high toughness
Mold steels	P	Low carbon, good fabricability
Hot work tool steels	H1-H19	Chromium types
	H20-H39	Tungsten types
	H40-H59	Molybdenum types
High speed tool Steels	M	Molybdenum types
	T	Tungsten types
Special-purpose tool steels	L	Low-alloy types

Source: Budinski, K., *Engineering Materials: Properties and Selection,* 2nd ed., Reston, 1983.

carbon and alloy steels even though their chemical compositions can overlap with these other steels.

The AISI system for identifying tool steels consists of a letter prefix to denote a class of tool steels and a number to denote a particular alloy within the class. Table 10–3 shows the AISI tool steel categories. These categories are based on intended use, although it is "not against the law" to use a tool steel from one category for an application outside of this category. The role of these steels in hardfacing is that some are used as hardfacing consumables. They are mostly available in bare rod for GTAW application, and they are most widely used for the rebuilding of tools and for small hardfacing applications. We will next discuss the various categories of tool steels that can be used as hardfacing consumables.

Water Hardening (W) The W series tool steels are essentially carbon steels with improved quality control in manufacture. The composition of the alloys of commercial importance is presented in Table 10–4. All these steels have low hardenability and require water quenching to achieve a hard surface; thus the class designation W. The alloy with chromium has a little better hardenability than the other two alloys, and the alloy with vanadium has a little better resistance to quench cracking, but as a class of steels these materials are very susceptible to cracking in hardening. They are the cheapest tool steels and they are often used for massive auto body panel dies. They are very shallow hardening; in heavy sections the surface gets hard and the core stays soft and tough. W series hardfacing alloys are most often used for repairing W series tools. These steels do not have any properties that would make them suitable as a general-purpose hardfacing.

TABLE 10–4 COMPOSITION OF W SERIES TOOL STEELS

	Nominal composition (wt. %)		
Type	C	Cr	V
W1	0.6–1.4	—	—
W2	0.6–1.4	—	0.25
W5	1.1	0.5	—

Cold Work Tool Steels Table 10–3 shows three classes of cold work tool steels. As their name implies, these steels are used for tools that work and shape other materials at room temperature or at temperatures lower than about 700°F (370°C). The O series steels (Table 10–5) are oil hardening, and they are characterized by a carbon content of about 1 percent and small percentages of other alloying elements. The O1 and O2 grades are widely used for short-run tools. The normal use hardness is about 60 HRC, and these steels are popular because they are low cost, easy to machine, and are not as prone to heat-treating distortion and cracking as are the W tool steels. The other O grades are used for special applications, and the hardfacing consumables that are predominately based on this class of steels are like the O2 alloy in properties.

The A series of tool steels (Table 10–6) are air hardening and they are characterized by carbon contents of about 1 percent, 1 percent molybdenum, and 5 percent chromium. Chromium and molybdenum increase the hardenability of these steels so that sections up to several inches (100 mm) thick will harden with air quenching from the hardening temperature. The A2 grade is almost a standard for general-purpose tool and die work. The other grades are modifications of the A2 grade for special purposes. For example, the A7 grade has high carbon and vanadium content to promote the formation of wear-resistant vanadium carbides in the structure. The A10 grade has a composition that produces free graphite in the microstructure for galling resistance.

Like the O tool steels, the welding consumables patterned after this class of steels are like the A2 grade in composition. These steels have sufficient alloy content to allow the formation of hard second phases in their microstructure. Carbides of chromium and other elements are present to promote abrasion and metal-to-metal wear resistance.

TABLE 10–5 COMPOSITION OF O SERIES TOOL STEELS

	Nominal composition (wt. %)					
Type	C	Mn	S_1	W	Mo	Cr
O1	0.90	1.00	—	0.50	—	0.50
O2	0.90	1.60	—	—	—	—
O6	1.45	0.80	1.00	—	0.25	—
O7	1.20	—	—	1.75	—	0.75

TABLE 10–6 COMPOSITION OF A SERIES TOOL STEELS

	Nominal composition (wt. %)							
Type	C	Mn	Si	W	Mo	Cr	V	Ni
A2	1.00	—	—	—	1.00	5.00	—	—
A3	1.25	—	—	—	1.00	5.00	1.00	—
A4	1.00	2.00	—	—	1.00	1.00	—	—
A6	0.70	2.00	—	—	1.25	1.00	—	—
A7	2.25	—	—	—	1.00	5.25	4.75	—
A8	0.55	—	—	1.25	1.25	5.00	—	—
A9	0.50	—	—	—	1.40	5.00	1.00	1.50
A10	1.35	1.80	1.25	—	1.50	—	—	1.80

The D series of tool steels (Table 10–7) are characterized by a composition of about 1.5 to 2 percent carbon and about 12 percent chromium. These steels are also called high-carbon/high-chromium tool steels. They are considered to be the most wear resistant standard steels for high-production tools like punches and dies. They are air hardening in section thicknesses as great as 6 in. (150 mm). All these steels contain chromium carbides in their microstructure for wear resistance (Figure 10–9). The standard high-carbon/high-chromium tool steel is type D2. Two grades of D tool steels have higher carbon for added wear resistance. Cobalt is added to one grade to produce better resistance to softening at elevated temperature, and the alloy with vanadium contains massive vanadium and chromium carbides in its microstructure. The hardfacing consumables that have compositions similar to this family of alloys are usually used for rebuilding tools made from the D2 alloy. A very important aspect of the A and D tool steels is the presence of carbide particles in the microstructure. Many other hardfacing alloys have been developed that use the concept of hard second phases to make them more wear resistant than competitive surface treatments.

TABLE 10–7 COMPOSITION OF D SERIES TOOL STEELS

	Nominal composition (wt. %)				
Type	C	Mo	Cr	V	Co
D2	1.50	1.00	12.00	1.00	—
D3	2.25	—	12.00	—	—
D4	2.25	1.00	12.00	—	—
D5	1.50	1.00	12.00	—	3.00
D7	2.35	1.00	12.00	4.00	—

Figure 10–9 Microstructure of D2 tool steel (400×). The white particles are chromium carbides.

Shock resistant Tool Steels (S) The common composition characteristic of this class of tool steels is medium carbon content (Table 10–8). These steels were developed for tools that are battered in use, such as chisels and riveting devices. The S2 alloy with silicon as the major alloy addition is the basic grade of this family of steels. Tungsten is added to the S1 grade for wear resistance; S5 and S6 have slightly different hardening characteristics than S2, but the S7 alloy is significantly different from the other S tool steels. It is the only grade that is air hardening; the others are oil hardening. The air-hardening capability reduces distortion in hardening, and this steel has excellent shock resistance even at a hardness of 56 HRC. The other grades are usually used at hardness levels in the upper forties (HRC). These steels are used when breakage problems

TABLE 10–8 COMPOSITION OF SHOCK-RESISTANT TOOL STEELS

	Nominal composition (wt. %)					
Type	C	Mn	Si	W	Mo	Cr
S1	0.50	—	—	2.50	—	1.50
S2	0.50	—	1.00	—	0.50	—
S5	0.50	1.80	2.00	—	0.40	—
S6	0.45	1.40	2.25	—	0.40	1.50
S7	0.50	—	—	—	1.40	3.25

arise with other grades of tool steels. Because of the low carbon content, they are not particularly wear and abrasion resistant; but their impact properties preclude this property deficiency for many applications. The S7 composition is used for hardfacing rods, and some of the other compositions are used for hardfacing rods for repair of S tool steel parts.

Mold Steels (P) The mold steels are (Table 10–9) intended for use as plastic injection mold cavities and mold bases. Most grades have very low carbon content and they cannot be direct hardened. The low-carbon grades are usually carburized. The various low-carbon grades have subtle differences in hardenability and properties, with the exception of P21, which is precipitation hardened instead of quench hardened. Type P20 at one time was the most popular steel for injection molding cavities. It is usually supplied by steel mills prehardened to 30 HRC and is used at this hardness. None of these compositions is suitable for hardfacing applications, but the basic P20 composition is sometimes used for buildup consumables. Bare rods are available to match the composition of some of the other grades, but repair welding cannot usually be done on P steel tools that have been carburized; carburized steels tend to crack when fusion welded.

TABLE 10–9 COMPOSITION OF P SERIES TOOL STEELS

	Nominal composition (wt. %)				
Type	C	Mo	Cr	Ni	Al
P2	0.07	1.20	2.00	0.50	—
P3	0.10	—	0.60	1.25	—
P4	0.07	0.75	5.00	—	—
P5	0.10	—	2.25	—	—
P6	0.10	—	1.50	3.50	—
P20	0.35	0.40	1.70	—	—
P21	0.20	—	—	4.00	1.20

Hot Work Tool Steels (H) Hot work tool steels were developed for forging and extrusion dies and other tools that work or process metals in the visible heat range. Like the shock-resisting steels, these steels have medium carbon content and most have chromium as a major alloying element (Table 10–10). There are more grades than those listed, which are only the most widely used grades. Of the group, types H11 and H13 are the most widely used. H13 tool steel is a standard for aluminum and zinc die casting tooling; H11 is also used for die casting cavities, but it is also considered to be an ultrahigh-strength steel. It is used for critical structural components in aircraft. Most of the other grades are used for forging tools and tools that work materials in the red heat range. The refractory metal additions to these grades contribute resistance to softening at high temperatures and to erosion resistance from scaled metal.

As a class of steels, these materials have excellent toughness; they have a maximum

TABLE 10–10 COMPOSITION OF HOT WORK TOOL STEELS

	Nominal composition (wt. %)					
Type	C	W	Mo	Cr	V	Co
Chromium types						
H10	0.40	—	2.50	3.25	0.40	—
H11	0.35	—	1.50	5.00	0.40	—
H12	0.35	1.50	1.50	5.00	0.40	—
H13	0.35	—	1.50	5.00	1.00	—
H14	0.40	5.00	—	5.00	—	—
H19	0.40	4.25	—	4.25	2.00	4.25
Tungsten types						
H21	0.35	9.00	—	3.50	—	—
H22	0.35	11.00	—	2.00	—	—
H23	0.30	12.00	—	12.00	—	—
H24	0.45	15.00	—	3.00	—	—
H25	0.25	15.00	—	4.00	—	—
H26	0.50	18.00	—	4.00	1.00	—
Molybdenum types						
H42	0.60	6.00	5.00	4.00	2.00	—

hardness capability of about 52 HRC. They can be used for battering applications, and they are extremely deep hardening. Sections up to 16 in. (40 cm) thick can be air hardened. There are hardfacing alloys patterned after these compositions, and there are consumables that can be used to rebuild these grades of steels.

High Speed Tool Steels (T&M) The EFe and RFe grades of hardfacing consumables are high speed steels. They will be discussed in the next section. Over 30 grades of high-speed steels are commercially available. Table 10–11 lists some of the more important grades. These steels have about 1 percent carbon, and the T series have tungsten as a major alloying element; the M series have molybdenum as a major alloying element. These steels were developed to have resistance to softening when they are used for cutting tools. Cutting steels at high speed often generates chips that are red hot and the tip of the tool that produced the chips is also hot. These steels contain refractory metal additions that provide resistance to softening at temperatures in excess of 1000°F (540°C).

These steels are very wear resistant. Working hardnesses are often in the range of 63 to 67 HRC, harder than most other steels. There are hardfacing consumables that match some of these alloys, but high-speed steel consumables are widely used for all sorts of hardfacing applications. They are favorites for die repair since single-layer deposits are almost always at least 60 HRC, which is not the case with many of the other tool steel consumables and hardfacing alloys.

One other class of tool steels is listed in Table 10–3, but these are special-purpose materials that are really no more than alloy steels made to tool steel quality requirements. They are not important hardfacing consumables. The tool steel classes that we have

TABLE 10–11 COMPOSITION OF HIGH SPEED TOOL STEELS

	Nominal composition (wt. %)					
Type	C	W	Mo	Cr	V	Co
M1	0.85	1.50	8.50	4.00	1.00	—
M2	1.00	6.00	5.00	4.00	2.00	—
M3-1	1.05	6.00	5.00	4.00	2.40	—
M3-2	1.20	6.00	5.00	4.00	3.00	—
M4	1.30	5.50	4.50	4.00	4.00	—
M7	1.00	1.75	8.75	4.00	2.00	—
M42	1.10	1.50	9.50	3.75	1.15	8.00
T1	0.75	18.00	—	4.00	1.00	—
T15	1.50	12.00	—	4.00	5.00	5.00

discussed are important as substrates that are frequently hardfaced, and their properties are such that they have been adopted by many hardfacing consumable manufacturers for general hardfacing applications.

Cemented Carbides

The welding consumables that are called ''composites'' by the AWS are based on technology related to cemented carbides. This group of engineering materials fits into the generic category of cermets; a cermet is a material that is part metal and part ceramic. Cemented carbides are ceramic materials, usually carbides, in particle form that are bonded together with a metal, usually cobalt or nickel. The carbides used in cemented carbides are compounds of carbon and tungsten, tantalum, titanium, columbium, and other metals. Their hardness varies, but all are harder than the hardest metal:

ABSOLUTE HARDNESS (kg/mm^2)

Hardest steel, 900

Tantalum carbide, 1750

Molybdenum carbide, 2000

Tungsten carbide, 2000

Columbium carbide, 2100

Titanium carbide, 2500

Vanadium carbide, 2600

Boron carbide, 2700

The most common cemented carbide composition is tungsten carbide with a cobalt binder. Tungsten carbide is made by arc melting tungsten and carbon or by carburizing tungsten powder. Once the carbide is produced, it is ball milled to make fine powder. In some processes, cobalt powder is put into the ball mill when the WC is being crushed.

The cobalt becomes mechanically plated onto the powder particles. The plated powders are then graded to the desired size fraction, compacted into a shape, and then sintered. The WC particles become bonded or cemented by diffusion bonding or liquation of the cobalt coating at the sintering temperature of about 2000°F (1090°C).

The carbide component can be any of those listed previously, and the binder content can vary from a few percent to about 30 percent. The higher binder contents produce higher toughnesses. Mixed carbides are used to produce different wear-resistance characteristics. The particle size and distribution also affect wear resistance and mechanical properties. The finer the particle size the greater the abrasion resistance; the larger the binder content, the lower the abrasion resistance. There are no agreed-to controls on the composition of cemented carbides in the United States. There are standard grades of cemented carbides, as shown in Table 10–12, but the C2 grade from one supplier may not have the same composition as the C2 grade from another supplier. The grades simply designate that a particular grade will perform a particular function. For example, C2 grades are suitable for machining cast irons and nonferrous metals.

The C2 composition will be adequate for most wear applications. A cemented carbide will almost always outwear any steel in most wear processes. They have excellent abrasion resistance, excellent metal-to-metal wear resistance, good erosion resistance, and good resistance to surface fatigue. They cannot take battering or bending, so they may not be suitable for gouging abrasion and impact wear systems. Corrosion-resistant

TABLE 10–12 CARBIDE GRADE CODE

Use category	Code	Recommended application	Example properties					
			Composition*				Hardness* RA	Transverse* rupture strength (MPa)
			WC	TiC	TaC	Co		
Machining of Cast iron, Nonferrous and non-metallic material	C-1	Roughing	94	—	—	6	91	2000
	C-2	General propose	92	—	2	6	92	1550
	C-3	Finishing	92	—	4	4	92	1520
	C-4	Precision finishing	96	—	—	4	93	1400
Machining of carbon, alloy, and tool steels	C-5	Roughing	75	8	7	10	91	1870
	C-6	General purpose	79	8	4	9	92	1650
	C-7	Finishing	70	12	12	6	92	1750
	C-8	Precision finishing	77	15	3	5	93	1180
Wear applications	C-9	No shock	94	—	—	6	92	1520
	C-10	Light shock	92	—	—	8	91	2000
	C-11	Heavy shock	85	—	—	15	89	2200
Impact applications	C-12	Light	88	—	—	12	88	2500
	C-13	Medium	80			20	86	2600
	C-14	Heavy	75			25	85	2750

* Composition and properties are averages from several manufacturers.

Source: Budinski, K., *Engineering Materials: Properties and Selection,* 2nd ed., Reston, 1983.

Figure 10–10 Cemented carbide microstructure, WC/cobalt binder (1100×).

grades are available for chemical applications (nickel or nickel/chromium binder). Because cemented carbides are made by press compaction of powders, the shapes that are available are those that can be made in a compaction die. An extremely large piece of carbide would be 2 in. (50 mm) in diameter by 36 in. (900 mm) long. In fact, this size is above the capability of many manufacturers. The high volume use of cemented carbides is for cutting tool inserts. Small wear parts are molded to shape, and dies and punches are made from standard shapes such as blocks and rounds. They cannot be machined; shapes must be generated by grinding or electrical discharge machining.

We will discuss the specific welding consumables that use carbide technology in the section on composite consumables, but essentially they are cermets, tungsten carbide in metal binders. The big difference is that the carbide concentration is much lower, often as low as 10 percent, and the binders are usually hard metals, alloy steels, or hard cobalt-base alloys. The carbide particles in composite welding consumables can be as large as ⅛ in. (3.1 mm) in diameter. The size range for carbide particles in a cemented carbide tool material is much smaller, from 1 to 3 μm (40 to 120 μin.), (Figure 10–10). In summary, cemented carbides are very useful for wear applications, and composite welding electrodes are patterned after these materials, but the carbide particles are much larger; their concentration is much lower and the binders are usually hard metals instead of soft cobalt.

Stainless Steels

The definition of a stainless steel is an alloy of iron, carbon, and chromium with at least 10.5 percent chromium and the ability for passive behavior. Passive behavior or passivity is the ability of a material to resist attack in an environment that is capable of attacking the material. Stainless steels are corrosion resistant in many environments because they have a passive film of chromium oxide that protects them from attack. If the film is removed by abrasion or some other cause, the material can be attacked. Most corrosion-resistant metals derive their corrosion resistance from passive surface films. These films are usually on the order of 100 angstroms (10×10^{-10} m) thick, and they usually form spontaneously in air. The phenomenon of passivity plays a role in hardfacing alloys; many hardfacing consumables contain more than 10.5 percent chromium, but they are not considered to be stainless steels because they usually have high carbon contents that inhibit the formation of a passive surface. The same thing is true of the high-carbon, high-chromium tool steels. They have 12 percent chromium in their composition, but they are not stainless because of the high carbon content. From the hard-facing user's standpoint, this means that one cannot always depend on obtaining corrosion resistance from the iron/chromium and other high chromium hardfacing consumables that produce high-hardness deposits.

As a class of materials, stainless steels are not noted for their wear resistance. There are a number of basic types of stainless steels but only a few alloys are suitable for wear applications:

Ferritic (400 series, low C): 60 to 95 HRB
Austenitic (300 series): 60 HRB to 45 HRC
Martensitic (400 series, high C): 80 HRB to 58 HRC
PH (precipitation hardening): 20 to 48 HRC

The classification system coincides with the microstructures of the alloys. The ferritic grades cannot be quench hardened, and they are alloys of iron, chromium, and carbon with carbon usually less than 0.2 percent. They are the lowest-cost stainless steels, and they do not have much utility in wear processes with the exception of some forms of erosion. The original austenitic stainless steel had a composition of 18 percent chromium and 8 percent nickel. The nickel in these alloys produces the austenitic structure. Carbon contents are usually less than 0.1 percent. They are very corrosion resistant, but notoriously poor in sliding wear systems. They are very prone to galling. They are not used for hardfacings, but they are frequently used as substrates for hardfacing. They have excellent weldability. Some grades of austenitic stainless steel have manganese additions to replace the normal nickel that is present (200 series of austenitic stainless steels). Some proprietary stainless steels with high manganese are used for wear applications; they work harden, and galling tendencies are better than for the 300 series alloys. These high-manganese compositions are not presently available as welding consumables.

The martensitic stainless steels are quench hardenable, but most alloys cannot achieve a hardness greater than the low forties HRC. Types 440 A, B, and C and type 420 can be hardened to above 50 HRC, and they have significant utility in wear systems. Type 440 C is almost a standard for knives and cutting devices that are used in corrosive environments. Type 420 stainless steel has a maximum hardness of about 52 HRC, and it is widely used for plastic injection molds. Type 440C is not used as a hardfacing consumable; it is unweldable, and deposits of this material would be very prone to cracking. On the other hand, type 420 stainless steel is used as a hardfacing consumable. It has excellent weldability, and it will form deposits with a minimum hardness of 50 HRC on most weldable substrates. It is used for fusion welding hardfacing of steel mill rolls and the like. Consumables are available in wire form for automatic welding machines (GTAW, PTA, GTAW, etc.).

The precipitation-hardening stainless steels (PH) are martensitic or semiaustenitic in structure, and they are hardened by precipitation of copper or aluminum. Some of the alloys can be hardened to as high as 48 HRC, but they are not abrasion resistant or resistant to metal-to-metal types of wear. Their carbon content is typically less than 0.1 percent, and the low-carbon martensite structure that they have in the hardened condition does not resist most forms of sliding wear. There are no hardfacing consumables based on this alloy system, but they have excellent weldability and are suitable substrates for fusion welding hardfacings.

In summary, stainless steels are not normally used for hardfacing consumables, but the weldable grades of stainless steel are often used as hardfacing substrates. Type 420 stainless steel is the exception; this stainless steel is an excellent hardfacing consumable for parts that require wear and corrosion resistance.

FUSION ALLOYS

The AWS specification system for hardfacing rods and electrodes uses an alphanumeric system with a prefix of ''E'' for electrodes and a ''R'' for bare wire or cast bare rods. Designating a consumable as a electrode means that it carries current in the deposition process. Rods do not normally carry current. Rods are used in OAW, GTAW, PAW, and the like; electrodes are used in SMAW, GMAW, SAW, and similar processes. The principal identifier in the system is a series of letters that are the chemical symbols for the major elements in the alloy. The designation RNiCr-A designates a bare rod (R), with nickel (Ni) and chromium (Cr) as the major elements comprising the hardfacing alloy. The ''A'' suffix denotes a particular range of chemical composition in a family of nickel/chromium alloys. That is, several nickel/chromium hardfacing alloys have been recognized in the AWS specification, and letter suffixes denote these individual alloys.

The AWS hardfacing designation system only applies to high-speed steel, iron/chromium alloys, manganese steels, cobalt/chromium alloys, nickel/chromium/boron alloys, copper-base alloys, and composites. Many other families of alloys, such as the tool steels and alloy steels, are commercially available. The Appendix III lists many of these by trade name. With regard to product form, the AWS letter designation system only covers rods and electrodes; fusion hardfacing alloys are also available in powder form for spray-and-fuse application. The following discussion covers the AWS alloys; Appendix III can be consulted for a more complete listing of what is available.

RFe5 and EFe5 High Speed Filler Metals

The term high speed steel is used in metallurgy to describe a family of steel alloys that are characterized by high hardness and exceptional resistance to softening at elevated temperatures. The term high speed comes from the days of the development of these alloys in the 1920s. Prior to the introduction of these alloys in industry, high-carbon steels were used as cutting tools. They were not very resistant to softening due to the frictional heat that arises in machining. The advent of this new family of alloys allowed machining at much higher speeds, thus the term high speed steels. The original high-speed steels contained up to 20 percent tungsten to promote resistance to heat softening. Tungsten is a refractory metal with a very high melting temperature; when in solid solution in steels, it promotes resistance to heat softening. Another effect of this alloy addition is to promote the formation of alloy carbides in the microstructure that improve wear characteristics.

During World War II, tungsten was relatively unavailable, and new high-speed steels were developed that had molybdenum, another refractory metal, as the principal alloying element. Most present-day high speed steels contain both tungsten and molybdenum for heat resistance, chromium for hardenability, other elements such as vanadium for wear resistance, and cobalt for heat resistance and other purposes. The carbon content varies in the different high-speed alloys, but most contain about 1 percent. Table 10–13 lists alloys that have an AWS designation.

Two-layer deposits of the alloys have hardness in the range of 55 to 60 HRC. They typically lose only about 10 points of hardness when heated to temperatures as high as 1200°F (650°C). Although they contain a large weight percentage of alloy, they are not considered to be rust, chemical, or oxidation resistant. They are brittle, and machining in the as-deposited condition is not possible. This family of hardfacing alloys has properties essentially similar to the previously mentioned M and T series

TABLE 10–13 AWS ALLOY DESIGNATIONS FOR HIGH SPEED STEEL HARDFACING ALLOYS*

	C	Mn	W	Cr	Mo	V	Si	Fe	Typical two-layer hardness (HRC)
RFe5-A	0.7–1.0	0.5	5–7	3–5	4–6	1.0–2.5	0.5	Rem.	55
RFe5-B	0.5–0.9	0.5	1.0–25	3–5	5–9.5	0.8–13	0.5	Rem.	55
EFe5-A	0.7–1.0	0.6	5–7	3–5	4–6	1.0–2.5	0.8	Rem.	55
EFe5-B	0.5–0.9	0.6	1.0–2.5	3–5	5–9.5	0.8–1.3	0.8	Rem.	55

* This table and subsequent tables of AWS hardfacing compositions are reproduced by permission of the American Welding Society from A5.13–80, Specification for Solid Surfacing Rods and Electrodes. For a complete copy of AWS A5.13–80, contact AWS, 550 N.W. LeJuene Rd., P.O., Box 351040, Miami, Florida, 33135. Telephone: 305-443-9353.

tool steels used for cutting tools (drills, lathe bits, etc.). They are available in small-diameter wires that are suitable for such things as die repairs.

Austenitic Manganese Steels

Austenitic manganese steels are medium-carbon steels that contain sufficient manganese to force the room temperature structure to be austenite instead of the ferrite/pearlite structure that is normal in medium-carbon steels. Austenite is soft and ductile, but it has a pronounced ability to work harden. Anyone who has ever tried to drill a hole in austenitic stainless steel with a dull drill can attest to the ability of austenitics to work harden. A hardness of 50 HRC can be produced in the center of an attempted drill hole in austenitic stainless steel (using a dull drill). The austenitic manganese steels respond in a similar fashion when used in a battering mode. The more they are impacted with rocks and the like, the harder they get. The chemical composition of the AWS austenitic manganese alloys are shown in Table 10–14.

The difference between these two alloys is subtle; the A alloy has better resistance to embrittlement than the B alloy, and the B alloy has a higher yield strength. Embrittlement can be a problem in these materials (Figure 10–11). The austenitic structure is produced by water quenching from the austenitizing temperature. Reheating these alloys (in the heat-affected zone or by preheating) can cause changes in the metastable austenite that lowers the impact strength. In other words, these alloys should be kept cool during welding. Embrittlement can occur in the B alloy when the material is reheated above about 600°F (315°C). The A alloy resists embrittlement at temperatures up to 800°F (425°C). Embrittlement can also occur when these alloys are put on carbon steels. The most suitable substrates are manganese steels or austenitic stainless steels.

The as-deposited hardness of these alloys is about 20 HRC; cold work can raise their hardness up to 50 HRC depending on the degree of cold work. Austenitic manganese steels are not corrosion resistant; because of the embrittlement problem, they should not be used at elevated temperature. As one might expect, these alloys are extremely difficult to machine. It is best to use these alloys as deposited.

The austenitic manganese hardfacing alloys are used for applications where battering is the best description of the intended service. They are widely used for gyratory rock crushers, for earth loader bucket teeth, and for many of the other types of equipment used in earth moving and mining. Railroad switch frogs are another typical application. In summary, these alloys are most suited to application where the hardfaced parts will

TABLE 10–14 AWS AUSTENITIC MANGANESE STEEL HARDFACING ALLOYS

	C	Mn	Ni	Cr	Mo	Si	P	Fe	Typical two-layer hardness (HRC)
EFeMn-A	0.5–0.9	11–16	2.75–6.0	0.5	—	1.3	0.03	Rem.	20
EFeMn-B	0.5–0.9	11–16	—	0.5	0.6–1.4	0.3–1.3	—	Rem.	20

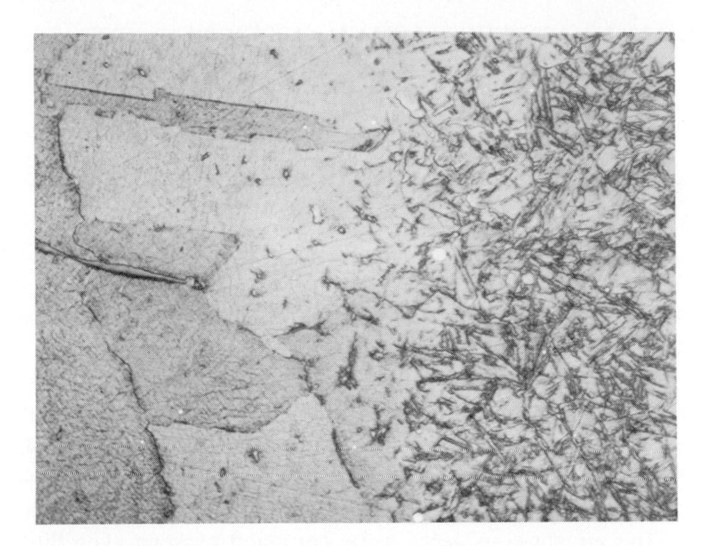

Figure 10–11 Manganese steel (800×). The area to the left is the normal austenitic structure. The needlelike structure to the right is untempered martensite caused by welding.

be impacted. The austenitic manganese steels work harden and minimize deformation and material removal. They are not well suited for metal-to-metal wear or for scratching abrasion. The user should also keep in mind that the lower-strength alloys will not work harden without plastic deformation. This means dimensional change. For example, if a $\frac{1}{4}$-in. (6.2-mm) thick deposit is applied to a substrate, it will not become hard until it is deformed to probably one-half of the deposit thickness. If your application cannot tolerate this type of dimensional change in cold working, these alloys may not be a proper choice.

High-Chromium Iron Alloys

This family of alloys is a by-product of cast-iron foundry technology. The cast-iron people have known for over 60 years that alloying cast iron with a significant amount of chromium will produce a casting with a large volume fraction of very hard chromium carbides. The matrix of iron/chromium irons can be austenite, pearlite, or martensite depending on the alloy and processing. Most of these alloys have a hardness in excess of 50 HRC, and they were used for a wide variety of wear-resistant castings. At some time in the genesis of hardfacing alloys, these types of materials were made into hardfacing consumables. The iron/chromium hardfacing alloys covered by AWS designations are only two, but many proprietary versions are available. The compositions of the AWS alloys are shown in Table 10–15. These alloys have an austenitic matrix with hard carbides of the Cr_xC_y type (Figure 10–12). The austenitic matrix provides better toughness than the other possible matrixes, but with a sacrifice of matrix strength. The martensitic

TABLE 10–15 AWS IRON/CHROMIUM HARDFACING ALLOYS

	C	Mn	Cr	Mo	Si	Fe	Typical two-layer hardness (HRC)
RFeCr-A1	3.7–5.0	2–6	27–35	—	1.1–2.5	Rem.	53
EFeCr-A1	3.0–5.0	4–8	26–32	2	1.0–2.5	Rem.	53

grades have the highest resistance to deformation, but they are more prone to cracking than the austenitic grades.

These alloys can be applied by GTAW, FCAW, SAW, and SMAW, and there are proprietary grades of iron/chromium alloys that can be applied by almost any of the fusion welding processes. The high chromium content of these alloys provides good scaling resistance for use at elevated temperatures, and they have good retention of their hardness at temperatures up to 1000°F (540°C). Because of the high carbon content, these alloys are not considered to be corrosion resistant even though they contain more chromium than many stainless steels. The high carbon prevents passivity. They do have a modicum of atmospheric rust resistance. With as-deposited hardness in excess of 50 HRC, iron/chromium alloys are not machinable. Some of the proprietary martensitic grades can be annealed, machined, and rehardened. For most applications, these alloys are used as deposited.

The AWS alloys combined with the proprietary grades of iron/chromium alloys make up the largest fraction of the total usage of hardfacing alloys (Figure 10–13).

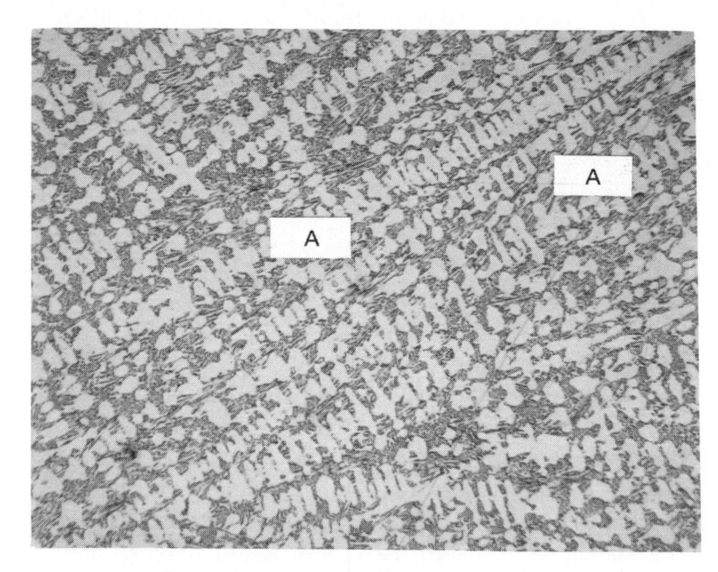

Figure 10–12 Microstructure of an FeCr hardfacing alloy; bulk hardness, 57 HRC (400×). White areas are dendritic austenite; dark areas contain fine Cr_7C_3 carbides.

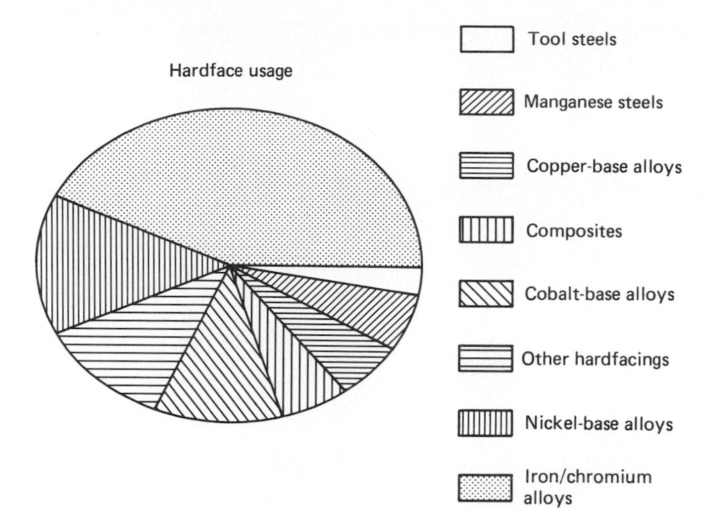

Figure 10–13 Estimated relative usage of fusion-type hardfacing consumables.

One of the most important reasons for this is that they are among the lowest-cost hardfacings. A second reason is their overall wear performance. They are applicable to a wide variety of wear situations, metal-to-metal wear, abrasion, erosion, and so on. They are good all-around hardfacing alloys, with good availability and good applicability.

Nickel-Base Alloys

This family of hardfacing alloys has no wrought counterpart, and its only other common application is for wear-resistant castings. Boron is the principal alloy strengthening and hardening element. The matrix in these alloys is solid solution strengthened nickel, and wear resistance is obtained by the formation of hard chromium boride phases in the microstructure (Figure 10–14). They are not hardenable by heat treatment and they are generally used as deposited. Many of these alloys have proprietary compositions, but the alloys covered by AWS alloy designations are shown in Table 10–16. They are almost identical in composition, but the seemingly small variations in carbon and boron are responsible for the change in bulk hardness from about 35 HRC in the A alloys to as high as 60 HRC in the C alloys. This is due to the solid solution strengthening and carbide effect of the carbon and boron on the nickel matrix. Even the softer A alloy still has a significant volume fraction of chromium borides to provide significant wear resistance; it has much better general wear resistance than a ferrous alloy hardened and tempered to the same hardness range.

The physical properties of nickel-base hardfacing alloys are similar to those of nickel. These alloys have corrosion characteristics similar to other nickel-base alloys. They are not particularly chemical resistant, but they will not rust in normal industrial environments and in outdoor service. They are not ferromagnetic.

These alloys retain their as-deposited hardness at use temperatures up to about 1000°F (540°C). They cannot be formed at room temperature, and in general they are

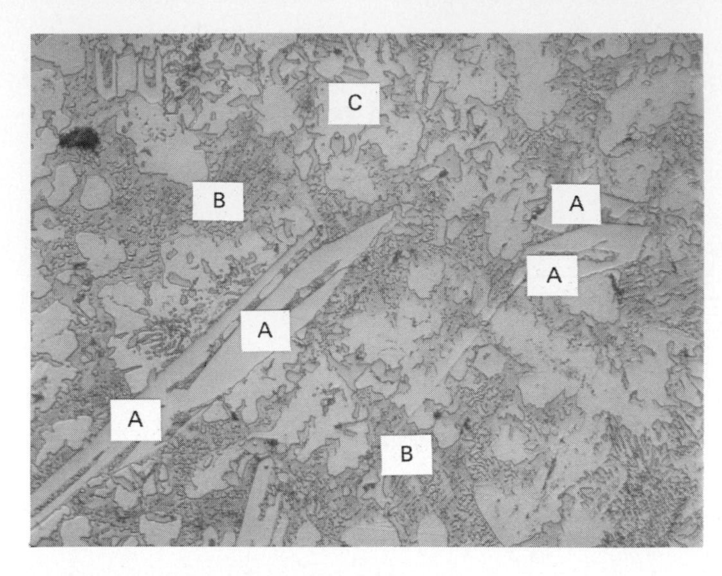

Figure 10–14 Microstructure of a NiCr-C hardfacing; bulk hardness, 54 HRC; hardness of A (Cr borides), 1300 HK_{50}; hardness of B (Ni eutectic), 700 HK_{50}; hardness of C (Ni boride phase), 570 HK_{50}, (400×).

considered to be very brittle. The C alloys exhibit considerable checking in two-layer deposits. A distinguishing characteristic of these alloys is their ease of application. Their melting points are near 2000°F (1090°C), and they wet ferrous and other substrates better than most other nonferrous hardfacing alloys (especially the cobalt alloys). Another signal property of these alloys is that they are more machinable than other hardfacings at a comparable hardness. Cam paths and similar sliding surfaces can be hardfaced and accurately machined with cemented carbide tools. There are, however, limitations. Drilling and tapping are essentially impossible. Machining can only be done with operations that involve positive chip removal.

TABLE 10–16 AWS NICKEL-BASE HARDFACING ALLOYS

	Composition							Typical two-layer hardness (HRC)	
	C	Co	Cr	Fe	Si	B	Ni	OAW	ARC
RNiCr-A	0.3–0.6	1.5	8–14	1.26–3.25	1.25–3.25	2–3	Rem.	35–40	24–35
RNiCr-B	0.4–0.8	1.25	10–16	3–5	3–5	2–4	Rem.	45–50	30–45
RNiCr-C	0.5–1.0	1.00	12–18	3.5–5.5	3.5–5.5	2.5–4.5	Rem.	56–62	35–56
ENiCr-A	0.3–0.6	1.50	8–14	1.25–3.25	1.25–3.25	2–3	Rem.		24–35
ENiCr-B	0.4–0.8	1.25	10–16	3–5	3–5	2–4	Rem.		30–45
ENiCr-C	0.5–1.0	1.00	12–18	3.5–5.5	3.5	2.5–4.5	Rem.		35–56

Nickel-base hardfacing alloys are available in coated rods for arc welding, in bare cast rods for arc and gas welding, and in powders for thermal spray and plasma application techniques. Some proprietary version of these alloys are available in sweat-on pastes. These pastes can be applied to a surface with a variety of techniques, including paint brush. They are then fused to the substrate with OAW. As a family, these alloys are suitable for abrasive wear and metal-to-metal wear, and they are widely used on machine components that require finishing of hardfaced areas. Typical examples of this type of application are extruder screws, pump and fan impellers, machine ways, and the like.

Cobalt-Base Alloys

Cobalt-base hardfacing alloys are well established in industries that require heat-resistant, corrosion-resistant, and wear-resistant surfaces. The high chromium concentration (over 20 percent) that is typical in these alloys produces alloys that are often more corrosion resistant than 300 series stainless steels. Like the nickel-base hardfacings, there are many proprietary alloys, and AWS only has designations for three rods and three electrodes (Table 10–17).

The first cobalt hardfacing alloys were developed by the Haynes Stellite Company, and the trade name Stellite is commonly misused as the generic name for these alloys. More recently, a second family of cobalt-base alloys was introduced under the trade name Tribaloy®. The Tribaloy alloys are cobalt- or nickel-based with molybdenum, silicon, and chromium as the major alloying elements. These alloys have many of the same use properties as the Stellite alloys, but they are very different from the metallurgical standpoint. The compositions are balanced so that the bulk of the structure is a hard, brittle laves phase (Figure 10–15). A laves phase is an intermetallic compound. The laves alloys have two-layer hardnesses usually in the range of 50 to 60 HRC. They are most suited to applications that require extreme chemical resistance.

The cobalt-base hardfacing alloys that are similar to Stellite alloys vary in carbon and tungsten content to produce alloys with different hardnesses. The matrix is essentially

TABLE 10–17 AWS COBALT-BASE HARDFACING ALLOYS

	Composition									Typical two-layer hardness
	C	Mn	W	Ni	Cr	Mo	Fe	Si	Co	(HRC)
RCoCr-A	0.9–1.4	1	3–6	3	26–32	1	3	2	Rem.	38–47
RCoCr-B	1.2–1.7	1	7–9.5	3	26–32	1	3	2	Rem.	45–49
RCoCr-C	2–3	1	11–14	3	26–33	1	3	2	Rem.	48–58
ECoCr-A	0.7–1.4	2	3–6	3	25–32	1	5	2	Rem.	23–47
ECoCr-B	1–1.7	2	7–9.5	3	25–32	1	5	2	Rem.	34–47
ECoCr-C	1.75–3	2	11–14	3	25–32	1	5	2	Rem.	43–58

® Registered trademark of the Stoody/Deloro Stellite Corp.

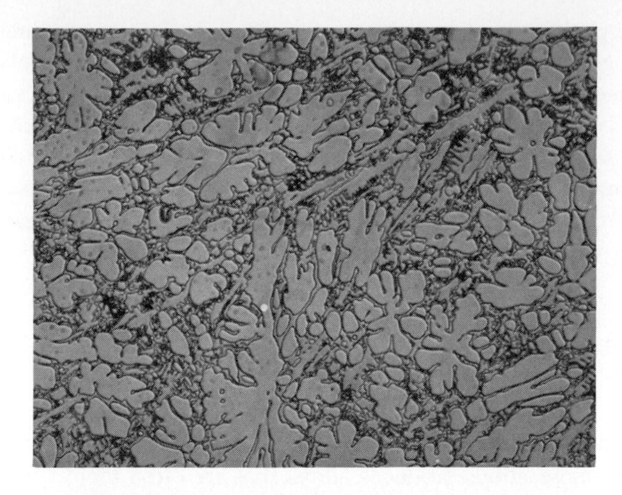

Figure 10–15 Microstructure of a laves phase hardfacing alloy (Tribaloy T800) (400×). The light areas are hard intermetallic compounds with a stoichiometric composition of AB_2.

cobalt strengthened by chromium and tungsten. The wear resistance of these alloys is enhanced by the formation of carbides. The high chromium content leads primarily to the formation of chromium carbides (Figure 10–16). The shape, distribution, and volume fraction of the alloy carbides is a function of the alloy composition and deposition technique.

The abrasion resistance of the three basic AWS alloys increases with carbon content and hardness. The harder alloys are more prone to cracking than the A alloys, and thus they are less used. The A alloys are widely used to produce wear-resistant areas on 300 series stainless steel chemical-processing equipment. The cobalt-base hardfacing

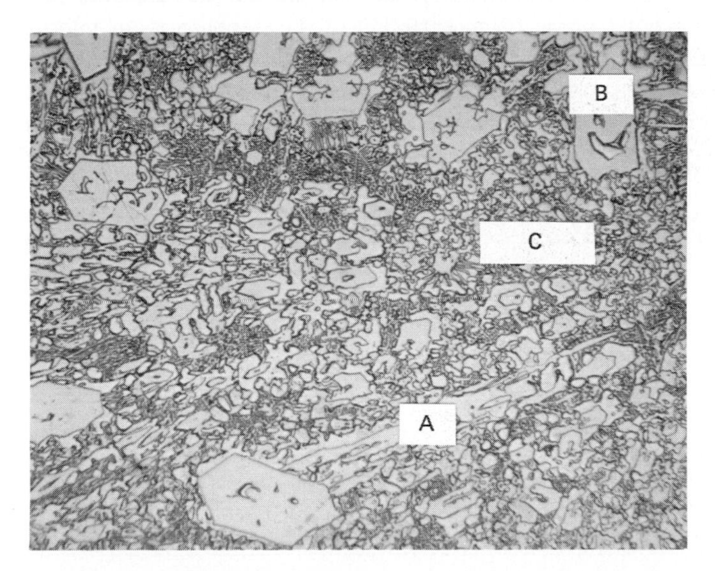

Figure 10–16 Microstructure of RCoCr-C alloy; bulk hardness 55 HRC, hardness of A, 1250 HK_{100}; hardness of B, 1320 HK_{100}; hardness of C, 600 HK_{100}.

alloys are not hardened by heat treatments, and most alloys do not respond to annealing. Their room temperature properties are maintained in use temperatures up to about 1200°F (650°C). These alloys are considered to be nonmachinable. The A type alloys can be lathe turned, but grinding is usually the only applicable machining process. Their high chromium content provides good oxidation resistance at temperatures up to 1800°F (980°C).

Cobalt-base hardfacing alloys have been in industrial use for over 50 years, and they probably will maintain their utility in the hardfacing field for many more years. Their biggest advantage over other hardfacing alloys is their corrosion and oxidation resistance. They are expensive compared to other surfacings; thus they are best used where their unique properties are really needed. They are available in bare rod for GTAW and OAW deposition, in coated electrode form for SMAW, in wire form for GMAW, FCAW, SAW, and PTAW, and in powder form for thermal spray processes. They have wide utility in the chemical process industries for wear- and corrosion-resistant parts and in the power industry for elevated-temperature applications.

Copper-Base Alloys

Copper-base alloys have been used for centuries for wear applications, but they only resist certain modes of wear. Copper alloys are not at all resistant to scratching abrasion, but they do perform satisfactorily in certain metal-to-metal wear systems, and they are resistant to certain forms of corrosive wear such as liquid erosion.

Copper-base hardfacing alloys generally fit into four alloy groups:

1. Brasses: copper/zinc alloys
2. Aluminum bronzes: copper/aluminum/iron alloys
3. Phosphor bronzes: copper/tin/phosphorus alloys
4. Silicon bronzes: copper/tin/silicon alloys

The brass hardfacing alloys are single-phase solid solutions of zinc in copper. They are not hardenable by heat treatment, and their hardness never reaches the start of the Rockwell C scale.

Aluminum bronzes can be quench hardened to a martensitic structure that can have a hardness as high as 36 HRC. The hardfacing alloys are usually not quench hardened. They are used as deposited, and the hardness varies with the alloy (Figure 10–17).

The phosphor bronzes are not hardenable by heat treatment, and their bulk hardness is similar to that of the brass alloys. The tin in phosphor bronzes is relatively insoluble in the copper matrix; when the tin concentration is in excess of about 3 percent, there is a tendency for the "extra" tin to form a second tin-rich phase that is much harder than the matrix. This is one reason why phosphor bronzes can produce wear on mating soft steel surfaces. When bronzes are used in metal-to-metal sliding systems, the counterface should have a hardness of about 60 HRC. This is also true with aluminum bronze.

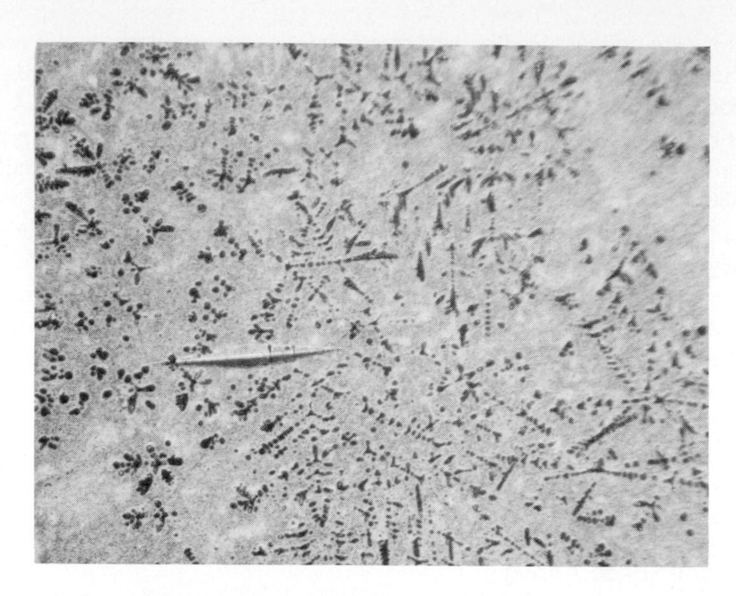

Figure 10–17 Microstructure of aluminum bronze hardfacing alloy (400×). The dark areas are aluminum/copper compounds. They strengthen but have little effect on bulk hardness (note indentation).

The silicon bronze alloys are usually single phase. They are not hardenable by heat treatment, and their wear properties are similar to the yellow brasses. They are most widely used in systems that involve seawater corrosion.

The copper base hardfacing alloys covered by an AWS alloy designation are listed in Table 10–18.

Copper alloy hardfacings are not really "hard." This precludes their use for applications involving abrasion. The brass alloys are commonly used for rebuilding such things as large cast-iron gears. In lubricated metal-to-metal applications, they will perform about the same as a soft gray iron. Aluminum bronzes are used for the same types of applications, but their higher hardness and strength allow higher loads and more severe operating conditions. Silicon bronzes are used for such things as rebuilding ship propellers. Phosphor bronzes are most widely used to rebuild bearings and the like.

None of the copper-base hardfacing alloys should be used at operating temperatures over 400°F (200°C); they are subject to significant oxidation at elevated temperatures in air. All these alloys are machinable. Their corrosion characteristics are similar to those of their wrought counterparts. The copper-base hardfacing alloys are most suited to metal-to-metal wear applications. They are also used in sliding systems where it is desired that one member of the sliding couple be sacrificial; for example, aluminum bronzes are often used for roller chain guides. The chain is very expensive, and if it rubs against a hard iron/chromium surfaced guide, chain wear will result. When copper-based hardfacings are used, the chain is protected.

Copper-base hardfacings are available in bare rod, small diameter wire, coated

TABLE 10–18 AWS COPPER-BASE HARDFACING ALLOYS

	Typical composition									Typical two-layer hardness
	Mn	Fe	Al	Zn	Si	Pb	Sn	P	Cr	
RCuZn-E	0.3	1.5	0.01	40	0.04–0.25	0.05	2–3	—	Rem.	<20 HRC
ERCuSi-A	1.5	0.5	0.01	1.5	2.8–4.0	0.02	1.5	—	94 min	<20 HRC
ERCuA1-A2	—	1.5	9–11	0.02	0.10	0.02	—	—	Rem.	<20 HRC (130–190 HB)
ERCuA1-A3	—	3–5	10–11	0.10	0.10	0.02	—	—	Rem.	<20 HRC (130–190 HB)
RCuA1-C	—	3–5	12–13	0.02	0.04	0.02	—	—	Rem.	<20 HRC (140–290 HB)
RCuA1-D	—	3–5	13–14	0.02	0.04	0.02	—	—	Rem.	20–40 HRC
RCuA1-E	—	3–5	14–15	0.02	0.04	0.02	—	—	Rem.	<20 HRC
ERCuSn-A	—	—	0.01	—	—	0.02	4–6	0.1–0.35	93.5 min	<20 HRC
RCuSn-D	—	—	0.01	—	—	0.05	9–11	0.1–0.30	88.5 min	<20 HRC
ECuSi	—	0.5	0.01	—	2–4	0.02	1.5	—	Rem.	<20 HRC
ECuA1-A2	—	0.5–5.0	7–9	—	1	0.02	—	—	Rem.	130–190 HB
ECuA1-B	—	2.5–5.0	8.4–10	—	1	0.02	—	—	Rem.	140–290 HB
ECuA1-C	—	3.0–5.0	12–13	0.02	0.04	0.02	—	—	Rem.	140–290 HB
ECuA1-D	—	3–5	13–14	0.02	0.04	0.02	—	—	Rem.	230–390 HB
ECuA1-E	—	3–5	14–15	0.02	0.04	0.02	—	—	Rem.	230–390 HB
ECuSn-A	—	0.25	0.01	—	—	0.02	4–6	0.05–0.35	Rem.	<20 HRC
ECuSn-C	—	0.25	0.01	—	—	0.02	4–6	0.05–0.35	Rem.	<20 HRC

electrodes, and powder. Thus they can be applied with most of the fusion processes. Application of these alloys is normally limited to rebuilding applications and for special metal-to-metal wear situations. They can be applied to ferrous, nickel, and copper substrates.

Composite Materials

This consumable category covers straight-length, bare, tubular rods and coated electrodes. The composite term arises from the fact that the rods and electrodes are metal tubes filled with wear-resistant particles, usually carbides. A composite by definition is a material made from different materials, and the properties of the composite are different and superior to the properties of the materials making up the composite. Low-carbon steel tubing, about 0.010 in. (0.25 mm) thick, is roll formed into a tube, and carbide particles are introduced into the tube at the point of forming. The finished rod is made by pinching closed the end of the filled tube at the point of cutoff. These filled tubes can then be coated with an appropriate coating to provide a shield gas during welding. The bare rods are used with oxyacteylene and GTAW welding, and the coated composite electrodes are used with SMAW. Tables 10–19 and 10–20 give chemical analyses of the components of composite rods, the steel sheath, and the carbide.

The AWS designations for composite rods and electrodes (Table 10–21) essentially coincide with the various sizes of tungsten carbide particles that are available for insertion into the hollow tube that will become the rod or electrode.

Figure 10–18 is an illustration of one method of construction of a composite rod. Figure 10–19 shows a typical cross section of a deposit made with a composite rod. It is obvious that this deposit is a true composite. Carbide particles that have a size determined by selection (by the user) are held in a metal binder. Since the tungsten carbide particles have a hardness of about 2000 HK compared with the hardest metal, with a hardness of about 900 HK, one would expect that the composite deposits would have wear characteristics akin to the hard metal carbides. They do, but the wear resistance is also helped by partial dissolution of the carbide during welding. Carbon and tungsten from the carbide particles go into solution in the mild steel, making its as-deposited

TABLE 10–19 CHEMICAL REQUIREMENTS FOR MILD STEEL TUBES FOR COMPOSITE CONSUMABLES

C	Mn	P	S	Si	Ni	Cr	Mo	V
0.10	0.45	0.02	0.03	0.01	0.30	0.20	0.30	0.03 %

TABLE 10–20 CHEMICAL REQUIREMENTS FOR TUNGSTEN CARBIDE GRANULES FOR COMPOSITE CONSUMABLES

C	Co	W	Ni	Mo	Fe	Si
3.6–4.2	0.3	94.0 min	0.3	0.60	0.5	0.8 %

TABLE 10–21 AWS DESIGNATIONS FOR COMPOSITE RODS AND ELECTRODES

AWS classification	Carbide mesh size range*	Weight percent of tungsten carbide
Rods		
RWC-5/8	Through 5 on 8	60
RWC-8/12	Through 8 on 12	60
RWC-12/20	Through 10 on 20	60
RWC-20/30	Through 20 on 30	60
RWC-30/40	Through 30 on 40	60
RWC-30	Through 30	60
RWC-40	Through 40	60
RWC-40/120	Through 40 on 120	60
Electrodes		
EWC-12/20	Through 12 on 20	60
EWC-20/30	Through 20 on 30	60
EWC-30/40	Through 30 on 40	60
EWC-40	Through 40	60
EWC-40/120	Through 40 on 120	60

* U.S. standard sieves.

hardness as high as 60 HRC. Thus the composite weld deposit consists of very hard particles in a moderate to high hardness steel matrix. There is a potential problem with dissolution of carbide particles to strengthen the matrix. If excessive welding currents are used, the wear-resistant carbide particles can be completely dissolved. Thus these consumables require the utmost care in application.

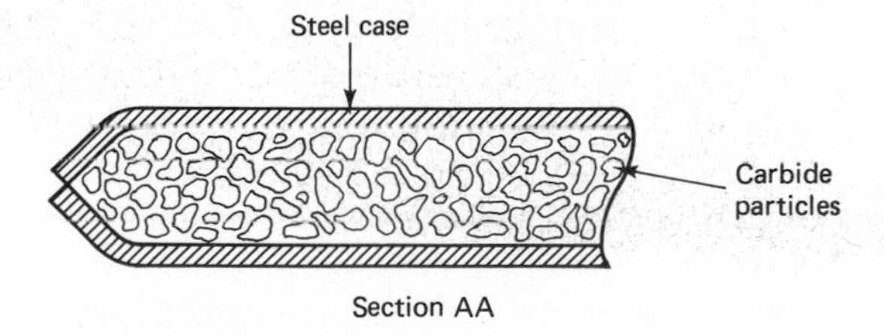

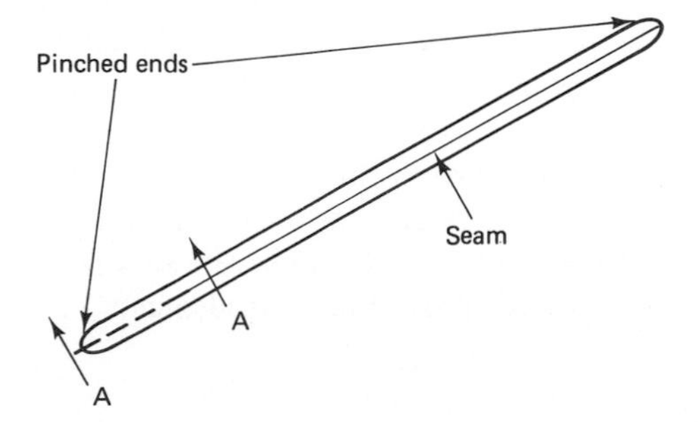

Figure 10–18 Typical composite bare rod.

Figure 10–19 Microstructure of composite hardfacing (100×). Hard carbide particles are standing proud from polishing wear.

Where are these materials used? If properly applied, it will not be possible to machine composite deposits. Thus these materials are used for applications where the deposit can be used as applied. Bucket teeth, sand augers, and coal-loading devices are typical applications for these materials. Since the matrix is essentially steel with some alloy pickup from the carbide, the matrix properties will be those of alloy steel; this precludes the use of these materials for elevated-temperature applications and for applications where corrosion resistance is required.

Composite deposits are not well suited to metal-to-metal wear applications. The softer metal matrix can be eroded, leaving the carbide particles standing proud. These hard particles will act as an abrasive lap to wear the metal counterface.

In summary, composite rods and electrodes with tungsten carbide particles as the wear-resisting additive are used for general-purpose abrasion applications where deposits can be left as applied. They are somewhat troublesome to use because of the risk of getting poor deposits (dissolved carbides) and because the operating characteristics of the conventional rods and electrodes are usually better. There are proprietary grades of these materials with vanadium and other carbides that are even harder than the tungsten carbides. Some proprietary grades have matrixes other than steel. Stellite types of materials and nickel/chromium matrixes are used where corrosion resistance is a factor. As a class, all the composite consumables are best suited to severe service applications where the more conventional consumables have proved to be inadequate.

NONFUSION MATERIALS

Thermal Spray Powders

Welding consumables in the powder form are used in oxyacetylene spray torches, plasma, and detonation gun equipment and in a variety of other welding equipment, such as plasma arc. The powders used in the former processes are not usually fused with the substrate. Powders used with transferred arc plasma and GTAW are fused to the substrate. It is the purpose of this section to discuss the characteristics of these powders and to categorize them to make selection and use easier. We will concentrate discussion on the powders used in the nonfusion processes because of their wider use.

To the occasional user of thermal spray techniques, the selection of an appropriate consumable can be an awesome task; any powder that can be melted without decomposing in the flame can be sprayed. Similarly, any substrate can be sprayed if it resists degradation at typical substrate temperatures, which are usually less than 330°F (150°C). Powders are usually made in the size range of 5 to 100 μm (0.2 to 0.4 mils). They are introduced into the thermal spray torch in a variety of ways, but a requirement of the process is that they melt and can be propelled at the substrate to be coated. The ease of melting depends on the torch design (where the powder is introduced), the nature of the powder (round, spongelike, blocky, etc.), and the flame temperature. An oxyacetylene torch has a flame temperature of about 5000°F (2760°C); plasma temperature can be as high as 50,000°F (2800°C), and d-gun explosions produce temperatures of about 7000°F (3870°C). Obviously, the plasma torch is capable of melting almost any powder. Some ceramics could have limited sprayability with the "cooler" processes. Powder coatings are formed by overlapping splat-cooled droplets of the melted powder, as shown in Figure 10–20. The splats usually contain some contaminants from reaction of the molten droplets with the medium that the droplets passed through on their way to the workpiece.

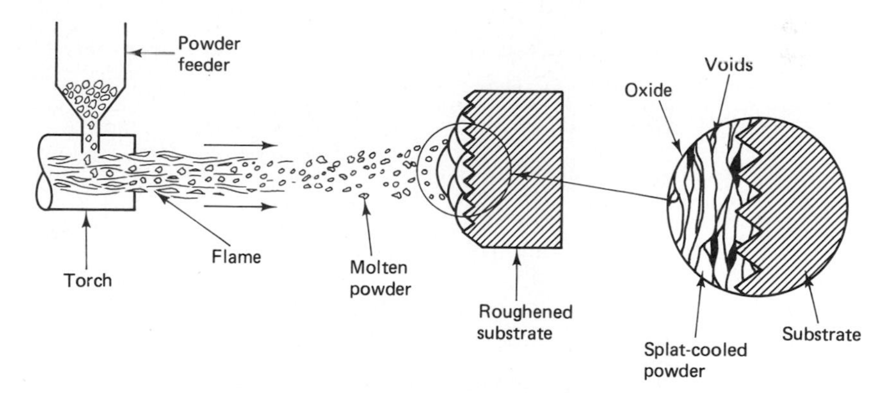

Figure 10–20 Schematic of typical powder spray morphology.

The area of impingement is usually a circle about 1 in. (25 mm) in diameter. There are usually some voids in the deposit from incomplete conformance of the splat-cooled droplets with each other. The porosity or void content is affected by the velocity of impingement of the molten droplets. Oxy/fuel spray velocities are usually less than 500 ft/s (152 m/s); standard plasma sprays have velocities in the range of 700 to 1200 ft/s (210 to 365 m/s), and high-velocity plasma processes can have particle velocities as high as 1800 ft/s (550 m/s). The detonation gun has a particle velocity in the range of 2000 to 2800 ft/s (610 to 635 m/s), and rocket torches can have velocities as high as 4500 ft/s (1370 m/s). The relative porosity of a nonfused thermal spray coating essentially coincides with impingement velocity (Figures 10–21 to 10–23).

The mechanism of bonding of nonfused powder spray coating is still a subject being addressed by researchers in the field, but most agree that a large amount of the adhesion is due to mechanical locking with the substrate. It is for this reason that the substrate to be sprayed is usually grit blasted to a surface roughness between 100 and 300 microinches prior to coating. The other factors contributing to bonding are diffusion, van der Waals forces, and chemical effects. The best powder spray coating will have adhesion comparable to a good electroplate on metal. Bond strengths of good plasma spray deposits are usually in the range of 5000 to 10,000 psi (34.5 to 69 MPa) in tension (a cylinder is sprayed on end, epoxied to another cylinder, and tensile tested).

Now that we have reviewed some of the common characteristics of all nonfusion thermal spray coatings, we will address the various types of powders that are commercially available. The hundred or so specific powders available from major manufacturers can roughly be categorized into seven groups based on material system. Figure 10–24 shows the basic powder categories and lists some specific types of powder systems in each of

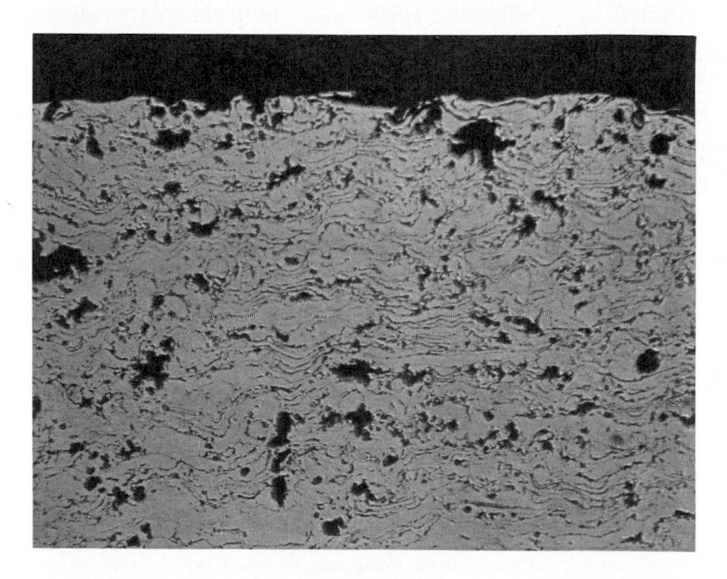

Figure 10–21 Microstructure of a flame-sprayed high-carbon steel (400×). Black areas are porosity.

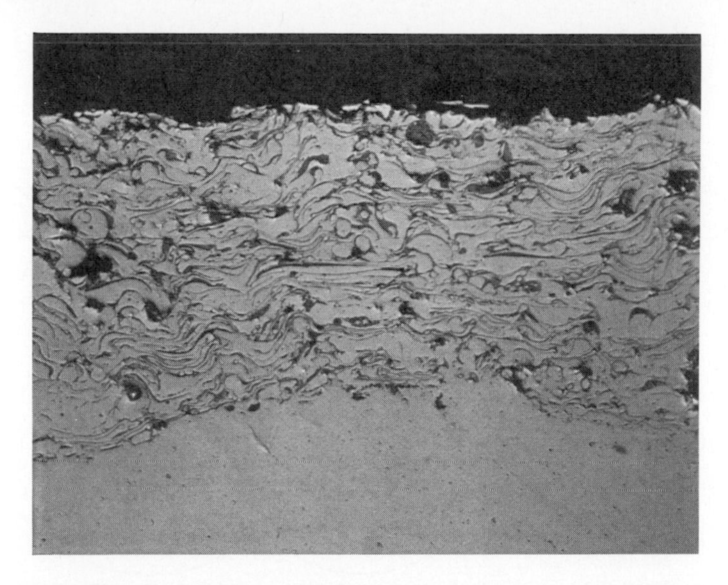

Figure 10–22 Microstructure of WC/Co applied by plasma arc spraying (400×).

these categories. Some of the powder categories are self-explanatory; most ceramics, pure metals, and metal alloys can be sprayed, and the ones listed are the most widely used. A cermet is a material composed of a metal and a nonmetal. The usual combination in cermets is a ceramic with a metal binder. A composite, as described in the previous section, is a material made from differing material systems, and the resulting composite has properties that are usually superior to the properties of the two components used to

Figure 10–23 Microstructure of WC/Co powder applied by detonation gun (400×).

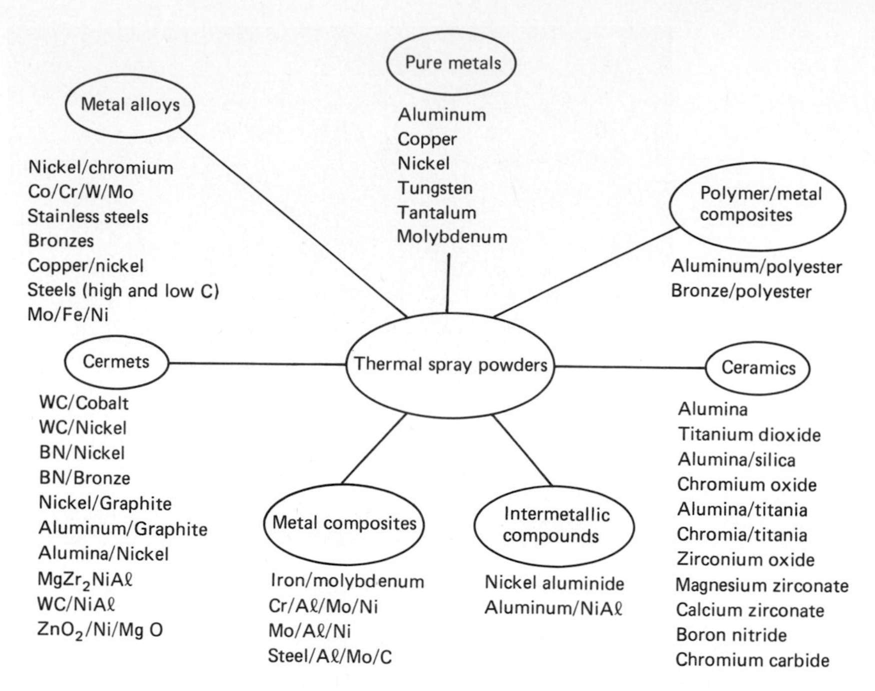

Figure 10–24 Spectrum of powders for thermal spray processes.

make the composite. Hardfacing powder composites are made from the physical mixture in the deposit of differing material systems. They are not alloys. The formation of an alloy requires fusion and solid solution of differing elements. The other composite system listed is a physical mixture of a relatively high melting point plastic, polyester, and a metal. The final category listed, intermetallic compounds, could be put under the category of ceramics because many intermetallic compounds have properties that are more like ceramics than metals. By definition, an intermetallic compound is a compound formed by covalent bonding of differing metal atoms. The intermetallic that is most important in thermal spraying is nickel aluminide, NiA1. It is a soft, ductile material with metallic appearance. It is widely used as a bond coat for spraying of other wear-resistant materials. It adheres to many substrates better than hard coatings, and hard coatings adhere better to the nickel aluminide than they do to difficult-to-coat substrates such as hard steels.

Most of the specific powder systems listed in Figure 10–24 are not a single powder but a family of powders. There are many nickel/chromium alloy powder formulations, many tungsten carbide formulations, many types of alumina (they vary in particle size and the like). There are simply too many specific powder consumables to discuss each separately. In addition, many of these are seldom used. To establish a repertoire of useful powder consumables, we will discuss only the most widely used powders.

Figure 10–25 is an estimate of the current relative usage of thermal spray powders for wear applications. The powder systems in the circles are the most widely used

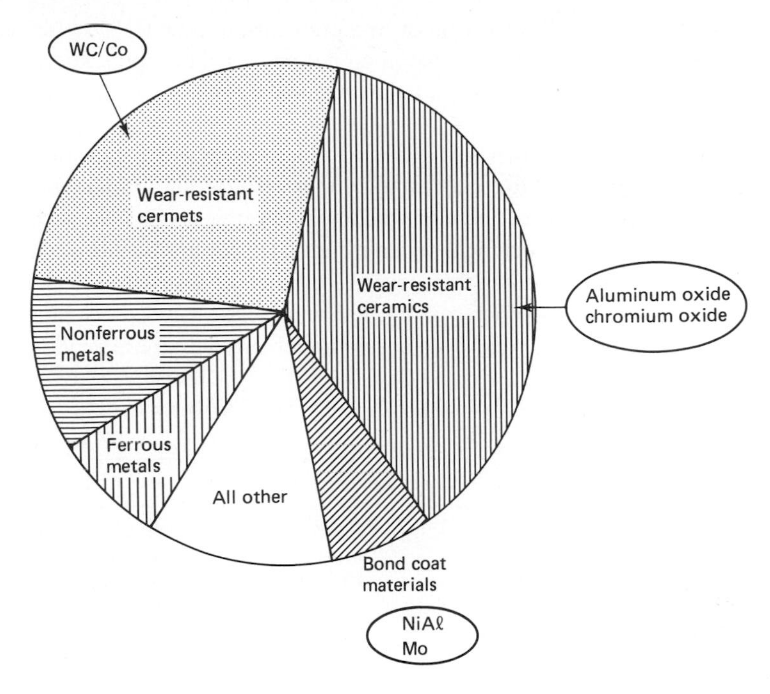

Figure 10–25 Estimated usage in the United States of thermal spray powders for wear applications.

systems in the most important categories. We will discuss these and make some comments on each of the other categories.

Metal Alloys Nickel/chromium alloys when applied as thermal spray coatings can have hardnesses in the range of hard electroplates, >50 HRC. They have utility as wear- and corrosion-resistant surfaces on such things as paper mill rolls. Cobalt alloys similar to Stellite are used for similar wear- and corrosion-resistant applications. Almost all types of stainless steels and corrosion-resistant alloys such as the Hastelloys can be sprayed. The usual application is for rebuilding equipment used in chemical service. In such cases it is desirable to match the composition of the substrate with a thermal spray rebuilding alloy. Bronzes and copper alloys are used for bearing applications. High- and low-carbon steels are sprayed for a wide variety of applications, but mostly for buildup of worn steel parts. The low-carbon steel deposits are machinable, and the high-carbon steel deposits usually require grinding.

Pure Metals Most of the pure metal consumables are used for corrosion or technical applications, that is, for special applications where the physical properties of the metal are of more interest than the wear resistance of the metal. For example, copper is sprayed on the inside of electronic equipment enclosures for RF shielding. One exception to this statement is molybdenum. As a thermal spray deposit, it has been learned that it has good metal-to-metal wear characteristics, and it is used in bearing applications.

Composites Almost without exception, these powder systems are used for technical applications. There are no general guidelines for their application. They would probably only be used upon recommendation from a thermal spray specialist.

Ceramics As shown in Figure 10–25, ceramics are the most widely used thermal spray consumables for wear applications. There are probably 50 additional powders available, but most of them are technical ceramics. Like the technical applications discussed previously, technical ceramics are used for applications where the user is looking for some property other than wear resistance (Table 10–22). Magnesium zirconate is used as a thermal barrier coating; hafnium oxide is used for neutron shielding and the like. The important wear-resistant ceramics are aluminum oxide, chromium oxide, and

TABLE 10–22 CERAMIC MATERIALS THAT CAN BE FLAME OR PLASMA SPRAYED

Material	Flame or plasma	Properties and uses
Aluminum oxide	Either	Erosion and heat resistant
Beryllium oxide	Either	Thermal conductivity, dielectric
Boron carbide	Either	Nuclear shielding
Calcium zirconate	Either	Dielectric, dense
Ceric oxide	Either	Refractory, emittance
Columbium carbide	Plasma	Refractory, hard
Chromium boride	Either	Hard
Chromium carbide	Plasma	Refractory, hard
Chromium oxide	Either	Dense, oxidation resistant
Hafnium carbide	Plasma	Refractory
Hafnium oxide	Either	Thermal insulation, refractory
Ilmenite	Either	Dense
Lithium aluminate	Flame	Used with oxides to aid melting
Magnetite	Either	Magnetic
Molybdenum boride	Plasma	Refractory, hard
Molybdenum carbide	Plasma	Refractory, hard
Molybdenum silicide	Plasma	Oxidation protection
Mullite	Either	Glassy, refractory
Nickel oxide	Either	Emittance
Rare earth oxides	Either	Refractory
Spinel	Either	Oxidation resistant
Tantalum carbide	Plasma	Refractory, hard
Titanium boride	Plasma	Refractory
Titanium carbide	Plasma	Hard
Titanium oxide	Either	Dense, dielectric, resists oxidation
Tungsten carbide	Either	Hard, wear resistant
Vanadium boride	Plasma	Refractory, hard
Yttrium oxide	Either	Refractory
Zirconium boride	Plasma	Hard
Zirconium carbide	Plasma	Hard
Zirconium oxide	Either	Thermal insulation, refractory

Source: *Materials Engineering,* © Penton IPC, November 1974.

mixtures of these with silica, titania, and other hard ceramics. All the hard ceramic sprays have good resistance to low-stress scratching and polishing abrasion. They may or may not work for bearing-type applications. If an application clearly involves low-stress abrasion of the type encountered in shaft packing or when paper slides over a surface, any of the ceramics mentioned will usually outperform even the hardest metal. To make selection easier, most thermal spray job shops recommend the use of chromium oxide or chromia/silica. These coatings seem to produce a better ground finish than aluminum oxide, but the latter will do the same job.

Cermets All the cermets listed in Figure 10–24 are intended for technical applications, with the exception of the tungsten carbide cermets. There are three main binder systems for tungsten carbide cermets: nickel, cobalt, and corrosion-resistant alloy binders like cobalt/chromium. For low-stress abrasion and for metal-to-metal wear applications, the tungsten carbide/cobalt powders predominate. This system produces wear characteristics approaching those of cemented carbides. Higher binder contents increase the toughness of the deposit, and lower binder contents lead to improved abrasion resistance. This system is the best to use for general wear applications where impact rules out the use of the wear-resistant ceramics.

In summary, powder thermal spray consumables are available for almost any application. There are no summary statements that can be made on properties since the properties of deposits made with these consumables are as diverse as the selection of materials that can be sprayed. The occasional user can successfully use these materials without knowing the properties of each specific powder. For rebuilding metals with metals, match the composition of the substrate if similar wear characteristics are desired. If improved wear characteristics are desired, use tungsten carbide or chromium oxide. For technical applications and for applications where it is known that the conventional powders do not work, it is recommended that powder manufacturers be consulted.

Thermal Spray Wires

Wire consumables for thermal spray torches can be made from almost any metal that is available in wire form and meets the requirements for torch spraying mentioned in the previous section. The applicability niche for wire spray processes in wear applications is usually thick buildups, coatings thicker than about 0.030 in. (0.75 mm). Thermal spray wires are not available in as wide a range of materials as powder consumables; ceramics, cermets and highly alloyed metals are difficult to make into wires. Thus the selection of wires is limited mostly to pure metals and metal alloys:

High-carbon steel
Low-carbon steel
Medium-carbon steel
Low-alloy steel

Nickel
Monel
Copper
Brass
Bronze
Aluminum
Nickel aluminide
Lead
Tin
Babbitt
Stainless steel

There are at least a few alloys available in each of these categories, but this selection does not contain any of the highly wear resistant materials that are available in other consumable systems. The steel wires are usually used for rebuilding parts with large amounts of wear. The low-carbon steels are machinable, and the high-carbon steel wires usually require grinding. The stainless steel wires are used for cladding roll surfaces and the like. The properties of the other spray metals listed are essentially the same as the properties of the wrought counterparts of these metals. Some of these wires are intended for arc guns, but most of these alloys can be sprayed in oxy/fuel torches or in arc torches.

If a shaft is worn 0.062 in. (1.5 mm) on a side, it is probably a good candidate for rebuilding with one of the preceding spray consumables. The composition of the shaft can be matched in the spray deposit, or another alloy can be applied. Most of the time, parts rebuilt with wire spray processes are rebuilt with a matching composition wire. Thus, from the selection standpoint, this category of hardfacing consumables is more for rebuilding rather than for improving the properties of a surface. They are useful for applying large quantities of material to a worn surface.

SUMMARY

In this chapter we have described the basic categories of hardfacing consumables and the more important specific materials in each of these categories. At this point it may seem that as a user you still have insufficient information to make an intelligent selection for a specific application. This is not without intent. We have not fully discussed where these consumables should be used. This will be done in Chapter 12 on selection. At this point it is sufficient to have an understanding of the material systems that are involved in making hardfacing consumables. Fusion materials include tool steels, copper alloys, Stellite-type alloys, hard nickel/boron alloys, manganese steels, iron/chromium alloys, and tungsten carbide composites. The nonfusion alloys include ceramics, cermets, intermetallic compounds, pure metals, and special composites. The other nonfusion materials are wires for thermal spray processes. These wircs can be made from most materials, but since these wires are used mostly for rebuilding, they are predominately ''ordinary metals'' and not wear-resistant materials.

The principal objective of this chapter was to present what is available; we showed several categories and many specific materials. We discussed the metallurgy of the materials; basically, any metal alloy can be used for rebuilding, but if improved wear resistance is desired for an application, then the other hardfacing consumables are needed. The hard metal alloys almost all contain hard second phases in a softer and more shock-resistant matrix. Manganese steels are different from all the other metal systems in that they are soft metals and their usefulness depends on their ability to work harden on impact. If a wear problem involves severe abrasion, the ceramics and cermets were presented for consideration. They are harder than the metal systems, and they have characteristics that cannot be matched by metals. Thus many consumables are available for hardfacing, but they fit into a relatively few categories that can be easily understood and remembered.

REFERENCES

Hardfacing Structures. New York: Welding Research Council, 1982.

MOCK, J. A. ''Ceramic and Refractory Coatings,'' *Materials Engineering,* vol. 80, no. 6, 1974, pp. 101–108.

RIDDIBOUGH, M. *Hardfacing by Welding,* 4th ed. London: Deloro Stellite Ltd., 1975.

Specification for Composite Surfacing Welding Rods and Electrodes, AWS A5.21–70. Miami: American Welding Society, 1970.

Specification for Solid Surfacing Welding Rods and Electrodes, AWS A5.13–80. Miami: American Welding Society, 1980.

Technical Bulletins. Westbury, N.Y.: Metco, Inc., June 29, 1984.

TUCKER, R.C. Plasma and Detonation Gun Deposition Techniques and Coating Properties. *Deposition Technologies for Films and Coatings*. Park Ridge, N.J.: Noyes Publications, 1982.

ZATT, J. H. ''A Quarter Century of Plasma Spraying,'' *Annual Review of Materials Science,* vol. 13, 1983, pp. 9–42.

ZUM GAHR, K. H. ''How Microstructure Affects Abrasive Wear Resistence,'' *Metal Progress,* Sept. 1971, pp. 46–49.

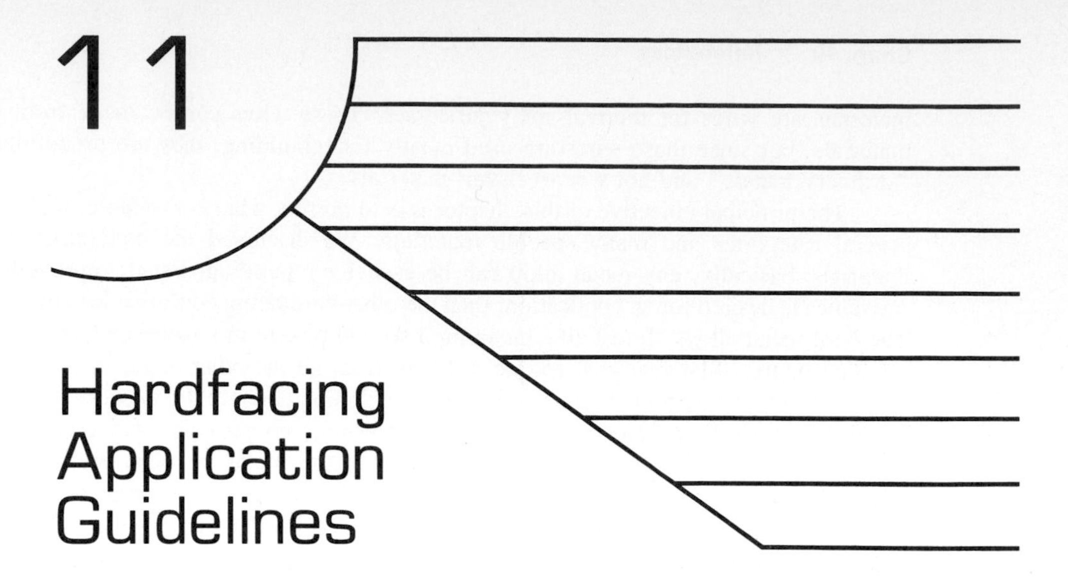

11 Hardfacing Application Guidelines

If we have done our job in the preceding chapters, the reader should now know how to select a hardfacing process and suitable consumable for a specific task. The next step in the chronology of hardfacing is to tell the welder how to apply this material. The details of how to use welding symbols for this will be dealt with in Chapter 13. In this chapter, we will address the things to consider in deciding if hardfacing is applicable to a part, details of how hardfacing materials should be applied to parts with varying geometries, weldability considerations, and finishing and postweld treatment of hardfacing deposits.

There are many ways that surfacing can be applied to various types of parts; how the hardfacing is applied is often as important as what is applied in determining if the job will be successful. As an example, if a worn shaft distorts when the worn area is rebuilt, there is likelihood that the shaft may still be unusable. The goal of a usable shaft will not have been obtained. Similarly, if the shaft cracks from the welding, the goal is not obtained. Finishing, on the other hand, can cause a new set of problems. Hardfacings can be applied that are just too hard to grind or machine. Again, the objective of a serviceable shaft is not obtained. Thus, in this chapter, we will address the factors that are concerned with the physical application of hardfacings.

HARDFACING IN NEW DESIGNS

If a designer is given the assignment for some mechanism that will be subject to some form of wear, how does he or she decide whether to use hardfacing on some of the parts? Where should the surfacing be applied? The answer to the first question is simply to weigh the advantages and disadvantages of competitive processes. If hardfacing compares favorably, then it should be used. In general, the circumstances that are ideal for the application of hardfacing are as follows:

1. The wear will be concentrated in a small area.
2. The parts that need a small wear-resistant area are weldable.

The requirement of a small area of potential wear means that the surfacing can be applied with a minimum of distortion when fusion processes are used, and the cost will be low when high alloys are used in the hardfacing process. The requirement of a weldable substrate is self-explanatory, but it is still necessary to consider this at a very early point in the design process. As an example, if a part is to be subjected to significant operating stresses, it may be desirable to use hardenable steels. Hardenable steels create problems of weldability with any of the fusion processes. With the nonfusion processes, weldability can still be a factor. If a part is very fragile, it may be subject to distortion just from the pretreatment (abrasive blasting, etc.) needed for plasma and similar nonfusion processes; the heat from thermal spray processes may be enough to cause distortion. Thus a prime factor to consider in designing new parts that employ hardfacing is distortion from the hardfacing, economics, and the weldability of the substrate and the application process.

HARDFACING FOR REPAIRS

The considerations that apply to the question of whether to use hardfacing for a wear repair are essentially the same as those that apply to new-part designs. In this instance, the question of where to apply hardfacing has already been answered—in the worn areas. The next item to consider is if the substrate is weldable. If the answer to this question is yes, hardfacing techniques should definitely be considered. Hardfacing should be weighed against such alternatives as replacing the worn part with a new one, rebuilding with electroplating techniques, or rebuilding by mechanical attachment (sleeving, bushings, bolt on plates, etc.). As mentioned in previous discussions, if the wear depth on a part is only several thousandths of an inch (75 μm), plating techniques may indeed be the most cost-effective approach. If the wear depth is deeper, hardfacing is a candidate.

The disadvantage of a mechanical repair of wear damage is usually that the wear may reoccur. If the material used for the mechanical attachment is similar in composition to the substrate that was damaged by wear, a repeat failure is guaranteed. One outstanding characteristic of hardfacing is that expensive and metallurgically complicated alloys can be considered for use because only a relatively small volume of these sophisticated materials will be needed. As an example, it would be out of the question to use cobalt-base alloys like Stellite for large parts such as a screw conveyer because of their cost. On the other hand, such alloys are commonly used for hardfacing the edges of the screw flights. Ten pounds (4.5 kg) of alloy may cover all the high-wear areas on a 1000-lb (450 kg) screw.

On repairs, hardfacing is almost always a cost-effective candidate. The decision to use hardfacing on wear repairs should be weighed against alternatives. Hardfacing usually compares favorably with the alternatives because it allows repairs to be made with alloys that are more wear resistant than the original material, and a repeat wear failure can be avoided.

HARDFACING WITH FUSION PROCESSES

Assuming that the decision has been made that hardfacing is a cost-effective technique for rebuilding a worn part or for the design of a new part, we will now present some guidelines on how to apply surfacing to parts.

Only three basic deposits are used in hardfacing with fusion processes; these are shown in Figure 11–1. The bead-on-plate deposit is the simplest form of hardfacing, the easiest to apply and the lowest in cost. It also has a significant advantage over the other types of deposits in that deposits produced in this fashion have the lowest cracking tendencies. Cracking or checking is highly likely with many of the harder surfacings. These alloys are brittle, and when they solidify they are subject to a volume contraction. If the weld bead is restrained when it is solidifying, as it would be in a groove weld, it is much more likely to crack than if a bead-on-plate deposit is used. Thus this type of deposit should be used whenever possible. It can be a single bead or complete coverage of a surface with multiple beads.

Groove deposits are different in shape than grooves used in weld joining. The groove should only be as deep as the desired thickness of wear-resistant material, and the ends of the groove should be radiused as shown in the illustration. The radius reduces cracking tendencies. A rule of thumb to observe on the thickness of a deposit is to make it no thicker than the wear that can be tolerated in the system. If, for example, you are hardfacing a shaft that runs in a plain bearing, the shaft could probably only tolerate perhaps 0.010 in. (0.25 mm) of wear before the shaft starts to whip and cause damage to other members of the mechanism. In such a case, it would be unwise to specify a hardfacing deposit $\frac{1}{4}$ in. (6 mm) thick.

If arc welding processes are used, it is common practice to use two layers of surfacing. The first layer gets diluted with the substrate alloy, and the full properties of the hardfacing are not realized. Thus, for arc welding, two layers should be specified, and the thickness of the individual layers can be in the range of $\frac{1}{16}$ (1.5 mm) to $\frac{1}{4}$ (6 mm) in. Oxy/fuel processes can have dilution levels as low as 10 percent, and it is

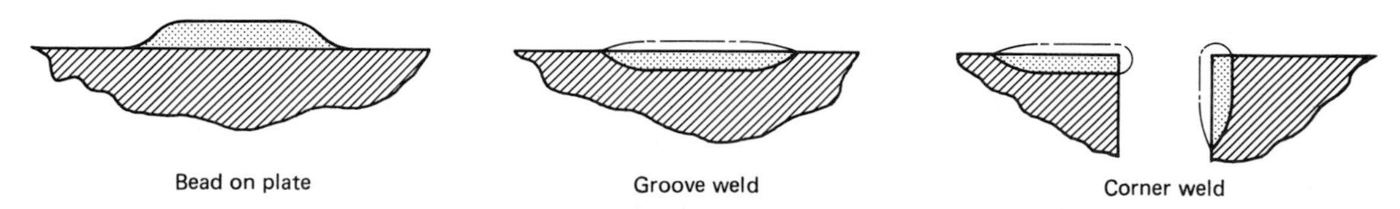

Figure 11–1 Basic types of hardfacing deposits.

common to achieve the desired deposit thickness with a single layer in the thickness range of $\frac{1}{16}$ to $\frac{1}{4}$ in. (1.5 to 6 mm). Spray and fuse consumables can have thicknesses as low as $\frac{1}{32}$ in. (0.75 mm) with minimal dilution.

Corner deposits are between bead-on-plate deposits and groove deposits in cracking and checking tendencies. The thickness criteria mentioned for the other two deposit configurations still apply.

Figure 11–2 to 11–7 present some hints on how to apply hardfacing to typical parts that could benefit from hardfacing. The following are some points to keep in mind when specifying fusion deposits:

1. Minimize the area and thickness of the deposit (to minimize distortion).
2. Try to use a bead-on-plate deposit (to minimize cracking).
3. Never apply more than two layers of high hardness (greater than 55 HRC). Use a buildup alloy under the hard deposits if the desired thicknesses of the hard welds are inadequate for a repair (to minimize cracking).
4. Never load hardfacing deposits in tension or in a fatigue mode. Where possible, load hardfacing deposits in compression.
5. Avoid fusion processes on parts that are distortion prone (long and slender, thin sections, etc.).
6. Consider porosity and flux inclusions. If these cannot be tolerated, tell the welder. They are likely to occur if the welder does not take extra care to prevent them.

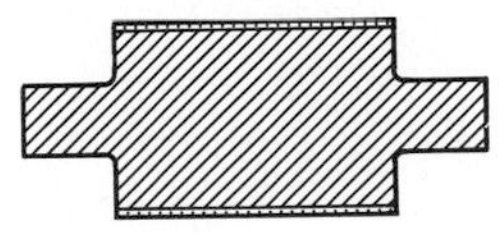

Rolls: Use automatic or semiautomatic processes (bulk welding, SAW, etc.)

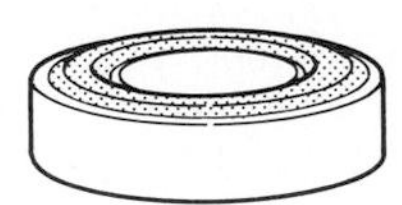

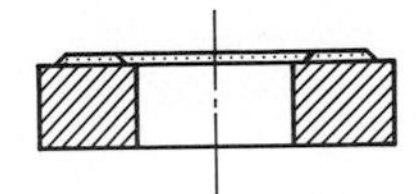

Face seals: Use bead on plate

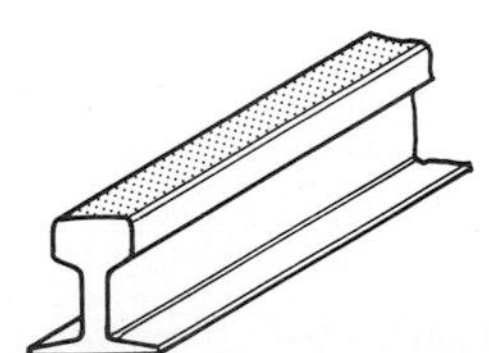

Tracks: Taper deposit

Figure 11–2 Hardfacing application suggestions.

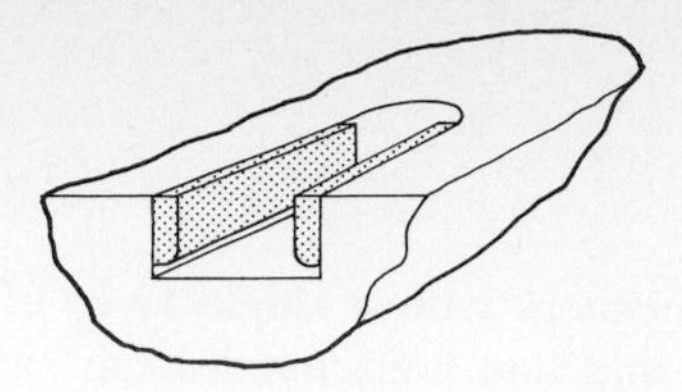

Keyway/cam follower groove: Do not weld into bottom of groove; minimizes cracking

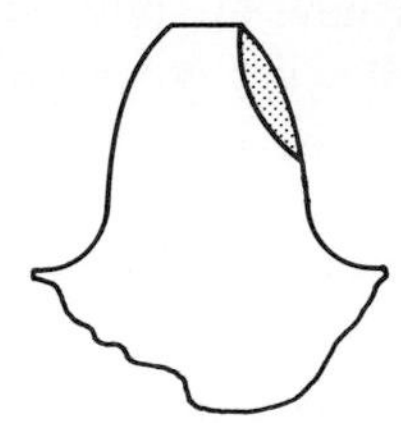

Gear/sprocket teeth: Taper deposit; weld only loaded side

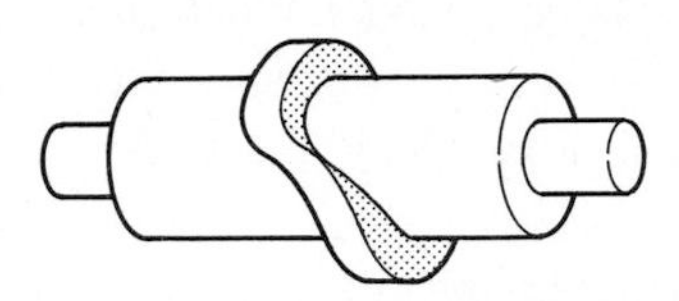

Barrel cam: Weld only loaded side; do not bring deposit to base of projection

Figure 11–3 Hardfacing application suggestions.

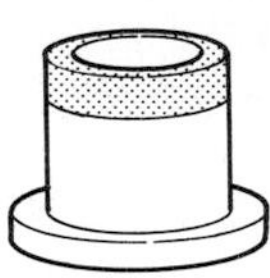

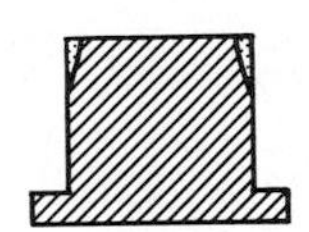

Punch: Do not coat entire end; heaviest deposit where wear will be highest

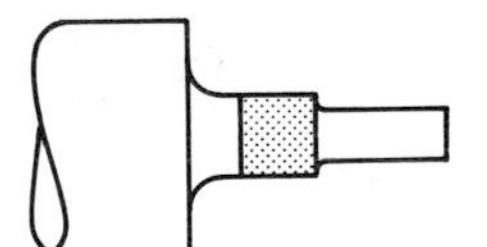

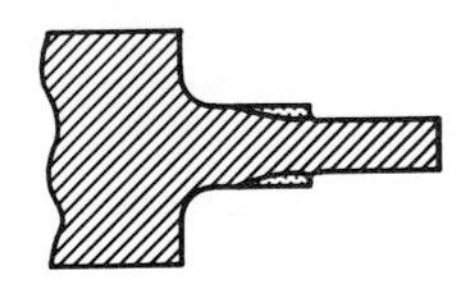

Journal: Taper deposit to minimize stress concentration

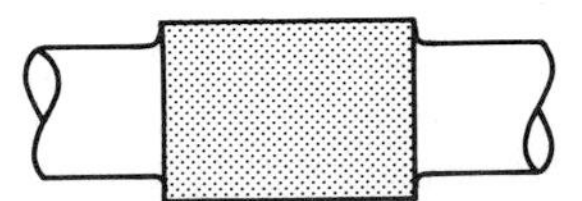

Shaft: Blend deposit; put on boss to prevent shaft weakening

Figure 11–4 Hardfacing application suggestions.

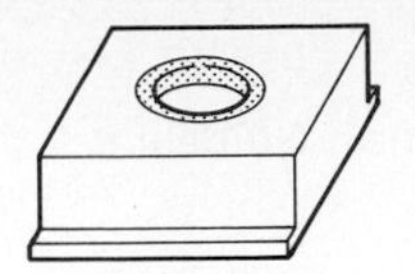
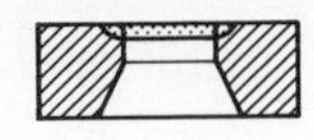

Blanking die: Use tapered horizontal groove for ease of deposition

Battering tool: Taper deposit at ends

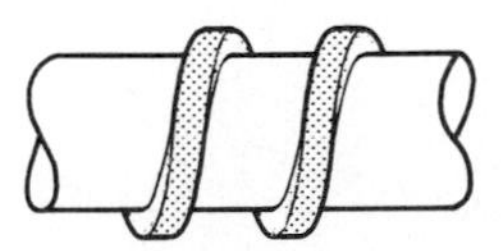
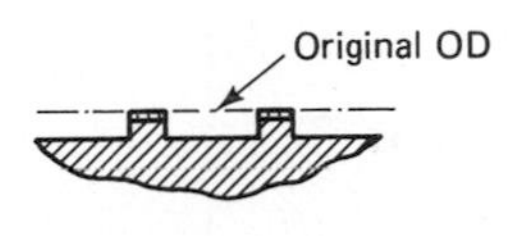

Extruder screw: Put groove in bar and deposit surfacing before machining flights

Figure 11–5 Hardfacing application suggestions.

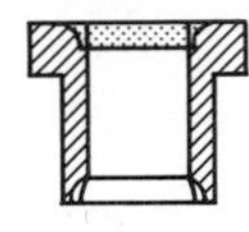

Guide bushings: Radius weld grooves

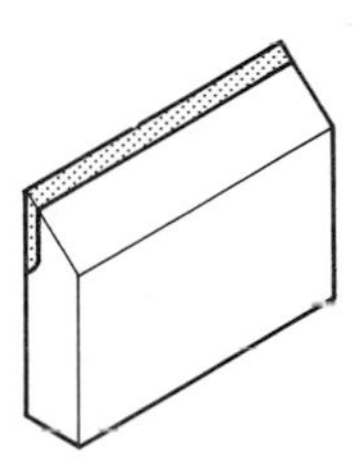
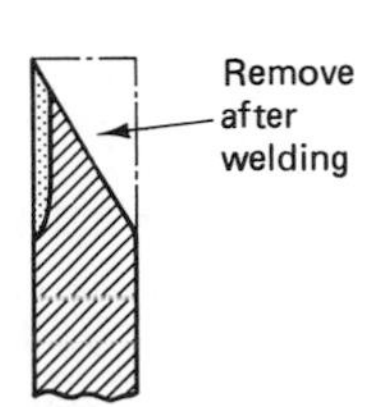

Knife edges: Prebow to reduce distortion; stress relieve and machine angle after

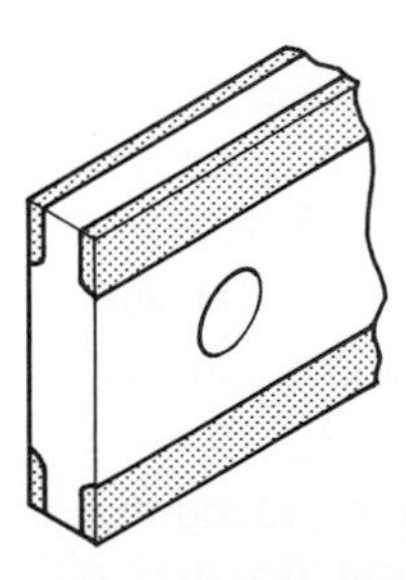

Shear blades: Welding of four edges minimizes distortion; stagger weld

Figure 11–6 Hardfacing application suggestions.

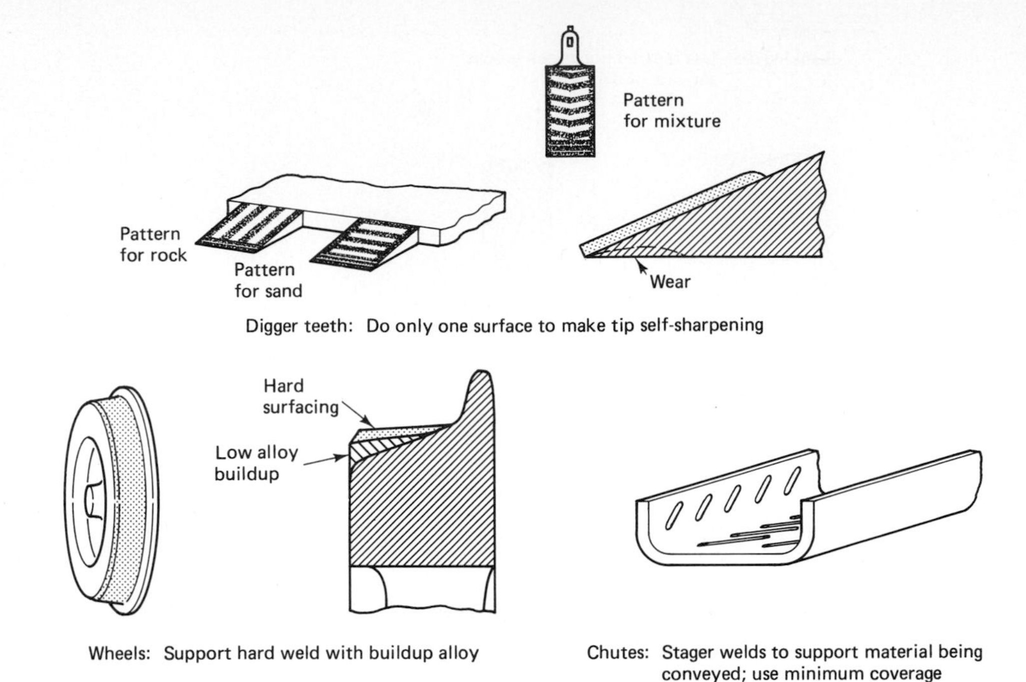

Figure 11–7 Hardfacing application suggestions.

NONFUSION DEPOSITS

Basically, three application processes are used with nonfusion deposits: plasma spray, detonation gun, and the electric or oxy/fuel processes; they all fall into the generic category of thermal spray processes. Each process has some idiosyncrasies from the design standpoint. The following are some guidelines to keep in mind when using these processes.

Plasma Spray

1. Thicknesses are normally in the range of 2 to 10 mils (50 to 250 μm). Deposits can be made much thicker (up to 60 mils [1.5 mm]), but the cost usually becomes prohibitive and there can be spalling problems with very thick deposits.
2. Sprays can only be applied to surfaces that are in a line of sight with the gun.
3. Masking is a hand operation; it is expensive and should be avoided if possible (remove overspray in finishing operations).
4. Substrates should be machined such that deposits are not ground to a featheredge in finishing. Deposits should be applied to an undercut that has radiused ends and that is the depth of the finished deposit thickness.

5. Sprayed hardfacing is easiest to apply to cylindrical surfaces. Flat surfaces are often done by manual manipulation of the bulky plasma torch. This can result in inconsistent thicknesses and coating quality. Large flat deposits can be difficult to successfully apply.
6. Sprayed coatings contain varying degrees of porosity; if this is undesirable, impregnation with polymers should be specified. Teflon dispersions, phenolic resins, or epoxy resins are the usual choices.
7. These coatings have a bond strength similar to electrodeposited platings; fatigue loading conditions should be avoided.

Figure 11–8 illustrates some things to do and not to do in designing for plasma spray deposits.

Detonation Gun

All the things mentioned for plasma coatings apply to d-gun deposits. The big difference between plasma coatings and d-gun coatings is the availability. D-gun is a proprietary process and it can only be done by one company in the United States. It is usual practice to limit the thickness of d-gun deposits, after finishing, to about 3 mils (75 μm). This is not the process limit; it is an economic limit. These coatings are more expensive than plasma coatings, and the cost is based on the geometry as well as the deposit thickness. One thing to keep in mind from the design standpoint is that this process is done in a soundproof room with a remote operator. Parts to be d-gun coated must be small enough to fit in these chambers, and they must be of such a configuration that they can be manipulated with some mechanism during coating. Very large items such as a 20-ft (6-m) extruder screw are done in special d-gun facilities, but this is not a suitable process for coating the tip of a bulldozer blade. The process is ideally suited to production surfacing of small areas on relatively small parts, such as turbine blades.

Other Thermal Spray Processes

The design hints suggested for plasma and d-gun apply to deposits produced with powder torches that use fuel gases for a heat source. They also apply to systems that are fed by a consumable in rod form (like the Rockide® process). The thermal spray process that requires special consideration from the design standpoint is wire metallizing. Conventional metallizing requires that the substrate be dovetailed to assist adhesion of the deposit. As shown in Figure 11–9, the normal sequence for part preparation is to turn grooves in a cylindrical part; the grooves are usually about 0.40 in. (1 mm) wide, 0.040 in. (1 mm) deep on a pitch of 0.080 in. (2 mm). Grooving is ideally performed with special tools available from suppliers of metallizing equipment. After grooving, the threads are deformed to produce a dovetail that is about 0.30 in. (0.75 mm) in depth. Two design constraints are obvious: (1) This process is essentially limited to cylindrical parts. (2) It is not applicable to thin deposits and thin-walled cylindrical parts. Metallized deposits are normally 0.060 in. (1.5 mm) thick or more. A very

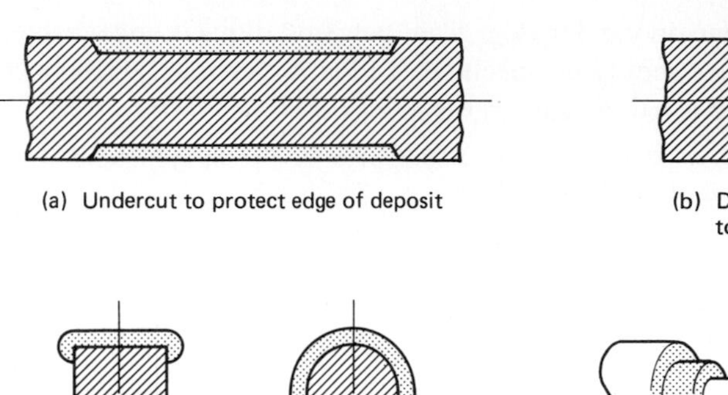

(a) Undercut to protect edge of deposit

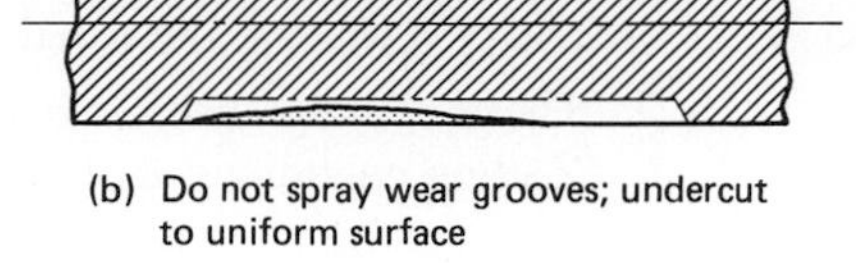

(b) Do not spray wear grooves; undercut to uniform surface

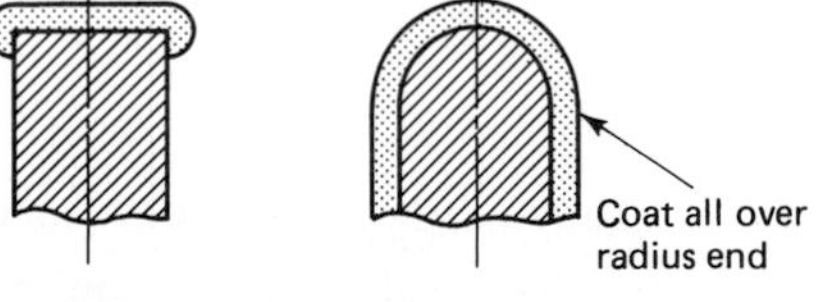

(c) Do not spray tips of punches, etc., with sharp edges

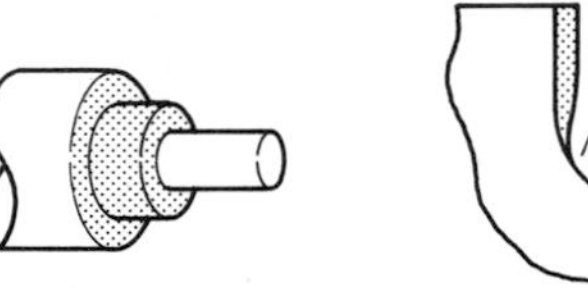

(d) Do not try to spray into corners

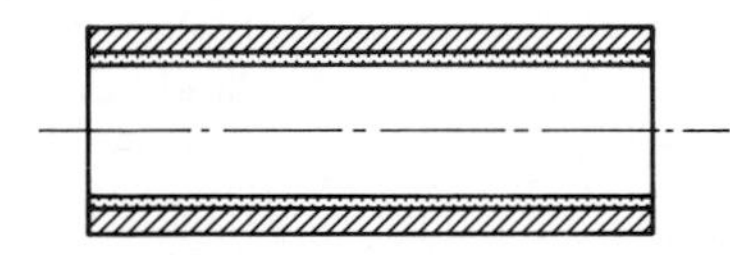

(e) The smallest bore that can be sprayed by most job shops is ~1.5 in. (10 cm). IDs smaller can be sprayed from ends to a depth equal to one diameter or by using special torches.

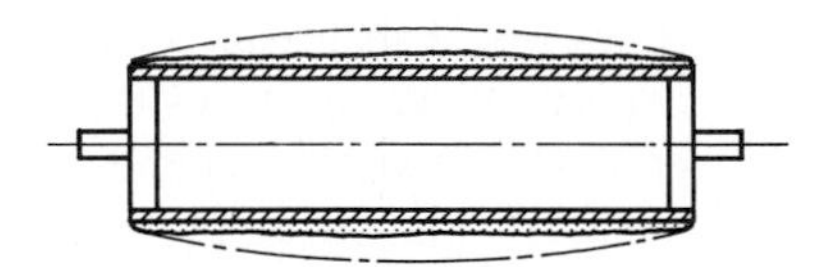

(f) Avoid spraying the OD of thin cylinders; they can distort to form a crown.

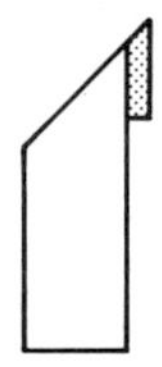

(g) Avoid grinding of sprayed coatings to form cutting edges; only use as-sprayed and very thin (0.0003 to 0.0005 in.; 7.5 to 12.5 mm).

Figure 11–8 Application hints for thermal spray processes.

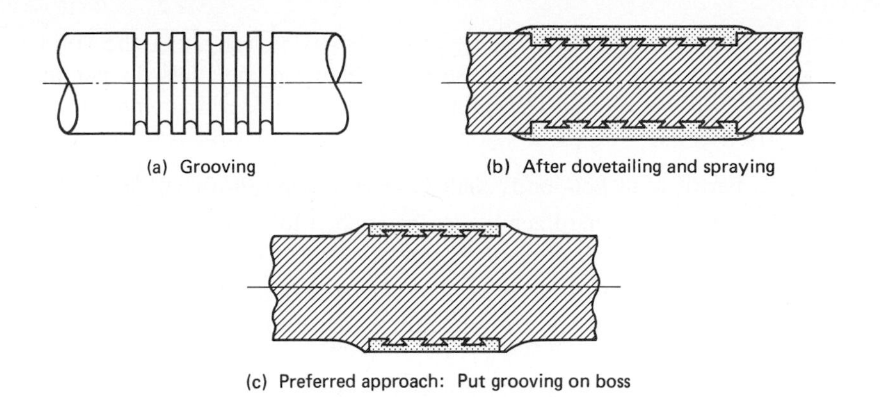

Figure 11–9 Grooving for wire spray processes.

important consideration in using the coatings on mechanical components is that the grooving operation produces severe geometric stress concentrations. To overcome this, it is advisable to put deposits on bosses that are larger in diameter than the remainder of the cylindrical part (see Figure 11–9c).

WELDABILITY CONSIDERATIONS

The nonfusion processes can, in general, be applied to any metal substrate. Thermosetting plastics can be suitable substrates for plasma and d-gun, but there is always a risk of overheating of the plastic. Heavy-wall thermosets are usually safe. Ceramic substrates are not normally sprayed because they often have wear resistance without hardfacing, but they could be done. Cast irons and free machining steels are sometimes troublesome. The graphite in cast irons can interfere with the bond. Similarly, the sulfur or selenium in free machining parts can have a detrimental effect on bond. One factor that affects almost all the thermal spray processes is substrate hardness. Abrasive blasting of the substrate is used in thermal spray hardfacing to enhance the bond. Hard metals are difficult to roughen with abrasive blasting. Soft metal bond coats can be used to somewhat override this problem, but, if possible, it is desirable to keep substrate hardness to 50 HRC maximum.

Thus there are very few substrate weldability considerations in using nonfusion hardfacing; but if a potential hardfacing job involves a deposit on a strange substrate, such as indium or lead, it is advisable to consult the spray vendor.

Fusion-welded hardfacing deposits involve all the weldability considerations that exist with any fusion welding process. There are two main considerations to be addressed when applying fusion hardfacing deposits:

1. Will the hardfacing deposit adhere to the substrate?
2. Will the hardfacing cause cracking that will propagate in the substrate?

The first question can be answered by investigating the solid solubility of the hardfacing material in the substrate. If the two materials do not have solubility, the deposit will not adhere. The following are some guides for common hardfacing situations:

1. Ferrous-, nickel-, and cobalt-base hardfacing alloys will fuse with ferrous substrates.
2. Copper-base hardfacings can be applied to copper-, nickel-, and ferrous-base substrates.
3. There are no fusion-type hardfacings that can be applied to aluminum, titanium, refractory metals, or low-melting-temperature metal alloys.

Copper-base alloys have poor solid solubility in ferrous substrates, but copper-base hardfacings adhere to ferrous substrates in the same way that brazing alloys adhere to ferrous materials. OAW is the preferred application technique; arc welding may lead to cracking due to liquid metal embrittlement. The nickel-base alloys are compatible with copper-base substrates; there is adequate solubility for good adhesion.

The second major fusion weldability consideration, metallurgical damage of the substrate, is the more formidable problem to be dealt with by the hardfacing user. Welding on any of the substrates that have the capacity for allotropic phase transformations can lead to cracking. Essentially, this weldability consideration is directed at ferrous substrates. If the carbon content or carbon equivalent of a ferrous substrate is in excess of 0.3 percent, the hardfacing user should take precautions to deal with the potential of the substrate cracking in the heat-affected zone under the hardfacing deposit. The cracking potential arises from the possible formation of hard, brittle martensitic structures when the substrate is cycled through hardening temperature ranges from the heat of the welding. The quench that is required for the formation of hardened structure can come from the mass effect of the substrate material. The accepted way of preventing formation of hard, crack-prone structure in hardfacing of ferrous substrates is to use preheating of the substrate. This slows down the quench and prevents the transformation to hardened structure. Table 11–1 (pp. 300–301) shows suggested preheat temperatures for welding of ferrous alloys. These temperatures can be used as a rough guide for preheat temperatures for hardfacing, but they apply mostly to cracking of the substrate. Much higher preheat temperatures are often required to keep the harder hardfacing alloys from cracking. If crack-free deposits are a design requirement, the hardfacing manufacturer should be consulted for recommended preheat temperatures. A firm rule is to use preheat whenever hardfacing is to be used on ferrous substrates that have the capacity for quench hardening.

The carbon equivalent term mentioned previously is a way of determining the effect of alloy additions on hardenability. Plain carbon steels can quench harden when their carbon content is greater than 0.3 percent, but alloy steels can be quench hardened if their carbon content is lower than 0.3 percent. The carbon equivalent is a way to take this into consideration. The following is the common expression:

$$\text{C.E.} = \%\text{C} + \frac{\%\text{Mn}}{6} + \frac{\%\text{Ni}}{15} + \frac{\%\text{Cr}}{5} + \frac{\%\text{Mo}}{4} + \frac{\%\text{V}}{4}$$

Thus, if preheating is specified everytime the carbon content or carbon equivalent is greater than 0.3 percent, the risk of substrate cracking is greatly reduced. Although not shown in Table 11–1, all tool steels require preheating during fusion surfacing. Suggested preheating temperatures are at least the tempering temperature (use the secondary tempering temperature where applicable).

Another common problem that arises in fusion surfacing of ferrous substrates is free machining steels. The sulfur or selenium in these steels causes hot short cracking on cooling from the welding temperature, and these alloys should be avoided as substrates for hardfacing deposits.

When depositing new hardfacing material over old, worn hardfacing, it is recommended that the old hardfacing be completely removed. If this is not possible, grind to sound deposit and apply the new hardfacing with a generous preheat. Do not mix alloy systems; the new hardfacing should be from the same alloy family as the worn deposit (e.g., do not use iron base alloys over nickel/chromium/boron alloys, etc.).

FINISHING CONSIDERATIONS

Each hardfacing alloy has machining parameters that are best obtained from the hardfacing alloy manufacturer. There are, however, some finishing precautions of a global nature that apply to hardfacing deposits:

1. Use gauges to assure adequate height of a deposit for machining (see Figure 11–10).
2. Design the deposit thickness such that the final wear surface is in an undiluted layer.
3. Avoid finishing of composite alloys (they are extremely difficult to machine).
4. Abrasive blast to remove flux prior to machining.
5. Stress relieve parts prior to machining wherever possible. (Caution: Do not stress relieve deposits such as FeCr and tool steels that will soften in stress relieving).

As mentioned in prior discussions, most hardfacing alloys are not really machinable. Most can only be finished by grinding. When softer hardfacing deposits are used, it is possible to do more complicated machining operations, but the hardfacing user should

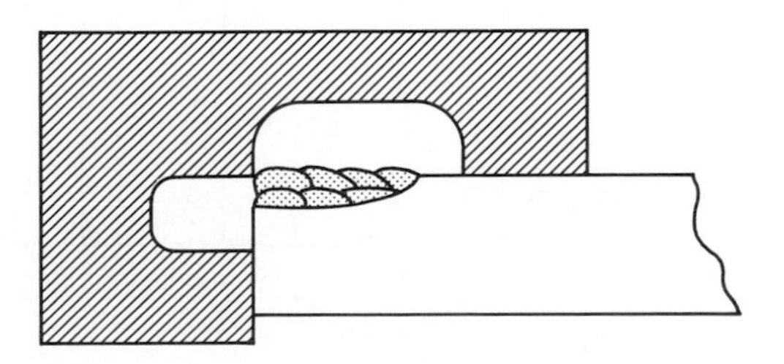

Figure 11–10 Gauge for determining adequate deposit height.

TABLE 11–1 SUGGESTED PREHEAT TEMPERATURES FOR WELDING FERROUS MATERIALS

Metal group	Metal designation	C	Recom-mended preheat (°F)
Plain carbon steels	Plain carbon steel	Below 0.20	Up to 200
	Plain carbon steel	0.20–0.30	200–300
	Plain carbon steel	0.30–0.45	300–500
	Plain carbon steel	0.45–0.80	500–800
Carbon moly steels	Carbon moly steel	0.10–0.20	300–500
	Carbon moly steel	0.20–0.30	400–600
	Carbon moly steel	0.30–0.35	500–800
Manganese steels	Silicon structural steel	0.35	300–500
	Medium manganese steel	0.20–0.25	300–500
	SAE T 1330 steel	0.30	400–600
	SAE T 1340 steel	0.40	500–800
	SAE T 1350 steel	0.50	600–900
	12% Manganese steel	1.25	*
	Manganese moly steel	0.20	300–500
	Manten steel	0.30 max	400–600
	Armco High-Tensile steel	0.12 max	Up to 200
	Mayari R steel	0.12 max	Up to 300
	Nax High-Tensile steel	0.15–0.25	Up to 300
	Cromansil steel	0.14 max	300–400

Metal Group	Metal designation	C	Recom-mended preheat (°F)
Medium nickel chromium steels	SAE 3115 Steel	0.15	200–400
	SAE 3125 Steel	0.25	300–500
	SAE 3130 Steel	0.30	400–700
	SAE 3140 Steel	0.40	500–800
	SAE 3150 Steel	0.50	600–900
	SAE 3215 Steel	0.15	300–500
	SAE 3230 Steel	0.30	500–700
	SAE 3240 Steel	0.40	700–1000
	SAE 3250 Steel	0.50	900–1100
	SAE 3315 Steel	0.15	500–700
	SAE 3325 Steel	0.25	900–1100
	SAE 3435 Steel	0.35	900–1100
	SAE 3450 Steel	0.50	900–1100
Moly bearing chromium and chromium nickel steels	SAE 4140 Steel	0.40	600–800
	SAE 4340 Steel	0.40	700–900
	SAE 4615 Steel	0.15	400–600
	SAE 4630 Steel	0.30	500–700
	SAE 4640 Steel	0.40	600–800
	SAE 4820 Steel	0.20	600–800

High-tensile steels (see also steels below)	Corten steel	0.12 max	200–400
	Yoloy steel	0.05–0.35	200–600
	Jalten steel	0.35 max	400–600
	Double-strength #1 steel	0.12 max	300–600
	Double-strength #1A steel	0.30 max	400–700
	Otiscoloy steel	0.12 max	200–400
	A. W. Dyn-El steel	0.11–0.14	Up to 300
	Cr-Cu-Ni steel	0.12 max	200–400
	Cr-Mn steel	0.40	400–600
	Hi-steel	0.12 max	200–500
Nickel steels	SAE 2015 steel	0.10–0.20	Up to 300
	SAE 2115 steel	0.10–0.20	200–300
	2½% nickel steel	0.10–0.20	200–400
	SAE 2315 steel	0.15	200–500
	SAE 2320 steel	0.20	200–500
	SAE 2330 steel	0.30	300–600
	SAE 2340 steel	0.40	400–700

Low chrome moly steels	2% Cr.–½% Mo. steel	Up to 0.15	400–600
	2% Cr.–½% Mo. steel	0.15–0.25	500–800
	2% Cr.–1% Mo. steel	Up to 0.15	500–700
	2% Cr.–1% Mo. steel	0.15–0.25	600–800
Medium chrome moly steels	5% Cr.–½% Mo. steel	Up to 0.15	500–800
	5% Cr.–½% Mo. steel	0.15–0.25	600–900
	8% Cr.–1% Mo. steel	0.15 max	600–900
Plain high chromium steels	12–14% Cr. type 410	0.10	300–500
	16–18% Cr. type 430	0.10	300–500
	23–30% Cr. type 446	0.10	300–500
High chrome nickel stainless steels	18% Cr., 8% Ni. type 304	0.07	
	25–12 type 309	0.07	Usually do not require preheat, but it may be desirable to remove chill
	25–20 type 310	0.10	
	18–8 Cb. type 347	0.07	
	18–8 Mo. type 316	0.07	
	18–8 Mo. type 317	0.07	

Courtesy of Tempil Corp.

* Preheating usually not required. When welding outdoors in extremely cold weather, 11 to 13 percent manganese steel parts should be warmed to 100° to 200°F. Under normal conditions 11 to 13 percent manganese steel should not be preheated, and welding temperatures over 500°F should be avoided for prolonged periods.

always keep in mind that machining is usually difficult. The copper-base hardfacing alloys can be readily machined; the nickel-base alloys can be milled, turned, and maybe drilled, but tapping and similar operations are almost out of the question.

Stress relieving after welding and prior to machining will minimize distortion of the surfaced part during machining. Stress relieving can affect the deposit properties; thus it is advisable to refer to the hardfacing manufacturer recommendations on this subject. Cobalt- and nickel-base alloys are usually unaffected by stress relieving and the usual stress relief temperature is 1200°F (650°C), but ferrous hardfacings such as iron/chromium alloys and tool steel deposits are usually softened and stress relieving is not recommended.

It is not common practice to stress relieve thermal spray coatings. Because of their porous nature and because they have not been subject to the solidification shrinkage stresses of fusion deposits, the deposit stress usually does not cause significant part distortion.

SUMMARY

In this chapter we have addressed what hardfacing deposits should look like on various parts and how they should be treated after they are applied. These are the design aspects of hardfacing. They are the things that the hardfacing user, not the welder, must specify; they must be addressed on the engineering drawing that specifies hardfacing or on the work order to the shop that will be applying hardfacing.

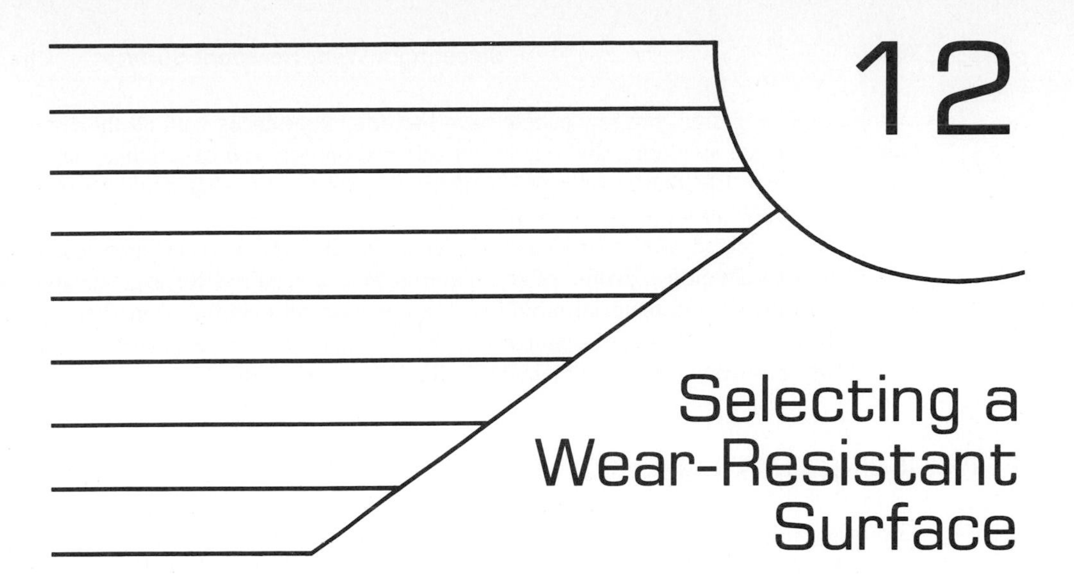

12 Selecting a Wear-Resistant Surface

There are at least 60,000 commercially available metal alloys in just the United States; there are hundreds of polymer families and 20 or so major families of engineering ceramics; there are limitless types of composites and scores of surface treatments and coating systems that may be used to produce a wear-resistant surface on parts that will be subject to some form of wear. How does a designer or material user select the optimum wear-resistant surface for a particular application? It is the purpose of this chapter to answer this question.

We will discuss the basic methodology of material selection and how wear processes enter into the selection process, and then we will describe when to use the more popular systems that are used to produce wear-resistant surfaces.

METHODOLOGY OF MATERIAL SELECTION

Everyone has a repertoire of engineering materials. Engineers are usually required to take a metallurgy or materials course in school, and on the job they learn by experience that certain materials work in certain places; machinists learn about the properties of various engineering materials when their job requires them to fabricate parts from these materials; maintenance and factory operators learn the properties of various engineering materials when they see what fails in service; and every homeowner learns what will work as protective coatings (housepaint, varnish, etc.), what lasts as an appliance housing, and what corrodes on the car. People develop a repetoire of ''good'' materials from their life experiences. When faced with the task of selecting a material for some application, the normal thing to do is to draw on our life experiences and select one of our ''good'' materials. This is fine, but few people have had life experiences with all the engineering materials that are available, and a material selected on personal experience may produce a useful service life that is only one-tenth of the service life that could be obtained if the person knew about XYZ material.

The suggested way to prevent this problem and the way to approach material selection is to develop a profile of the material that is required for a given application and then to match this material profile with a database on material properties, cost, and availability. The best material choice will be the material whose properties, cost, and availability best match the material profile that has been established.

ESTABLISHING A MATERIAL PROFILE

The basic steps in establishing a material profile are outlined in Figure 12–1. The first step is to firmly establish the service requirements of the part. What does the part have to do? Does it transmit torque? Does it just support another member? Is it part of a tribosystem? What is the required reliability? What are the ramifications of breakage or wear? How important is long-term stability? If we take the example of a blade for a rotary lawn mower, the service requirements may be that it has to be capable of cutting grass; it must resist bending and braking when it hits a foreign object; it must be capable of resisting fracture under conditions of vibration and high centrifugal forces; it must have moderate resistance to solid particle abrasion and abrasion from mineral types of materials that may be clinging to the grass; and it must be resharpenable and replaceable at moderate cost.

These factors are the "wants." The next step is to translate the "wants" into measurable properties of materials. Figure 12–2 can be used as a checklist of properties that may be important for a particular application. Chemical, mechanical, and physical properties are self-explanatory. For some applications, a chemical property such as composition may be a prime selection criterion. For example, in the nuclear industry it is often imperative that materials be cobalt-free. Cobalt can become radioactive in these types of installations, and the replacement of a radioactive part could present health hazards and a disposal problem. Sometimes a physical property such as thermal conductivity can be very important. Heads used to heat seal containers are often subjected to abrasive wear from pigments in packaging and the like, and a hard surface is needed to resist this wear; but the surface must also be a good conductor of heat because it transmits the heat for sealing of the package. Most machine components need certain mechanical properties, and it is the designer's responsibility to determine which of these properties are really important to the serviceability of the part and then, where possible, to put some specific limits on the required properties.

The last class of properties listed in Figure 12–2, dimensional properties, are not legitimate properties by most standards; but this classification was included to force consideration of things such as the degree of dimensional accuracy that will be required of a part. For example, many coating processes and surface treatments involve size changes by their very nature. If a part cannot tolerate, for example, a buildup of 1 μm

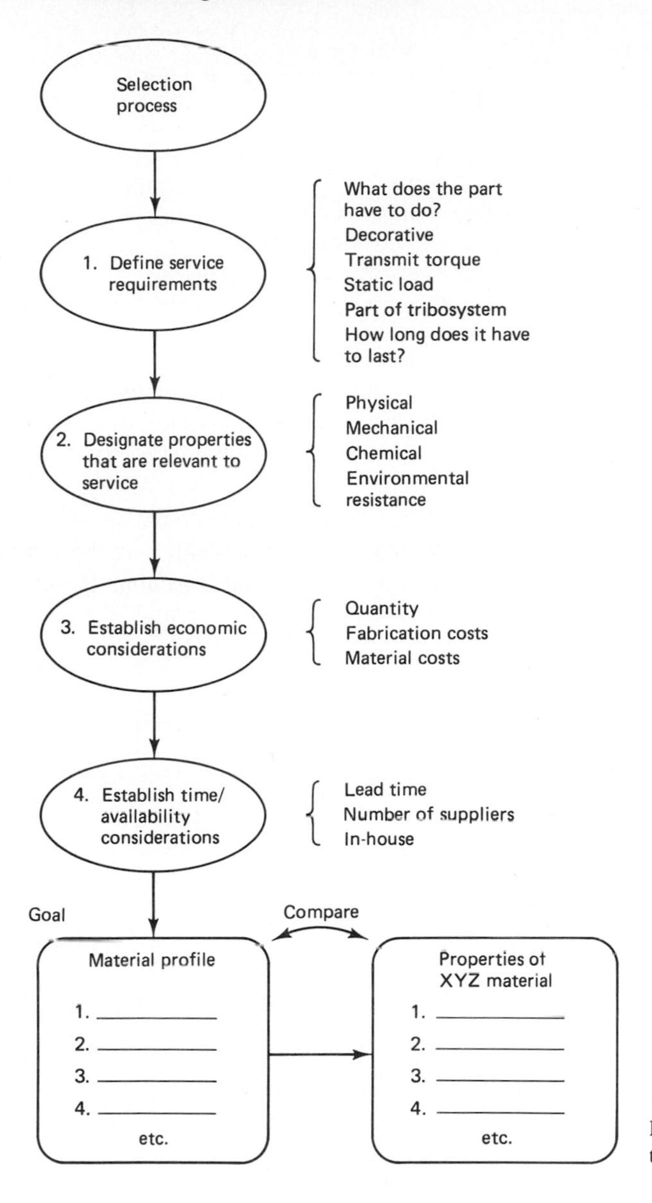

Figure 12–1 Steps in the material/treatment selection process.

(40 μin.) on a surface, it will not even be possible to apply many thin-film coatings. If no distortion can be tolerated in applying a surface treatment, most of the surface hardening processes will be ruled out. In fact, if no size change can be tolerated in the application of a surface treatment, it may be that surface treatments are not even a candidate for the part.

Using our example of selecting a material for a lawn-mower blade, the important properties for this application will probably resemble the following:

Service goal	Related property
Wear resistance	Surface hardness Suitable microstructure
Impact resistance	Impact strength Fracture toughness
Rust resistance	Atmospheric corrosion resistance
High strength	Yield strength > 100 ksi (689 MPa)
High stiffness	Modulus of elasticity = 30,000 ksi (210 GPa)

This list of material requirements already indicates that a hardenable steel is the prime candidate for this application. In fact, only a few metals meet the modulus of elasticity requirements: steels, nickel-base alloys, cobalt-base alloys, uranium, some ceramics, and boron-fiber-reinforced epoxy composites. Intuitively, steels will be the lowest in cost of these materials.

The next step in the selection process is to establish the economic constraints for the part under design. The most important economic fact to establish is the required quantity. The quantity often has a profound effect on material selection. If only one

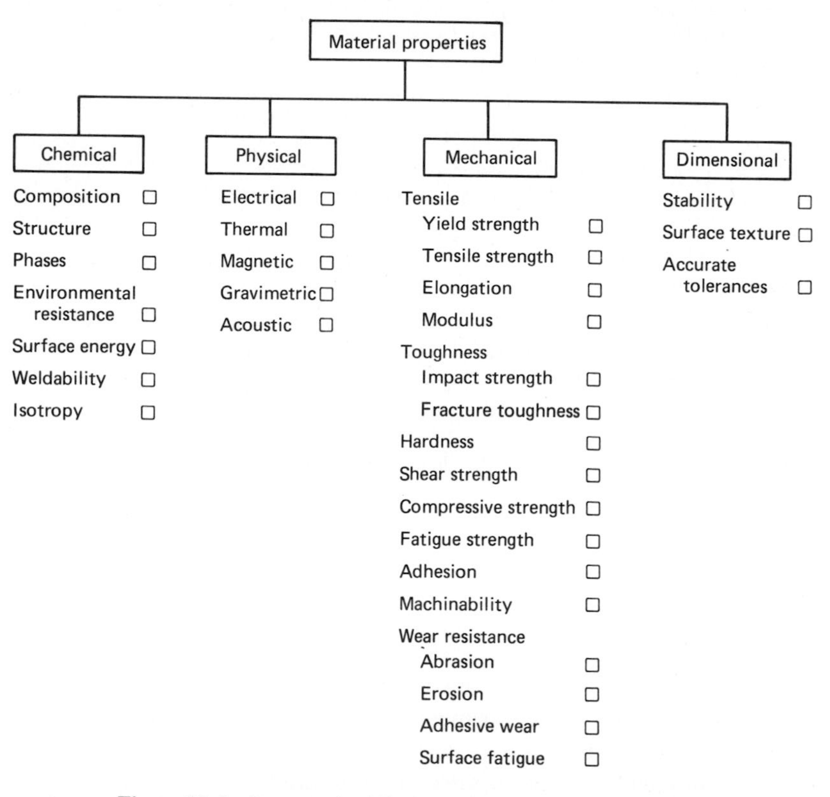

Figure 12–2 Property checklist for establishing a material profile.

blade is required, the use of an expensive cobalt-base alloy may be possible. This may still be true if the number of parts required goes as high as 100. But when the quantity required gets into thousands, the cost of the material may become the major selection factor. In our lawnmower example, thousands will be required; it is for a mower that will be sold at the average discount store. This makes steel the first choice of those materials that meet the stiffness requirement. Fabrication costs should be considered in this step. How difficult is the material to fabricate? If we were to use a cobalt- or nickel-base alloy, the machinability, formability, and blanking characteristics would undoubtedly be more difficult than for steel.

Associated with the economic considerations is the intended service life. If the lawn-mower blade that we are designing is guaranteed not to require sharpening for the life of the mower, it may be necessary to use some superduty material or a steel with some very wear-resistant surface treatment. This would result in an extra cost, and it would affect the economic step in the selection process. Service life is usually always a judgment decision on the part of the designer. Stationary power-generating facilities are usually designed for a service life of 30 years, about 25,000 hours; recreational marine engines are designed for a life of only about 500 hours. Designing for too long a service life can make a part cost so much that nobody can afford to buy it. A realistic life for the lawn-mower blade that we have been discussing may be 250 hours, 1/hour per week for 250 weeks. This type of service life will mean that the blade will not have to have the abrasion resistance of a material such as cemented carbide.

The last category of selection considerations in Figure 12–1, availability, concerns how long it is going to take to obtain the material that we select from the supplier. It can be easy to glance over this consideration, but, like cost, it can often be more important than mechanical properties. There are always time constraints on manufacturing any part. Once the part is designed, there is always a deadline for production of the part. If a material is selected that requires some special production run at the supplier, it may require a lead time of 18 months. In the early 1980s, it was almost impossible to get cobalt for hardfacings and wrought alloys. Aside from world situation availability problems, all materials are subject to warehouse versus mill-run constraints. For example, if we are going to make 15,000 lawn-mower blades in a year, our annual use of the blade material may be only 15,000 lb (6800 kg). Many steel mills (in good times) have minimum quantities for a mill run of 30,000 lb (14,000 kg). Thus either we have to buy a lot more material than we need or we have to get our material from a warehouse where we can buy monthly quantities of 1000 lb (450 kg) or thereabouts. This is one way that availability can affect material selection. Another way is the available number of suppliers. Many times it can be unwise to select a material that is only available from a single supplier. If it is a critical material and the supplier has a labor strike or a disaster of some sort, the manufacture of your product may cease. For our example lawn-mower blade, the production quantity of 15,000 lb (6800 kg) of material per year would probably dictate that we select a material that is available from a warehouse. We do not have sufficient quantity to have a material made special for our blade.

To summarize our selection example, our original rough material profile has been fine tuned to the point where we can now list a material profile with more specifics:

MATERIAL PROFILE FOR A ROTARY MOWER BLADE

1. Tensile strength of at least 100 ksi (689 MPa)
2. Impact strength of at least 20 ft lb (27 J)
3. Modulus of elasticity of 30,000 ksi (210 GPa)
4. Surface hardness of 45 to 47 HRC
5. Resistant to solid-particle impingement and sand abrasion
6. Service life of 250 hours
7. Capable of being flat blanked and formed
8. Machinability comparable to carbon steels
9. Available from a warehouse
10. Cost less than $0.5 per pound ($1.10/kg)
11. Resistant to atmospheric corrosion for at least 6 months in storage before sale
12. Available from at least two suppliers

This firmed-up material profile is now compared with our material database to make a final selection. The database may be only mental; an inexperienced material user may rely on handbooks for the database information, but the process is the same. We would choose a 1050 to 1060 carbon steel with a zinc phosphate chemical conversion coating. The blade would be hardened and tempered to 45 to 47 HRC, and the rust-protecting chemical conversion coating would be applied last. Other choices could be made, and these choices could give comparable service. There are always a number of materials that will meet a given service need, but the steps taken in the selection process should be the same. We selected a through-hardening material because of the core strength requirement, and we did not apply a wear-resistant coating because the service life requirement did not dictate one. The material profile requirements can be met by coatings and surface treatments, and in this instance the material database that is compared with the material profile will contain property information on coating and substrates, as well as on material that could be used without a coating or surface treatment.

IDENTIFYING A WEAR MODE

If the material profile generated in the selection process shows that a part under consideration is part of a tribosystem, it is necessary to determine which form of wear will predominate. In the chapter on wear processes, we defined the different forms of wear and we presented examples to help recognize these forms of wear on parts that have failed in service. On new parts, it is not possible to diagnose a wear mode from worn parts, so the designer should anticipate the type or types of wear that will occur in the tribosystem that is under design. The forms of wear that should be considered are abrasion, erosion, adhesion-related processes, and processes that are more obviously fatigue related. Once it is decided which of these broad categories applies to a particular tribosystem, the wear mode should be fine tuned to one or two of the 17 specific

TABLE 12–1 EXAMPLES OF VARIOUS MODES OF WEAR

Wear mode	Potentially occurs in:
Abrasion	
Low stress	Agriculture equipment, handling minerals
High stress	Comminution equipment
Gouging	Earth moving, mineral benefication
Polishing	Handling solids containing mineral fillers
Erosion	
Solid particle	Sandblast equipment
Fluid impingement	Pipe elbow, rain on aircraft
Cavitation	Pumps, mixing impellers, ultrasonic devices
Slurry Erosion	Oil drilling, pumping, mineral benefication
Adhesion	
Fretting	Bolted-together machine components
Adhesive wear	Gears, cams, slides, bushings
Seizure	Dry sliding systems
Galling	Valves, sliding surfaces, bushings
Oxidative wear	Hard-metal sliding systems
Surface fatigue	
Pitting	Gear teeth, rolling element bearings
Spalling	Surface-treated parts
Impact	Riveting tools, hammers
Brinelling	Static overload or rolling element bearings

forms of wear that we discussed in Chapter 2. Table 12–1 presents some information on tribosystems that are typically subjected to these different forms of wear.

If it is not clear what type of wear will occur in a particular tribosystem, some general statements can be made as a guide to what usually happens in some common tribosystems:

1. When two metals are sliding together and it is likely that they will not be hydrodynamically lubricated (there will not be sufficient speed and lubricant present to form film separation of the parts), metal-to-metal wear will occur. If the loads are very high as in point or line contact loading, severe wear, a form of adhesive wear, may occur. If both metals are above 50 HRC and the loads are moderate, oxidative or mild wear will occur.
2. In ceramic-to-metal systems, the metal member will wear, usually by a combination of abrasion from the ceramic member and oxidative wear from oxide detritus that will be formed from the relative sliding.
3. Polymers with inorganic fillers will almost always cause low-stress abrasion on mating metal surfaces. High-hardness metals or ceramics are the preferred counterfaces.
4. Equipment that handles hard inorganic substances will almost always be subject to one of the forms of abrasive wear.

5. Polishing wear occurs in most machine components that guide products through some piece of equipment.
6. Impingement and cavitation are potential forms of wear in most liquid-handling or liquid-mixing devices. If the liquids contain inorganic particles, slurry erosion is a potential mode of wear.
7. Fretting damage is almost inevitable in bolted-together systems that are subject to cyclic strains that are measurable.
8. Galling is a potential mode of wear in most threaded devices and in devices that require intermittent, unlubricated, slow-speed sliding under high normal forces.
9. Adhesive wear, in its worst form, severe wear, occurs in unlubricated metal-to-metal systems with soft metal as one or both surfaces.
10. Surface fatique in the form of pitting or spalling of coatings is likely in tribosystems that involve hertzian types of loading.

Other general statements can be made about the types of wear that can occur in a tribosystem, but the statements listed here apply to many systems. The main point to be emphasized is that, in selecting a material or a surface for an application, the material profile that has been generated should include specifics on the anticipated form of wear. An engineering material, no matter how exotic it may be, will not resist all forms of wear. It can be resistant to some, but not to all, forms of wear. Thus, in the selection process, it is essential to pinpoint the form of wear that is likely to occur.

COATING/SURFACE TREATMENT VERSUS A MONOLITHIC MATERIAL

The first step in selecting a surface coating or surface treatment including welding hardfacing is to decide whether or not one of these has benefits that outweigh the alternative of through hardening a part for wear resistance or using a material that is wear resistant as manufactured. What are the options? Assuming that it has been established that a part under consideration needs to be protected from one of the forms of wear that we have previously described, the choices that are available for a material that will resist wear fall into three categories: (1) materials that can resist some forms of wear in their as-manufactured condition, (2) metals that can be heat treated to enhance their wear characteristics, and (3) a metal that can be given a surface treatment to make it wear resistant. Figure 12–3 illustrates some of the materials that fit into the first two categories. Figure 12–4 shows the spectrum of processes that can be used for the third choice to alter the surface characteristics of a metal to enhance its wear characteristics. Which category is right for a particular part? The answer to this question is that the designer must scan all these material and process options. The designer should know something about each candidate in each category, and compare a material profile with the available properties from each of these candidates. This sounds like a job for a computer; it could be, but doing this is not all that awesome. The answers to a few

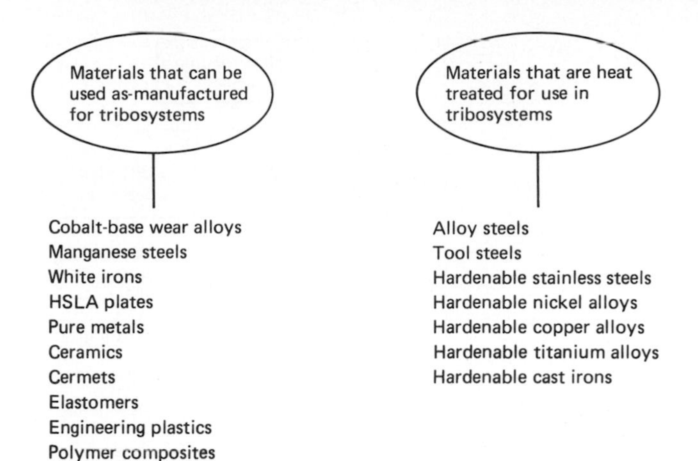

Figure 12–3 Spectrum of wear-resistant materials.

questions will put you into one of three categories: materials with utility in tribosystems as manufactured, heat-treatable material, or surface treatments. These questions are as follows:

1. Is the part made from a ferrous or nonferrous alloy?
2. Does the part need high strength?
3. Can the part tolerate welding or high-temperature treatments?
4. What is the permissible depth of wear in service?

The first question addresses the fact that hardfacing and many surface processes that we have discussed do not apply to a number of metal systems. As shown in Table 12–2, fusion hardfacing processes do not apply to aluminum, magnesium, titanium, and zinc alloys. Through hardening for wear resistance also does not apply. Some titanium, nickel, and copper alloys can be precipitation hardened to about 40 HRC, but this level of hardness on these alloys does not have a significant effect on the response of these materials to most wear processes. Thus, if a part must be made from one of these nonferrous alloys, the only way to produce wear-resistant surfaces on these materials is to use surface treatments.

The answer to the question on part strength must be quantified; does the part need a yield strength of 10 or 100 ksi (68 or 689 MPa)? If the strength, toughness, and stiffness dictate a ferrous material, through hardening becomes a candidate for wear protection. Hardenable steel and cast irons can be made wear resistant to different degrees with through hardening, but for applications such as tools, only tool steels and cemented carbides usually have properties that suffice. A definition of a "tool" is a device used to work, shape, or cut other materials. If the strength requirements do not dictate through hardening materials, hardfacing and the hardfacing alternatives apply.

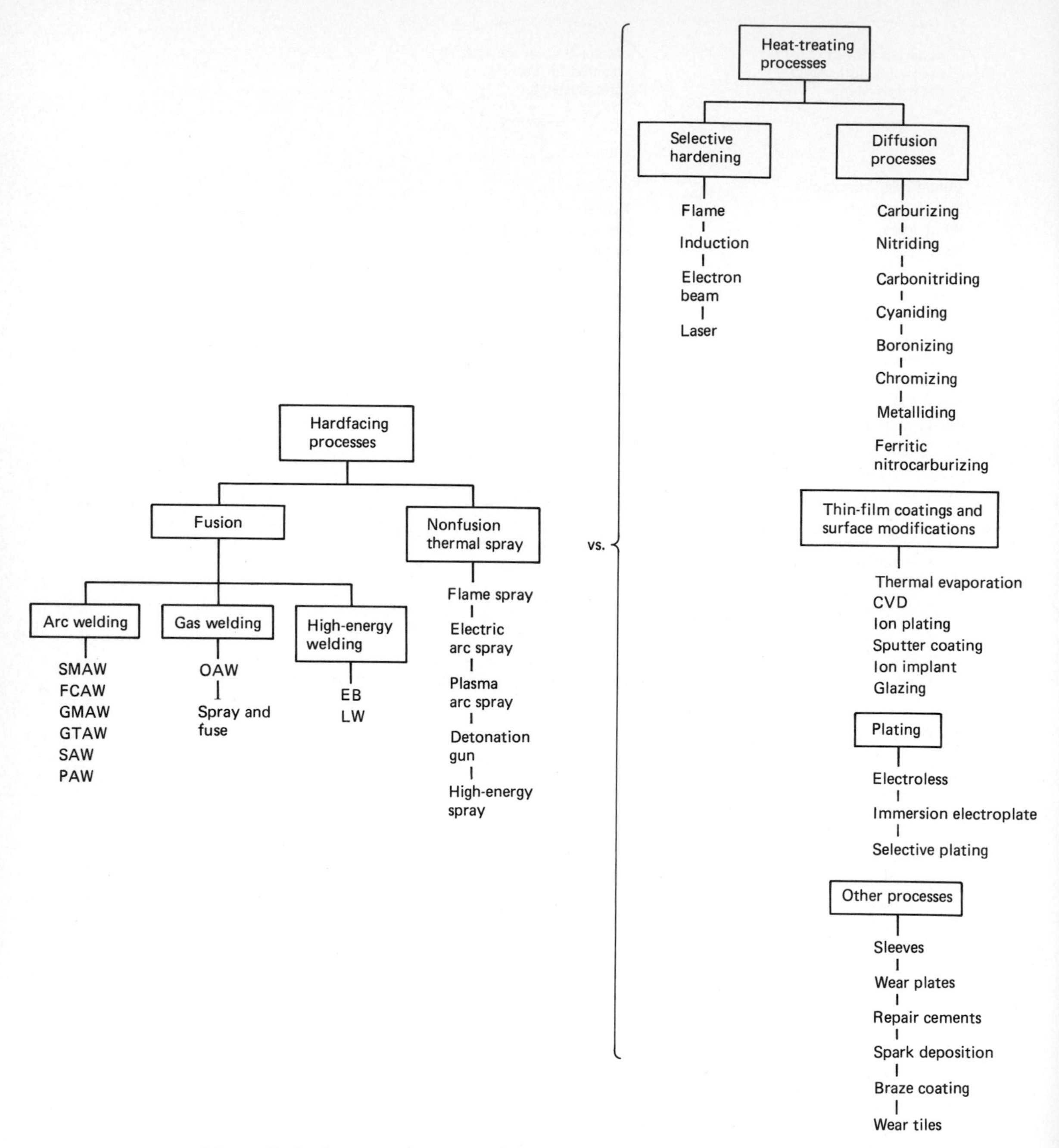

Figure 12–4 Spectrum of processes that can be used to improve the wear resistance of a metal surface.

TABLE 12–2 COMPATIBILITY OF VARIOUS SURFACE-HARDENING PROCESSES WITH IMPORTANT METAL SYSTEMS

Substrate	Applicable fusion hardfacings	Applicable other processes
Aluminum alloys	None	Anodizing, electroplate, PVD sputter coatings, thermal spray
Copper alloys	CuZn CuAl NiCrB	Plating, PVD, CVD coatings, ion plating, thermal spray
Low-carbon steels	All	Diffusion processes, thin-film coatings, platings, thermal spray
Alloy steels (hardenable)	All, but weldability concerns	Selective hardening, plating, thin-film coatings, high-energy modifications, nitriding
Cast irons	Poor arc weldability OAW: Bare rods	Selective hardening, plating, thermal spray
Martensitic stainless steels	Poor arc weldability OAW: Bare rods	Selective hardening, plating, thin films, nitriding, thermal spray
Austenitic stainless steels	All	Nitriding, thin films, plating, thermal spray
Magnesium alloys	None	Anodizing, PVD sputter coating, thermal spray
Nickel alloys	All	Thin films, plating, thermal spray
Titanium alloys	None	Anodizing, thin films, thermal spray
Zinc alloys	None	Plating, thermal spray
Tool steels	None	Thin films, plating, thermal spray

The third process screening question, can the part tolerate welding or high-temperature treatments, is aimed at determining the susceptibility to distortion. We discounted fusion hardfacing in our lawn-mower blade example because the blade shape would make the part prone to distortion. There will be many parts where this distortion concern exists. In fact, it is the major reason for not using hardfacing on some machine components. Fusion welding has the potential for causing distortion on almost any section thickness if the amount of welding, the length of weld or area of weld, is large with respect to the part section thickness.

The geometries that are prone to distortion in welding are thin sections and long slender parts. These same types of parts cause problems in quench hardening. If a part has a geometry that looks like it would produce tolerable distortion in high-temperature processing, hardfacing and the high-temperature alternative processes are candidates for improving tribological properties. If the answer is no, the part is a candidate for the nonfusion hardfacing processes and their low-temperature alternatives.

The last question, allowable wear, is very important since the answer to this question determines if the thin coating processes and surface treatments are applicable. It also determines if it is necessary to use some of the "super-wear-resistant" materials. For example, if it has been established that a part cannot tolerate a wear of greater than 0.001 in. (25 μm) after 1 year of service, it may be necessary to consider cemented carbides or ceramics. The part can be made from a solid carbide or ceramic, or the thermal spray processes can be used to apply a coating of these materials. At the other extreme, if the part can tolerate a wear loss of $\frac{1}{4}$ in. (6 mm) in a year, it may be a candidate for fusion hardfacing or materials such as austenitic manganese steel or HSLA steels that can be used in tribosystems as manufactured. Thin-film coatings such as sputtered TiN usually have a thickness of less than a few micrometers; these types of coatings may not be appropriate for use on parts that can tolerate more than this amount of wear.

In summary, before getting into the process of selecting a surface treatment for wear resistance, it is necessary to establish that a coating provides some advantage over the alternative of a material that is usable as manufactured or over a material that can be hardened by heat treatment to make it wear resistant. The four questions that we recommended are intended to guide the way to answering the question, is it more appropriate to use a through-hardened material or will a material combined with a surface-improvement process meet the service requirements?

WHEN TO USE PLATINGS

Some distinguishing characteristics of platings are their ability to be applied to a wide variety of substrates and the low temperatures at which they are applied. Most electrodeposits are applied from baths that are room temperature, and the highest bath temperature will be less than 100°C (212°F). They can be applied to hardened tool steels, to many nonferrous metals, and even to plastics. This means that a hard surface can be applied to most metals without regard for the metallurgical details of solid solubility and the like that exist is some of the welding processes. The following is a list of some of the substrates that are often plated with one of the electroless processes or with electroplating:

Tool steels
Alloy steels
Carbon steels
Stainless steels
Nickel-base alloys
Copper-base alloys
Cast irons
Zinc alloys

Anodizing is the electrochemical conversion coating that can be used like platings to produce a wear-resistant surface on aluminum, magnesium, and titanium. Hardcoating, the heavy anodizing that is used on aluminum, is the most widely used anodizing process for wear applications. The metal platings that have the most utility in producing wear-resistant surfaces are chromium, hard nickels, electroless nickel, and, as lubricating precious metal deposits, silver and gold. There is no technical limit to their thickness, but they are usually most cost effective for coatings in the thickness range of 0.0001 to 0.010 in. (2 to 250 μm).

The low-temperature aspects of these processes sometimes makes them the only coatings that can be applied to distortion-prone substrates. For example, if it is desired to apply some durability to a large machine platen (1 m^2 or 10.7 ft^2) that is machined to close tolerances, it will not be possible to surface this platen with a fusion welding process without distortion. It is expensive to coat large areas with the nonfusion hardfacing processing, and they involve some heat that could cause distortion. The diffusion hardening processes similarly will involve significant heat and possible distortion, as will surface hardening by flame and induction. The part is too large for most thin-film coating facilities. Even the special processes such as wear tiles are not very practical for large surfaces unless the anticipated wear is severe. Thus plating with hard chromium or nickel is the most practical approach.

Platings are widely used for rebuilding worn parts, mainly because they do not usually cause distortion. When a part needs rebuilding because of service wear, it is finish machined; there is no extra stock to remove from bearing diameters to accommodate distortion. Plating seldom produces distortion, and the coating can be applied only to the area that is worn.

Plating and hardcoatings have one other advantage over most of the other surface processes: they can be applied in small holes and recesses that are difficult to do with other coating processes. In fact, any surface that gets wet will be coated with the electroless plating processes. Solutions can even be pumped through holes to plate, for example, the inside surfaces of a heat exchanger. Selective plating may be the appropriate process to use on parts that are too large to fit in most plating tanks, on parts that must be rebuilt in situ, and for parts that may require significant masking in bath plating.

In summary, platings should be used when the goal is a wear-resistant surface of moderate thickness on parts that are troublesome geometries for the other processes. A troublesome geometry may be a very large surface, a distortion-prone part, or a cavity that cannot be easily reached with line-of-sight processes. Chromium is the most abrasion resistant of the hard platings; it is more abrasion resistant than hardened nickels and hardcoating. Chromium and the other hard platings can slide against hard metals, but the optimum couple may require testing. Chromium is the most corrosion resistant of the hard platings, and it can have resistance to some of the erosion processes that involve oxidizing corrodents. However, it is not advisable to rely on platings as a barrier to corrodents since they are likely to have pinholes or cracks that may penetrate to the substrate. A hard plating over a chemical-resistant substrate can be a usable combination. Finally, conventional platings should not be used for applications that clearly involve surface fatigue. For example, they are not wise choices for applications that involve point and line contact stresses of the type that occur in rolling element bearings and the like.

WHEN TO USE DIFFUSION TREATMENTS

The most popular diffusion process for producing wear-resistant surfaces is carburizing, followed by carbonitriding and nitriding. Diffusion of boron and other elements to produce wear-resistant compound layers on surfaces can produce some unique surface properties, but the commercial availability of these specialty diffusion treatments is limited. Thus the selection of diffusion processes for surface hardening is almost entirely limited to processes that diffuse carbon or nitrogen or combinations thereof into surfaces.

Carburizing and quench hardening are one of the most economical ways of imparting a wear-resistant surface to low-carbon steels. It can be done by many different processes (pack, gas, vacuum, and others). Hardening of the carburized surface layer can yield a hardness of 60 HRC with a case depth of up to 60 mils (1.5 mm) or deeper. There are two main limitations to this process: (1) The liquid quench required to produce the hard surface can cause part distortion. (2) There are substrate limitations. The distortion in carburizing comes from all the problems that can occur on heating a steel to its austenitizing temperature and from the size change that must occur on changing from a soft steel structure to a hardened martensitic structure. The substrate limitations are that it is primarily applicable to low-carbon steels and to carburizing grades of alloy steels. Low-carbon steels have limited hardenability even with the high-carbon surface that is produced in the carburizing operation. From the practical standpoint, this means that carburized heavy sections will not quench fast enough even with a water quench to form a hard surface. A heavy section in round bar is anything over 1 in. (25 mm) diameter. In plate material, a 1 ft^2 (0.3 m^2) plate $\frac{3}{4}$ in. (19 mm) thick may be the thickness limit that will respond to quench hardening. The hardenability problem is overcome by using alloy carburizing steels such as 4615 and 8620. The use of these steels may increase the applicable bar diameter to 2 or 3 in. (50 or 75 mm), but thicker sections may still be a problem. Thus the ideal place to use carburizing to produce a wear-resistant surface is on small parts that can be batch processed, and the parts should not have critical dimensions or they should be finish ground on critical surfaces after carburizing and hardening.

Carbonitriding has a similar type of applicability; the same distortion and substrate limitations apply, but some carbonitriding processes can be done at a slightly lower temperature to reduce quenching distortion, and the diffusion of both hardening species, carbon and nitrogen, can somewhat enhance the wear characteristics.

Nitriding requires special substrates; it only works on alloy steels, and some special processes will work on stainless steels. The big advantage of this process is the low process temperature of about 900 to 1100°F (480 to 590°C), and a liquid quench is not required. A second significant advantage over the other diffusion processes is that when it is used on special nitriding steels of the Nitralloy type a case hardness of 60 to 70

HRC can be obtained. The case depth can be as deep as 0.030 in. (0.75 mm) or more; it is one of the hardest surfaces that can be applied to a steel.

Besides the substrate limitations, a disadvantage of this process is the size limitation caused by the necessity of doing this process in a sealed retort or in a vacuum chamber (for ion nitriding). Sometimes it is difficult to find a heat treat shop with a chamber suitable for large parts. If the limitations can be overcome, nitriding will produce a surface that is very resistant to metal-to-metal wear processes and with good abrasion resistance. Both nitriding and carburizing have good resistance to surface fatigue, and they are quite widely used for gear and cam types of applications (a heavy case is necessary).

This process is more costly than carburizing because of the long times that are required to produce cases. It can only be justified on high-volume production parts if the application warrants the harder surface that it produces compared to carburizing. It is most applicable to special applications that need this very hard surface.

All the diffusion hardening processes that are capable of producing surfaces of 60 HRC or better are suitable for abrasion and metal-to-metal wear applications. They are generally most applicable to small parts that can be batch processed.

WHEN TO USE SELECTIVE HARDENING

Flame and induction hardening are limited to certain families of steels: medium-carbon steels, medium-carbon alloy steels, some cast irons, and the lower-alloy tool steels. There is no size limit to parts that can be flame hardened since only the portion of the part to be hardened need to be heated for the hardening process. The size limit on induction hardening is that the part or area to be hardened usually must fit within an inductor coil. Localization of the heating for hardening is the biggest advantage of these processes; distortion is minimized. The hardened surface layer produced will have the properties of the substrate heat treated by conventional processes, and, in general, the hardnesses are comparable but on the lower side of the recommended working hardness range. Flame- and induction-hardened parts are used for a wide variety of applications, but the tendency is to use these surfaces for metal-to-metal types of tribosystems, gears, cams, rolls, large bearing races, shafts, and journals. Flame hardening has been successfully applied for many years to tillage and mineral benefication equipment where the mode of wear is most often low-stress abrasion.

Flame hardening is usually used for very heavy cases, in the range of about 0.06 to 0.25 in. (1.2 to 6 mm). Thin case depths are difficult to control because of the nature of the heating process. The equipment required for flame hardening is minimal, and most heat treating shops can do flame hardening. There are many nuances to the proper control of the flame and quenching equipment, and critical flame-hardening jobs should be directed to a company that specializes in these processes. Flame hardening can be adapted to production hardening of large quantities of parts, but the more common area of application is on low-quantity parts and one-of-a-kind large parts.

Induction hardening requires putting the area of the part to be hardened in an inductor coil, or with the newer processes a coil is put in proximity with the surface to be hardened. It is much more suited to production hardening of small parts than is flame hardening. It can be more precisely controlled than flame hardening, and cases as thin as 0.010 in. (0.25 mm) can be obtained. Both flame and induction hardening require a liquid quench, but quenching distortion is minimal if small areas are hardened.

In tribosystems, flame hardening is well suited for heavy cases on wheels, gears, cams, mineral-handling equipment, and similar parts that benefit from a wear-resistant surface and a lower-hardness core. Induction hardening is used for smaller parts, mass production parts, and thinner cases. Many automotive parts are induction hardened.

WHEN TO USE THIN-FILM COATINGS

In 1987, the only widely available thin-film coating was titanium nitride, TiN or graded coating of TiN, TiC, and Al_2O_3. It is available from every major cutting tool manufacturer, and there are scores of coating companies that apply this coating as a service. The cutting tool companies usually only apply TiN to high-speed steel and cemented carbide substrates. The coating services will apply TiN to most substrates, but they usually do not guarantee the serviceability. These coatings can vary in composition and morphology depending on the deposition process, but most have a hardness between 1000 and 2000 HK. The deposition techniques vary from planar diode sputtering, to magnetron sputtering, to cathodic arc sputtering and other techniques. The coating can be produced by reaction or from a TiN target. One can expect some property differences depending on the application process, but an important point to keep in mind is that some of these processes can raise the substrate temperature to as high as 800°C (1500°F). Some of the processes keep the part temperature to about 300°C (570°F), but the higher temperature processes can soften tool steel substrates. An inquiry should be made about process temperature whenever a temperature-sensitive substrate is coated. Similarly, when these coatings are applied in thicknesses greater than 1 μm (40 μin.) they tend to produce a matte surface roughness. If a part has a mirror surface before coating (less than 0.1 μm 4 μin.), it may have a 5-μm (200 μin.) roughness after coating and nodules (macros) may be present. The latter usually make coated substrates abrasive to softer surfaces. If this is undesirable, they should not be used unless the supplier can guarantee "no macros."

TiN coatings are usually applied in the thickness range of 2 to 3 μm (80 to 120 μin.). The beneficial effects of the coating are allegedly increased abrasion resistance and increased resistance to sliding wear processes. The effectiveness of TiN in improving service life has been well documented in the area of cutting tools. Pieces per tool are often increased several fold on high speed steel and cemented carbide cutting tools. The coating usually quickly wears off at the cutting edge, but it stays on in the areas of the tool where the chips slide on the tool rake or flank faces. Apparently, the coating minimizes welding of chips to the tool. Thus a proven place to apply TiN is on cutting tools.

At least 20 other thin-film coatings are used for tribological applications, but the average person looking for a wear-resistant surface would be advised against using these without some type of test program. They are not developed to the point of the TiN coatings and, very simply, it is not known if they will work on any tribosystem other than the one for which the coating was developed. Most of these noncommercial coatings have been developed for some special application. For example, silicon photoreceptors are protected from polishing wear by 1 or 2 μm thick (40 or 80 μin.) silicon nitride coatings produced by plasma-assisted CVD. They offer wear protection on silicon substrates, but it is anybody's guess if this same coating will reduce abrasion on a punch press die. A similar situation exists on coatings such as diamondlike carbon, molybdenum disulfide, silicon dioxide, titanium diboride, and scores of other coatings. They are simply not commercially developed coatings.

In general, for tribological applications, thin-film coatings are competitive with the thin metallic platings such as chromium and hard nickels. Titanium nitride is harder than chromium, and in tribosystems that benefit from a hard surface, it may outperform metallic platings. It is generally accepted as beneficial to the service life of cutting tools. Other applications of TiN or any of the thin-film coatings may require testing to determine if they produce improved service life over uncoated substrates.

WHEN TO USE HIGH-ENERGY SURFACE MODIFICATIONS

High-energy surface modifications by our definition are those surface treatments that involve high concentrations of energy, such as 5 kJ/cm^2. The specific processes considered in this category are electron-beam and laser hardening and glazing and ion implantation. Laser and electron-beam surface hardening are well developed processes, and they are used in the same places as flame and induction hardening. They do the same thing as flame and induction hardening, but the hardened surface layer is usually in the thickness range of a few mils to 0.030 in. (50 μm to 0.75 mm), and the beam has to be made to cover the area to be hardened by electronic rastering or by manipulation of the part. The applicable materials are the same materials as those used in flame hardening. These processes usually produce less distortion than flame and induction hardening because of the lower heat input and mass quenching. Thus these processes are used where it is desired to obtain a wear-resistant surface in, for example, the inside of a bore or on a small projection or cutting edge. The hardened surface area and depth can be very accurately controlled.

Laser and EB hardfacing are still in the developing stages, but they can be used to overlay surfaces much like the GTAW and PTA processes. A powder feed tube follows the beam and an alloy is melted into the surface. Thermal spray coatings that have a mechanical bond to the substrate can be fused to the substrate with these processes, but the commercialization of this technique is still under development.

Laser and electron-beam glazing to produce amorphous surfaces and ion implantation have applicability that is similar to some of the thin-film coatings. They have been used successfully in some tribosystems, but these processes are not developed to the point where the average machine designer can freely specify them on a drawing. Where they work and do not work is still being investigated. Ion implantation is more developed than surface glazing. A number of coating job shops will treat a wide variety of parts, but they usually recommend trying a number of species and implantation parameters and a test program to evaluate the success in service. Since the effect of surface modification will typically be in the range of 0.1 to 1 μm (4 to 40 μin.) from the surface, these processes should only be considered for tribosystems that require essentially zero wear. Most tribosystems can tolerate substantially more than this amount of wear, and it is often more realistic to use surface processes that produce wear surfaces that have much greater depth or thickness.

WHEN TO USE HARDFACING

Hardfacing applies to surfaces that can tolerate a substantial surface buildup. It produces coatings on the surface of substrates, not a penetration like the case-hardening processes, and it produces much thicker deposits than thin-film coating and metallic platings. Plasma spray and related thermal spray processes produce the thinnest hardfacing deposits, and the thinnest that these coatings are usually applied is about 0.0003 in. (8 μm). The preferred minimum coating thickness is 0.003 in. (75 μm). There is no upper thickness limit to hardfacing; fusion processes can be used to apply 12 in. (75 cm) of hardfacing if an application required it. Thus hardfacing is the process that is best suited for heavy surface deposits. It is well suited for rebuilding worn parts because it can produce any type of coating, and hardfacings can be applied to almost any substrate. The fusion processes require weldable substrates, but the thermal spray processes will work on almost any substrate that can take the moderate application temperatures.

Because of the wide range of consumables that are available, hardfacing can be used to address all the modes of wear. There is a hardfacing alloy that will meet the needs of almost any tribosystem. The high-carbon, high-chromium white irons have well-documented abrasion resistance; the cobalt-base alloys have well-documented resistance to metal-to-metal types of wear such as galling and sliding wear; this same family of alloys has significant utility in erosive wear processes that involve corrosion. The austenitic manganese steels are widely used for resistance to surface fatigue on rails. The list of testimonials can go on indefinitely; it is very well established and hardfacing processes are widely available. Any shop that has welding facilities can apply the fusion hardfacings. The thermal spray processes are available in hundreds of welding shops, and the more sophisticated thermal spray processes such as plasma spraying are available in many shops that specialize in coatings.

Some of the factors that can limit the applicability of hardfacing are the following:

1. It is not well suited to batch processing.
2. Parts usually require finishing after deposition.
3. Part distortion can occur in processing.

Case-hardening processes can be used to produce wear-resistant surfaces on hundreds or even thousands of parts at the same time if they are small enough to be processed in baskets for batch processing. Parts to be hardfaced have to be treated one at a time. Occasionally, some parts can be batch treated by the thermal spray processes; for example, thin carbide plasma spray coatings have been applied to the cutting edges of knives with the parts fixtured such that multiple edges were covered with the spray pattern, but the normal method is to apply the hardfacing to each part separately. Single part processing is dealt with by automation of the deposition and handling processes, but individual part processing may be less cost effective than some of the other processes on small parts.

Fusion hardfacing deposits are composed of overlapping beads of weld deposit that would not be a suitable counterface for most metal-to-metal tribosystems. They must be finish machined or ground. They can only be used as deposited on applications like shovel teeth and earth-moving equipment. This is a significant part of hardfacing, but one of the intents of this book is to promote the use of hardfacing for machine applications. For these types of applications, finishing costs become a selection consideration to be dealt with.

The last major consideration in using hardfacing, part distortion, exists in the fusion processes from weld shrinkage. This shrinkage cannot be eliminated, but it is dealt with by machining after the hardfacing operation. The thermal spray coatings are often used as deposited. For example, if these coatings are to be used for one of the erosion forms of wear, the surface roughness that is indigenous to thermal spray processes is usually of no consequence; the erodent is usually some type of fluid and surface roughness usually does not affect rates of erosion processes. For conforming surface tribosystems, these coatings usually require machining. Thermal spray processes usually do not involve heating the substrate to temperatures in excess of 300°F (150°C). This temperature will not cause distortion on most substrates. The user of thermal spray coatings, however, should be aware of the distortion potential from the abrasive blasting operation that is used to prepare the substrate for coating. To prevent this type of distortion, it is usually sufficient to tell the coater that distortion cannot be tolerated and to use mild blasting or a bond coat in place of the normal blasting operation. Thus there are some disadvantages of hardfacing compared to competitive processes for generating wear-resistant surfaces. These potential problems can be dealt with, but these are factors to be considered in process selection.

In summary, hardfacing is only suited to tribocomponents that can tolerate a surface buildup. It applies to most forms of wear because hundreds of different consumables can be applied. It is ideal for parts that do not require batch processing and for parts that can be used as-hardfaced. If this is not the case, the cost of postprocessing finishing must be made a selection factor. The risk of part distortion must be considered, especially with the fusion processes. Like all the processes that we have discussed, hardfacing has limitations, but these can be dealt with by proper process specification. In comparison to all the processes that we have discussed, hardfacing probably has the most versatility, and it offers wear-resistant surfaces that can work in almost any application.

PROCESS COMPARISONS

We have recommended that the selection process be started by establishing a material profile to define what is needed; we presented some general comments on when particular processes are most likely the best choices, and we will conclude this discussion on the selection of wear-resistant surfaces by making some additional comparisons of hardfacing processing and the 25 or so competitive processes that have been discussed.

Important Selection Factors

Some years ago, 50 experienced machine designers in a very large design engineering department were given a list of material properties and asked to rank the five most important properties that they consider in selecting a material for machine applications. The property choices include such things as specific mechanical and physical properties; various economic considerations such as machinability, material cost, and the like; availability considerations included the material is on hand, the material is available locally, or the material must be ordered from the manufacturer. Surprisingly, the ranking came out as follows:

1. Corrosion characteristics
2. Availability (on hand)
3. Weldability
4. Machinability
5. Hardening characteristics

One might expect that a mechanical property such as yield strength would have made it into the top five list; none of the mechanical or physical properties made it. This ranking undoubtedly would be different if designers at another type of industry were surveyed, but the results of this survey suggested that, in this plant, freedom from corrosion was the most important selection factor; the manufacturing processes in the industry surveyed involved handling chemicals, so the corrosion factor was not too surprising. But the second choice was not even a property consideration; it was an availability concern: was the material on hand in the stock room? A conclusion that can be made from this survey is that designers often consider availability and economics to be more important selection factors than the ''high-tech'' properties that are stressed so much in engineering school. It also shows that designers prioritize based on the worst thing that can happen to a part. Corrosion characteristics won in this survey in the chemical process industry; the designers felt that corrosion was the worst thing that can happen. Apparently, these designers were not worried about parts breaking in service; they probably overdesigned to the point where breakage was never anticipated, but they were concerned about the costs to machine parts and the ability to join parts (weldability).

In selecting a process for a wear-resistant surface, the five most important selection factors will probably include the following:

1. Process availability
2. Wear resistance
3. Cost
4. Distortion or size change tendencies
5. Thickness attainable

We will conclude this discussion of selection surface treatments by comparing the processes that we have discussed in each of these selection factor areas.

Availability

This selection factor is often given high priority in selecting materials. This is especially true if a person has a time limit on getting a part made. Even without time constraints, designers tend to use materials and processes that are available in house or locally. The availability of the surface treatments that are alternatives to hardfacing vary from probably in house to available only at a few places in the United States.

Selective hardening processes, flame, induction, EB, and laser hardening, vary in availability. Most heat-treating job shops have flame- and induction-hardening facilities, but most of these jobs shops can only handle simple jobs and small parts. Large parts, rolls, extruder screws, gears, and the like are best left to heat-treating firms that specialize in these processes. There are such firms in all areas of the United States, but not in every city. Most cities with a large manufacturing base have companies that do laser hardening on a job-shop basis. Electron beam-equipped shops are fewer in number than laser treating shops, but the availability is about comparable to that of specialty flame- and induction-hardening shops.

Most diffusion hardening processes (carburizing, carbonitriding, cyaniding, ferritic nitrocarburizing and nitriding) can be done by any heat-treat shop. Each shop, however, will have size limitations. Almost any shop can handle parts that are less than 18 in. (0.45 m) in diameter and less than 2 ft (0.6 m) long. Five foot (1.5 m) long shafts and similar large parts may require searching for heat-treat shops that have the required furnace capacity. The special diffusion processes, boronizing, chromizing, metalliding, and the like, have to be done at companies that specialize in these processes. They are available, but not in every city.

Thin-film coatings are not done in the average heat-treating firm. These processes are done by tool manufacturers or in companies that specialize in these coatings. TiN coatings are available through almost every machine tool supplier. Since these coatings are widely used on cutting tools, tool supply houses usually act as manufacturer's representatives and parts can be coated through them. Tool supply houses are in every city. Other PVD and CVD coatings are available from some of the same companies that supply TiN coatings. Ion implantation is done by companies that specialize in this process and by large corporations that have in-house units that can be used by others on a contract basis. Surface glazing with EB and laser is not widely used, and this process can probably only be obtained by a special arrangement with laser and EB job shops.

Platings for wear resistance can be applied at most plating shops, and most cities have such shops. All the nickel electroplates, chromium, electroless nickel, and the soft platings (tin, copper, gold, silver, and cadmium) are done at the average plating shop. Some companies offer proprietary chromiums and nickel platings. In general, these platings are not different from chromium or nickel done by other suppliers, but these shops have become adept at putting these platings on functional surfaces, as opposed to surfaces for decorative purposes. They may do a better job for a wear application than a standard plating shop. Selective plating is offered by some standard plating shops, but there are companies that specialize in this process and may even come into a plant and do an in situ coating on a large piece of machinery.

The other hardfacing alternatives that we discussed, wear plates, sleeving, and repair cements, can be done in any shop or plant. One only has to buy the materials and apply them.

Fusion hardfacing processes are available in any shop that does welding. The welder simply has to buy the appropriate consumables. The thermal spray processes are available in countless job shops, and the equipment is commercially available for use in captive shops. The consumables are available from any welding supply house. Figure 12–5 is a tabulation of the types of alloys that are available for the various hardfacing processes. Below the tabulation of processes is a tabulation that shows what forms of wear these consumables apply to. Table 12–3 is a more complete list of the hardfacing processes that are available and where they should be used.

The availability situation can be summarized by saying that the processes that involve surface hardening by allotropic transformation or by diffusion of species into a host substrate can be done in almost any company that has heat-treat facilities or by the heat-treat shops that exist in most large cities. The same statement applies to platings. The thin-film coatings, special platings, and special diffusion treatments may only be available from firms that specialize in these processes. The hardfacing processes probably have the best availability since any shop with an arc welder or oxyacetylene torch can apply some of the hardfacings, and the consumables are as available as the nearest welding supplier. All the processes that we have described in this text are available, but all may not be available locally. If this creates a lead-time problem or some other problem, then this is a factor to consider in process selection.

Wear Resistance

The selection of a surface process or a bulk material for a wear application is probably the least developed part of the selection process. Tribologists who have worked in the field for many years are stumped on a daily basis because, as we mentioned in our introductory remarks, wear is not a material property; it is the product of a tribosystem. If a tribologist is presented with a tribosystem that he or she has never seen before, he

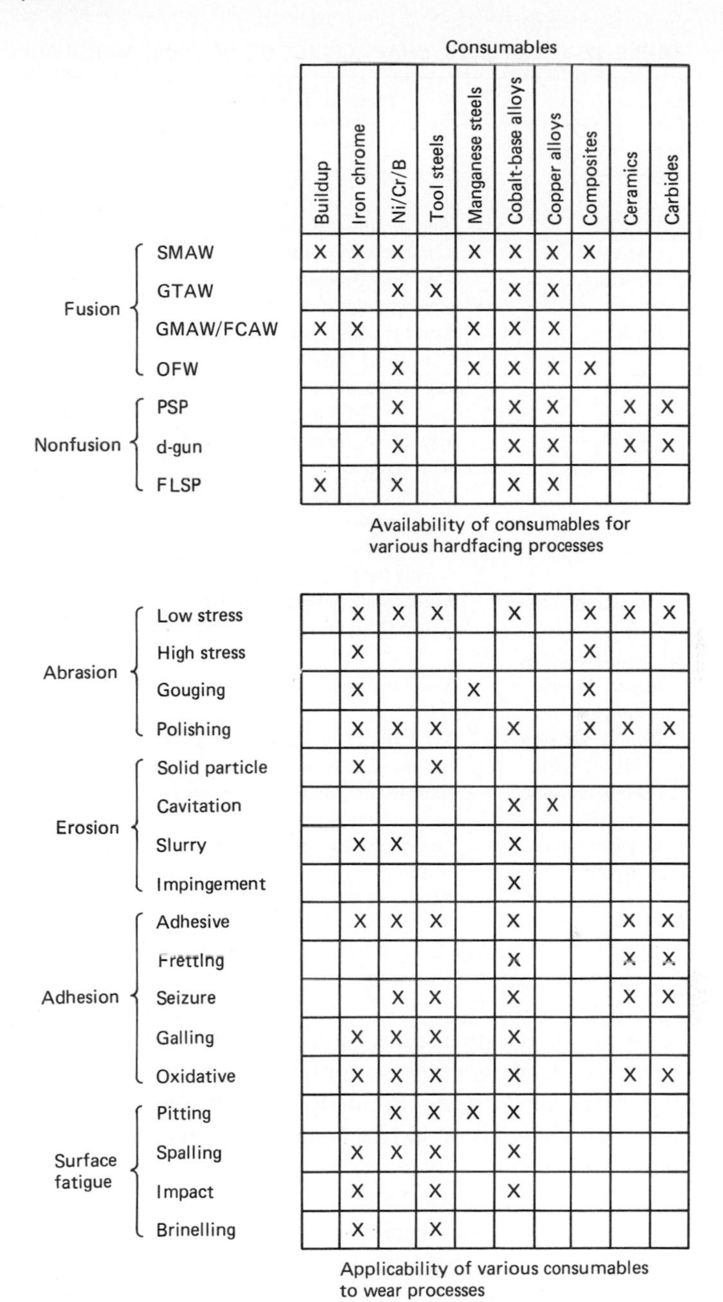

Consumables

		Buildup	Iron chrome	Ni/Cr/B	Tool steels	Manganese steels	Cobalt-base alloys	Copper alloys	Composites	Ceramics	Carbides
Fusion	SMAW	X	X	X		X	X	X	X		
	GTAW			X	X		X	X			
	GMAW/FCAW	X	X			X	X	X			
	OFW			X		X	X	X	X		
Nonfusion	PSP			X			X	X		X	X
	d-gun			X			X	X		X	X
	FLSP	X		X			X	X			

Availability of consumables for various hardfacing processes

		Buildup	Iron chrome	Ni/Cr/B	Tool steels	Manganese steels	Cobalt-base alloys	Copper alloys	Composites	Ceramics	Carbides
Abrasion	Low stress		X	X	X		X		X	X	X
	High stress		X						X		
	Gouging		X			X			X		
	Polishing		X	X	X		X		X	X	X
Erosion	Solid particle		X		X						
	Cavitation						X	X			
	Slurry		X	X			X				
	Impingement						X				
Adhesion	Adhesive		X	X	X		X			X	X
	Fretting						X			X	X
	Seizure			X	X		X			X	X
	Galling		X	X	X		X				
	Oxidative		X	X	X		X			X	X
Surface fatigue	Pitting			X	X	X	X				
	Spalling		X	X	X		X				
	Impact		X		X		X				
	Brinelling		X		X						

Applicability of various consumables to wear processes

Figure 12–5 Availability of hardfacing consumables for various welding processes and their applicability to various forms of wear.

TABLE 12–3 GENERAL CHARACERISTICS OF SOME HARDFACING PROCESSES

	Principal limitations	Best suited for:
Fusion processes		
SMAW	Low deposition rate	Small jobs, out of position, wide variety of consumables
FCAW	Not all alloys available	Heavy deposits
GMAW	Limited alloys available	Large jobs
GTAW	Very low deposition rate	Small jobs, tool repairs
SAW	Limited alloys, flat position	Heavy deposits
PAW	Limited alloys (must use powder or wire), expensive equipment	Mechanized production jobs
OAW	Low deposition rate	Cobalt and nickel base alloys, small jobs, field welding
LASER	Expensive, techniques not well developed	Special production jobs
EB	Poor availability of equipment, techniques not well developed	Special production jobs
Electro slag	Poor availability of equipment, only for alloys in bare wire form	Heavy deposits
Furnace braze	Equipment availability, only suited to a few consumables	Placement of carbide tiles
Nonfusion processes		
Flame spraying (FLSP)	Significant porosity	Heavy deposits > 0.040 in. of metal alloys
Electric arc spraying (EASP)	May require dovetailing, only for consumables in wire form	Heavy deposits > 0.040 in. of metal alloys
Plasma arc spraying (PSP)	Expensive equipment, thin deposits < 0.040 in.	Cermics, cermets, thin metal alloy deposits
Detonation gun, d-gun	Proprietary equipment (one vendor)	Thin deposits < 0.005 in. of ceramics, carbides, and high alloys
Jet Kote	Relatively high gas consumption	Thicker deposits (0.010 to 0.020 in.) of tungsten carbides and high-alloy metal powders

or she cannot be certain about what is the most wear-resistant couple or surface. There are only a few standard wear tests for materials, and no tribologist has tested all materials with all the available tests. Similarly, no tribologist has had experience with all the available bulk materials or surface treatments. What can be used as a basis for selection? The best that one can do is to use past experiences of yourself and others and test data developed by yourself and others.

Figure 12–6 shows where a variety of bulk materials and some surface treatments have been successfully used for various modes of wear. This illustration is based on personal experiences with the specific wear processes and materials that are listed. There could be a fourth ring of labeled hardfacings on this illustration. As mentioned previously, there are so many different hardfacing consumables that there is one or several that will work for almost any wear process.

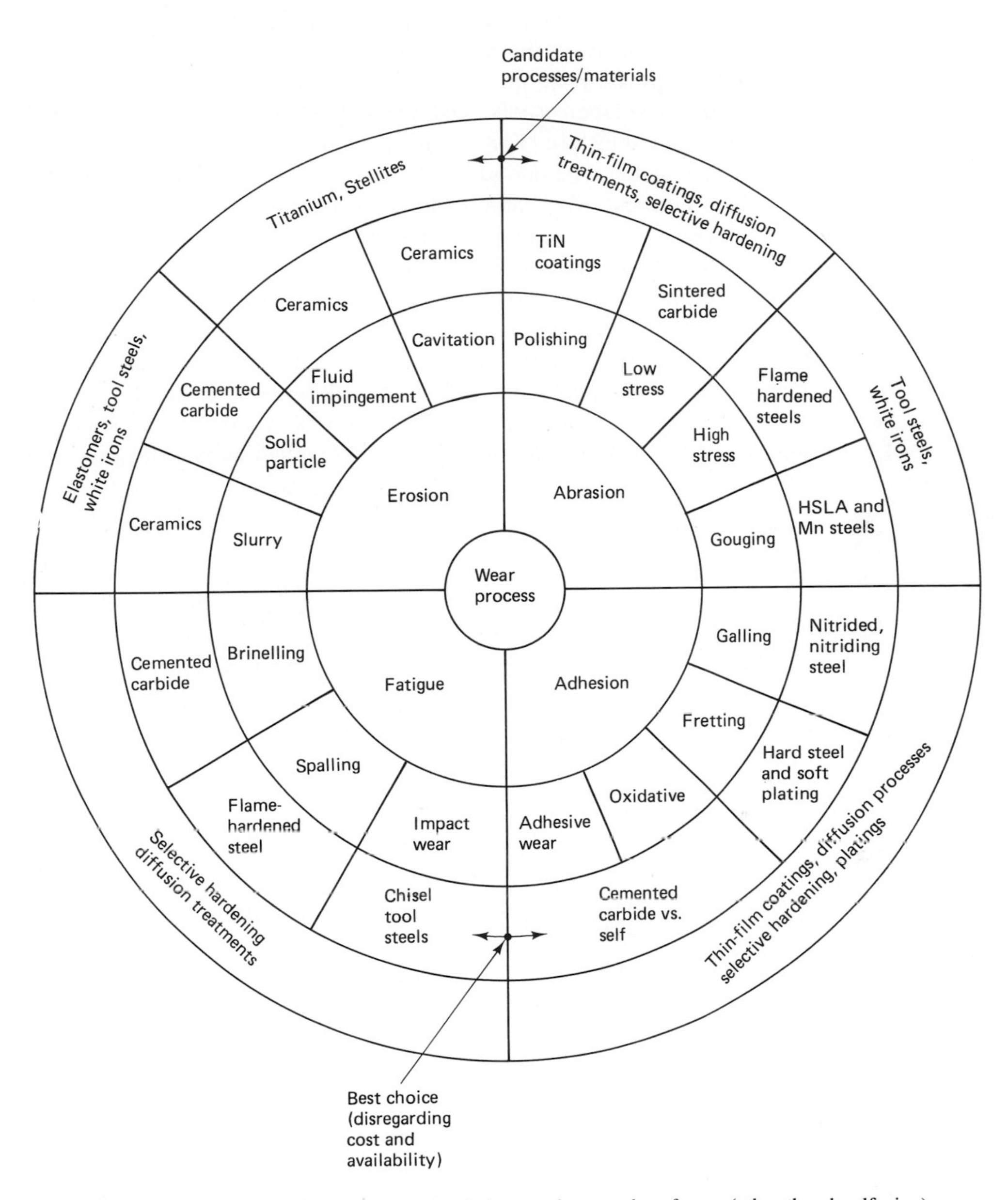

Figure 12–6 Wear modes and material solutions to these modes of wear (other than hardfacing).

Although we have stated in our discussion of wear processes that there is not a direct correlation of wear resistance to hardness for many wear processes, hardness does have a role in selecting coatings and surface treatments. Figure 12–7 shows typical ranges in hardness for many of the processes that we have discussed. All the surface treatments and coatings shown in this illustration have hardnesses greater than ordinary construction steel, low-carbon steels, and the like. The surface-hardening processes that rely on martensitic transformations all have comparable hardness, and the diffusion treatments that can produce harder surfaces are nitriding and nontraditional diffusion processes such as boriding. The hardest metal coating is chromium plate. There is no other plating with its hardness capability. Hardened electroless nickel can have a hardness approaching that of chromium, but in most abrasion tests chromium outlasts electroless nickel by a significant factor. The surfaces that are harder than chromium are ceramics, cermets, or surfaces that are coated or reacted so that they are ceramics or cermets. They are nitrides, carbides, borides, or similar compounds. The popular solid ceramics for wear applications, aluminum oxide, silicon carbide, silicon nitride, and the like, generally have hardnesses in the range of 2000 to 3000 kg/mm^2. As shown in Figure 12–7, when materials such as aluminum oxide are applied as plasma spray and other

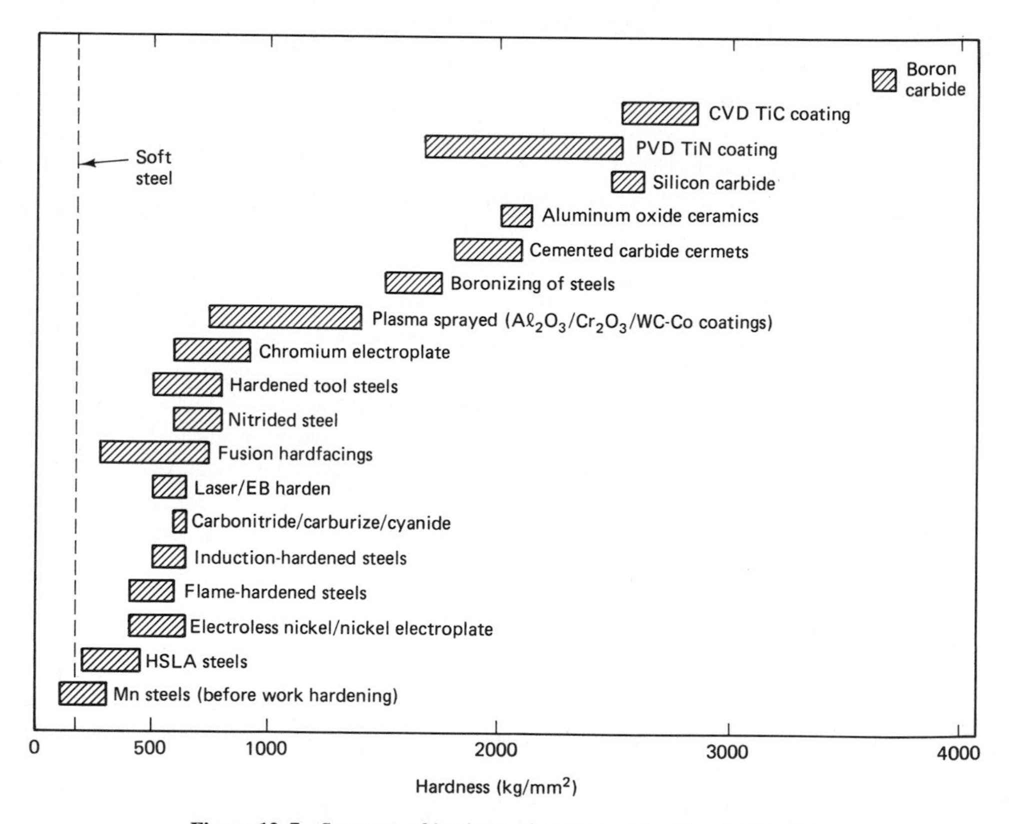

Figure 12–7 Spectrum of hardness of coatings and surface treatments.

thermal spray processes, they have hardnesses that are less than the same material in solid sintered form. This is because the sprayed materials contain porosity and oxides that are not present in the sintered form. The other hard surface for tools, cemented carbide, has a hardness of about 2000 kg/mm^2, about twice as hard as the hardest metal. Some vacuum-deposited carbides and similar compounds have hardnesses that can be harder than the traditional ceramics.

How should hardness information be used in the selection of a surface for a wear application? The wear processes that are usually mitigated by the use of hard surfaces are low-stress abrasion, wear in systems involving relative sliding of conforming solids, fretting wear, galling, and, to some extent, solid-particle erosion. Unfortunately, there are many caveats to this statement. For example, a 3000 Å thick coating of TiN can have a hardness of 2000 kg/mm^2, but it will not resist abrasion from sand with a hardness of 800 kg/mm^2 when it is put on a soft substrate. The coating will not have enough mechanical strength to resist hertzian stress loading from grains of sand, and it will probably be quickly removed by a fracture mechanism. The same thing is true for all the thin, hard coatings. In fact, any coatings that are adhered with a mechanical bond will probably not work in three-body abrasion even with a hard substrate. Thin, hard coatings and surface treatments should be applied to a hardened substrate, and they work best in polishing abrasion types of systems and in conforming solid sliding systems.

Using solid, hard materials is not as simple as it may appear. For example, cemented carbide and sintered aluminum oxide are near the top of our list in hardness, but a sliding couple of aluminum oxide versus aluminum oxide will probably gall. Because of their chemical compatibility, there will be a tendency for solid-state bonding. Self-mating of some grades of cemented carbide will do the same thing. In relative sliding of conforming solids, it is often necessary to perform compatibility tests, and one should not rely entirely on hardness as a selection criterion.

Some designers hold the misconception that sliding systems involving conforming solids should have one member softer than the other, for example, a 60 HRC steel versus a 40 HRC steel. This produces high system wear, and the wear on the harder member will be higher than if the mating steel was also 60 HRC; the hard member is not protected. The reason for this is that adhesion tendencies are higher with one member soft, and this adhesion produces damage to the hard member as well as to the soft member.

To summarize, hardfacings can have hardnesses that range from that of tin/lead babbitts to approaching that of cemented carbides. The coatings and surface treatments that are alternatives to hardfacing can have hardnesses comparable to all the hardfacing processes, and in some cases they can be harder. Most sliding wear processes and low-stress abrasion processes show reduction in rates with high-hardness surfaces, but hardness should not be considered by itself. It is necessary to consider the thickness of the hardened surface layer, stresses on hard surfaces, and compatibility of hard surfaces with mating materials.

Figure 12–8 presents some laboratory wear data on the metal-to-metal and low-stress abrasion resistance of a variety of hardfacing alloys. Based on the results of

Alloy†	Nominal Composition*													Hardness, Rockwell C
	C	Si	Cr	W	Co	Aℓ	Mn	B	Fe	Ni	Cu	Other	Other	
Fe-1 △	0.18	0.30	2.90				1.02		Bal					35
Fe-2 △	0.68	0.47	6.80				1.48		Bal					20
FeCr-1 △	2.20	0.90	30.00				1.30		Bal			3.8 Mo	1.50 Ti	50
FeCr-2 △	0.44	1.20	29.60				1.70		Bal			4.0 Mo		60
FeCr-3 △	4.00	1.00	23.00						Bal			0.5 Mo	7.50 Cb	60
FeCr-4 △	1.00	3.00	13.00				0.70	3.00	Bal					60
FeCr-5 △	6.00	1.00	13.00				2.70		Bal			5.2 Ti		62
FeMn-1 △	0.44	0.52	14.10				16.60		Bal					20
FeMn-2 △	0.64	0.26	0.47				13.60		Bal					20
Co-1 +	1.20	2.06	30.00	4.50	Bal		2.00		3.00	3.00		1.5 Mo		40
Co-2 +	1.80		29.00	9.00	Bal									50
Co-3 +	2.00	0.85	30.90	13.80	Bal				2.30					55
Co-4 +	2.50		32.50	17.50	Bal									58
Co-1A +	0.95	1.20	27.40	5.00	Bal				1.90					30
Co-1C △	1.10		29.00	4.50	Bal					3.00				40
Co-7 △														36
Cu-1 +						14.00			4.00		Bal			20
NiCr-4 +	0.45	2.25	10.00					2.00	2.50	Bal				35
NiCr-5 +	0.65	3.75	11.50					2.50	4.25	Bal				50
NiCr-6 +	0.75	4.25	13.50					3.00	4.75	Bal				56
NiCr-C +	0.04	0.86	15.50	4.10				0.24	5.70	Bal		16 Mo	0.34 V	30
COM-1 +									Bal			60 WC		62
COM-2 △									Bal			60 WC		62
COM-3 +	0.48	0.80	12.00	1.80	10.00		0.80		1.20	1.20		60 WC		60

*Fe, iron base; FeMn, iron manganese; FeCr, iron chromium; Co, cobalt base; NiCr, nickel chromium; Cu, copper base; COM, composite.

† +, rod; △, electrode.

(a) Compositions of Hard-Surfacing Alloys

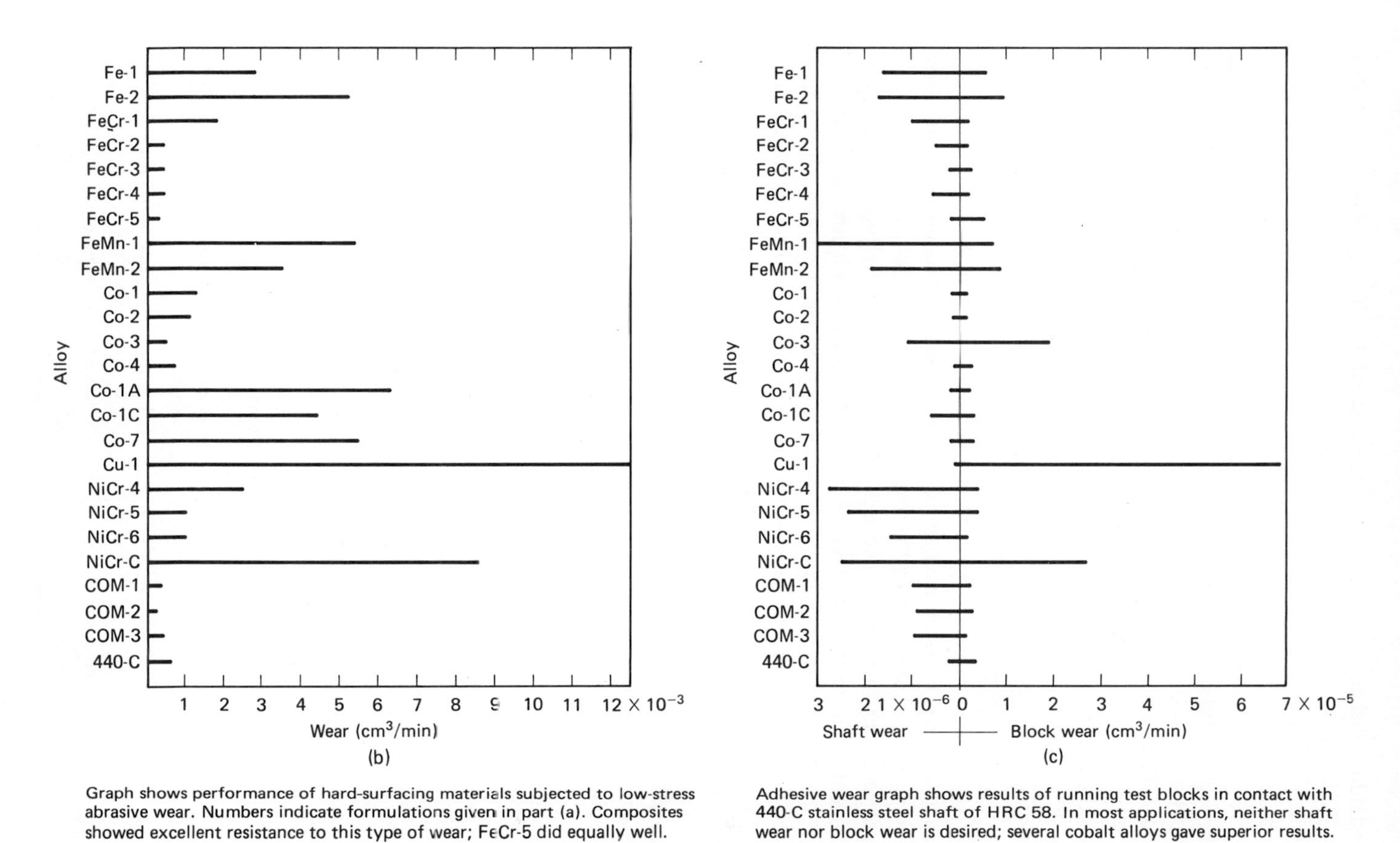

Graph shows performance of hard-surfacing materials subjected to low-stress abrasive wear. Numbers indicate formulations given in part (a). Composites showed excellent resistance to this type of wear; FeCr-5 did equally well.

Adhesive wear graph shows results of running test blocks in contact with 440-C stainless steel shaft of HRC 58. In most applications, neither shaft wear nor block wear is desired; several cobalt alloys gave superior results.

Figure 12–8 Laboratory wear test data on a range of hardfacing consumables.

these tests and additional tests, it was concluded that the hardfacing consumables that are shown in Table 12–4 will meet most hardfacing applications; if a designer becomes familiar with this handfull of consumables, he or she will be able to address most wear problems. Some hardfacing wear data that were generated in a different laboratory are shown in Table 12–5. Some of the same types of alloys were tested, but the tests were different and the role of process type was evaluated. These types of data can be reviewed by potential hardfacing users as selection aids.

Essentially, fusion hardfacings have wear properties that correspond to the alloy systems that they are based on. For example, there are bronze consumables; they have wear characteristics that are the same as wrought bronzes. The same is true for the tool steel consumables, the high speed steel consumables, the cobalt-base alloys, and so on. The porosity of the nonfusion deposits makes their wear properties somewhat different from the wear properties of the same materials in bulk form, but their wear characteristics are often similar or better than the bulk materials on which they are based. For example, it has been shown that plasma and d-gun applied aluminum bronzes have better sliding wear characteristics that the same alloys in the wrought form.

The wear characteristics of the surfaces that are generated by processes that are competitive to hardfacing vary appreciably. Chromium plating and hardened electroless nickel are the most widely used platings for wear applications. Chromium has better abrasion resistance than the nickels, but some of the nickels have application advantages. Many of the diffusion surface-hardening processes, such as carburizing, carbonitriding, and cyaniding, are the same in wear characteristics; only the thicknesses and application parameters vary.

Surfaces of hardenable steels that have been hardened by flame, induction, laser, or electron beam have the same wear characteristics as the same material hardened by conventional bulk hardening techniques; the hardened layer varies in thickness and the

TABLE 12–4 SIMPLIFIED CONSUMABLE SELECTION

	Consumable	Application
Fusion	R or E Fe5-B (60 HRC) High speed steel	Small repairs on tool steels
	R or E FeCr-A1 (58 HRC) Iron/chromium alloy	High-stress abrasion
	R or E NiCr-C (35–56 HRC)	Low-stress abrasion/metal-to-metal wear and for deposits that require machining
	R or E CoCr-A (38–47)	Metal-to-metal wear and low-stress abrasion; use for elevated temperatures and for corrosive environments
	R or E WC 20/30 (60 HRC)	High-stress and gouging abrasion; to be used as-deposited
Nonfusion	Chromium oxide	PSP for low-stress abrasion
	Low-alloy steel	FLSP, heavy spray rebuilds

unhardened core provides toughness that is usually superior to that of the bulk hardened material.

The wear characteristics of the thin-film coatings are still being documented. It is known that titanium nitride deposits of a few micrometers offer improvements to the wear life of high speed steels and cemented carbide cutting tools, but similar statements cannot be made about most of the other coatings. They are still in the development stage. These coating will always be limited to tribosystems that can only tolerate very small amounts of wear. They will never be candidates for bulldozer blades.

There are so many special surfacing processes that it is not realistic to make general statements on their wear resistance, but they should be considered when it looks like none of the hardfacings or more common surface processes will meet service needs.

In summary, we do not have quantitative wear data on all the hardfacing consumables applied with all the welding processes; we do not have quantitative wear data on all the processes that are competitive with hardfacing. Selecting for wear resistance is a judgment process involving many factors, some of which do not relate to wear. All the processes that we have discussed will provide optimum wear resistance for some applications. Thus, in selecting a wear-resisting surface for a particular application, hardfacing and all the alternative processes should be considered as candidates, and the final choice should be determined by matching the wear properties of these surfaces with the wear requirements that were arrived at in your material profile.

Cost of Surface Treatments

The cost of application of a surface treatment to a particular part requires a quotation from candidate suppliers, but some general statements can be made about costs. Probably the most important factor that relates to costs of producing a wear-resistant surface on a part is part quantity. Treating many parts usually allows economies in treatment and finishing; but if many parts are involved, it also follows that the cost per part must be low. In evaluating the cost aspects of candidate surface treatments, the importance of this factor varies if the part is one of a kind, a low-production part, a high-production part, or a repair job.

One-of-a-kind parts usually mean that most surface treatments can be considered as candidates. Special processes such as chromizing and sputtered thin films may become practical. Repair jobs are similar in nature; the cost of a surface treatment may be a lesser consideration than getting the part back in service in a timely manner. Again, most surface treatments are candidates. The exception to this statement is if the one-of-a-kind part is large. A roll 3 ft (0.9 m) in diameter and 12 ft (3.6 m) long is expensive to coat with almost anything. There are some critical sizes for each surface-treatment process above which the cost of obtaining the treatment may be high. Table 12–6 presents some critical-size comments on the various hardfacing and alternative processes. Arc welding, gas welding, selective plating, and similar processes can be done on any size part because the machine can be brought to the part and the machines are available almost everywhere. Electron-beam surfacing and hardening and vacuum

TABLE 12–5 WEAR COEFFICIENTS FOR SELECTING HARDFACING ALLOYS

Composition and hardness of iron-base alloys							
		Nominal hardness		Sliding wear* coefficients		Density	
Alloy	Composition	DPH	Rc	Unlubricated	Abrasive wear† coefficients	gm/cc	lb/in.^{3}
Pearlitic steels	Fe-2Cr-1Mn-0.2C	318	32	6.6×10^{-5}	5.6×10^{-4}	7.85	0.283
	Fe-3.5Cr-2Mn-0.2C	446	45	6.9×10^{-5}	5.8×10^{-4}	7.26	0.262
	Fe-1.7Cr-1.8Mn-0.1C	372	38	9.9×10^{-5}	1.1×10^{-3}	7.6	0.274
Austenitic steels	Fe-14Mn-2Ni-2.5Cr-0.6C						
	As-deposited	188	Rb88	2.8×10^{-5}	5.1×10^{-4}	7.86	0.284
	Work-hardened	458	46				
	Fe-15Cr-15Mn-1.5Ni-0.2C						
	As-deposited	230	18	2.5×10^{-5}	8.2×10^{-4}	7.84	0.283
	Work-hardened	485	48				
Martensitic steels	Fe-5.4Cr-3Mn-0.4C	544	52	9.8×10^{-5}	9.3×10^{-4}	7.6	0.274
	Fe-12Cr-2Mn-0.3C	577	54	6.7×10^{-5}	1.1×10^{-3}	7.69	0.278
High-alloy irons	Fe-16Cr-4C	595	55	7.9×10^{-5}	2.5×10^{-4}	7.61	0.275
	Fe-26Cr-2.5C	544	52	1.3×10^{-5}	8.8×10^{-4}	7.72	0.279
	Fe-26Cr-4.6C	633	57	1×10^{-5}	2.4×10^{-4}	7.17	0.259
	Fe-29Cr-3C-3Ni	697	60	1.8×10^{-5}	2.6×10^{-4}	7.56	0.273
	Fe-30Cr-4.6C	560	53	5.3×10^{-5}	2.6×10^{-4}	7.33	0.265
	Fe-36Cr-5.7C	633	57	2.4×10^{-5}	2.4×10^{-4}	7.69	0.277

* Unlubricated wear tests conducted against 4620 rings on Dow Corning LFW-1 tests at 150 lb for 2000 revolutions at 80 rpm.

† Dry sand rubber wheel abrasion tests using AFS 50/70 sand, 30-lb loads, and 1.44×10^6 mm sliding distance.

Composition and hardness of nickel-base alloys

Alloy	Nominal composition	Nominal macro hardness DPH	Nominal macro hardness Rc	Self-mated threshold galling stress (ksi)	Sliding wear coefficient unlubricated	Abrasive wear oxyacetylene	Coefficient tungsten arc
Boride-containing alloys	Ni-14Cr-4Si-3.4B-0.75C (E/R NiCr-B)	633	57	>72	6×10^{-6}	0.4×10^{-3}	0.3×10^{-3}
	Ni-12Cr-3.5Si-2.5B-0.45C (E/R NiCr-A)	530	51	18	4×10^{-6}	0.4×10^{-3}	0.3×10^{-3}
Carbide-containing alloys	Ni-16.5Cr-17Mo-0.12C (C)	200	(Rb-95)	18	2×10^{-6}	—	1.1×10^{-3}
	(Ni + Co)-27Cr-23Fe-10(W + Mo)-2.7C	405	41	36	1×10^{-6}	0.4×10^{-3}	0.6×10^{-3}
	Ni-10Co-26Cr-32Fe-3W-3Mo-1.1C	315	32	18	1×10^{-6}	0.4×10^{-3}	1.1×10^{-3}
	Ni-17Cr-17Mo-4W-0.4C	315	32	18	2×10^{-6}	—	1.1×10^{-3}
Laves-phase alloys	Ni-32Mo-15Cr-3Si	470	45	36	5×10^{-6}	—	1.2×10^{-3}

Composition and hardness of cobalt-base alloys

Alloy	AWS specification	Nominal composition	Nominal macro hardness DPH	Nominal macro hardness Rc	Self-mated‡ threshold galling stress (ksi)	Wear coefficient unlubricated (0–150 lb) (0–68.2 kg)	Wear coefficient unlubricated (180–400 lb) (81.180 kg)	Abrasive wear oxyacetylene	Coefficients† tungsten arc
Carbide-containing alloys									
A		Co-27Cr-5Mo-0.5C [21]	255	24	72	6.6×10^{-5}	3.3×10^{-4}	2×10^{-3}	0.9×10^{-3}
B	E/R CoCr-A	Co-28Cr-4W-1.1C [6]	424	42	72	6.6×10^{-5}	3.7×10^{-4}		
C	E/R CoCr-B	Co-29Cr-8W-1.35C [12]	471	47	72	6.6×10^{-5}	5.6×10^{-4}	0.8×10^{-3}	1.2×10^{-3}
D	E/R CoCr-C	Co-30Cr-12W-2.5C*	577	54	72	1.1×10^{-5}	1.1×10^{-5}		1.4×10^{-3}
E		Co-32Cr-17W-2.5C	653	58	72	1.1×10^{-5}	1.1×10^{-5}		0.9×10^{-3}
Laves-phase alloys									
F		Co-28Mo-8Cr-2Si (T-400)	580	55	72	3.3×10^{-5}	3.3×10^{-5}		2.2×10^{-3}
G		Co-28Mo-17Cr-3Si (T-800)	653	58	72	3.3×10^{-5}	3.3×10^{-5}		0.9×10^{-3}

* Coefficients calculated from wear tests conducted on Dow Corning LFN-1 against 4620 steel ring at 80 rpm for 2000 revolutions varying the applied load.

† Coefficients calculated from dry sand rubber wheel abrasion tests. Tested for 2000 revolutions at a load of 30 lb (13.6 kg) using a 9 in. (229 mm) diameter rubber wheel and AFS test sand.

‡ Stress out which pin and block show visual signs of material transfer (galling) after one revolution of the pin in unlubricated condition.

Source: K. Bhansali, ''Wear Coefficients of Hard Surfacing Material,'' *Wear Control Handbook*, eds. M. B. Peterson, W. Winer, ASME, 1980, pp. 380–81.

TABLE 12–6 SIZE AND PROCESS CONSIDERATIONS FOR SELECTED SURFACE TREATMENTS

Process	Size limitations	Processing limitations
Arc welding	None	One at a time
Gas welding	None	One at a time
EB treatment	A very large unit would have a vacuum chamber with dimensions of 5 × 5 × 5 ft (3.5 m^3)	One at a time
Laser treatment	Work areas are usually less than 3 × 3 ft (0.9 × 0.9 m)	One to a few at a time
Flame spray, arc spray	None	
Plasma spray, d-gun	Part or gun movement must be mechanized	One at a time
Flame hardening	None	One at a time
Induction hardening	Need 8 kW/in.^{2} and coil to comform to part OD	One at a time
Diffusion treatments	Furnace size	Batch process
Thin films	Parts must fit in vacuum chamber or retort	Batch
Plating	Tank size	One at a time or batch (barrel plating)
Selective plating	None	One at a time
Wear plates	None	
Repair cements	None	

coatings require that the part fit into the work zone of a vacuum chamber. The cost of vacuum equipment goes up exponentially with chamber volume, and there are very few EB units that have a work chamber larger than 100 ft^3 (2.8 m^3). When you want to use this process for a very large part, it may be necessary to conduct a search to find a company with a large enough unit. When an adequate unit is found, it is likely that the shop will charge a lot more for a job than for the same job on a small part because it has to amortize the cost of the large machine. There are out-of-vacuum EB units, but they are not widely available for surface treatments, since job shops tend to buy the more flexible hard vacuum units. A similar situation on vacuum chamber size exists in using thin films that are applied by vacuum technology. There is the same chamber size dilemma, as well as the problem of fixturing the parts to get coating coverage. CVD coatings can have the vacuum chamber size problems, and the ones not done in vacuum can have retort size problems. These retorts tend to be small, and if a very large retort is found, the cost of using this special piece of equipment may add considerable cost to the job.

Even simple carburizing and nitriding operations can get costly if you have a large part. Carburizing can be done in any furnace with adequate temperature capability and atmosphere control, but most heat-treating shops do not have furnaces with work zones longer than about 6 ft (1.8 m). Experience has shown that diffusion treatments on 6 ft (1.8 m) long shafts gets very expensive simply because of the equipment size limitations. Thus a thing as simple as part size can become a significant cost consideration. The comments in Table 12–6 can be used to determine if a part under consideration may get expensive to treat just because few suppliers have facilities that can handle the part.

When the volume of parts becomes any more than a few, the cost of surface treatment can be a critical process-selection factor. Some processes can treat a number of parts for essentially the cost of treating one part. As shown in Table 12–6, diffusion treatments and some of the thin-film coatings can be batch processes. If a part has a shape that does not allow nesting, like a bushing, thousands of these can be put jumbled in a basket and carburized in one furnace operation. In such a case, the cost of the hardening process may only be 10 cents per pound. If you only had one of the same bushings, the cost to carburize it may be $50, the minimum charge. Most heat-treat shops have such a minimum. Thin-film treatments can be done as batch processes, but it is not as simple as the diffusion processes. Some CVD treatments can be done on jumbled parts, but others require individual support of parts like many electroplating operations; each part must be handled. PVD coatings and other line-of-sight coating processes can be done as batch processes, but the parts must fit into the vacuum chamber; each must be fixtured and only the surfaces that face the source or target get coated. Thus, a batch vacuum coating process is not going to be as low cost as a batch diffusion process where jumbled parts can be processed.

The other processes listed in Table 12–6 require one-at-a-time treatment. This means individual handling and the cost of this handling is part of the cost of the treatment.

Table 12–7 presents some process cost comments to consider along with the part size and volume factors. The determination of fusion welding costs is fairly straightforward; each welding process is capable of applying so much metal per hour, the consumable costs so much per pound, and the machining and finishing costs are estimated in the same way that they are for any machining operation. The labor cost includes the welder's and machinist's time times their labor rate (maybe $25 to $50/hour). Low-alloy consumables cost a few dollars per pound; high alloys (cobalt base, etc.) may cost as much as $25/lb ($55/kg). The cost of the metallizing types of thermal spray processes follows the same formula. Figure 12–9 presents an estimate of the cost of various hardfacing processes. The approximate cost of hardfacing consumables is shown in Figure 12–10.

Plasma and the high-energy spray processes are often costed by a rule of thumb of \$0.50 to \$3/in.^{2} (\$0.07 to 0.15/cm^2) of coating per mil (25 μm) of thickness. Finishing costs are added to this figure, and if ceramics or cermets are used, diamond grinding may be required. This is much more expensive than conventional grinding, since the machine shop raises the price to amortize the cost of the diamond wheels, which can be as high as $2500 per wheel.

TABLE 12–7 GUIDELINES FOR DETERMINING THE COST OF PROCESSES FOR MAKING WEAR-RESISTANT SURFACES

Process	Cost comment
Fusion welding hardfacing	SMAW: 5lb/h; GTAW: 2 lb/h; FCAW/GMAW: 10lb/h; SAW: 30 lb/h; PAW: 10 lb/h (Welding cost) + Finishing costs; OAW: 2 lb/h
Flame spray	2 lb/h + finish costs
Electric arc spray	4 lb/h + finish costs
Plasma arc spray	\$0.5 to \$3/in.^{2}/mil + finish costs
Selective hardening	Usually set up and/or tool costs + time per part; induction is faster than flame
Diffusion heat treatments	In quantity, parts are charged by weight: \$0.25–\$2/lb
PVD, ion plate	TiN thin coatings are low cost, but others may be expensive
Plating	Costs depend on labor for masking, racking, and square inches to be plated

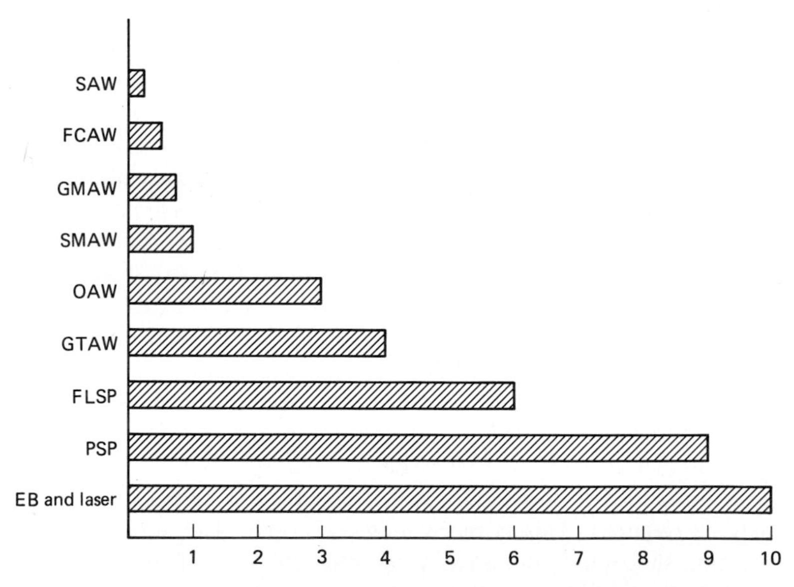

Figure 12–9 Relative hardfacing application costs (based on pounds of alloy deposited per hour).

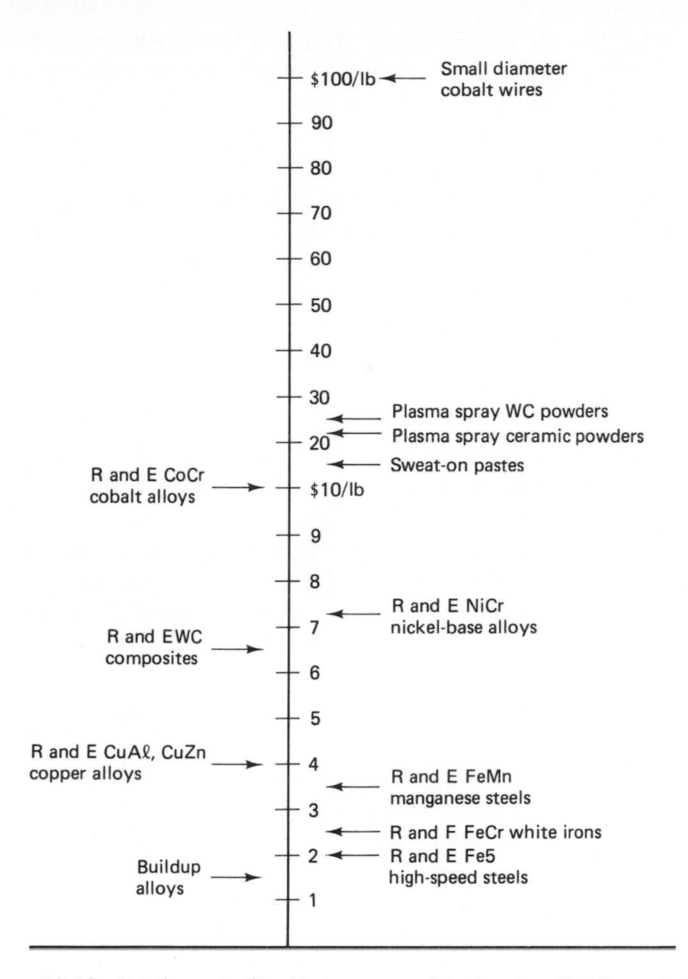

Figure 12–10 Relative costs of hardfacing consumables (based on 1984 *Strategic Materials Survey*).

Plating is almost never as low cost as heat treating, and the cost is mostly based on the labor requirements for fixturing, masking, and inspection. The energy costs and the solution costs are often added to the labor rate as burden, so the labor rates will be higher than that for machine shops; they may range from $50 to $125 per hour. An idea of the cost of electroplating a part can be obtained by estimating how much labor it would take to handle a part through cleaning, masking, and fixturing. The customer usually does not pay for time in the bath. If precious metals are plated, the practice that is sometimes used is to analyze the gold content of a bath before and after plating, and the customer pays for the amount of gold used times the current cost of gold. Selective plating involves mostly labor costs, and coating costs are estimated using labor guidelines like those used in tank plating. Plating of small parts like bolts and nuts can be done by barrel plating. This is a batch process; jumbled parts are put into a barrel and the barrel rotates submerged in the plating bath. This is probably the most economical plating process for small parts.

The cost of thin-film coatings varies significantly with the nature of the part. The only vacuum coating that is widely available is TiN applied by sputtering or some ramification of this process. Tool bits and drills can be coated for pennies per piece because many suppliers are set up to do many parts in one cycle. The same situation exists for razor blades and small cutting devices. Larger parts such as gear hobs may cost one hundred to several hundred dollars, especially if the part must be done alone. In general, if the part is larger than a lathe tool bit insert, it is wise to obtain a quotation from a number of suppliers.

We would like to summarize this discussion of costs by presenting a cost in dollars per unit area for every process. Unfortunately, because of the factors mentioned previously, it is not possible to do this with any great degree of accuracy. We have presented some quantitative cost information where possible, but part size, geometry, and volume affect various surface-treatment processes in different ways, and quotations are the only accurate way of determining part costs. In general, the low-cost processes are conventional diffusion treatments, fusion welding, and selective hardening, probably in that order. The processes that could be expensive are electroplating, thin-film coatings, and high-energy surface modifications; they require a quotation.

Distortion Tendencies

A major reason why alternative processes are used over welding hardfacing is the potential distortion from some of the welding processes. We have already addressed the distortion aspects of welding, but some of the processes that have been presented as alternatives to welding also have the potential for causing distortion. A common misconception by many potential users of all types of hardening processes is that heat causes distortion; it does not. Distortion can occur when something is heated, but it is not caused by the heat and the volume expansion of a material that occurs when the item is heated. Distortion arises from alteration of residual stresses by heating, from plastic deformation when some portions of a part are heated or cooled differentially, from nonuniform support during heating, from expansion stresses caused by part restraint during heating, and finally from volume changes that occur in surface hardening operations. If a cube of low-carbon steel is completely free of residual stresses, it could be hung in a furnace and heated to 2000°F (1090°C) and cooled to room temperature a thousand times with no distortion—if the heating and cooling was always the same on all faces of the cube, and if the block was small enough so that it did not sag from its own weight, and if the furnace atmosphere caused no oxidation, carburization, or decarburization, and if the support wire did not stress only one face. Obviously, nobody heats and cools parts in such a way that the heating process produces no stresses, and few parts are completely free of residual stresses; one way to minimize distortion in surface-treating processes is to not heat the part or to use a process that only requires heating to a low temperature.

Figure 12–11 shows the surface temperatures that are encountered in various surface-hardening processes.

Processes are shown in two groups; one group of processes is likely to produce negligible part distortion, and the other group contains processes that have varying potentials for causing distortion. The discriminating temperature in Figure 12–11 is 1000°F (540°C). This temperature limit really only applies to ferrous metals; obviously, a temperature of 1000°F (540°C) can melt a number of nonferrous metals, and it would cause distortion on metals such as aluminum and magnesium. However, this process temperature information can be used to compare the heating that will be required for a particular process, and this temperature can be compared with the stress-relieving tempera-

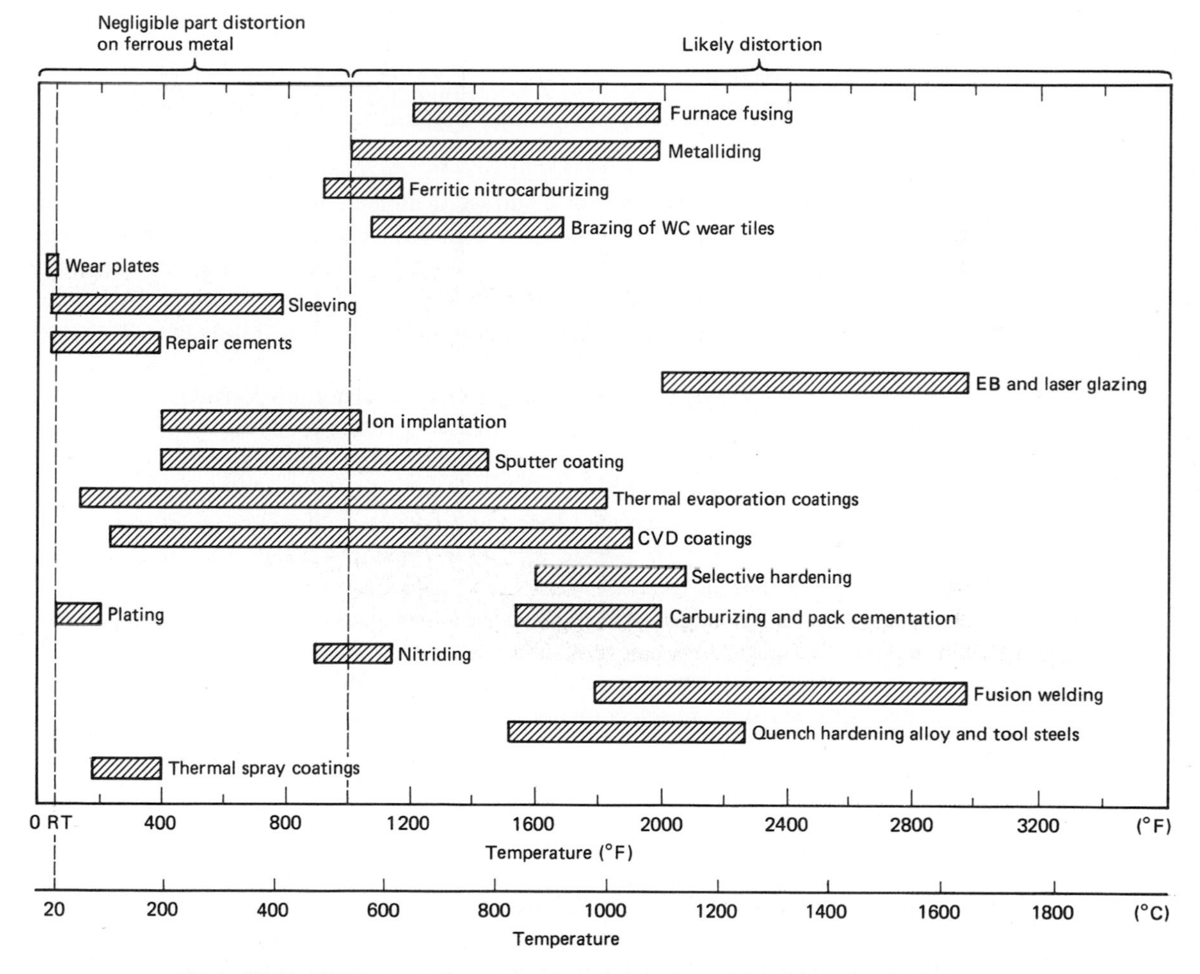

Figure 12–11 Maximum surface temperatures that can be anticipated in various surface treatment and coating processes.

TABLE 12–8 STRESS-RELIEVING TEMPERATURE RANGES FOR VARIOUS ALLOY SYSTEMS

Alloy system	Stress-relieving temperature (°F)
Aluminum	300–500
Copper	400–650
Low-carbon steel	1100–1200
Cast iron	1000–1100
Alloy steel	1100–1200
Austenitic stainless steel	1600–1700
Tool steels	1100–1200
Martensitic stainless steel	1100–1200
Magnesium	300–800
Nickel	900–1300
Titanium	1000–1400

ture of the metal that is a candidate for a surface treatment. If the process temperature is below the stress-relieving temperature for a metal, chances are good that heating below this temperature will not cause significant distortion. Typical stress-relieving temperatures for some metal systems are shown in Table 12–8. Potential users of surface treatments should always inquire about the application temperature of the specific process under consideration.

The other very important consideration in rating the distortion potential of a surface treatment on a particular substrate is if the process by itself produces a change in the volume of the treated surface. Nitriding causes a surface growth of up to 0.001 in. (25 μm) per surface. Carburizing and similar processes produce a volume expansion of several percent in the treated surface. This same surface volume expansion occurs in induction, flame, laser, and EB surface hardening. Fusion welding leads to distortion because the deposited metal shrinks several percent on cooling from the molten condition. Thus some processes are almost certain to cause some distortion because by their very nature they involve alteration of the part shape (the volume of the treated surface region).

If a part could benefit from a surface treatment, but distortion in treatment cannot be tolerated, it is advisable to select one of the processes that requires minimal heating.

Thickness

One of the most distinguishing characteristics of coatings and surface coatings is thickness. In selecting one for an application, the thickness capability should be matched with the wear depth requirement that we just discussed. If a part can tolerate 0.010 in. (0.25 mm) of wear, it is logical to select a process that can provide a surface that is treated to a depth of about 0.020 in. (0.5 mm). Figure 12–12 shows the typical thickness/penetration capabilities of various coatings and surface treatments compared with welding hardfacing processes. This illustration immediately points out one of the biggest differences

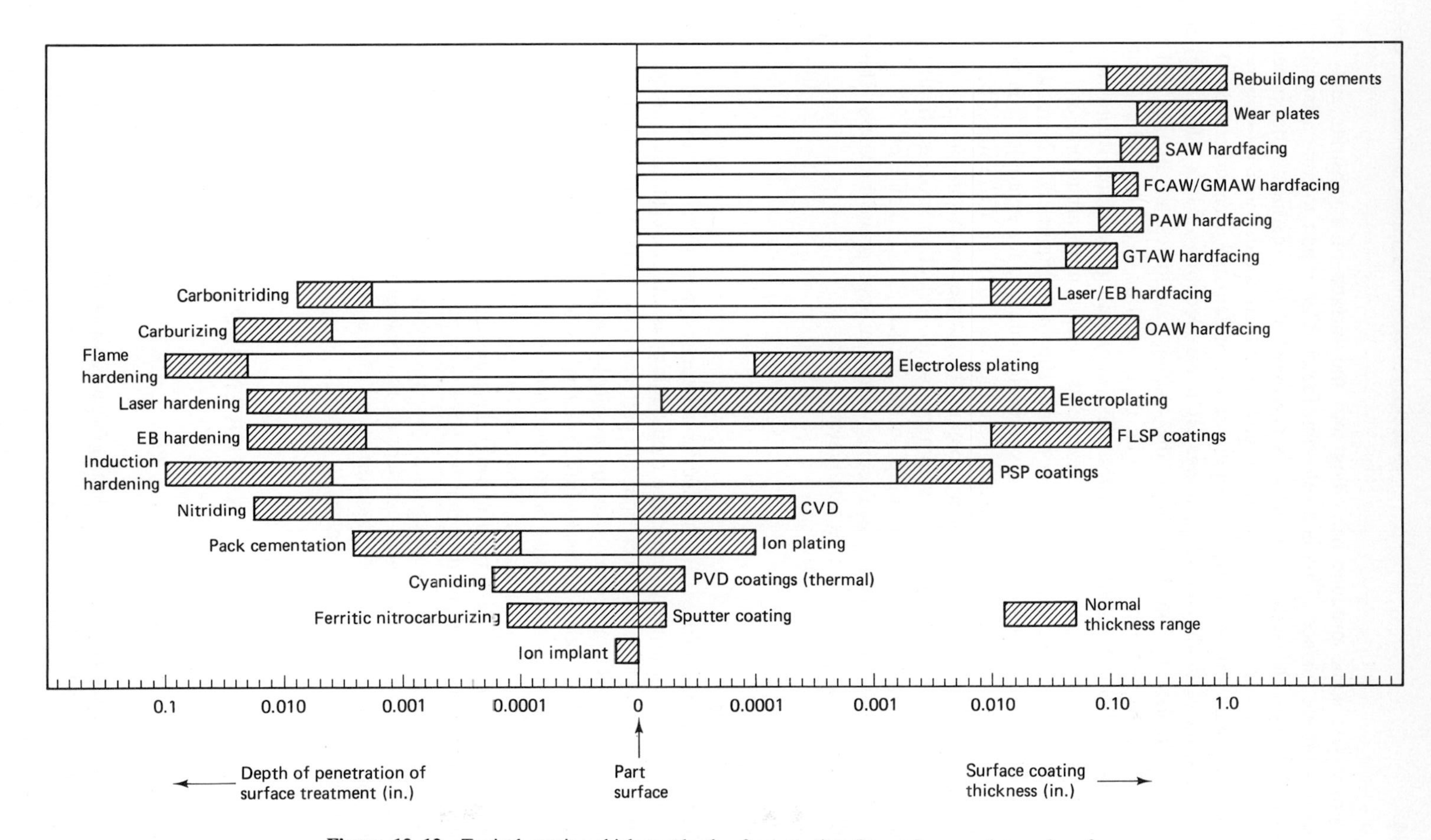

Figure 12–12 Typical coating thickness/depth of penetration for various coating and surface-hardening processes.

in processes; some produce hardening into the surface of a substrate and some are coatings on the surface. This is a selection factor. Can a part tolerate a buildup on the surface? If not, the selection process is narrowed to the treatments that penetrate into the surface. The potential user of these processes then decides which of the penetrating treatments will provide adequate depth for the anticipated wear tolerance. The same thing is done with the surface coatings. If a wear coating of less than 0.001 in. (25 μm) is adequate for an application, the coating candidates are plating and vacuum coatings. If the part can tolerate 0.010 in. (0.25 mm) of wear, the welding hardfacings and heavy plating become candidates. The very thin penetration treatments are ion implantation and the thin diffusion treatments, ferritic nitrocarburizing and cyaniding. The coating thickness and penetration depths shown in Figure 12–12 are not technical limits, but rather typical limits. With special techniques, for example, it is possible to apply thermal evaporation PVD coatings as thick as 0.030 in. (0.75 mm), but at typical coating rates of about 500 Å/hour, coating thicknesses with this process are seldom greater than 2000 or 3000 Å. In summary, in the selection process, think about the use conditions and decide if a coating on the surface is acceptable or if it is desirable to use a penetration treatment. Then decide what is an appropriate thickness. This exercise will drop out many processes as candidates.

SUMMARY

The selection of a process for producing a wear-resistant surface is the critical part of addressing wear in tribosystems. The methodology is simple: establish a material profile that describes what you expect from the part and how much you can afford to pay for the part; screen the processes that are available for producing wear-resisting surfaces and consider the option of using bulk materials; then decide which of these options best meets the part needs that were uncovered in your material profile. Obviously, doing this is not a trivial matter. There are so many ways that wear-resistant surfaces can be obtained. We have tried to fully describe the important processes for creating wear-resistant surfaces; the data that we have presented should serve as the database to use in your cost/benefit comparison. If you go though all the selection steps that we have outlined and you are still not sure of the best process to use, make a judgment decision. Selecting a wear-resistant surface is often a judgment decision for even experienced metallurgists and tribologists.

REFERENCES

BUDINSKI, K. G. ''Guide to Hardfacing, *Plant Engineering*, Vol. 44, No. 19, 1974, pp. 181–183.

LUTES, W. L., and H. F. REED. ''Selecting Wear Resistant Materials,'' *Chemical Engineering*, Vol. 6, 1956, pp. 243–247.

Matthews, S. J., R. D. Zordan, and P. Crook. "Laboratory Solutions to Unlubricated Wear Problems," *Metal Progress,* Vol. 44, No. 8, 1984, pp. 60–63.

Price, L. H. "Fighting Wear in Agricultural Off Road Equipment," *Metal Progress,* Vol. 124, No. 3, 1983, pp. 21–27.

Riddebough, M. "Stellite as a Wear Resistant Material," *Tribology,* Vol. 3, 1970, pp. 4–8.

Stoody Hardfacing Guidebook. Whittier, Calif.: Stoody Company, 1966.

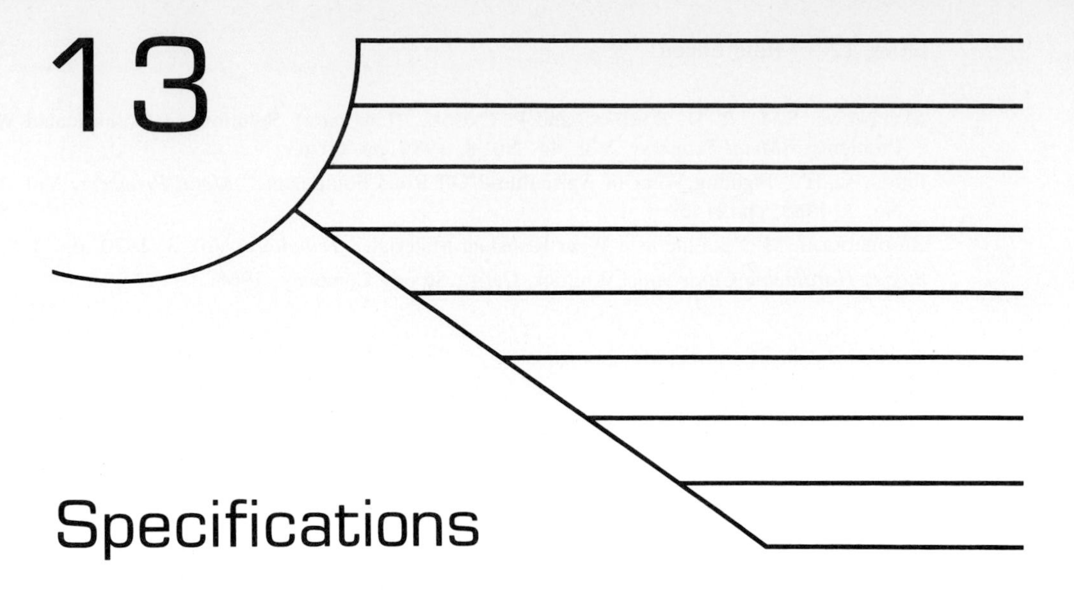

13 Specifications

In the preceding chapters, we discussed the important processes that are used for producing wear-resistant surfaces. The original purpose of this text was to discuss only welding hardfacing processes, but it was expanded to include the processes that compete with welding. This was done because a decision on the use of a welding process should be based on weighing its costs and benefits compared with alternative processes. We have presented sufficient information to allow this comparison to be made. At this point, we will address the last step in the process of producing wear-resistant surfaces, how to specify these processes on engineering drawings and on purchasing specifications. How do you properly specify a plating, a diffusion treatment, thin-film coatings, or hardfacings? This step is not to be belittled. If the process is not done right, the surface treatment or coating may not work and money and time will be wasted.

PLATINGS

We have discussed electroplatings, electroless platings, chemical conversion coatings, and electrochemical conversion coatings. Their role in producing wear-resistant surfaces can be briefly summarized as follows:

1. Hard chromium electroplate is the hardest metal; it has utility for thin coatings on tools and for rebuilding in heavy thicknesses.
2. Electroless nickel and similar electroless coatings typically are used in thicknesses of less than 2 mils (50 μm). They can have hardnesses approaching that of chromium, and they are used for enhancing the wear resistance of tools, not for rebuilding.

3. Selective electroplating has substantial utility in rebuilding shafts and worn parts. Hard nickel/nickel alloy platings are the most popular platings used for this purpose.
4. Hard anodizing is about the only way to enhance the wear properties of aluminum surfaces. The normal thickness limit is 2 mils (50 μm). Anodized surfaces have fair abrasion resistance, but they may wear mating materials in sliding contact applications.
5. Soft metal platings such as silver and gold can act as lubricative surfaces to reduce sliding wear and fretting damage. In these applications, thickness as low as 0.1 mil (0.5 μm) may be adequate. Heavy coatings may lead to adhesive wear.

Many other statements can be made about platings, but the preceding list covers the most commonly used plating systems. The elements of a proper plating specification are the following:

1. Area to be plated.
2. Special masking instructions (holes, threads, etc.).
3. Plating thickness; if the plating is to be finished after plating, the finished thickness requirement must be clearly noted.
4. Postplating treatments; baking to reduce hydrogen embrittlement is the common postplating requirement for hardened steel substrates.
5. Complete identification of substrate (alloy, heat treatments, etc.).

If proprietary platings are to be used, the drawing or purchasing specification must show who the trade name belongs to. Many times, 5 year old drawings call for plating with something such as "Waskally," and the person who made the drawing is long gone when the part is to be remade. This can result in an unnecessary 8-hour hunt to relate the trade name to a company. You cannot send a quotation for a plating to a trade name; you need a company name and address.

DIFFUSION TREATMENTS

Most machine designers and shop personnel are very familiar with carburizing, but their familiarity with the other processes that we described, nitriding, carbonitriding, ferritic nitrocarburizing, and diffusion CVD, is often a bit fuzzy. The successful use of many of these processes depends on the substrate material. If a design constraint is "lowest possible" cost, this usually means the use of low-carbon steel as the substrate. The processes that apply in this instance are carburizing (all types), ferritic nitrocarburizing, carbonitriding, and some of the special diffusion treatments, such as boriding and some CVD treatments. In using these treatments on low-carbon steels, it must be kept in mind that the processes such as carburizing and carbonitriding that require quenching to achieve a hard surface layer may not work if the part has a heavy cross section. Low-carbon steels with a thickness over 1 in. (25 mm) may not quench fast enough to

TABLE 13–1 COMPARISON OF SOME COMMONLY USED DIFFUSION TREATMENTS

Process	Suitable substrates	Typical case thickness	Quench required
Carburizing	Low-carbon and alloy steels	5 to 50 mils (0.12 to 1.25 mm)	Yes
Carbonitriding	Low-carbon and alloy steels	2 to 10 mils (50 to 250 μm)	Yes
Ferritic nitrocarburizing	Low-carbon and alloy steels	0.1 to 0.5 mil (2.5 to 12.5 μm)	Yes
Nitriding	Nitriding steels and some alloy steels	2 to 20 mils (50 μm to 0.5 mm)	No

form hard structure. Carburizing grades of alloy steel must be used. Table 13–1 is an abbreviated version of the table that was presented on the properties of diffusion treatments in Chapter 4. This table shows the thickness and suitable substrates for the most commonly used diffusion treatments. For low-cost, low-carbon steel substrates, three processes could be used to produce a surface with enhanced wear resistance, but the thickness capabilities vary with the process.

A nitrided surface is one of the finest wear-resistant surfaces for steels; if nitriding steels are used, surface hardnesses of 70 HRC can be obtained and distortion is low because the process is done at a low temperature (about 1000°F or 540°C), and a liquid quench is not required. Similar high-hardness cases (but thinner) are possible on stainless steels and tool steels with proprietary processes. If the lowest-cost substrate is not a design requirement, the nitriding process is really an exceptional tool for surface hardening. Boriding, chromizing, and special pack cementation processes are best reserved for special applications where it is known that the more conventional diffusion processes do not do the job. In general, these processes produce hard surfaces that have a thickness range of about 1 to 5 mils (25 to 125 μm). Chemical vapor deposition processes that involve diffusion produce cases that are usually less than 1 mil (25 μm) in thickness, and they often involve process temperatures as high as 1800°F (980°C). They can produce some very wear resistant surface compounds, such as titanium carbide, but the high process temperatures often preclude their use on parts that cannot tolerate distortion. The high process temperatures usually produce minor distortion by grinding, and the coatings are not thick enough to remove the distortion without penetrating the coating. Not all CVD coatings require processing at temperatures that are high enough to produce diffusion of the reacting species into the substrate, but any time a CVD process is used the details of the processing temperatures should be reviewed.

The preceding are some general statements reviewing what was previously discussed about these processes. Assuming that one of these processes is the surface-hardening process that offers the most favorable cost/benefit ratio, the proper specification for one of these treatments should contain the following elements:

1. Type of substrate (alloy, condition of heat treatment).
2. Areas to be treated.
3. Specific diffusion process desired (gas carburizing, ferritic nitrocarburizing, etc.); do not simply specify "case harden."
4. Depth of treatment; give an allowable range (e.g., 0.005 to 0.007 in. or 125 to 175 μm). If this is a critical parameter, describe how you are going to check the depth, microhardness, fracture, and so on.
5. Post treatment processing (e.g., carburized cases should be hardened and tempered).
6. Expected surface hardness.

If proprietary diffusion processes are to be used, the statement that was made on proprietary platings applies. State the trade name of the process and the name and address of the company who performs this proprietary process.

SURFACE HARDENING

Flame- and induction-hardening processes are used to produce hard surfaces on hardenable steels by local austenitizing and quenching. Flame hardening can produce the heaviest cases; induction hardening is usually more suited for production operations, and it can be used for thin cases; flame hardening cannot. A major consideration for users of these processes is potential for success. These are not easy processes to control, but they can be made routine if sufficient quantities of parts are treated to allow refinement of equipment and process variables. If these processes are considered for one of a kind or for a few parts, it is advisable to use a heat treater that specializes in these processes.

The applicable steels are medium-carbon steels, medium-carbon alloy steels, cast irons, and some tool steels and stainless steels. High-alloy steels are troublesome because of the lack of austenitizing soak. They often do not respond. If one of the "easy to selective harden" steels is to be used (1040 to 1050 steels, 4140, 4340 steels), the specification of this process may only involve the following:

1. Specification of the steel and condition of heat treat.
2. Areas to be hardened.
3. Desired depth of hardening and hardness level (0.050 in. or 1.25 mm depth of hardening over 50 HRC).
4. Surface hardness range.
5. Tempering requirements.
6. Inspection technique (for hardness, depth of hardening, cracking, etc.).
7. Post hardening surface treatments (chemical conversion coating, plating, etc.).

If the steel to be selective hardened is one that a heat treater has not done before, it may be necessary to sit down with the heat treater and agree on a procedure and the processing variables.

One of the most common problems that occurs with flame and induction hardening is verification of the depth of hardening. If the end of a shaft is to be flame hardened to a depth of 0.100 in. (2.5 mm), how does one know if this has been achieved without cutting a sample out of the finished part? The fact of the matter is that cutting a sample from the part is the only positive technique. The compromise technique that is used, where possible, is to run the hardened surface to an edge and apply a chemical etch to bring out the transition from soft to hard structure (in situ metallography). After etching, the hard structure depth is measured with a loupe and scale or by some similar technique. This technique is not foolproof, but experience has shown that it works most of the time. A second technique is to ask for a setup sample for cross sectioning and case measurement. If the setup sample is of similar mass and geometry as the part to be hardened, this is a very good technique to ensure the desired hardening, and it helps the heat treater set up his equipment. Setup samples are highly recommended on jobs that are one of a kind or a few parts. Setup samples are almost a must if the material to be selective hardened is not one of the easy-to-harden alloys.

One last precaution in specifying this type of hardening is that on large surfaces the hardening is often done by progressive movement of the flame head or by several cycles in an induction coil. It should be kept in mind that in spiral hardening of, for example, a roll surface the flame can be softening the case on a preceding spiral. Large surfaces should always be hardened by the progression of a heating pattern that covers the full width of the area to be hardened. In the case of a roll surface, the roll should be austenitized by translation through a toroid and 360-degree quench ring. If some other technique is used, a proper specification will require enough hardness readings to ensure that there are no soft spots in the hardened area.

THIN-FILM COATINGS

Thermal evaporation, sputter coating, and ion plating are the most commonly used thin-film coating processes, and the predominating use of these processes is for technical applications such as coatings for electronics. The only widely used thin-film coating for wear applications is titanium nitride and variations thereof. This coating is usually applied in thicknesses of less than 3 μm (120 μin.). The properties of the coating depend on the deposition technique, but most coaters claim a coating hardness in excess of 1500 kg/mm^2. This hard skin on tools has lengthened the service life on many machine tools and on tools such as taps, drills, and tool inserts. It is available from many tool manufacturers and from job shops. The most common substrates are high-speed tool steels and cemented carbides. A substrate consideration to be dealt with in specifying this coating on tools is the effect of the processing temperatures on substrate hardness. Some coatings are applied by plasma-assisted CVD processes that can involve substrate temperatures as high as 1800°F (980°C). Obviously, such a high process temperature will soften any hardened steel substrate, and the coating performance will likely be affected. There are other processes that keep the substrate temperatures below 1000°F (540°C). Even this temperature will affect many steel substrates, but this temperature will not affect most high speed steels. The point to be made is that, when specifying a thin-film coating, determine the process temperature and make sure that your substrate will not be adversely affected by this temperature.

A second process-related effect is surface roughening. Often these coatings are used to enhance the wear properties of some cutting device (knives, dies edges, etc.). If the deposition process produces a nodular coating, the cutting edges can be adversely affected; they can be dulled by the coating buildup and nodules.

The optimum thickness of these coatings for a particular application is not that well known. Thicker is not always better. A 2000 Å (2×10^{-7} m) thickness may have better adhesion and use characteristics than the same material applied to a thickness of 6000 Å (6×10^{-7}m). The recommended thickness for an application may best be determined by consultation with the firm that will be doing the coating. Thin-film coaters are very familiar with the proper thickness for cutting tools, but other types of parts usually require collaboration with the coater.

In addition to the popular TiN coating for tools, most thin-film coating facilities can apply a wide range of other coatings. Many metals can be ion plated with chromium, nickel, aluminum, copper, gold, and the like. Ion-plated coatings can have better adhesion than electroplated coatings of the same metal. This technique can be used when there is some special problem that warrants this special process. Some of the other thin-film coatings that are used for tribological applications are aluminum oxide, titanium carbide, and silicon carbide. They can be applied by sputter coating using targets of these materials or by reactions produced on the substrate from active species that are introduced into the vacuum chamber. Diamondlike carbon is another tribological coating that is produced by plasma-assisted surface reactions. Hydrocarbon gases such as ethylene are introduced into the plasma, and an amorphous carbon coating with a hardness as high as 5000 kg/mm^2 can be formed. At the present time, these coatings are not in wide use and their usefulness in tribosystems is still being studied.

Thin-film coatings are not to be used as freely as electroplates and hardfacings, but the proper specification of one of these coatings should contain many of the same elements:

1. Thickness range
2. Surface roughness after coating
3. Areas to be coated
4. Species to be coated
5. Physical and chemical property requirements
6. Permissible process temperature

A typical thickness range for the commonly used TiN coatings is 2 to 3 μm (80 to 120 μin.). The roughness of the substrate should not be altered by a proper coating.

In specifying coverage of coatings, it should be kept in mind that these coatings are done in a vacuum chamber; the coater must have a chamber that your part will fit in. Most vacuum coating operations are line of sight; the coater must have manipulation facilities to coat the areas that you specify. The composition of a coating may be difficult to confirm. TiN coatings usually have the proper stoichiometry when they have a gold color. Other colors suggest an off-analysis coating. Physical and chemical properties of thin-film coatings are difficult to measure. If these are important to an application, agreement should be established with the coater on how these properties will be confirmed.

In summary, thin-film coatings are increasing in importance in tribology every year, but successful coatings require specifications that ensure that you get what you want.

HARDFACING

In the preceding chapters, we have tried to present sufficient information to decide if hardfacing is the best process to use to produce a wear-resistant surface for a particular application. We have described the modes of wear that occur, the processes that compete with hardfacing, the methodology for selecting hardfacing processes and consumables, and how to design for hardfacing. In this section, we will discuss the final step in the process, the specification of hardfacing on an engineering drawing or on a process specification. The basic elements of hardfacing specification include the following:

1. Process (which welding technique; use the AWS acronym)
2. Consumable (by AWS type or trade name)
3. Area to be hardfaced (and surface preparation, if any)
4. Deposit thickness
5. Deposit pattern, sequence, and so on
6. Pre- and postweld treatments
7. Deposit hardness

The following discussion will present some details on each of these elements of a hardfacing specification.

Process

Figure 13–1 summarizes the methodology in the selection of a hardfacing process. After the decision has been made to use hardfacing on a part, the first step is to decide whether to use a fusion or a nonfusion process. This is a very critical step; fusion processes applied to a fragile substrate can cause distortion that is so severe that the part will be rendered useless; a nonfusion process used on a part under the wrong kind of service stresses and strains will fail by spalling. In previous discussions, we cited many examples to provide guidelines for this step, but the fact of the matter is that it

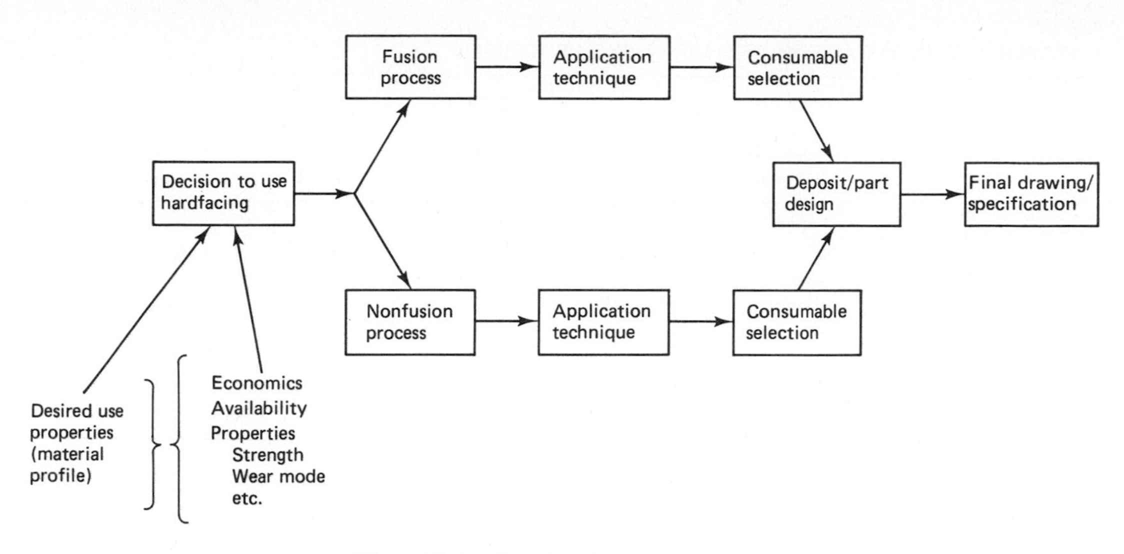

Figure 13–1 Flow chart for specification of hardfacing.

is still a judgment decision. For people who do not have wide experience with fusion welding, a conservative criterion for using fusion versus nonfusion techniques is that, if the part will only see relative sliding from a conforming surface and if the apparent contact stress is below about 1 ksi (6890 MPa), nonfusion welding processes will have adequate bond strength. On the other hand, if a part is clearly too fragile to survive melting of the surface to apply a fusion consumable, then a nonfusion process must be used regardless of the nature of the tribosystem. Sometimes the type of consumable desired dictates a fusion or nonfusion process. For example, if a design requires a nonmetal surface for electrical insulation or as a thermal barrier, there is no alternative but to use a nonfusion process. Ceramics are the only hardfacings that meet these criteria, and they can only be applied by nonfusion thermal spray processes.

After deciding on a fusion or nonfusion process, a proper hardfacing specification will designate the welding process that will be used. The proper way to list the welding process is to use the AWS process acronyms, the name in words, or, if a proprietary process is used, the process name should be used along with the name and address of the company that has the process. The process choices that are available are shown in Table 13–2 along with some process discriminators.

Each fusion and nonfusion process has its niche, something that it does better than the others. These are described in Chapter 9, but the most popular processes for fusion hardfacing are oxyacetylene and shielded metal arc welding and plasma spray for nonfusion coatings. The application comments in Table 13–2 are really an opinion on where a particular process might be used over other processes. All the process limits shown are "violated" by users at some time or another. People have found ways to apply plasma spray coatings as thin as 0.0001 in.; GTAW deposits have been made 1 in. thick and so on. If an application requires some special deposit requirements,

TABLE 13–2 COMPARISON OF HARDFACING PROCESSES

Process	Applicable consumables	Normal application method	Approx. min. deposit thickness, (in.)	Approximate deposition rate (lb/h)*	Application comments
Fusion					
OAW: oxyacetylene welding	Bare rod, wire, tubes	Manual	0.03 (0.75 mm)	1–5	Lowest dilution
SMAW: shielded metal arc welding	Coated electrodes	Manual	0.08 (2 mm)	2–6	Easiest to use process
GTAW: gas tungsten arc welding	Bare rod, wire, powder	Manual	0.04 (1 mm)	1–4	Best for very small deposits
GMAW: gas metal arc welding (open arc)	Wire, tube	Manual	0.08 (2 mm)	4–20	Good for manual deposits on large areas
SAW: submerged arc welding	Bare wire, tube strip	Semiautomatic	0.08 (2 mm)	4–150	Heavy deposits, large area must be done in shop
FCAW: flux cored arc welding	Cored tubular wire	Manual	0.08 (2 mm)	4–20	Good for manual deposits on large areas, neater deposit than open arc
PAW: plasma arc welding/transferred arc PTA	Bare wire, powder	Automatic	0.01 (0.25 mm)	1–15	Good for production hardfacing
LBW: laser beam welding	Bare wire, powder	Automatic	0.01 (0.25 mm)	1–5	Good for production hardfacing
EBW: Electron beam welding	Bare wire, powder	Automatic	0.01 (0.25 mm)	1–5	For special applications only
ESW: Electro slag welding	Bare rod, wire strip	Automatic	0.80 (2 mm)	120–850	For massive deposits, must be done in shop
Nonfusion					
FLSP: wire flame spray	Wire	Semiautomatic	0.01 (0.25 mm)	12–50	For heavy deposits (0.040 to 0.100 in.)
FLSP: powder/rod flame spray	Powder, rod	Semiautomatic	0.01 (0.25 mm)	1–5	General shop work
EASP: electric arc spray	Wire	Semiautomatic	0.01 (0.25 mm)	4–100	For large areas, mostly used for soft metals
PSP: plasma arc spray	Powder	Semiautomatic	0.003 (75 μm)	1–5	Best general-purpose process
High energy PSP: d-gun, Jetkote, etc.	Powder	Automatic	0.003 (75 μm)	1–5	For applications where PSP quality is not adequate

* Multiply by 2.2 to convert to kilograms.

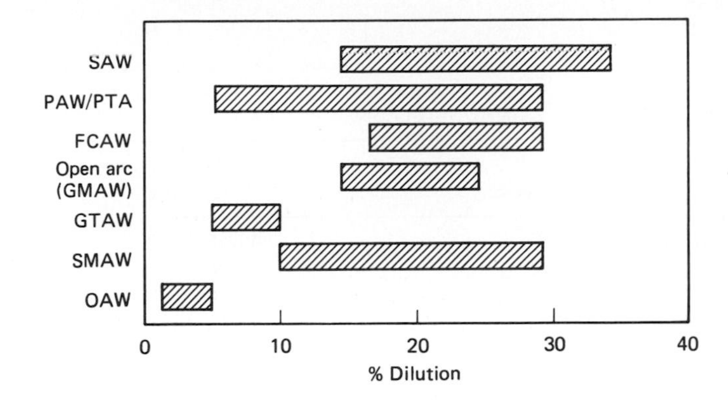

Figure 13–2 Dilution ranges for conventional fusion welding processes. A second layer of hardfacing may be necessary if dilution in the first layer is high.

some process modification can be made to make it work. Table 13–2 indicates normal use.

The importance of specifying the welding process is that it controls the results of the hardfacing job. Specifying a low-heat process can reduce distortion; as shown in Figure 13–2, dilution of the hardfacing with the substrate varies with the process; process also determines the cost of the job. So there are many advantages to proper process specification. Now that we have told you how to specify the process and why it is important, we will summarize this discussion with the following statement: if in doubt as to the best welding process, leave it out of the specification; just specify "hardface" and the other items on the specification list, consumable, area to be surfaced, thickness, and so on. This may sound like a heretical statement after all that was said about processes; but if a user is not firmly familiar with these welding processes, it may be better to leave the process selection up to the welding contractor. This is especially true if you are dealing with a firm that specializes in hardfacing. The exception to this recommendation is the thermal spray hardfacings. In this case, if you are not familiar enough with all the processes to comfortably specify a process, specify plasma arc spray (PSP). This process can be used to apply almost all the hardfacings that are available, and the bond will be very good. We make this recommendation because some thermal spray processes do not give the same quality of deposit. The bond strength and porosity of a flame sprayed wire will not be the same as that of a plasma spray; the same thing is true of oxy/fuel powder spray deposits and electric arc sprays. It is also safe to specify one of the high-energy processes such as d-gun, but the cost may be higher. Thus the "default" for a welding process specification is not to state the process for fusion welding (but specify consumable, hardness, thickness, etc.), but to specify plasma arc spray or d-gun (or equivalents) for nonfusion processes.

Consumables

The appropriate hardfacing material for a particular part is selected by matching the service requirements for the part with the properties that are available from various consumables. A material profile is generated for the desired material. This profile is

TABLE 13–3 COMMONLY USED FUSION WELDING CONSUMABLES

Type	Composition range	Hardness range (HRC)
Tool steels, oil hard	1/1.5 C, 0/1 Mn, 0.5/1 Cr, bal. Fe	40–55
air hard	1/2 C, 1/1.5 Mo, 1/12 Cr, bal. Fe	40–60
Alloy steels, 4130, etc.	0.3 C, 0.9/0.6 Mn, 0.8/1.1 Cr, 0.15/0.25 Mo, bal. Fe	20–40
High-speed steels (E/R Fe5)	0.5/1.1 C, 0.5 Mn, 1/7 W, 3/5 Cr, 4/9 Mo, 1/2.5 V, 0.5/Si, bal. Fe	50–60 + carbides
Austenitic manganese steels (E/R FeMn)		
Nickel type		
Moly type	0.5/1 C, 11/16 Mn, 2.75/6 Ni; 0.5 Cr, 1.3 Si, bal. Fe	
	0.5/1 C, 11/16 Mn, 0.6/1.4 Mo, 0.3/1.3 Si, bal. Fe	20–25
Low chromium martensitic iron	3.5/4 C, 1 Mn, 4/15 Cr, 2/4 Mo, bal. Fe	50–60 + carbides
Low-chrominum austenitic iron	2/4 C, 2/2.5 Mn, 12/16 Cr, 0/8 Mo, bal. Fe	50–60 + carbides
High-chromium martensitic iron	2.5/4.5 C, 1/1.5 Mn, 26/30 Cr, bal. Fe	40–60 + carbides
High-chromium austenitic iron (E/R FeCr)	3/5 C, 2/8 Mn, 26/35 Cr, bal. Fe	40–60 + carbides
Low-carbon nickel/chromium/boron (E/R NiCr-A)	0.3/6 C, 1.5 Co, 8/14 Cr, 1.25/3.25 Fe, 1.25/3.25 Si, 2/3 B, bal. Ni,	30–40
Medium-carbon nickel/chromium/boron E/R NiCr-B)	0.4/0.8 C, 1.25 Co, 10/16 Cr, 3/5 Fe, 3/5 Si, 2/4 B, bal. Ni	45–50
High-carbon nickel/chromium/boron (E/R NiCr-C)	0.5/1 C, 1 Co, 12/18 Cr, 3.5/5.5 Fe, 3.5/5.5 Si, 2.5/4.5 B, bal. Ni	50–60 + carbides
Low-carbon cobalt-base alloys (E/R CoCr-A)	0.9/1.4 C, 1 Mn, 3/6 W, 3 Ni, 26/32 Cr, 1 Mo, 3 Fe, 2 Si, bal. Co	38–47
Medium-carbon cobalt-base alloys (E/R CoCr-B)	1.2/1.7 C, 1 Mn, 7/9.5 W, 3 Ni, 26/32 Cr, 1 Mo, 3 Fe, 2 Si; bal. Co	45–59
High-carbon cobalt-base alloys (E/R CoCr-C)	2/3 C, 1 Mn, 11/14 W, 3 Ni, 26/33 Cr, 1 Mo, 3 Fe, 2 Si, bal. Co	48–58 + carbides
Aluminum bronze (E/R CuAl-D)	3/5 Fe, 13/14 Al, bal. Cu	20–40
Composite tubular rods (R/E WC)	Tube: 0.1 C, 0.45 Mn, 0.3 Ni, 0.2 Cr, 0.3 Mo, bal. Fe Carbides: 3.6/4.2 C, 94 W min.	50–65 + various size carbides

matched with the properties that are available from hardfacings and alternative processes. If hardfacing is clearly the process of choice for generating a wear-resistant surface, the chapter on consumables will supply information that can be matched with the material profile. When it is possible, the specific consumable is specified by the AWS designation (EFeCr, etc.). The use of these generic consumable designations allows the welder options in purchasing the consumable; on large jobs this approach may allow competitive bids from a number of suppliers and lower costs to you, the customer. The welding contractor may have a preference from the operating standpoint for a particular manufacturer's consumables, and the use of the generic designation may allow him to use a consumable from his "favorite" supplier. When it is not possible to use a generic consumable designation, trade names or generic classes of materials can be specified:

Preferred consumable designation: EFeCr
Alternatives: Waskaloy 230, Wasso Corp., Batavia, N.Y.
400 series stainless steel (generic alloy)

Appendixes III and IV list over 800 hardfacing consumables by trade name. These consumables are categorized by alloy system. There are many different alloys, and it may seem like an awesome task to select only one from so many. As shown in Tables 13–3 and 13–4, there are not really that many different types of hardfacing consumables. Table 13–3 lists 16 classes of consumables that can be used for fusion welding deposits, and AWS alloy designation systems apply to most. The appendixes list quite a number of consumables in a catchall category entitled iron-base alloys. Some of these are buildup consumables, some are proprietary materials that do not fit the AWS categories, and some match the composition of commonly used wrought tool materials. If, for example, you would like to have a fusion deposit with the properties of a chisel steel (S series tool steel), you will find a number of consumables of this type in this grouping.

Nonfusion consumables fall into about ten different material categories, but only two of these categories are widely used for producing wear-resistant deposits; the other materials, such as composites, pure metals, polymers, and the like, are used for technical applications and not wear applications. Most wear problems can be addressed with three or four ceramic or cermet materials. Most metals can be applied by thermal spray

TABLE 13–4 COMMONLY USED NONFUSION THERMAL SPRAY CONSUMABLES

Ceramic/cermets	Aluminum oxide
	Chromium oxide
	Chromium carbide
	Tungsten carbide/cobalt
Metals	400 Series stainless steels
	Cobalt-base alloys
	Nickel/chromium/boron alloys

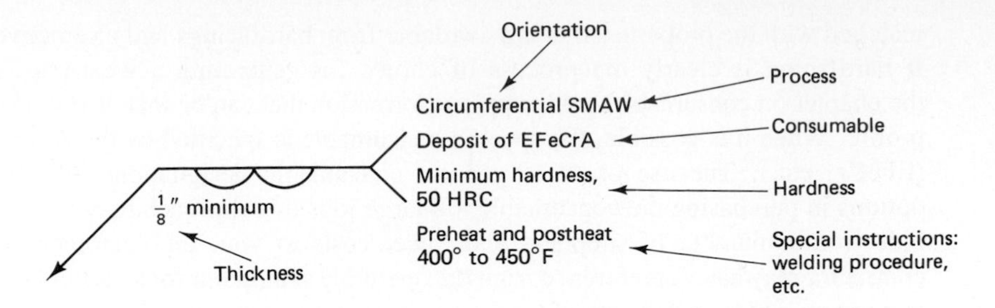

Figure 13–3 Hardfacing weld symbol showing process and consumable details.

processes; if a material profile points toward the use of a cobalt-base alloy, you can specify only the nominal chemical composition of the alloy that you desire, and the welder will find a powder or wire manufacturer who makes one that matches your composition. If you want a stainless steel deposit, specify the AISI or generic type: 18–8 austenitic stainless steel, martensitic stainless steel, or 420 stainless steel, or 410 stainless steel, and so on. Specifying a consumable from the hundreds that are available is not that complicated.

Area to Be Hardfaced

In the United States, the accepted technique for specification of weld deposits is to use the standard symbols that have been developed by the American Welding Society and the American National Standards Institute (ANSI/AWS A2.4–86).* The welding symbol for specification of hardfacing deposits is the same as the symbol for surfacing, which is two "bumps" (which represent weld beads) placed under an arrow that points to the area to be surfaced. An arrow with all the additions for process and other details is shown in Figure 13–3. Surfacing is applying weld deposit over an area to increase a dimension of a substrate; hardfacing is doing the same thing with a weld deposit that is intended to improve the wear resistance of the surface. The bumps on the weld symbol always go on the underside of the arrow; the thickness of the deposit is specified by a dimension placed to the left of the bumps. If the deposit is to be done in layers of differing thickness, this is denoted by a second line on the arrow, as shown in Figure 13–4a. If the deposit is to have a particular lay or orientation, this is designated by a note in the tail of the arrow, as shown in Figure 13–4b. Additional details about the deposit, such as the welding process and type of consumable, are placed in the tail of the arrow (Figure 13–3).

Figure 13–5 shows how the location of a fusion hardfacing deposit would be specified on a variety of parts. If an entire area is to be covered, as shown in the first example in Figure 13–5, the weld symbol could be put on the part drawing. If only selected parts of an area are to be covered, as shown in the other examples in Figure

* "Symbols for Welding and Nondestructive Testing." For a complete copy, contact AWS, 550 N.W. LeJeune Rd., P.O. Box 351040, Miami, Florida, 33135. Telephone: 305-443-9353.

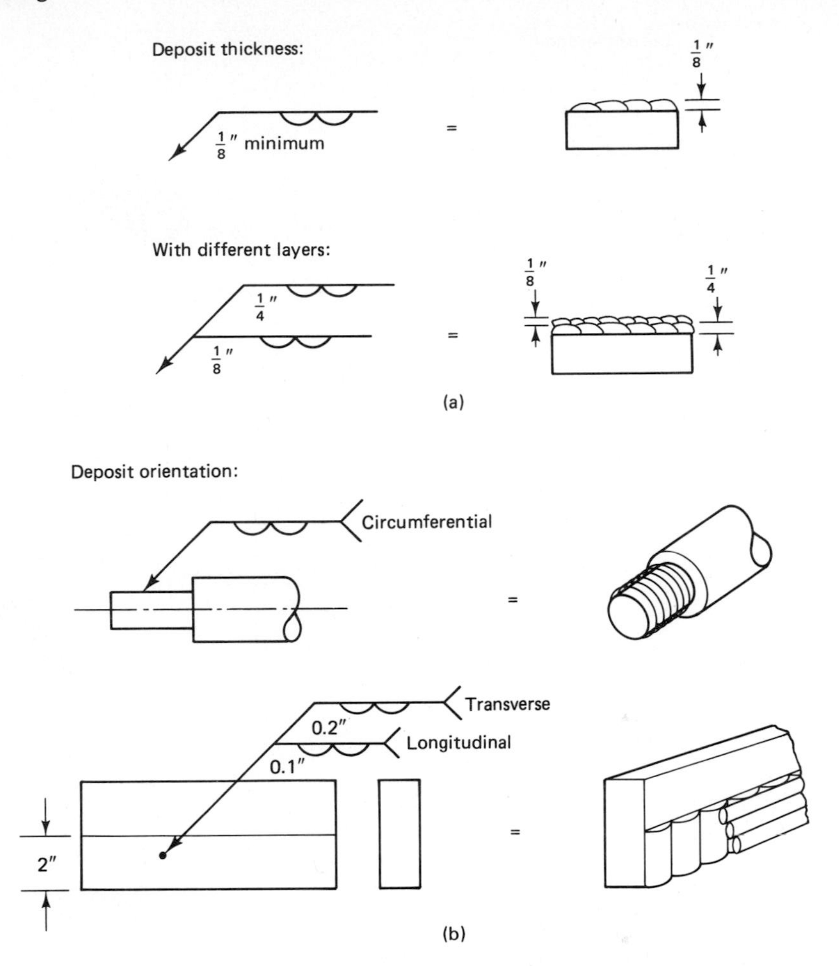

Figure 13–4 AWS system for specifying hardfacing thickness and orientation.

13–5, it is recommended that a drawing or sketch be made of the part with no other dimensions other than those that are associated with the hardfacing deposit. This will prevent confusion between before-welding part dimensions and after-finishing dimensions on the deposit. Figure 13–6 shows how welding symbols are used to designate deposit location on parts that are to be hardfaced with nonfusion processes.

If the part to be hardfaced requires significant machining before the welding operation (to prepare the weld area), three drawings will prevent mixups. A drawing is made to machine the part to prepare for the hardfacing; a drawing is made to specify the areas to be hardfaced and the details of the deposit; and the final drawing shows the finish machined part, including the finishing required on the deposit. This sequence of drawings is illustrated in Figure 13–7. The part that was made with this sequence of drawings is shown in various stages of completion in Figures 13–8 to 13–11. Obviously, this much detail is not always warranted, but a key part of the successful use of hardfacing is to

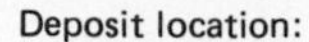

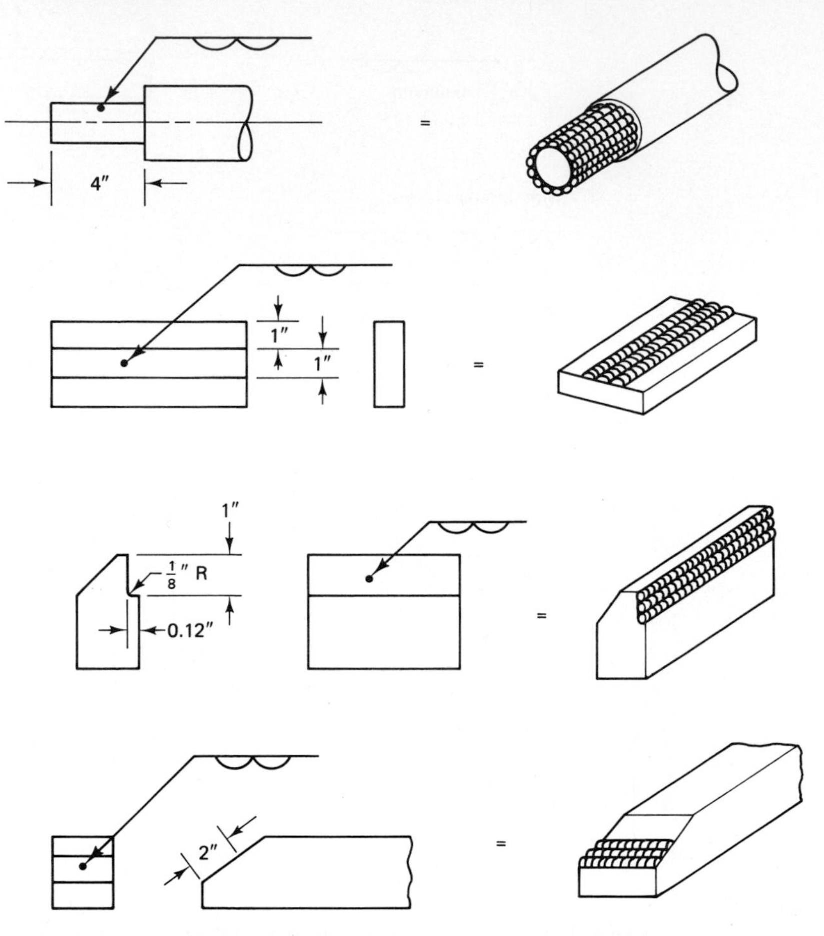

Figure 13–5 AWS system for specifying hardfacing deposit location; only dimension the deposit of the preparation for the deposit on the drawing.

adequately specify what you want. Good drawings and specifications on where the deposit is to be applied are essential.

Deposit Thickness

We have shown how deposit thickness is specified using welding symbols. The following are general guidelines on the proper thickness for hardfacing deposits.

The thickness ranges that can be obtained with the various nonfusion processes are listed in Table 13–2. The flame spray processes that require grooving of the substrate must be thick because of the groove depth, but in most other cases the thickness of the deposit should be no greater than about twice the allowable wear. If the serviceability of the part will be lost if the surface wear is 0.002 in. (50 μm), a coating thickness of

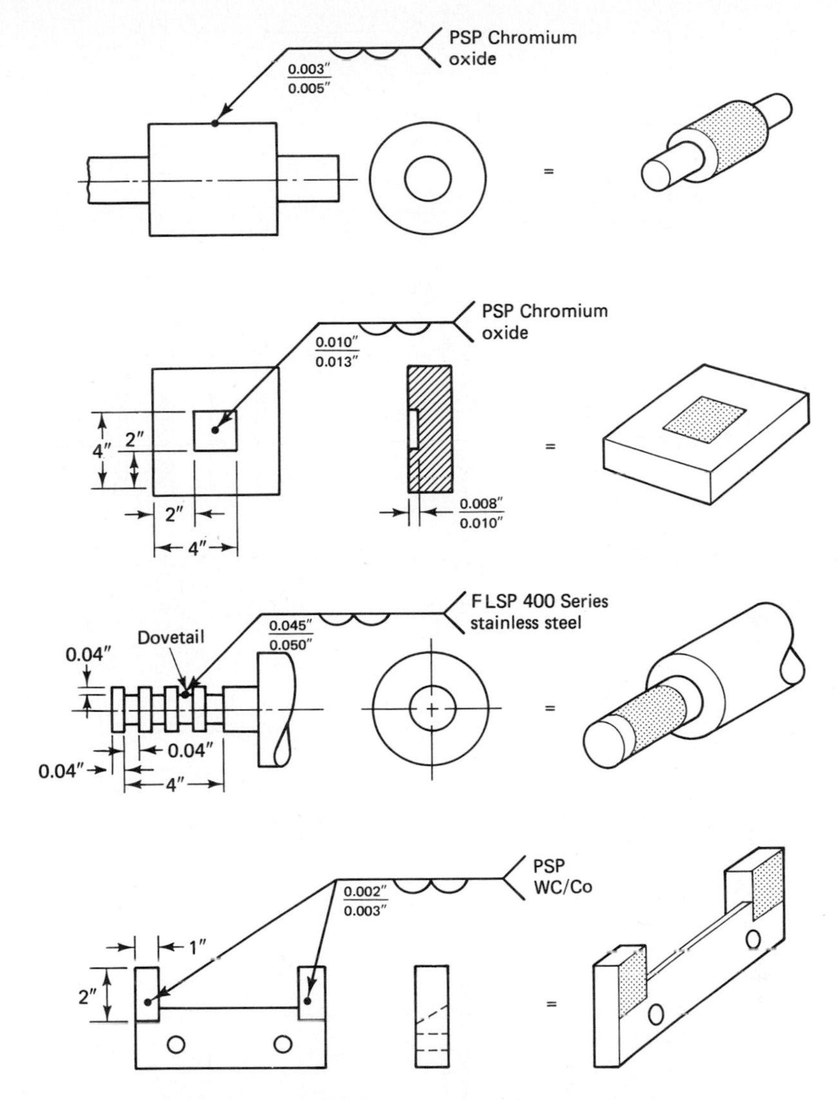

Figure 13–6 Designation of nonfusion hardfacing deposits.

3 to 5 mils (75 to 125 μm) will be adequate. Extra coating thickness increases the part cost, and it serves no purpose to have a 10-mil (250 μm) deposit when the part will be removed from service when it is worn 2 mils (50 μm). Another reason for minimizing deposit thicknesses is that plasma spray and similar thermal spray deposits may have poorer bond strength in heavy thicknesses than in thin deposits. Grinding heavy deposits is more likely to cause deposit cracking than grinding thin deposits. This is especially true with ceramics. They are poor conductors of heat, and the heat of grinding can cause abnormal stresses and cracking that would not occur on thin deposits, which are

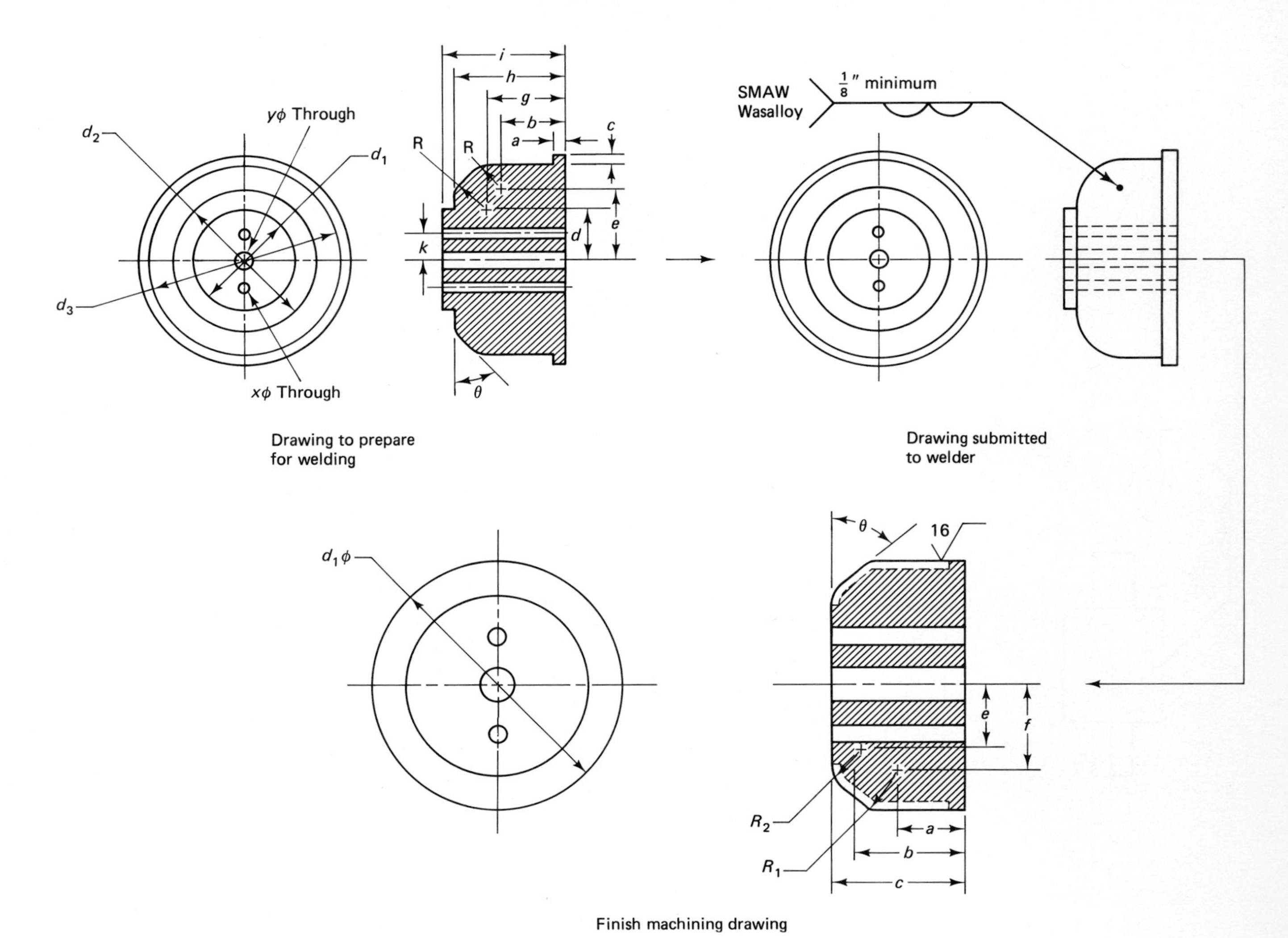

Figure 13–7 Drawing sequence for specifying hardfacing on a new part.

Figure 13–8 Part prepared for hardfacing.

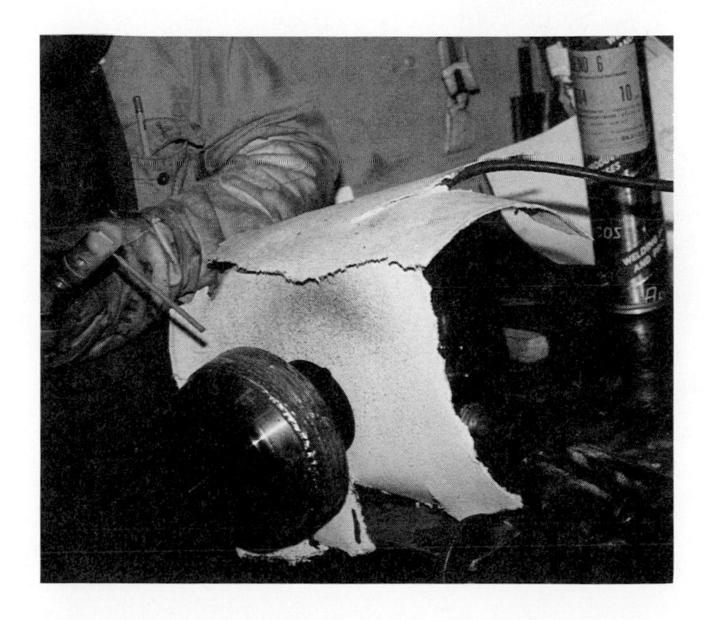

Figure 13–9 Hardfacing of part shown in Figure 13–8.

Figure 13–10 Hardfaced part.

Figure 13–11 Hardfaced part after final machining.

Figure 13–12 Cross checking on a hardfaced knife blank.

less prone to heat buildup in grinding. Thus the recommendation on specification of thickness of thermal spray deposits is to minimize the deposit thickness; use a thickness that is commensurate with the allowable system wear.

Thicker is not necessarily better with fusion hardfacings. In fact, some formidable problems can arise from too thick deposits; they can spall, and cracking can cause service problems. The harder hardfacings, those with a bulk hardness above 50 HRC, often cross check after welding (Figure 13–12). This is normal, and it is the way that these relatively brittle materials stress relieve themselves. Cross checks occur about every 1/2 in. or inch (12.5 or 25 mm) transverse to the deposit; they are tight and they are only visible on smooth deposits. The cracks that develop with too heavy deposits are large enough to put your fingernail into, and if the cracks are close together, pieces of the deposit can spall. The harder the deposit, the thinner it should be:

Deposit hardness 62 HRC and up: do not use more than one layer
55 to 62 HRC: do not use more than two layers
50 to 55 HRC: do not use more than three layers
40 to 50 HRC: do not use more than five layers
Buildup materials 20 to 40 HRC: no limit

These thickness guidelines apply mostly to the E/R FeCr consumables. The cobalt/chromium alloys and the nickel/chromium/boron alloys are so brittle that it is advisable never to use more than two layers of these materials. They are relatively brittle even at hardnesses as low as 40 HRC. Figure 13–13 shows spalling of a heavy deposit. In this instance, welding over worn surfacing contributed to the failure; this practice should be avoided. Old surfacing and wear should be ground off before applying new hardfacing.

At the other extreme, the thinnest hardfacing deposit that is commonly used is 0.060 in. (1.5 mm). Deposits this thin (with low dilution) can be achieved with oxyacety-

Figure 13–13 Spalling of a heavy hardfacing deposit (see arrow).

lene processes and some of the more sophisticated processes such as laser surfacing. The thickness of fusion deposits should also be commensurate with the allowable system wear. If a system can tolerate substantial amounts of wear, processes such as SAW or ESW should be considered. These processes can be used to produce heavy deposits without using many layers.

Deposit Pattern

Some of the hardfacing deposit patterns that are used with fusion welding techniques can have an affect on serviceability. Examples of the patterns that are used on digger teeth and dozer buckets were illustrated in Chapter 11. We will not discuss the patterns that have been found to be successful in these types of applications because many suppliers of fusion hardfacing consumables have excellent illustrations of these patterns in their hardfacing manuals. A general rule is to space deposits $\frac{1}{4}$ to $1\frac{1}{2}$ in. (6 to 37 mm) apart transverse to the motion of rocks and abrasive solids on the wear surfaces. Dot weld patterns on centers from about $\frac{3}{4}$ in. (18 mm) to about $1\frac{1}{2}$ in. (37 mm) produce a wear-resistant pattern and minimize distortion of the substrate; heat input is less because there is less weld than in covering an entire area. Many deposit patterns that are recommended for earth-moving equipment and mineral benefication equipment have been empirically determined. By experimentation, hardfacing users have found that, for example, a herringbone deposit pattern works well in digging iron ore, and a dot pattern works well on digger teeth in rocky soil. There are often no theoretical explanations as to why particular patterns work, but patterns that tend to keep the flow of abrasive materials from impinging on exposed substrate are intuitively preferred.

The use of girth welds reduces the tendency for distortion on cylindrical parts. Again, this is just common sense. A weld down the long axis of a cylinder will tend to distort the cylinder concave to the side that is being welded, eventually yielding a "banana." Girth welds produce circumferential stresses that are no different than putting a very strong hose clamp on the part; there is less tendency for distortion.

In summary, the deposit pattern is not usually a part of the specification of nonfusion hardfacing deposits, but it can be important on fusion deposits that are used as-deposited. A general rule is to pattern the deposit to minimize distortion and to make beads transverse to the flow of the abrasive. On machinery parts that are hardfaced and subsequently machined, the specification of a deposit pattern is usually unnecessary.

Deposit Hardness

It has been stated many times throughout this text that the hardness of a hardfacing deposit is not the sole predictor of wear resistance. Most hardfacing deposits are composed of softer matrixes and hard microconstituents. The nature, size, and volume fraction of these microconstituents are more important than bulk hardness in determining wear resistance. However, the hardness of fusion weld deposits should be specified and checked, since every fusion deposit is supposed to have some as-deposited hardness. This hardness is listed in the consumable manufacturer's catalog. If you do not achieve this specified hardness, there is a possibility that the deposit is excessively diluted (with substrate material) and that the welding procedure was improper. Conventional Rockwell C hardness tests can be used on two-layer deposits where each layer is about $\frac{1}{8}$ in. (3 mm) thick. The lighter-load Rockwell superficial test, 15N, is applicable when the layers are thinner; but even this test does not give a valid indication of hardness if the deposit is less than $\frac{1}{16}$ in. (1.5 mm) in thickness.

Rockwell hardness has no significance on nonfusion thermal spray deposits. They are too thin; Rockwell penetrators will simply go through the deposit and measure the hardness of the substrate. The hardness of these materials can only be measured by microhardness techniques, and then these hardnesses are not even adequate predictors of wear characteristics since the porosity and oxide inclusions in thermal spray deposits can lower the penetration hardness test results. Thus it is not necessary to specify the hardness of a nonfusion thermal spray deposit. If an application requires assurance of a proper thermal spray deposit, the recommended approach is to ask the hardfacing contractor to supply a test sample that was sprayed with the part, and microhardness and metallographic studies can be made on this test coupon (an extra part can serve as the destructive test coupon). The microhardness of the deposit can be compared with the consumable supplier's specified hardness range.

Even though hardness is not an indicator of the wear resistance of a hardfacing consumable, hardness should be specified and checked on all fusion hardfacing deposits to make sure that the consumable is applied properly. The hardness specification should show the manufacturer's range of hardness for multiple-layer deposits. If the consumable manufacturer shows a one-layer deposit hardness of 50 to 55 HRC and you are using one layer, your welding symbol should look like Figure 13–14.

In summary, hardness is not normally specified on thermal spray deposits, but deposit quality can be checked by a laboratory microhardness test on an extra part or

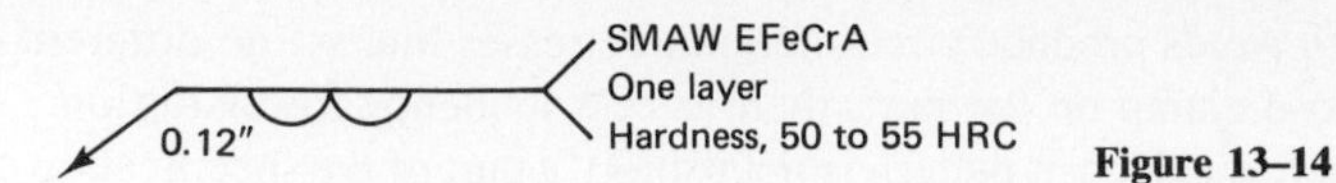

Figure 13–14

on a test coupon sprayed with the parts. The hardness of fusion hardfacing deposits should always be specified since hardness is an indicator of proper application technique.

Special Instructions

The tail of the hardfacing welding symbol is the proper place to put special instructions relating to a particular deposit. Some of the things that may need to be specified are the following:

FUSION HARDFACING

Peening
Stress relieving
Tempering
Straightening
Nondestructive testing
Preheating and postheating

NONFUSION HARDFACING

Impregnation
Surface roughness

Peening Peening can be used to minimize distortion and cross checking. The distortion and cracking that can occur with some fusion hardfacing operations arise from the shrinkage of the weld deposit on solidification and cooling. Laws of physical properties dictate that the weld must shrink, and one way to prevent this is to peen the deposit as it cools from the welding temperature. Peening is simply battering the deposit with a blunt-nosed hammer. It stretches the weld (by compression) and this lowers distortion. This process must be done while the weld is still hot, preferably over 1000°F (540°C), and to be effective, it must be done with a pneumatic hammer such as a chipping hammer. Peening cold could cause deposit damage such as cracking and spalling, so it is essential that it be done immediately after welding. In manual processes, the welder will make a deposit of about 6 in. (150 mm), put down his torch, and peen the weld. This makes overlaying a tedious process. Peening is only used for special applications, such as for deposits that cannot tolerate cross checking. When it is used, the specification would state something like "peen weld while hot (>1000°F/540°C)."

Stress Relief Stress relieving is not normally performed on fusion or nonfusion deposits, because it can soften fusion deposits and it can oxidize or do some other form of damage to nonfusion deposits. Some fusion welding consumables such as the cobalt-base alloys can tolerate a stress relief at 1200°F (650°C) without affecting the properties of the deposit. If these alloys are used for hardfacings that will be machined and held to accurate tolerances, it is advisable to stress relieve after welding. The weld symbol would show "stress relieve after welding" in the tail. If it is likely that the heat treaters are unfamiliar with stress-relieving weldments, the temperatures and times should be specified. Consumable manufacturers' catalogs should be consulted to determine if a stress relief will harm a particular deposit.

Tempering Tempering does not apply to nonfusion deposits, but it does apply to fusion deposits. Tempering of hardened steels in the temperature range of 300° to about 1100°F (150° to 590°C) is an essential part of the hardening process. The tempering process significantly increases the toughness of the hardened structure. Since tool steel deposits are part of hardfacing, it is recommended to temper deposits of these materials. The tempering temperatures and times are the same as those that would be used in hardening and tempering the alloy in wrought form. The specification would read "temper after welding (2 hours at 600°F, 315°C)," and so on. Tempering is not normally done on the other hardfacings, and postheating is absolutely not recommended on manganese steels. They can embrittle.

Straightening Straightening can be used to reduce the distortion effects of welding. It can be done by peening, press straightening, or local heating. All these things can cause damage to a fusion or nonfusion deposit. Spalling may occur on nonfusion deposits. The potential damage on fusion deposits is usually cracking. In general, straightening should only be done on the softer hardfacings, such as buildup materials. Where possible, straightening should be followed by a thermal stress relief.

Nondestructive Testing Nondestructive testing of hardfacing deposits is always appropriate. Fusion welds should be checked for hardness and, where necessary, cracking. Nonfusion deposits can be checked for porosity and bond strength. A microscope can be used to determine if porosity is excessive, and a simple bond-strength test is to apply a piece of duct tape and briskly pull it from the surface. Both of these will detect very bad deposits. When nondestructive tests will be used for the acceptance of a hardfacing job, it should be made clear on the part drawing what tests will be used and what will be the basis for part rejection. For example, if cracks and cross checks cannot be tolerated on a part, the drawing or hardfacing specification would show "hardfacing deposit to be free of cracks when checked with dye penetrant." Do not overinspect. If cross checking will not affect part serviceability, do not ask for a deposit to be free of them. Only ask for deposit requirements that are essential.

Thermal Treatments Preheating and postheating can be the most important parts of a fusion hardfacing process. Hardfacing deposits may spall by underbead cracking if hard deposits are put on quench hardenable steels without proper preheat and postheat procedures. We discussed this briefly in Chapter 11, but this procedure is so important it warrants reiteration.

The first step in even considering the use of fusion hardfacing should be to determine the weldability of the substrate. Essentially, all steels that are quench hardenable require precautions. Know the substrate material. Never weld on an unidentified substrate. Assuming that the substrate identification has been established and it is a hardenable material, the appropriate preheat and postheat temperatures need to be established and put onto the weld specification.

As mentioned in our discussions of hardfacing metallurgy, every hardenable steel requires a certain cooling rate to achieve a hard structure. Some alloys need a water quench; some alloys will harden with an air quench. Usually, the higher the alloy content, the easier it is to harden a steel. On quenching, the hardening temperature steels do not instantaneously transform to hard structure. They transform progressively over a temperature range. The transformation starts at a temperature that is different for each alloy. This temperature is called the martensite start temperature, M_s, and it can be anywhere from about 300°F (150°C) to about 1000°F (540°C). The temperature at which the martensite transformation is complete is called the M_f temperature (martensite finish), and this temperature may be below room temperature. Transformation to hard structure will not start on quenching until the steel reaches the M_s temperature, and it will not be complete until the steel temperature reaches the M_f temperature. Welding on a hardenable steel is the same as an austenitizing and quenching operation. The heat of welding raises the heat-affected zone into the austenitizing temperature range, and the mass of the part can produce a rapid quench. If this happens during the hardfacing process, hard, brittle martensite will form in the heat-affected zone, and there is a strong possibility that this hard, brittle structure will crack from the weld shrinkage stresses.

The theory behind preheating is to keep the part temperature above the M_s temperature during welding, and postheating is performed to soak the part at a temperature above the M_s temperature to form soft structure. If a tool steel deposit or some other hardfacing is being applied to a hardened steel, the properties of the part may be impaired if it is preheated and postheated to a temperature above the M_s if this temperature is above the tempering temperature that was used in hardening the part; the part hardness will be reduced. In this situation, the proper preheat temperature is the original tempering temperature. The part should be brought to room temperature after welding and retempered at the original tempering temperature. These thermal cycles are illustrated in Figure 13–15. M_s temperatures and normal tempering temperatures for a variety of steels are shown in Table 13–5. A proper thermal cycle for a hardened 01 tool steel that is to be hardfaced is as follows:

> Preheat to 400°F (200°C); slow cool to room temperature; retemper 2 hours at 400°F (200°C).

If the 01 steel was in the annealed condition (for subsequent hardening), the welding procedure would be as follows:

> Preheat to 500°F (260°C); maintain preheat during welding, and postheat for 2 hours at 500°F (260°C); slow cool to room temperature.

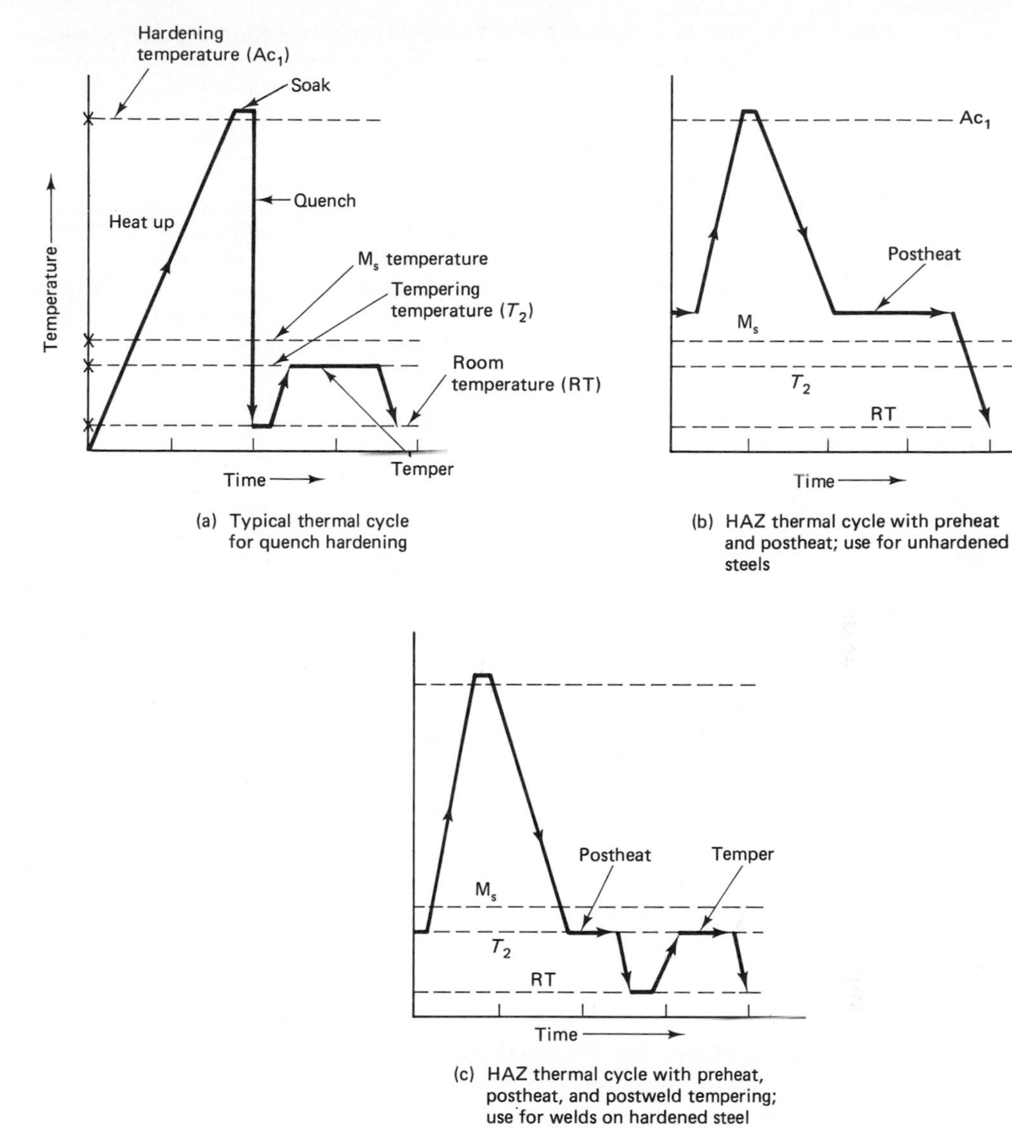

Figure 13–15 Thermal cycles for hardening of steel compared with pre- and postheating cycles.

This latter procedure will produce a soft weld and no cracking. The proper preheat temperature is 100°F (37°C) above the M_s temperature. The former procedure will produce a hard weld with no cracking. Thermal treatment specifications such as this are put as a general note on the hardfacing drawing or, if they fit, into the tail of the welding symbol. It is not necessary to preheat and postheat nonfusion thermal spray deposits, because the temperatures involved in deposition do not raise the surface temperature of the substrate to the point where quench hardening is possible.

TABLE 13–5 THERMAL TREATMENT TEMPERATURES FOR HARDENABLE STEELS

	Steels	M_s, °F (°C)	Normal tempering temperature, °F (°C)	Normal working hardness (HRC)
Stainless steels	440C	370 (188)	Preheat, postheat, and temper at 800°F (427)	
	420	500 (260)	400 (205)	52–54
	414	525 (274)	Preheat, postheat, and temper at 500 °F (260)	
	416	(Sulfurized, do not weld)		
	410	600 (316)	500 (260)	38–40
	17–4	(Weld in aged condition and reage after welding)		
Tool steels	D2	430 (222)	Preheat, postheat, and temper at 800°F (427)	
	A2	350 (177)	500 (260)	60–61
	H13	600 (316)	Preheat, postheat, and temper at 500°F (260)	
	M1	400 (205)	1030 (555)	63–65
	M2	370 (188)	1050 (566)	60–62
	01	400 (205)	400 (205)	61–62
	02	300 (149)	400 (205)	59–60
	S1	600 (316)	400 (205)	56
	S7	600 (316)	400 (205)	58
	W1	400 (205)	350 (177)	60–62
	Viscount 44*	(Sulfurized, do not weld)		
Alloy steels	4150	(Sulfurized, do not weld)		
	4140	660 (349)	500 (260)	50–52
			1150 (621)	28–32
	4340	570 (299)	400 (205)	52–54
			1200 (649)	28–32
	Nitralloy*	590 (310)	(Do not weld nitrided material)	
Carbon steels	1040	740 (393)	(Shafting; not usually hardened)	

* Proprietary steels

Special Instructions for Thermal Spray Processes The most common postprocessing specification used on thermal spray deposits is some type of impregnation. If thermal spray hardfacings are to be used for applications where chemical resistance is required, an impregnation of the porosity with an impregnant that is unaffected by the chemical environment will reduce the chances of the environment penetrating the coating. Typical impregnants are furfural alcohols, epoxies, fluorocarbons, and cyanoacylates. Impregnation is usually performed in a vessel designed to do impregnation. The parts are placed in a basket that can be raised and lowered within a pressure vessel. The lower portion of the vessel is filled with the impregnant; the vessel is sealed and evacuated with the parts suspended above the impregnant. When all the gases have been removed (by the vacuum) from the coating, the basket containing the parts is lowered into the impregnant. The vacuum is then turned off and air pressure is applied. This helps to push the impregnant into the pores of the deposit. The vessel is then brought to room conditions, and the excess impregnant is removed from the parts and the parts are air dried or baked to set up the impregnant. This process is only 100 percent effective if the porosity is completely interpenetrating to the surface. If it is not, there is a chance of penetration of corrodent to the substrate. For this reason, it is advisable to use substrates under thermal spray deposit that have environmental resistance of their own. The impregnating operation in this case will prevent crevice corrosion, and similar effects that can arise from having a porous surface in a liquid corrodent.

PTFE and similar fluorocarbons are a common impregnant for thermal spray coatings that are intended for unlubricated sliding systems. The fluorocarbon acts as a dry film lubricant.

The other common specification note that is added to drawings for thermal spray coatings is a control of surface roughness. The roughness of thermal spray deposits (as-deposited) can vary from about 100 microinches R_a to as high as 400 microinches R_a (2.5 to 10 μm). Many wear applications and chemical-resistance applications require a certain surface roughness to perform as intended. If this is the case, the deposit topography should be specified in the drawing notes or in notes in the tail of the weld symbol.

The final phase in the specification of hardfacing is to make sure that all the details are in place. We have discussed many factors that may or may not apply, and it may seem that hardfacing specification is difficult to do right. It is not. Use of the AWS welding symbols really makes the task quite easy. There are provisions on the symbol for all aspects of hardfacing specification; even the preheating and postheating details for hardenable steels can be put in the tail. Hardfacing specification will be complete and accurate if you consider the list of items that we presented and if you supply the welder with a sketch or drawing that only shows the details of deposition. Hardfacing is an easy welding process; if the welder knows what you want through a proper specification, he will make it right.

SUMMARY

This text was originally intended as a simple guide to the use of hardfacing, prompted by the observation that many design engineers do not know how to use this process. It was expanded to include information on competitive processes when it was pointed out that some engineers felt that they did not need another surface coating process when they already knew about plating and the other common surface-hardening processes. We have tried to show how to compare hardfacing with these other processes so that a decision can be made on using the most cost-effective process, with hardfacing as a candidate.

Right or wrong, the single factor that seems to make designers reluctant to use hardfacing is that they do not know what consumable to use. We sincerely hope that we have answered this question. We have shown the different categories of consumables and the welding processes that are used to apply them, and we have even tabulated the trade names and manufacturers of over 800 consumables in the appendixes. The information necessary to use hardfacing on a daily base is in this text and if we have done our job, you, the potential hardfacing user, will consider this process everytime that a design requires a wear-resistant surface. It is a phenomenal process and it deserves more emphasis than it gets. It is hoped you will become an advocate and loyal user.

REFERENCES

Symbols for Welding and Nondestructive Evaluation, AWS A2.4–86. Miami: American Welding Society, 1986.

Welding Handbook, Vol. 2, 7th ed. Welding Processes—Arc, Gas Welding, and Cutting. Miami: American Welding Society, 1980.

Welding Terms and Definitions, AWS A3.0–80. Miami: American Welding Society, 1980.

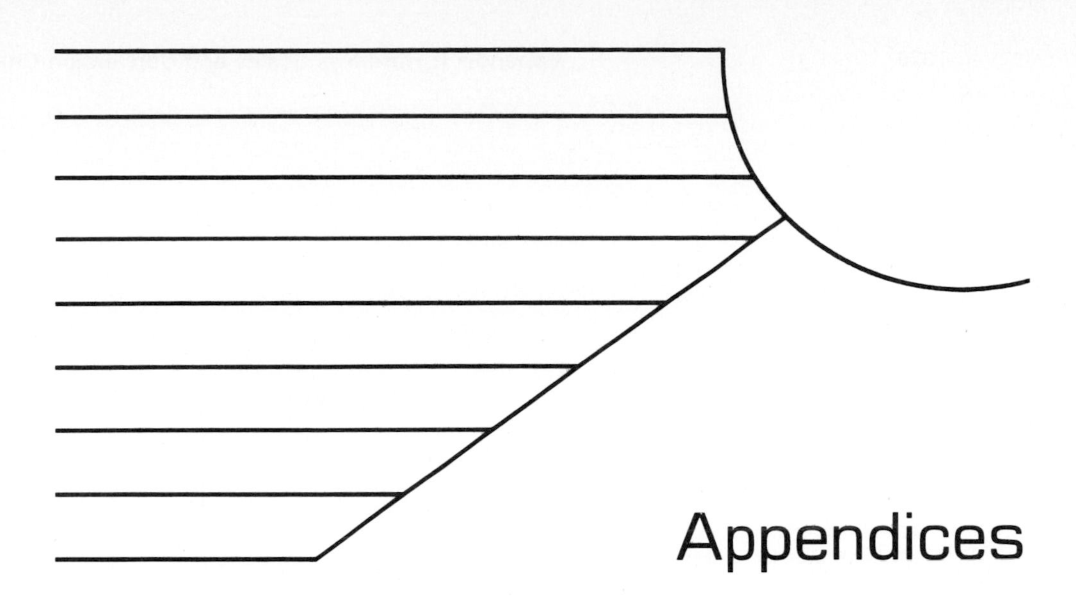

Appendices

APPENDIX I: HARDNESS SCALES AND CONVERSION CHARTS

Description of Hardness Code

Specify hardness according to the code described in Figure A-1.

This code is in agreement with the method of designation used by the following standards organizations:

1. American Society for Testing and Materials (ASTM)
2. American National Standards Institute (ANSI)
3. International Standards Organization (ISO)

EXAMPLES

1. 50–60 HRC means: a hardness value of 50 to 60 using the Rockwell C scale.
2. 85 HR15T MAX means: a maximum hardness value of 85 using the Rockwell superficial 15T scale.
3. 185–240 HV_{10} means: a hardness value of 185 to 240 using the Vickers hardness tester and a test load of 10 kilogram-force.
4. 500 HK_{200} MIN means: a minimum hardness value of 500 using the Knoop hardness tester and a test load of 200 grams-force.

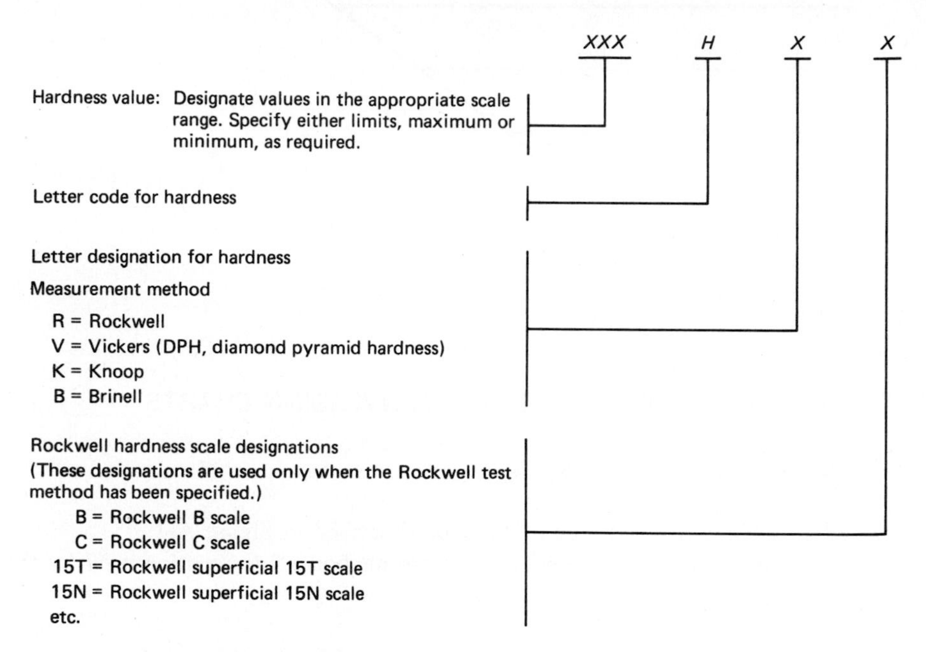

Figure A–1 Specification of hardness numbers for metals. (K. G. Budinski, *Engineering Materials: Properties and Selection*, 2nd ed. Reston, VA: Reston, 1983).

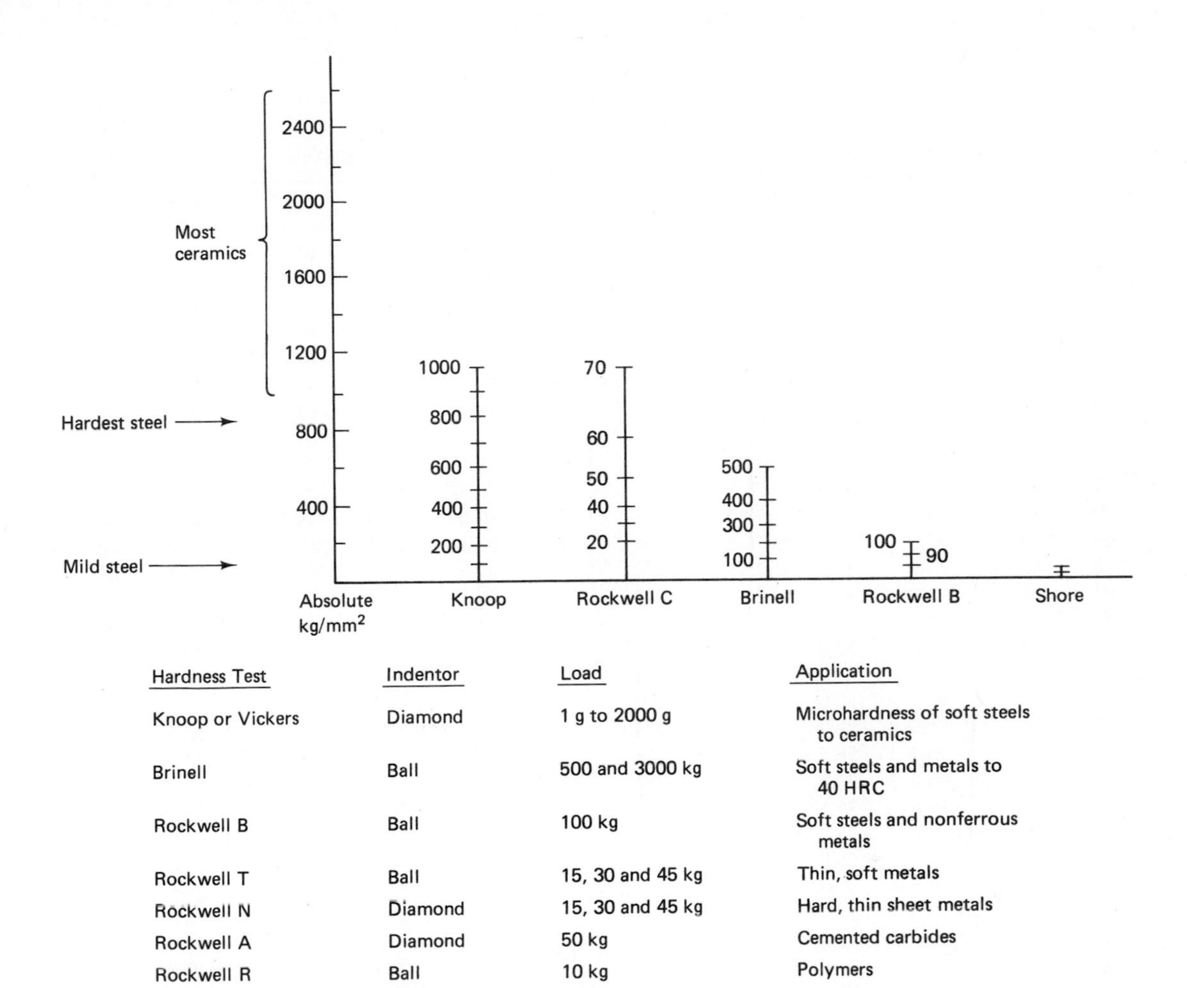

Hardness Test	Indentor	Load	Application
Knoop or Vickers	Diamond	1 g to 2000 g	Microhardness of soft steels to ceramics
Brinell	Ball	500 and 3000 kg	Soft steels and metals to 40 HRC
Rockwell B	Ball	100 kg	Soft steels and nonferrous metals
Rockwell T	Ball	15, 30 and 45 kg	Thin, soft metals
Rockwell N	Diamond	15, 30 and 45 kg	Hard, thin sheet metals
Rockwell A	Diamond	50 kg	Cemented carbides
Rockwell R	Ball	10 kg	Polymers
Shore durometer	Needle	Spring	Elastomers
Rockwell C	Diamond	150 kg	Hardened metals (thick)

Figure A–2 Comparison of hardness tests (K. G. Budinski, *Engineering Materials: Properties and Selection*, 2nd ed. Reston, VA: Reston, 1983).

TABLE A-1 HARDENED STEEL AND HARD ALLOYS

Rockwell scale					HK,* 500 g and over	Rockwell scale					HK,* 500 g and over
C	A	15N	30N	45N		C	A	15N	30N	45N	
80	92.0	96.5	92.0	87.0	—	49	75.5	85.0	67.5	54.0	526
79	91.5	—	91.5	86.5	—	48	74.5	84.5	66.5	52.5	510
78	91.0	96.0	91.0	85.5	—	47	74.0	84.0	66.0	51.5	495
77	90.5	—	90.5	84.5	—	46	73.5	83.5	65.0	50.0	480
76	90.0	95.5	90.0	83.5	—	45	73.0	83.0	64.0	49.0	466
75	89.5	—	89.0	82.5	—	44	72.5	82.5	63.0	48.0	452
74	89.0	95.0	88.5	81.5	—	43	72.0	82.0	62.0	46.5	438
73	88.5	—	88.0	80.5	—	42	71.5	81.5	61.5	45.5	426
72	88.0	94.5	87.0	79.5	—	41	71.0	81.0	60.5	44.5	414
71	87.0	—	86.5	78.5	—	40	70.5	80.5	59.5	43.0	402
70	86.5	94.0	86.0	77.5	972	39	70.0	80.0	58.5	42.0	391
69	86.0	93.5	85.0	76.5	946	38	69.5	79.5	57.5	41.0	380
68	85.5	—	84.5	75.5	920	37	69.0	79.0	56.5	39.5	370
67	85.0	93.0	83.5	74.5	895	36	68.5	78.5	56.0	38.5	360
66	84.5	92.5	83.0	73.0	870	35	68.0	78.0	55.0	37.0	351
65	84.0	92.0	82.0	72.0	846	34	67.5	77.0	54.0	36.0	342
64	83.5	—	81.0	71.0	822	33	67.0	76.5	53.0	35.0	334
63	83.0	91.5	80.0	70.0	799	32	66.5	76.0	52.0	33.5	326
62	82.5	91.0	79.0	69.0	776	31	66.0	75.5	51.5	32.5	318
61	81.5	90.5	78.5	67.5	754	30	65.5	75.0	50.5	31.5	311
60	81.0	90.0	77.5	66.5	732	29	65.0	74.5	49.5	30.0	304
59	80.5	89.5	76.5	65.5	710	28	64.5	74.0	48.5	29.0	297
58	80.0	—	75.5	64.0	690	27	64.0	73.5	47.5	28.0	290
57	79.5	89.0	75.0	63.0	670	26	63.5	72.5	47.0	26.5	284
56	79.0	88.5	74.0	62.0	650	25	63.0	72.0	46.0	25.5	278
55	78.5	88.0	73.0	61.0	630	24	62.5	71.5	45.0	24.0	272
54	78.0	87.5	72.0	59.5	612	23	62.0	71.0	44.0	23.0	266
53	77.5	87.0	71.0	58.5	594	22	61.5	70.5	43.0	22.0	261
52	77.0	86.5	70.5	57.5	576	21	61.0	70.0	42.5	20.5	256
51	76.5	86.0	69.5	56.0	558	20	60.5	69.5	41.5	19.5	251
50	76.0	85.5	68.5	55.0	542						

* Knoop hardness conversion: These values are approximate only since they were determined on a limited number of tests and samples. They are only for loads of 500 g or heavier.

From various sources. Detailed hardness conversion tables can be found in ASTM publication E140–83.

TABLE A-2 SOFT STEEL, GRAY AND MALLEABLE CAST IRON, AND MOST NONFERROUS METALS

Rockwell scale				HK,* 500 g and over	HB, 500 kg (10-mm ball)	HB, 3000 kg	Rockwell scale				HK,* 500 g and over	HB, 500 kg (10-mm ball)	HB, 3000 kg
B	15T	30T	45T				B	15T	30T	45T			
100	93.0	82.0	72.0	251	201	240	49	76.5	49.0	22.0	106	82	—
99	92.5	81.5	71.0	246	195	234	48	—	48.5	20.5	105	81	—
98	—	81.0	70.0	241	189	228	47	76.0	47.5	19.5	104	80	—
97	92.0	80.5	69.0	236	184	222	46	75.5	47.0	18.5	103	—	—
96	—	80.0	68.0	231	179	216	45	—	46.0	17.5	102	79	—
95	91.5	79.0	67.0	226	175	210	44	75.0	45.5	16.5	101	78	—
94	—	78.5	66.0	221	171	205	43	74.5	45.0	15.5	100	77	—
93	91.0	78.0	65.5	216	167	200	42	—	44.0	14.5	99	76	—
92	90.5	77.5	64.5	211	163	195	41	74.0	43.5	13.5	98	75	—
91	—	77.0	63.5	206	160	190	40	73.5	43.0	12.5	97	—	—
90	90.0	76.0	62.5	201	157	185	39	—	42.0	11.0	96	74	—
89	89.5	75.5	61.5	196	154	180	38	73.0	41.5	10.0	95	73	—
88	—	75.0	60.5	192	151	176	37	72.5	40.5	9.0	94	72	—
87	89.0	74.5	59.5	188	148	172	36	—	40.0	8.0	93	—	—
86	88.5	74.0	58.5	184	145	169	35	72.0	39.5	7.0	92	71	—
85	—	73.5	58.0	180	142	165	34	71.5	38.5	6.0	91	70	—
84	88.0	73.0	57.0	176	140	162	33	—	38.0	5.0	90	69	—
83	87.5	72.0	56.0	173	137	159	32	71.0	37.5	4.0	89	—	—
82	—	71.5	55.0	170	135	156	31	—	36.5	3.0	88	68	—
81	87.0	71.0	54.0	157	133	153	30	70.5	36.0	2.0	—	67	—
80	86.5	70.0	53.0	154	130	150	29	70.0	35.5	1.0	—	—	—
79	—	69.5	52.0	151	128	147	28	—	34.5	—	—	66	—
78	86.0	69.0	51.0	158	126	144	27	69.5	34.0	—	85	—	—
77	85.5	68.0	50.0	155	124	141	26	69.0	33.0	—	—	65	—
76	—	67.5	49.0	152	122	139	25	—	32.5	—	—	64	—
75	85.0	67.0	48.5	150	120	137	24	68.5	32.0	—	—	—	—
74	—	66.0	47.5	147	118	135	23	68.0	31.0	—	82	63	—

TABLE A-2 ***(Continued)***

Rockwell scale				HK,* 500 g and over	HB, 500 kg (10-mm ball)	HB, 3000 kg	Rockwell scale				HK,* 500 g and over	HB, 500 kg (10-mm ball)	HB, 3000 kg
B	15T	30T	45T				B	15T	30T	45T			
73	84.5	65.5	46.5	145	116	132	22	—	30.5	—	—	—	—
72	84.0	65.0	45.5	143	114	130	21	67.5	29.5	—	—	62	—
71	—	64.0	44.5	141	112	127	20	—	29.0	—	—	—	—
70	83.5	63.5	43.5	139	110	125	19	67.0	28.5	—	79	61	—
69	83.0	62.5	42.5	137	109	123	18	66.5	27.5	—	—	—	—
68	—	62.0	41.5	135	107	121	17	—	27.0	—	—	60	—
67	82.5	61.5	40.5	133	106	119	16	66.0	26.0	—	—	—	—
66	82.0	60.5	39.5	131	104	117	15	65.5	25.5	—	76	59	—
65	—	60.0	38.5	129	102	116	14	—	25.0	—	—	—	—
64	81.5	59.5	37.5	127	101	114	13	65.0	24.0	—	—	58	—
63	81.0	58.5	36.5	125	99	112	12	64.5	23.5	—	—	—	—
62	—	58.0	35.5	124	98	110	11	—	23.0	—	73	—	—
61	80.5	57.0	34.5	122	96	108	10	64.0	22.0	—	—	57	—
60	—	56.5	33.5	120	95	107	9	—	21.5	—	—	—	—
59	80.0	56.0	32.0	118	94	106	8	63.5	20.5	—	71	—	—
58	79.5	55.0	31.0	117	92	104	7	63.0	20.0	—	—	56	—
57	—	54.5	30.0	115	91	103	6	—	19.5	—	—	—	—
56	79.0	54.0	29.0	114	90	101	5	62.5	18.5	—	69	55	—
55	78.5	53.0	28.0	112	89	100	4	62.0	18.0	—	—	—	—
54	—	52.5	27.0	111	87	—	3	—	17.0	—	—	—	—
53	78.0	51.5	26.0	110	86	—	2	61.5	16.5	—	68	54	—
52	77.5	51.0	25.0	109	85	—	1	61.0	16.0	—	—	—	—
51	—	50.5	24.0	108	84	—	0	—	15.0	—	67	53	—
50	77.0	49.5	23.0	107	83	—							

* Knoop hardness conversion—These values are approximate only since they were determined on a limited number of tests and samples. They are only for loads of 500g or heavier.

From various sources. Detailed hardness conversion tables for metals can be found in ASTM publication E140–83.

APPENDIX II: CHEMICAL SYMBOLS AND PROPERTIES OF SOME ELEMENTS

Element	Symbol	Melting point, °F	(°C)	Density, g/cm^3	Resistivity at 20°C, 10^{-6} ohm-cm	Linear coefficient of thermal expansion,[f,g] 10^{-6} in./in. °F[e]	Thermal conductivity at 25°C, W/cm/°C[h]
Silver	Ag	1761	(960)	10.49	1.59	10.9	4.29
Aluminum	Al	1220	(660)	2.70	2.66	13.1	2.37
Gold	Au	1945	(1062)	19.32	2.44	7.9	3.19
Beryllium	Be	2345	(1285)	1.84	4.20	6.4	2.01
Carbon	C	6740[e]	(3726)	2.25	75.00	0.3–2.4	2.1
Calcium	Ca	1564	(851)	1.54	4.60[a]	12.4	2.01
Columbium	Cb[d]	4474	(2467)	8.57	14.60	4.06	0.537
Cerium	Ce	1463	(795)	6.66	75.00[a]	4.44	0.113
Cobalt	Co	2719	(1492)	8.90	5.68[b]	7.66	1.00
Chromium	Cr	3407	(1825)	7.19	12.80	3.4	0.939
Copper	Cu	1981	(1082)	8.94	1.69	9.2	4.01
Iron	Fe	2795	(1535)	7.87	10.70	6.53	0.804
Germanium	Ge	1717	(936)	5.32	60×10^6	3.19	
Hafnium	Hf	4032	(2222)	13.29	35.5	3.1	0.230
Mercury	Hg	−70	(−38)	13.55	95.78		0.083
Iridium	Ir	4370	(2410)	22.42	5.30[b]	3.8	1.47
Lanthanum	La	1688	(892)	6.17	57.00[a]	3.77	0.134
Magnesium	Mg	1204	(651)	1.74	4.46	15.05	1.56
Manganese	Mn	2271	(1244)	7.44	185	12.22	0.078
Molybdenum	Mo	4730	(2610)	10.22	5.78[a]	2.7	1.38
Nickel	Ni	2646	(1452)	8.90	7.8	7.39	0.909
Osmium	Os	5432	(3000)	22.50	9.5[b]	2.6	0.076
Lead	Pb	621	(327)	11.34	22	16.3	0.353
Palladium	Pd	2826	(1552)	12.02	10.3	6.53	0.718
Platinum	Pt	3216	(1768)	21.40	10.58	4.9	0.716
Plutonium	Pu	1183	(639)	19.84	146.45[b]	30.55	0.067
Rhenium	Re	5755	(3179)	21.02	19.14	3.7	0.480
Rhodium	Rh	3560	(1960)	12.44	4.7[b]	4.6	1.50
Silicon	Si	3520	(1382)	2.33	15×10^6[c]	1.6–4.1	1.49
Tin	Sn	449	(232)	7.30	11.5	13	0.668
Tantalum	Ta	5425	(2996)	16.60	13.6[a]	3.6	0.575
Thorium	Th	3182	(1750)	11.66	18[a]	6.9	0.540
Titanium	Ti	3035	(1668)	4.54	42	4.67	0.219
Uranium	U	2070	(1132)	19.07	30	3.8–7.8	0.275
Vanadium	V	3486	(1918)	6.11	24.8	4.6	0.307
Tungsten	W	6170	(3410)	19.30	5.5	2.55	1.73
Yttrium	Y	2748	(1508)	4.47	65		0.172
Zinc	Zn	787	(419)	7.13	5.75	22	1.16
Zirconium	Zr	3366	(1852)	6.45	44	5.8	0.227

[a] At 25°C.
[b] At 0°C.
[c] At 300°C.
[d] Also called niobium, Nb.
[e] Sublimes.
[f] Multiply by 1.8 to convert to cm/cm °C.
[g] At 68°F (20°C).
[h] Multiply by 55.7 to convert to Btu/ft/°F.

APPENDIX III: FUSION HARDFACING CONSUMABLES

The consumables in Table A-4 are grouped by AWS type or by type of basis metal. The information in Table A-4 and Appendix IV on thermal spray consumables was obtained from a survey sent in June 1985 to 50 welding consumable manufacturers that were listed in AWS A5.0—83 Filler Metal Comparison Charts (Miami: AWS, 1983). The respondents were asked to supply trade name, chemical composition, deposit hardness, and product forms. This list contains information obtained with permission from the survey respondents; it does not list all the consumables that are available since only a fraction of those surveyed responded, but it is intended as a sample of the types of consumables that are commercially available. Some manufacturers consider chemical composition proprietary; in such cases, this information lists the word description of the consumable, and grouping was based on this description. Table A-3 contains the key for the manufacturers' codes.

No claim is made or intended concerning the properties or merits of the consumables that are listed or those that are not listed. Questions on the properties and applicability of specific welding consumables should be directed to the manufacturer. Consumable manufacturers are invited to update their listed information, and additional manufacturers can add their consumables to this listing in subsequent revisions of this text by contacting the publisher, Prentice-Hall, Inc.

TABLE A-3 KEY FOR CONSUMABLE MANUFACTURERS CODE IN TABLE A-4

No. 1	Stoody Deloro Stellite, Inc. 610 W. Ash Street San Diego, CA 92101
No. 2	Amsco Welding Products Division Stoody Deloro Stellite, Inc. 610 W. Ash Street San Diego, CA 92101
No. 3	Stultz-Sickles Steel Co. 929 Julia Street Elizabeth, NJ 07201
No. 4	Weld Mold Company 750 Rickett Road Brighton, MI 48116
No. 5	Metallurgical Technologies, Inc. 14435 Max Road Pearland, TX 77581
No. 6	Specialty Powders, Union Carbide Corp. 1550 Polco Street Indianapolis, IN 46224
No. 7	Wall Colmonoy Corporation 30261 Stephenson Highway Madison Heights, MI 48071–1650
No. 8	Cerro Metal Products Co. P.O. Box 388 Bellefonte, PA 16823
No. 9	Alloy Metals, Inc. 1972 Meijer Drive Troy, MI 48084
No. 10	Lincoln Electric Company 22801 St. Clair Avenue Cleveland, OH 44117
No. 11	Metco Division of Perkin Elmer 1101 Prospect Avenue Wesbury, NY 11590
No. 12	Postle Industries, Inc. 3615 Superior Avenue Cleveland, OH 44114
No. 13	Norton Company 1 New Bond Street Worcester, MA 01601
No. 14	Metallurgical Industries 1 Coldstream Way Tinton Falls, NJ 07724
No. 15	Arcos Corp. Routes 54 and 61 Mt. Carmel, PA 17851
No. 16	Welding Equipment & Supply 5225 E. Davison Detroit, MI 48212
No. 17	Teledyne McKay 850 Grantley Road P.O. Box 1509 York, PA 17405–1509
No. 18	Inweld Manufacturing Co. P.O. Box 631 Greenwich, CT 06836
No. 19	Hobart Brothers Corp. Filler Metal Division 600 W. Main Street Troy, OH 45373
No. 20	Powder Alloy Corp. P.O. Box 429415 Cincinnati, OH 45242

TABLE A-4 FUSION HARDFACING CONSUMABLES

Consumable tradename	Manufacturer code*	Deposit hardness	Forms†	Nominal chemical composition‡ (wt. %)
		Cobalt-Base Fusion Welding Consumables		
Cobalt No. 1	2	52 HRC	R	2.5C, 31Cr, 13W, 50Co
Cobalt No. 6	2	41 HRC	R	1C, 28Cr, 5W, 60Co
Cobalt No. 12	2	45 HRC	R	1.5C, 30Cr, 9W, 55Co
Cobalt 1B	15	53–54 HRC	R	2/3C, 1Mn, 26/33Cr, 3Ni, 1Mo, 11/14W, 3Fe, Rem. Co
Cobalt 6B	15	33–40 HRC	R	0.9/1.4C, 1Mn, 2Si, 26/32 Cr, 3Ni, 1Mo, 3/6W, 3Fe
Cobalt 1C	15	40–42 HRC	CE	1.75/3C, 2Mn, 25/33Cr, 3Ni, 1Mo, 11/14W, 5Fe, Rem. Co
Cobalt 6C	15	37–40 HRC	CE	0.7/1.4C, 2Mn, 2Si, 25/32Cr, 3Ni, 1Mo, 3/6W, 5Fe, Rem. Co
Cobalt 7C	15	40–42 HRC	CE	0.2/0.4C, 1.5Mn, 1Si, 25/29 Cr, 2/4Ni, 5/6Mo, 2Fe, Rem. Co
Cobalt 6B A-3	15	32–40 HRC	R	1.2/1.4C, 0.5Mn, 1.2/1.4Si, 28/32Cr, 1.5Ni, 1Mo, 5/6W, 1.25Fe, Rem. Co
Cobalt 12B	15	50 HRC	R	1.2/1.7C, 1Mn, 2Si, 26/32Cr, 3Ni, 1Mo, 7/9.5W, 3Fe, Rem. Co
Cobar 1MC	15	—	W	1.75/3C, 2Mn, 2Si, 25/33Cr, 3Ni, 1Mo, 11/14W, 5Fe, Rem. Co
Cobar 1FC	15	—	WFC	1.75/3C, 2Mn, 2Si, 25/33Cr, 3Ni, 1Mo, 11/14W, Rem. Co
Cobar 6FC	15	—	WFC	0.8/1.4C, 2Mn, 2Si, 25/32Cr, 3Ni, 1Mo, 3/6W, 5Fe, Rem. Co
Cobar 6MC	15	—	W	0.7/1.4C, 2Mn, 2Si, 25/32Cr, 3Ni, 1Mo, 3/6W, 5Fe, Rem. Co
Cobar 7FC	15	26–35 HRC	WFC	0.2/0.4C, 1.5Mn, 1Si, 25/29Cr, 5/6Mo, 2Fe, Rem. Co
Cobar 7MC	15	26–30 HRC	W	0.2/0.4C, 1.5Mn, 1Si, 25/29 Cr, 2/4Ni, 5/6 Mo, 2Fe, Rem. Co
Cobar 12FC	15	50 HRC	WFC	1/1.7C, 2Mn, 2Si, 25/32 Cr, 3Ni, 1Mo, 7/9.5W, 5Fe, Rem. Co

* See Table A-3 for key to manufacturers' codes.
† Key: CE, coated electrode; R, bare rod, W, wire; WFC, flux core wire; P, powder.
‡ Rem. = remainder.

TABLE A-4 FUSION HARDFACING CONSUMABLES (*Continued*)

Consumable tradename	Manufacturer code	Deposit hardness	Forms	Nominal chemical composition (wt. %)
Cobend 1	15	45–46 HRC	CE	1.75/3C, 2Si, 2Mn, 25/33 Cr, 3Ni, 1Mo, 11/14W, 5Fe, Rem. Co
Cobend 6	15	46–48 HRC	CE	0.7/1.4C, 2Mn, 2Si, 25/32Cr, 3Ni, 1Mo, 3/6W, 5Fe, Rem. Co
Cobend 7	15	26–35 HRC	CE	0.2/0.4C, 1.5Mn, 1Si, 25/29Cr, 2/4Ni, 5/6Mo, 2Fe, Rem. Co
Cobend 12	15	39 HRC	CE	1/1.7C, 2Mn, 2Si, 25/32Cr, 3Ni, 1Mo, 7/9.5W, 5Fe, Rem. Co
Cobend L-605	15	—	CE, W	0.05/0.15C, 1/2Mn, 1Si, 19/21Cr, 9/11Ni, 14/16W, 3Fe, Rem. Co
Eureka 1	16	52–53 HRC	CE, R	Co/Cr/W alloy
Eureka 6	16	42–44 HRC	CE, R	Co/Cr/W alloy
Eureka 12	16	45–47 HRC	CE, R	Co/Cr/W alloy
Eurekalloy X	16	30 HRC	CE	Cobalt-base alloy
Hotwork No. 2	2	32 HRC	R	0.3C, 27Cr, 5.5W, 2.8Ni, 62Co
Polyface 601	4	46–55 HRC	CE, R, W	High alloy, cobalt base
Polyface 606	4	38–47 HRC	CE, R, W	High alloy, cobalt base
Polyface 612	4	42–52 HRC	CE, R, W	High alloy, cobalt base
Polyface 621	4	34–36 HRC	CE, R, W	High alloy, cobalt base
Polymatic 601 FC	4	50–55 HRC	W	ECoCr-C
Polymatic 606 FC	4	40–45 HRC	W	ECoCr-A
Polymatic 612 FC	4	40–45 HRC	W	ECoCr-B
Polymatic 612 FC	4	30–35 HRC	W	AMS 5385
Postalloy 2506-SPL	12	40–44 HRC	W	1.1C, 3.75W, 0.6Mn, 27Cr, 4Fe, 0.25Ni, 0.8Si, Rem. Co
Stellite No. 1	1	46–54 HRC	R, CE	30Cr, 2.5C, 12W, Rem. Co
Stellite No. 4	1	33 HRC	R	33Cr, 1C, 12W, Rem. Co
Stellite No. 6	1	34–49 HRC	R, CE, W	27/28Cr, 1.1/1.4C, 0/1Si, 0/2Mn, 0/5Fe, 4/5W, Rem. Co
Stellite No. 12	1	42–47 HRC	R, CE, W	27/29Cr, 1.4/1.9C, 0/1Si, 0/1.5Mn, 0/5Fe, 8W, Rem. Co
Stellite No. 20	1	58 HRC	R	33Cr, 2.5C, 18W, Rem. Co
Stellite No. 21	1	20–48 HRC	R, CE, W	27Cr, 0.25C, 5Mo, 0/5Fe, 2.8Ni, Rem. Co
Stellite No. 190	1	54 HRC	R	26Cr, 3.25C, 0.85Si, 0.5Mn, 1Mo, 3Fe, 3Ni, 14.5W, Rem. Co

TABLE A-4 FUSION HARDFACING CONSUMABLES (*Continued*)

Consumable tradename	Manufacturer code	Deposit hardness	Forms	Nominal chemical composition (wt. %)
Stellite No. 228	1	20 HRC	CE	0.1/C, 26Cr, 3Mo, 20Fe, Rem. Co
Stellite No. 306	1	36 HRC	W	0.55C, 24Cr, 1Si, 7Fe, 5Ni, 3W, 7Cb, Rem. Co
Stellite No. 506	1	36 HRC	W	1.8C, 34Cr, 1Si, 2Mn, 5Fe, 6.5W, Rem. Co
Stellite No. 694	1	50 HRC	R	0.85C, 28Cr, 1Si, 1Mn, 3Fe, 5Ni, 0.04B, 19.5W. IV, Rem. Co
Stellite No. 2006	1	44 HRC	R, CE	1.3C, 31Cr, 1.2Si, 8Mo, 18Fe, 8Ni, Rem. Co
Stellite No. 2012	1	49 HRC	R, CE	1.7C, 33Cr, 1.2Si, 10Mo, 15Fe, 8Ni, Rem. Co
Stellite F	1	43 HRC	R	1.7C, 25Cr, 22Ni, 12W, Rem. Co
Stoodex 1	1	53 HRC	R, CE	Cr/W/C/Si/Mn/Fe
Stoodex 6	1	56 HRC	R, CE	Cr/W/C/Si/Mn/Fe
Stoodex 12	1	56 HRC	R	Cr/W/C/Si/Mn/Fe
Tribaloy T-400	1	51–58 HRC	R, P	3max Ni+Fe, 28.5Mo, 8.5Cr, 2.6Si, 0.8C, Rem. Co
Tribaloy T-800	1	54–60 HRC	R, P	3max Ni+Fe, 28.5Mo, 17.5Cr, 3.4Si, 0.08C, Rem. Co
Wallex No. 1	7	50–55 HRC	R, P	2.25C, 1.25Si, 30Cr, 12.5W, Rem. Co
Wallex No. 6	7	39–44 HRC	R, P	1C, 1.25Si, 29Cr, 4.5W, Rem. Co
		Iron/Chromium Fusion Welding Consumables		
Abrasoweld	10	28–53 HRC	CE	Alloy steel
1000Agro Alloy	1	56 HRC	CE	4.2C, 31.5Cr, 1.3Si, 2.4Mn, Rem. Fe
Amsco 40	2	53–68 HRC	CE	3.5C, 2Cr, 5Mo, Rem. Fe
Amsco X-53	2	50–54 HRC	CE	3.5C, 16Cr, 1Mo, Rem. Fe
Amsco 77	2	45–50 HRC	CE	2.5C, 1Mn, 0.5Si, 19Cr, 1Mo, Rem. Fe
AW 73	2	57–60 HRC	W	4.5C, 3.5Mn, 2Si, 27Cr, Rem. Fe
AW Thermalloy 400	2	20–25 HRC	W	0.6C, 3Mn, 1.5Si, 17Cr, 5Mo, 8Ni, Rem. Fe
Colmonoy No. HC 240	7	60 HRC	W	30Cr, 1.5Mn, 5C, 1.5Si, Rem. Fe
Delcrome No. 11	1	55 HRC	CE	4.1C, 15.6Cr, 1.9Si, 1Mn, 0.6Mo, Rem. Fe

TABLE A-4 FUSION HARDFACING CONSUMABLES (*Continued*)

Consumable tradename	Manufacturer code	Deposit hardness	Forms	Nominal chemical composition (wt. %)
Delcrome No. 11–0	1	48 HRC	W	3.4C, 15.6Cr, 0.9Si, 1Mn, 0.4Mo, Rem. Fe
Delcrome No. 17–0	1	43 HRC	W	2.6C, 11Cr, 1.35Si, 1.8Mn, 1.5Mo, Rem. Fe
Delcrome No. 23–0	1	44 HRC	W	3C, 15.5Cr, 1Si, 1.2Mn, 4.5Mo, Rem. Fe
Delcrome No. 90	1	52 HRC	CE	2.5C, 26Cr, 0.25Si, 0.95Mn, Rem. Fe
Delcrome No. 90G	1	51 HRC	W, P	2.5C, 25Cr, 0.4Si, 1Mn, Rem. Fe
Delcrome No. 90S	1	46 HRC	W	2.6C, 26Cr, 1Si, 1.4Mn, Rem. Fe
Delcrome No. 91	1	57 HRC	CE	4.6C, 26.4Cr, 1.3Si, 0.35Mn, 3.4Mo, Rem. Fe
Delcrome No. 91–0	1	52 HRC	W	2.9C, 22.1Cr, 0.3Si, 0.3Mn, 3Mo, Rem. Fe
Delcrome No. 94	1	57 HRC	CE	3.5C, 31.5Cr, 0.45Si, 0.9Mn, 1.3Mo, Rem. Fe
Delcrome No. 946	1	60 HRC	W	3C, 28.2Cr, 1.5Si, 1.4Mn, 3.6Ni, 0.8B, Rem. Fe
Delcrome No. 94–0	1	56 HRC	W	3.5C, 28.8Cr, 1.1Si, 0.9Mn, 2.6Ni, 0.7B, Rem. Fe
Delcrome No. 94-S	1	57 HRC	W	2.9C, 27Cr, 1.7Si, 2Mn, 3.5Ni, 0.7B, Rem. Fe
Delcrome No. 100–0	1	52 HRC	W	4.3C, 28Cr, 0.8Si, 1.7Mn, 1.4Mo, Rem. Fe
Delcrome No. 103-S	1	53 HRC	W	4.6C, 30Cr, 0.9Si, 4.9Mn, Rem. Fe
Delcrome No. 911-S	1	57 HRC	W	5.7C, 36.2Cr, 1.15Si, 0.95Mn, Rem. Fe
Faceweld 1	10	50–58 HRC	CE	Fe/Cr alloy
Faceweld 12	10	50–58 HRC	CE	Fe/Cr alloy
Hardalloy 55	17	55 HRC	CE	4.6C, 27Cr, 3.5Mo, Rem. Fe
Hardalloy 140	17	55 HRC	CE	3C, 28Cr, Rem. Fe
Inweld #520	18	—	CE	6C, 22Cr, 7Mo, 5W, Rem. Fe
Linweld L-60/H560	10	58–61 HRC	W	Cr carbide in flux
Multipass No. 92	1	60 HRC	P	3.75C, 1.5Cr, 1Si, 1Mn, 10Mo, 1.75Co, Rem. Fe
Polywear 353	4	57–60 HRC	CE	High C, Cr iron
Polywear 357	4	60–64 HRC	CE	Cr/C martensitic iron
Polywear 358	4	30–36 HRC	CE	Cr/Mn/C iron base
Polywear 359	4	30–36 HRC	CE	Cr/Mn/C iron base

TABLE A-4 FUSION HARDFACING CONSUMABLES (*Continued*)

Consumable tradename	Manufacturer code	Deposit hardness	Forms	Nominal chemical composition (wt. %)
Postalloy 282-SPL	12	46–52 HRC	W	3C, 1.25Mo, 1.4Si, 15Cr, 1.2Mn, Rem. Fe
Postalloy 283-SPL	12	56–60 HRC	W	4.5C, 0.75Mo, 0.4Si, 26Cr, 1Mn, Rem. Fe
Postalloy 284-SPL	12	62–65 HRC	W	4.3C, 0.65B, 0.4Si, 24.5Cr, 1Mn, Rem. Fe
S/A MLA	2	50–54 HRC	WFC	3C, 16Cr, 1Mo, Rem. Fe
S/A MLA	2	45 HRC	W	2.6C, 1.4Mn, 1.4Si, 12Cr, 1.8Mo, Rem. Fe
S/A Super 20	2	60–65 HRC	WFC	5.2C, 1.2Mn, 0.5Si, 21Cr, 6Mo, 4W, Rem. Fe
S/A Superchrome	2	58–61 HRC	WFC	4.8C, 1.5Mn, 1.5Si, 28Cr, Rem. Fe
S/A Toughwear	2	40–55 HRC	WFC	2C, 1.4Mn, 1.4Si, 12Cr, 1.8Mo, Rem. Fe
S/A Tung-Chrome	2	56–64 HRC	WFC	4.5C, 1.5Mn, 0.5Si, 22Cr, 13W, Rem. Fe
Super 20	2	60–65 HRC	CE	6C, 22Cr, 7Mo, 5W, Rem. Fe
Superchrome	2	56–61 HRC	CE	4.5C, 2Si, 30Cr, Rem. Fe
Tube-Alloy 255–0	17	57 HRC	W	4.5C, 26.5Cr, Rem. Fe
Tube-Alloy 255S	17	53 HRC	W	4.5C, 29Cr, Rem. Fe
Tube-Alloy 263–0	17	63 HRC	W	6C, 24.3Cr, Rem. Fe
Tube Chromeface	2	60 HRC	R	4C, 1.5Mn, 2Si, 30Cr, Rem. Fe
Tufanhard 550	19	51–54 HRC	CE	3.5C, 2Mn, 1.5Si, 12Cr, 0.4Mo, 0.6V, Rem. Fe
Tufanhard 580	19	53–65 HRC	CE	6C, 2Mn, 2Si, 26Cr, 2V, Rem. Fe
		Iron/Manganese Fusion Welding Consumables		
AW CMS	2	220 HB	W	0.3C, 15Mn, 1.5Si, 15Cr, 1Ni Rem. Fe
AW Nicro Mang	2	200 HB	W	0.8C, 15Mn, 4Cr, 3.5Ni. Rem. Fe
Chrome-Mang	17	24 HRC	CE	0.4C, 14.5Mn, 14Cr, 1Ni, 1.5Mo, 0.55V, Rem. Fe, work hardens
Hardalloy 118	17	18 HRC	CE	0.8C, 16.5Mn, 5Cr, 0.3Ni, Rem. Fe, work hardens
Hardalloy 119	17	21 HRC	CE	1C, 20Mn, 0.5Ni, 4Cr, Rem. Fe, work hardens
Hardfacing #12-S/A	33	—	W	—
Inweld #536	18	235 HB	CE	0.9C, 17.5Mn, 0.5V, Rem. Fe

TABLE A-4 FUSION HARDFACING CONSUMABLES (*Continued*)

Consumable tradename	Manufacturer code	Deposit hardness	Forms	Nominal chemical composition (wt. %)
Inweld #538	18	200 HB	CE	8Co, 14.5Mn, 3.5Ni, Rem. Fe
Manganel	33	—	Bare rod	Austenitic manganese steel
Manganese-XL	33	—	CE	14Mn, Ni, Cr, austenitic steel
Manganese-XL-S/A	33	—	W	Buildup for FeMn
Mangjet	10	17–20 HRC	CE	Mn steel
Nicro-Mang	2	200 HB	CE	0.8C, 14.5Mn, 4Cr, 3.5Ni, Rem. Fe
Ni-Mang	1	99 HRB	CE	0.6C, 2.5Cr, 0.2Si, 13.5Mn, 3.5Ni, Rem. Fe
Ni-Mang-O	1	88 HRB	W	0.65C, 3Cr, 0.4Si, 15.5Mn, 1.7Ni, Rem. Fe
Ni-Mang-OA	1	88 HRB	W	0.85C, 3Cr, 0.6Si, 14Mn, 4Ni, Rem. Fe
Postalloy 285-SPL	12	15–20 HRC	W	0.8C, 3Ni, 13Mn, 2Cr, Rem. Fe
S/A CM-O	2	220 HB	WFC	0.3C, 16Mn, 16Cr, 1Ni, Rem. Fe
S/A Manganese	2	235 HB	WFC	0.9C, 17.5Mn, 0.5V, Rem. Fe
S/A Nicromang	2	200 HB	WFC	0.8C, 15Mn, 4Cr, 3.5Ni, Rem. Fe
S/A Roll-Build	2	180–230 HB	W	0.9C, 17.5 Mn, Rem. Fe
S/A Trackmang	2	235 HB	W	1C, 20Mn, 0.5Si, 4Cr, 0.5V, Rem. Fe
S/A-Z9OB	2	220–240 HB	W	0.3C, 16Mn, 16Cr, 1Ni, Rem. Fe
Sta-Mang	1	27 HRC	CE	0.6C, 15Cr, 0.3Si, 15Mn, 0.3Mo, 1.5Ni, Rem. Fe
Sta-Mang O	1	95 HRB	W	0.1C, 10.5Cr, 0.2Si, 13Mn, 6.9Ni, Rem. Fe
Sta-Mang OA	1	19 HRC	W	0.3C, 15Cr, 0.3Si, 15Mn, 1Ni, Rem. Fe
Stulz #24-S/A	3	—	—	Austenitic Mo/Mn/Cr deposit
Trackwear	2	235 HB	CE	0.9C, 17.5Mn, 0.5V, Rem. Fe
Tube-Alloy AP-O	17	22 HRC	W	0.42C, 16.5Mn, 13Cr, Rem. Fe
Tufanhard 150	19	10–18 HRC	CE	0.7C, 14Mn, 0.5Si, 4Cr, 4Ni, Rem. Fe
		Composite Fusion Welding Consumables		
AC-DC Tube Borium	1	2350–2700 HK (particles),	CE	Fe/WC
Acetylene Tube Borium	1	2350–2700 HK (particles),	R	Fe/WC

TABLE A-4 FUSION HARDFACING CONSUMABLES (*Continued*)

Consumable tradename	Manufacturer code	Deposit hardness	Forms	Nominal chemical composition (wt. %)
Bare Borod	1	—	R	Fe/WC
Coated Tube Stoodite	1	56 HRC	CE	Cr/Mn/B/Si/C/Fe
Electric Tube Borium	1	—	CE	Fe/WC
Haystellite Composite Rod No. 1	1	91 HRA (particles),	R	60WC, 40Hastelloy B
Haystellite Composite Rod No. 2	1	—	R	60WC, 40Stellite 6
Haystellite No. 38	1	—	W	38WC, Rem. Fe
Haystellite No. 38CC	1	—	CE	38WC, Rem. Fe
Haystellite No. 50	1	—	W	50WC, Rem. Fe
Haystellite No. 55	1	—	W	55WC, Rem. Fe
Haystellite No. 55 CC	1	—	CE	50WC, Rem. Fe
Haystellite No. 60	1	—	W	60WC, Rem. Fe
Haystellite No. 60 A	1	—	W	60WC, Rem. Fe
Haystellite No. 60 CC	1	—	CE	60WC, Rem. Fe
Horseshoe	2	—	R, CE	60WC (8Mesh), Rem. Fe
Horseshoe Alloy	1	—	W	WC/Fe
Horseshoe Borium	1	—	R	WC/Fe
Inweld #507	18	—	R, CE	60WC, Rem. Fe
S/A Tungstite	2	—	WFC	WC/Fe matrix
SRB	2	—	R	38WC, Rem. Fe
Tube Tungsite	2	—	R, CE	60WC, Rem. Fe
Tung-Chrome	2	56–64 HRC	W (tube),	15WC, 3.7C, 0.6Mn, 0.75Si, 27Cr, Rem. Fe
Tungfine	1	—	W, CE	61WC, Rem. Fe
Tungfine A	1	—	W	61WC, Rem. Fe
Tungrod	2	—	R, CE	60WC, Rem. Fe
Tungsmooth	1	—	CE, W	60WC, Rem. Fe
Vancar	1	—	R, W, P	VC/W/C/B/Mn/Si/Fe
		Copper-Base Fusion Welding Consumables		
Manganese Bronze W-78	8	60 HRB	R	59Cu, 0.04Mn, 0.8Sn, 0.8Fe, 0.1Si, Rem. Zn
Silicon Bronze W-133	8	—	R	95.9 Cu, 0.8 Mn, 3.3Si
Naval Bronze W-21	8	63–65 HRB	R	59Cu, 0.75Sn, 0.08Si, Rem. Zn
Naval Bronze W-60	8	55–58 HRB	R	60Cu, 0.75Sn, Rem. Zn
Penn Bronze W-16	8	57–60 HRB	R	58.5Cu, 0.02/0.1Mn, 0.06Si, 0.02/0.07Fe, Rem. Zn
Polyrod 26-C	12	120–160 HB	R	Cu/Zn/Ni
Polyrod 42-C	12	110–120 HB	R	Cu/Zn/Sn

TABLE A-4 FUSION HARDFACING CONSUMABLES (*Continued*)

Consumable tradename	Manufacturer code	Deposit hardness	Forms	Nominal chemical composition (wt. %)
Polyrod 56-B	12	—	R	Silicon bronze
Polyrod 85-C	12	150–220 HB	R	Cu/Zn/Ni
Silicon Bronze W-133	8	—	R	95.9Cu, 0.8Mn, 3.3Si
Wearwell Bronze W 46	8	83 HRB	R	57.5Cu, 2.4Sn, 0.17Si, Rem. Zn
		Nickel-Base Fusion Welding Consumables		
Colmonoy No. 4	7	35–45 HRC	R, CE, P	0.4C, 2.3Si, 10Cr, 3Fe, 2.1B, Rem. Ni
Colmonoy No. 5	7	45–50 HRC	R, CE, P	0.55C, 3.7Si, 12Cr, 3.7Fe, 2.5B, Rem. Ni
Colmonoy No. 6	7	56–61 HRC	CE, R, P	0.65C, 4.35Si, 13.5Cr, 4Fe, 3B, Rem. Ni
Colmonoy No. 21	7	26–31 HRC	R, P	0.25C, 3.25Si, 5Cr, 1Fe, 1.25B, Rem. Ni
Colmonoy No. 56	7	50–55 HRC	R, P	0.56C, 3.7Si, 12Cr, 3.7Fe, 2.5B, Rem. Ni
Colmonoy No. 70	7	52–57 HRC	R, P	0.65C, 3.4Si, 12.5Cr, 3.7Fe, 2.5B, 15W, Rem. Ni
Colmonoy No. 72	7	57–62 HRC	R, P	0.65C, 3.6Si, 12.8Cr, 3.8Fe, 2.7B, 12W, Rem. Ni
Colmonoy No. 84	7	40–45 HRC	R, P	1.1C, 1.95Si, 39Cr, 2Fe, 7.5W, 1.3B, Rem. Ni
Deloro No. 22	1	22 HRC	P	0.1C, 2.5Si, 1Fe, 1.3B, Rem. Ni
Deloro No. 35	1	35 HRC	P	0.15C, 3.7Cr, 3.5Si, 0.9Fe, 1.5B, Rem. Ni
Deloro No. 40	1	42 HRC	R	0.45C, 11Cr, 2.25Si, 2.25Fe, 1.5Co, 2.5B, Rem. Ni
Deloro No. 50	1	51 HRC	R	0.35C, 12Cr, 3.5Si, 3Fe, 2.5B, Rem. Ni
Deloro No. 60	1	57 HRC	R	0.75C, 15Cr, 4Si, 4Fe, 3.5W, Rem. Ni
Eurekalloy C	16	250 HB	CE	Mo/Cr/W, Rem. Ni
Haynes No. N-6	1	28–37 HRC	W	1.1C, 14Cr, 1.5Si, 1Mn, 5.5Mo, 3Fe, 3Co, 0.6B, 2W, Rem. Ni
Haynes No. 711	1	40–43 HRC	R, CE	2.7C, 27Cr, 1Si, 1Mn, 8Mo, 23Fe, 12Co, 3W, Rem. Ni
Haynes No. 716	1	24–32	R, CE	1.1C, 26Cr, 1.5Si, 1Mn, 3Mo, 29Fe, 11Co, 0.5B, 3.5W, Rem. Ni

TABLE A-4 FUSION HARDFACING CONSUMABLES (*Continued*)

Consumable tradename	Manufacturer code	Deposit hardness	Forms	Nominal chemical composition (wt. %)
Haynes No. 721	1	22 HRC	R, CE	0.4C, 17Cr, 1Si max, 1Mn max., 17Mo, 5.5Fe, 6.5Co, 4.5W, Rem. Ni
Ni-65	2	56–62 HRC	R	0.75C, 4.5Si, 15Cr, 4.5Fe, 3.5B, 1Co, Rem. Ni
Ni-Rod 55	3	—	R	—
Ni-Rod 99	3	—	R	—
Polyface 673	4	15–20 HRC	CE, R, W	High alloy, nickel base
Polyface 673-C	4	15–20 HRC	CE, R, W	High alloy, nickel base
Polyface 676-C	4	20–23 HRC	CE, R, W	Ni/Mo/Co
Postalloy 2814-2PL	12	55–60 HRC	W	0.75C, 3B, 6Fe, 17Cr, 4.5Si, Rem. Ni
Stoody 6N	1	50 HRC	R	Cr/C/Si/B/Fe
Tribaloy T-700	1	42–48 HRC	R, CE, P	3 max. Co+Fe, 32.5Mo, 15.5Cr, 3.4Si, 0.08C, Rem. Ni
		Iron-Base Fusion Welding Consumables		
AW 72	2	52 HRC	W	0.25C, 2Mn, 1Si, 6Cr, 0.8Mo, 1.5W, Rem. Fe
AW 79	2	44 HRC	W	0.15C, 2.2Mn, 0.7Si, 3.5Cr, 0.8Mo, Rem. Fe
AW 84	2	30 HRC	W	0.13C, 2.2Mn, 0.8Si, 0.6Mo, Rem. Fe
AW 87	2	41 HRC	W	0.14C, 2.2Mn, 0.8Si, 2.7Cr, 0.6Mo, Rem. Fe
AW 420	4	45 HRC	W	0.16C, 1.25Mn, 0.7Si, 12Cr, Rem. Fe
Bare Tube Stoodite	1	57 HRC	R	Composite Cr/Mn/B/Si/C/Fe
Buildup	33	—	CE	Low C, Cr/Ni
Buildup	2	22–30 HRC	CE	0.1C, 1.5Mn, 1.1Cr, 0.45Mo, Rem. Fe
Build-Up LH	1	29 HRC	CE	Cr/Mn/Si/C/Fe
Colmonoy No. 1	7	58–63 HRC	CE	0.9C, 1.75Si, 11Cr, 0.4Mn, 2.5B, Rem. Fe
Colmonoy Special No. 1	7	60–65 HRC	CE	1C, 3Si, 13Cr, 0.7Mn, 3B, Rem. Fe
Drawalloy	16	190–460 HB	CE	Austenitic matrix
Dynamag	1	15 HRC	W, CE	Mn/Ni/Cr/Si/C/Fe
Eureka 2	16	61–63 HRC	R, W	M42 high-speed steel
Eureka 31A	16	50–52 HRC	R	H13 tool steel
Eureka 35	16	32–35 HRC	CE	Hot-work tool steel

TABLE A-4 FUSION HARDFACING CONSUMABLES (*Continued*)

Consumable tradename	Manufacturer code	Deposit hardness	Forms	Nominal chemical composition (wt. %)
Eureka 45/45N	16	38–42 HRC	CE	Cr/Mo/W hot-work tool steel
Eureka 45-N	16	32–36 HRC	R, W	Medium C hot-work tool steel
Eureka 70A	16	—	R	01/02 tool steel
Eureka 71-M/70-W	16	58–62 HRC	CE	0 series tool steel
Eureka 72	16	52–56 HRC	CE	Hot-work tool steel
Eureka 72A	16	54–56 HRC	R, W	H12 tool steel
Eureka 73	16	56–58 HRC	CE	Cr/Mo/W hot-work tool steel
Eureka 73A	16	52–54 HRC	R, W	H21 tool steel
Eureka 74A	16	57–58 HRC	R	S7 tool steel
Eureka 75XA	16	—	R	W1/W2 tool steel
Eureka 78-A	16	58–60 HRC	R, W	High C, W-free, hot-work tool steel
Eureka 88	16	56–57 HRC	CE	Hot-work tool steel
Eureka 88-A	16	54–56 HRC	R	H19 tool steel
Eureka 130A	16	38–42 HRC	R, W	4130 steel
Eureka 130/145	16	32–36 HRC	CE	4130/6145 steel
Eureka 145 A	16	50–54 HRC	R	6145 steel
Eureka 350-A	16	38–40 HRC	R, W	Chromo-N
Eureka 400-C	16	45–47 HRC	CE	Buildup like W series tool steel
Eureka 1215	16	60–62 HRC	CE	5%Cr air-hardening tool steel
Eureka 1215A	16	—	R	A2 tool steel
Eureka 1216	16	60–64 HRC	CE	M1/M2 high speed steel
Eureka 1220A	16	38–40 HRC	R	D2 tool steel
Eureka 8510/75X	16	56–60 HRC	CE	W1/W2 tool steel
Eureka Air-40 & GTA	16	59–60 HRC	CE	Low air-hardening tool steels
Eurekamatic 5	16	40–46 HRC	WFC	Medium high C, Cr/Si/Mo alloy
Eurekamatic 145	16	50–54 HRC	W	6145 steel
Eurekamatic 1216	16	59–60 HRC	W	High speed tool steel
Eurekamold	16	31–32 HRC	R	P6 tool steel
EXP-10	16	38–40 HRC	CE	Cr-base alloy for castings
EXP-20	16	—	CE	Special Cr alloy
Frogalloy	17	20 HRC	CE	0.4C, 19.2Cr, 9.5Ni, 1.4Mo, Rem. Fe
Hammerweld	16	38–42 HRC	CE	Hot-work tool steel
Hardalloy 32	17	26 HRC	CE	0.18C, 0.7Cr, 0.3Mo, Rem. Fe
Hardalloy 42	17	40 HRC	CE	0.17C, 1.8Cr, 0.7Mo, 0.3V, Rem. Fe
Hardalloy 40TIC	17	45 HRC	CE	3C, 8.2Cr, 1.5Ti, Rem. Fe
Hardalloy 48	17	40 HRC	CE	1.8C, 30Cr, 3Ni, 1.5Mo, Rem. Fe
Hardalloy 55	17	55 HRC	CE	4.6C, 27Cr, 3.5Mo, Rem. Fe

TABLE A-4 FUSION HARDFACING CONSUMABLES (*Continued*)

Consumable tradename	Manufacturer code	Deposit hardness	Forms	Nominal chemical composition (wt. %)
Hardalloy 58	17	58 HRC	W	0.6C, 5.5Cr, 0.5Mo, Rem. Fe
Hardalloy 61	17	60 HRC	CE	0.8C, 4Cr, 8Mo, 1.1V, 1.3W, Rem. Fe
Hardalloy 120	17	20 HRC	CE	0.7C, 23.5Cr, 9.7Ni, Rem. Fe, work hardens
HW-T	2	55 HRC	W	0.4C, 5Cr, 1.5Mo, 1.25W, O.4V, Rem. Fe
Inweld #566 A.H.	18	58–60 HRC	R, CE	Air-hardening tool steel
Inweld #568 HSS	18	60–63 HRC	R, CE	High speed tool steel
Inweld #564 O.H.	18	58–60 HRC	R, CE	Oil-hardening tool steel
Inweld #560 W.H.	18	58–60 HRC	R, CE	Water-hardening tool steel
Inweld #505	18	58–60 HRC	CE	—
Inweld #562	18	52–55 HRC	R, CE	Hot-work tool steel
Jet-LH BU90	10	23–28 HRC	CE	Buildup
Lincore 33	10	28–38 HRC	W, WFC	Ferritic/martensitic structure
Lincore 50	10	48–52 HRC	W	Semiaustenitic structure
Lincore 55	10	50–59 HRC	W, WFC	Semiaustenitic structure
Lincore 30/801	10	27–32 HRC	W	Ferritic/martensitic structure
Lincore M	10	12–18 HRC	W, WFC	Austenitic structure
Linweld L-60/A96S	10	52–54 HRC	W	Like 420 stainless steel
Linweld L-60/A100	10	30–32 HRC	W	Like 420 stainless steel
Linweld L-60/385	10	34–46 HRC	W	Alloy in flux
Linweld L-60/H550	10	32–60 HRC	W	Semiaustenitic structure, alloy in flux
Marweld 250	16	30–32 HRC	R	250-ksi maraging steel
McKay GP	17	20 HRC	CE	0.6C, 26.5Cr, 9Ni, Rem. Fe
McKay C	17	95 HRB	CE	0.04C, 15.5Cr, 16Mo, 3.8W, Rem. Fe
McKay CG	17	90 HRB	W	0.03C, 15.5Cr, 16Mo, 3.8W, Rem. Fe
McKay C-S	17	90 HRB	W	0.03C, 15.5Cr, 16Mo, 3.8W, Rem. Fe
McKay C-T1	17	90 HRB	W	0.03C, 15.5Cr, 16Mo, 3.8W, Rem. Fe
Multipass No. 4	1	28 HRC	CE	0.1C, 2Cr, 0.4Si, 0.9Mn, Rem. Fe
Multipass No. 4–0	1	18 HRC	W	0.15C, 0.75Cr, 1.35Si, 1.9Mn, 0.3Mo, Rem. Fe
Multipass No. 4-S	1	30 HRC	W	0.1C, 0.6Si, 1.9Mn, 0.4Mo, Rem. Fe

TABLE A-4 FUSION HARDFACING CONSUMABLES (*Continued*)

Consumable tradename	Manufacturer code	Deposit hardness	Forms	Nominal chemical composition (wt. %)
Multipass No. 5	1	45 HRC	CE	0.2C, 2.3Cr, 0.8Si, 2Mn, 0.3Mo, Rem. Fe
Multipass No. 5–0	1	44 HRC	W	0.1C, 3.5Cr, 0.3Si, 2Mn, 0.2Mo, Rem. Fe
Multipass No. 5-S	1	33 HRC	W	0.15C, 3Cr, 0.6Si, 2Mn, 0.4Mo, Rem. Fe
Multipass No. 5 Mod.-S	1	43 HRC	W	0.28C, 5.35Cr, 1.3Si, 2.65Mn, 1Mo, Rem. Fe
Multipass No. 58-S	1	37 HRC	W	0.09C, 2.6Cr, 0.5Si, 2.3Mn, 0.4Mo, Rem. Fe
Multipass No. 7-S	1	20 HRC	W	0.1C, 1.7Cr, 0.5Si, 1.8Mn, 0.5Mo, Rem. Fe
Multipass No. 22	1	57 HRC	CE	0.7C, 5.5Cr, 0.4Si, 0.9Mn, Rem. Fe
Multipass 22-S	1	35 HRC	W	0.4C, 5.4Cr, 0.9Si, 3Mn, 0.8Mo, Rem. Fe
Polyforge 9630	4	38–43 HRC	CE	Medium C, high Cr, hot work tool steel
Polyforge 9650	4	38–42 HRC	CE, W	Medium C, Ni/Cr/Mo steel
Polymatic 325	4	35–38 HRC	CE	Ni/Mn alloy steel
Polymatic 325 FC	4	35–38 HRC	WFC	Ni/Cr/Mn steel
Polymatic 329 FC	4	54–58 HRC	WFC	High C, Cr/Si steel
Polymatic 356 FC	4	28–34 HRC	WFC	C/Cr/Cu, iron base
Polymatic 964 FC	4	50–55 HRC	WFC	Medium to low C, High Cr, hot work tool steel
Polyshock 943-L	4	54–57 HRC	CE, R, W	S7 tool steel
Polywear 354	4	52–58 HRC	CE	High C, Cr/Mo steel
Polywear 356	4	33–40 HRC	CE	C/Cr/Cu alloy steel
Postalloy 387-SPL	12	33–38 HRC	W	0.1C, 0.5Si, 1Mo, 1.75Mn, 2.4Cr, Rem. Fe
Postalloy 288-SPL	12	55–59 HRC	W	0.6C, 0.5Mo, 0.4Si, 8Cr, 0.5V, 0.4Mn, Rem. Fe
Postalloy 289-SPL	12	60–65 HRC	W	2.2C, 1Mo, 0.9Si, 7.4Cr, 0.6Mn, 0.95B, Rem. Fe
Postalloy 2885-SPL	12	50–55 HRC	W	0.35C, 13Cr, 0.5Si, 0.5Mn, Rem. Fe
Resistwear	2	58 HRC	CE	0.4C, 6Cr, 0.6Mo, Rem. Fe
S7-T	17	58 HRC	W	0.5C, 3.3Cr, 1.5Mo, Rem. Fe
S/A Roll-Face	2	50–54 HRC	W	0.3C, 16Cr, 1Mo, Rem. Fe
S/A T-40	2	20–25 HRC	WFC	0.6C, 3Mn, 0.5Si, 22Cr, 0.5Mo, 8Ni, Rem. Fe

TABLE A-4 FUSION HARDFACING CONSUMABLES (*Continued*)

Consumable tradename	Manufacturer code	Deposit hardness	Forms	Nominal chemical composition (wt. %)
Self Hardening	1	47 HRC	CE	Cr/Mn/Si/C/Fe
Stoody Build-up Wire	1	26 HRC	W	Cr/Mn/Si/Mo/C/Fe
Stoody C	1	14 HRC	CE	Cr/Mn/Si/Mo/C/VC/W/Fe
Supreme No. 4	16	To 67 HRC	R	T15 high speed tool steel
Thermalloy 400	2	20–25 HRC	CE	0.6C, 2Mn, 1Si, 22Cr, 0.5Mn, 8Ni, Rem. Fe
Tristelle TS-1	1	38 HRC	R, W, P	30Cr, 10Ni, 12Co, 5Si, 1C, Rem. Fe
Tristelle TS-2	1	45–46 HRC	R, W, P	35Cr, 10Ni, 12Co, 5Si, 2C, Rem. Fe
Tristelle TS-3	1	47–51 HRC	R, W, P	35Cr, 10Ni, 12Co, 5Si, 3C, Rem. Fe
Tube-Alloy 218–0	17	18 HRC	W	15C, 15Mn, 3.1Cr, Rem. Fe
Tube-Alloy 219–0	17	20 HRC	W	1C, 20Mn, 4.5Cr, Rem. Fe
Tube-Alloy 236-S	17	42 HRC	W	0.12C, 5.5Mo, 5.3Ni, Rem. Fe
Tube-Alloy 240–0	17	49 HRC	W	3.2C, 15.5Cr, Rem. Fe
Tube-Alloy 242–0	17	40 HRC	W	0.14C, 2.5Cr, Rem. Fe
Tube-Alloy 242-S	17	39 HRC	W	0.16C, 1.6Cu, 1.6Cr, 0.6Mo, 0.22V, Rem. Fe
Tube-Alloy 244–0	17	40 HRC	W	2.4C, 9.5Cr, 1.5Mo, 0.6Cu, Rem. Fe
Tube-Alloy 250-S	17	48 HRC	W	0.2C, 11Cr, Rem. Fe
Tube-Alloy 252-S	17	45 HRC	W	0.17C, 3.5Cr, Rem. Fe
Tube-Alloy 258–0	17	57 HRC	W	0.45C, 6Cr, 1.5Mo, 1.5W, Rem. Fe
Tube-Alloy 258-S	17	52 HRC	W	0.33C, 5.9Cr, 1.5Mo, 1.4W, Rem. Fe
Tube-Alloy 258 TIC–0	17	58 HRC	W	1.8C, 7.5Cr, 1.5Mo, 4.7Ti, Rem. Fe (contains TiC)
Tube-Alloy 821-S	17	45 HRC	W	0.15C, 6Cr, 1.5Mo, 1W, Rem. Fe
Tube-Alloy 829–0	17	28 HRC	W	2C, 13.5Mn, 3.2Cr, 3.5Ti, Rem. Fe
Tube-Alloy BU-0	17	18 HRC	W	0.06C, 1Cr, 0.5Ti, Rem. Fe
Tube-Alloy BU-S	17	27 HRC	W	0.13C, 0.8Cr, Rem. Fe
Tufanhard 160	19	15–20 HRC	CE	0.75C, 4Mn, 0.5Si, 20Cr, 9Ni, Rem. Fe
Tufanhard 250	19	23–26 HRC	CE	0.2C, 0.12Mn, 1.05Cr, Rem. Fe
Tufanhard 320	19	26–34 HRC	CE	0.2C, 1Mn, 0.4Si, 0.9Cr, 0.3Ni, 0.4Mo, Rem. Fe

TABLE A-4 FUSION HARDFACING CONSUMABLES (*Continued*)

Consumable tradename	Manufacturer code	Deposit hardness	Forms	Nominal chemical composition (wt. %)
Tufanhard 375	19	29–40 HRC	CE	0.23C, 0.69Mn, 0.23Si, 2.3Cr, 0.18Mo, Rem. Fe
Tufanhard 450	19	46–52 HRC	CE	2C, 1.9Mn, 0.5Si, 2Cr, Rem. Fe
Tufanhard 600	19	52–58 HRC	CE	0.29C, 0.84Mn, 4.3Cr, 0.28Mo, Rem. Fe
Ultra-Hard	33	60 HRC	CE	Cr/Si/C alloy
Universal Hardface	3	53 HRC	CE	—
Wearweld	10	50–55 HRC	CE	Alloy steel
Weldmold 919	4	42–46 HRC	CE, WFC	4340 steel
Weldmold 920	4	25–30 HRC	CE, W	AISI 8620 steel
Weldmold 922	4	32–36 HRC	CE, R, W	AISI 4230 steel
Weldmold 925	4	56–60 HRC	CE, R	W1 tool steel
Weldmold 927	4	32 HRC	CE	4140 steel
Weldmold 935	4	57–60 HRC	CE, R	O1 tool steel
Weldmold 937	4	56–58 HRC	CE, R	A2 tool steel
Weldmold 938	4	38–42 HRC	CE, R	D2 tool steel
Weldmold 954	4	38–42 HRC	CE, R, W	H series tool steel
Weldmold 957	4	54–57 HRC	CE	H12 tool steel
Weldmold 958	4	54–57 HRC	CE, R, W	H12 tool steel
Weldmold 9580	4	55–57 HRC	CE, W	H12 tool steel
Weldmold 959-L	4	54–57 HRC	CE, R, W	W-free, Cr hot work tool steel
Weldmold 966	4	60–63 HRC	CE, R	M2 high speed steel
Weldmold 988	4	54–57 HRC	CE, R	W, Cr, Co, hot work tool steel

APPENDIX IV: THERMAL SPRAY CONSUMABLES

TABLE A-5 THERMAL SPRAY CONSUMABLES

Consumable tradename	Manufacturer code*	Description/ chemical composition
	Wire Thermal Spray Consumables	
Metco 70	11	tin/zinc alloy, 70Sn, 30Zn
Metco 402	11	nickel/aluminum/iron alloy, 52 Ni, 20A1, 20Fe, 4Cr, and others
Metco 405	11	nickel aluminide, 80Al, 20Ni
Metco 470 AW	11	iron/chromium/nickel alloy, 25Fe, 15Cr, Rem. Fe
Metco Aluminum	11	aluminum
Metco Cadmium	11	cadmium
Metco Copper	11	copper
Metco Lead	11	lead
Metcoloy #1	11	18–8 stainless steel, 18Cr, 8Ni, 2Mn, 0.75Si, Rem. Fe
Metcoloy #2	11	stainless steel, 13Cr, 0.5Si, 0.5Ni, 0.35Mn, 0.35C, Rem. Fe
Metcoloy #5	11	stainless steel, 18Cr, 8.5Mn, 5Ni, 1Si, 0.15C, Rem. Fe
Metcoloy #4	11	stainless steel, 17Cr, 12Ni, 2.5Mo, 2Mn, 1Si, 0.8C, Rem. Fe
Metcoloy #33	11	nickel/chromium iron, 60Ni, 16Cr, 1.5Si, Rem. Fe
Metco Monel	11	Monel, 67Ni, 1.5Fe, 1Mn, Rem. Cu
Metco Nickel	11	nickel, 1Co, 0.6Fe, 0.25C, 0.25Cr, Rem. Ni
Metco SF Aluminum	11	94Al, 4Si
Metco Tin	11	99+% tin
Metco Zinc	11	99+% zinc
Metco Zinc-AL	11	aluminum/zinc alloy, 85Zn, 15Al
Spraybond Wire	11	99+% molybdenum
Spraybabbitt A	11	babbitt, 7.5Sb, 3.5Cu, 0.25Pb, Rem. Sn
Spraybrass	11	brass, 66Cu, 34Zn
Spraybronze	11	aluminum bronze, 90Cu, 9Al, 1Fe
Spraybronze C	11	red brass, 90Cu, 10Zn
Spraybronze P	11	bronze, 95Cu, 5Sn
Spraybronze TM	11	copper/zinc alloy, 58.2Cu, 0.8Sn, 0.75Fe, 0.25Mn, Rem. Zn

*See Table A-3 (Appendix III) for key to manufacturers' codes.

Note: All thermal spray consumables listed are in powder form unless otherwise specified.

TABLE A-5 (*Continued*)

Consumable tradename	Manufacturer code	Description and chemical composition
Spraysteel 10	11	carbon steel, 0.5 Mn, 1C, Rem. Fe
Spraysteel 80 AW	11	alloy steel, 2Mn, 1.9Cr, 1.1C, 0.4Si, 0.2Ti, Rem. Fe
	Ceramic Thermal Spray Consumables	
Amdry 304	9	chromium carbide (fine)
Amdry 306	9	chromium carbide (coarse)
Amdry 118	9	zirconium oxide
Amdry 125	9	chromium oxide
Amdry 125F	9	chromium oxide
Amdry 126F	9	chromium oxide/silica
Amdry 141	9	94% zirconium oxide, 6% yttrium oxide
Amdry 142	9	92% zirconium oxide, 8% yttrium oxide
Amdry 144	9	88% zirconium oxide, 12% yttrium oxide
Amdry 146	9	80% zirconium oxide, 20% yttrium oxide
Amdry 170	9	titanium oxide
Amdry 171	9	70% titania, 30% alumina
Amdry 180	9	aluminum oxide
Amdry 181	9	50% alumina, 50% titania
Amdry 182	9	97% alumina, 3% titania
Amdry 183	9	87% alumina, 13% titania
Amdry 184	9	60% alumina, 40% titania
Amdry 186	9	alumina/magnesia spinel
Amdry 187C	9	alumina (coarse)
Amdry 187F	9	alumina (fine)
Amdry 333	9	magnesium zirconate
Colspray 130	7	alumina/titania
Metco 70C-NS	11	chromium carbide
Metco 101	11	90% alumina, 2.5% titania, 2% silica, 1% iron oxide, Rem. other oxides
Metco 101B-NS	11	94% alumina, 2.5% titania, 2% silica, 1% iron oxide
Metco 101 FP	11	97% alumina, 3% titania
Metco 102	11	titania
Metco 105	11	98.5% alumina, 1% silica
Metco 106	11	96% chromium oxide, 2% titania, Rem. other oxides
Metco 110	11	50% alumina, 50% titania
Metco 136F	11	5% silica, 3% titania, Rem. chromium oxide
Metco 201	11	93% zirconia, 5% calcia, 0.5% alumina, 0.4% silica, Rem. other oxides

TABLE A-5 (*Continued*)

Consumable tradename	Manufacturer code	Description and chemical composition
Metco 202 NS	11	80% zirconia, 20% yttria
Metco 210	11	24% magnesia, Rem. zirconia
Metco 211	11	calcium zirconate, 31% calcia, Rem. zirconia
Metco 212F-NS	11	magnesia/alumina spinel, 72% alumina, 28% magnesia
MT 1010	5	aluminum oxide
MT 1011	5	94% alumina, 6% titania
MT 1020	5	titanium oxide, black
MT 1021	5	50% alumina, 50% titania
MT 1030	5	87% alumina, 13% titania
MT 1031	5	60% alumina, 40% titania
MT 1040	5	chromium oxide
MT 1050	5	92% chromium oxide, 5% silica, 3% titania
MT 1070	5	5% calcia stabilized zirconia
MT 1080 M	5	6% yttria stabilized zirconia
MT 1081	5	8% yttria stabilized zirconia
MT 1082 M	5	12% yttria stabilized zirconia
MT 1083 M	5	20% yttria stabilized zirconia
MT 1150 M	5	magnesium zirconate
MT 2050 C	5	chromium carbide
MT 2100 M	5	tungsten carbide
PAC 123	20	chromium carbide, 35–40 HRC
PAC 133	20	chromium carbide, 45–50 HRC
PAC 701	20	alumina
PAC 702	20	titania
PAC 705	20	alumina
PAC 710	20	alumina/titania
PAC 732F	20	chromium oxide, silica
PAC 801	20	zirconia, lime stabilized
PAC 801 B-1	20	zirconia, calcium carbonate
PAC 810	20	Mg stabilized zirconia
PAC 812	20	35% Ni/Cr, 65% zirconia
PAC 1106	20	chromium oxide
PAC 2020	20	20% yttria, Rem. zirconia
Rokide #106	13	alumina, 13% titania
Rokide #107	13	alumina, 13% titania
Rokide #108	13	alumina, 40% titania
Rokide #109	13	alumina, 40% titania
Rokide #110	13	alumina, 3% titania

TABLE A-5 (*Continued*)

Consumable tradename	Manufacturer code	Description and chemical composition
Rokide #112	13	alumina, 3% titania
Rokide #117	13	alumina, 3% titania
Rokide #150	13	alumina
Rokide #170	13	alumina, 28% magnesia
Rokide #183	13	alumina
Rokide #184	13	alumina
Rokide #233	13	zirconia, 18/25% magnesia
Rokide #235	13	zirconia, 5% calcia
Rokide #252	13	zirconia, 5% calcia
Rokide #272	13	zirconia, 30% calcia
Rokide #281	13	zirconia, 8% yttria
Rokide #324	13	chromium oxide
Rokide #328	13	chromium oxide
Rokide A	13	alumina rod
Rokide C	13	90% chromium oxide
Rokide MA	13	alumina, 30% magnesia rod
Rokide MBA	13	alumina rod
Rokide MBAT 60/40	13	alumina, 40% titania rod
Rokide MBAT 87/13	13	alumina, 13% titania rod
Rokide MBAT 97/3	13	alumina, 3% titania rod
Rokide MBC	13	83% chromium oxide
Rokide MZ	13	zirconia/magnesia rod
Rokide SA	13	97% alumina rod
Rokide YZ/8	13	zirconia, 8% yttria rod
Rokide Z	13	zirconia, 5% calcium oxide
Rokide ZS	13	zirconia, silica rod
UCAR A10-101	6	98% aluminum oxide
UCAR A10-105	6	3% titania, Rem. aluminum oxide
UCAR A10-108	6	3% titania, Rem. aluminum oxide
UCAR A10-117	6	99% aluminum oxide
UCAR CrO-131	6	chromium oxide
UCAR LA-2	6	99% aluminum oxide (d-gun)
UCAR LA-7	6	60% alumina, 40% titania (d-gun)
UCAR LC-4	6	99% chromium oxide
UCAR LC-19	6	70% chromium oxide, 30% alumina
UCAR ZrO-103	6	MgZrO
UCAR ZrO-110	6	$20Y_2O_3$, Rem. ZrO
UCAR ZrO-112	6	$8Y_2O_3$, Rem. ZrO
UCAR ZrO-113	6	$8Y_2O_3$, Rem. ZrO
UCAR ZrO-114	6	$8Y_2O_3$, Rem. ZrO

TABLE A-5 (*Continued*)

Consumable tradename	Manufacturer code	Description and chemical composition
		Cermet Thermal Spray Consumables
Amdry 301	9	WC, 12% Co
Amdry 302	9	WC, 12% Co
Amdry 305	9	75% chromium carbide, 25% Ni/Cr
Amdry 307	9	75% chromium carbide, 25% Ni/Cr
Amdry 308	9	85% chromium carbide, 15% Ni/Cr
Amdry 336	9	65% magnesium zirconate, 35% Ni/Cr/Al
Amdry 337	9	65% magnesium zirconate, 35% Ni/Cr alloy
Amdry 341	9	65% magnesium zirconate, 35% Ni/Al alloy
Amdry 364	9	93% chromium carbide, 7% Ni/Cr alloy
Amdry 367	9	90% chromium carbide, 10% Ni/Cr alloy
Amdry 731C	9	35% WC, 65% Amdry 761
Amdry 732C	9	80% WC, 20% Amdry 761
Amdry 734	9	50% WC/Co, 50% Amdry 761F
Amdry 736C	9	35% WC, 65% Amdry 761
Amdry 737C	9	50% WC, 50% Amdry 761
Amdry 855	9	30% WC/Co, 70% Amdry 850
Amdry 927	9	WC/12Co
Amdry 928	9	75% WC, 25% nickel alloy
Amdry 983	9	WC, 7% Co aggregate
Amdry W56	9	WC, Amdry 956 and 761
Metco 31C-NS	11	tungsten carbide, nickel binder, 46Ni, 35WC, 11Cr, 2.5Fe, 2.5Si, 2.5B, 0.5C
Metco 32C	11	tungsten carbide/nickel binder, 80WC, 14Ni, 3.5Cr, 0.8B, 0.8Fe, 0.8Si, 0.1C
Metco 34F	11	fine tungsten carbide/nickel, 50WC, 33Ni, 9Cr, 3.5Fe, 2Si, 2B, 0.5C
Metco 34FP	11	WC/nickel matrix, 50WC, 33Ni, 9Cr, 3.5Fe, 2Si, 2B, 0.5C
Metco 35C	11	WC/Ni-Co binder, 30WC, 28Co, 19Cr, 4Mo, 2.5Si, 2B, 1.9FE, 0.1C
Metco 36C	11	WC/nickel matrix, 35WC, 11Cr, 2.5B, 2.5Fe, 2.5Si, 0.5C, Rem. Ni.
Metco 71VF-NS	11	WC/cobalt binder, 12Co, 1Fe, Rem. WC
Metco 71NS	11	WC/cobalt binder, 12Co, 1Fe, Rem. WC
Metco 73SF-NS	11	WC/cobalt binder, 17Co, Rem. WC
Metco 80NS	11	chromium carbide/NiCr mixture, 85 chromium carbide, 12Ni, 3Cr
Metco 81-NS	11	chromium carbide/Ni/Cr, 75 chromium carbide, 20Ni, 5Cr

TABLE A-5 (*Continued*)

Consumable tradename	Manufacturer code	Description and chemical composition
Metco 82VF-NS	11	93% chromium carbide, 7% 80/20 nickel-chromium
Metco 83VF-NS	11	chromium carbide/NiCr, 50% chromium carbide, 50% Ni/Cr alloy
Metco 301C-NS	11	boron nitride cermet, 14Cr, 8Fe, 5.5BN, 3.5Al, Rem. Ni
Metco 303-NS	11	50% zirconium oxide, 28% nickel, 15% magnesium oxide, 7% chromium
Metco 304-NS	11	aluminum bronze/boron nitride, 93Al/Cu, 7BN
Metco 312-NS	11	nickel alloy/Bentonite, 4Cr, 4Al, 21 Bentonite, Rem. Ni
Metco 351	11	high-carbon iron/iron oxide, 5 iron oxide, 3C, 0.3Mn, Rem. Fe
Metco 404/404NS	11	nickel aluminide, 20 Al, Rem. Ni
Metco 410/410NS	11	alumina/nickel aluminide, 70% aluminum oxide, 30% nickel aluminide
Metco 411	11	alumina/nickel aluminide, 30% alumina, 70% nickel aluminide
Metco 412	11	zirconia/nickel aluminide, 60% zirconia, 35% nickel aluminide
Metco 421NS-1	11	65% magnesium zirconate, 35% nickel aluminide
Metco 430 NS	11	chromium carbide/nickel aluminide, 48Cr, 28Ni, 6C, 2Al, 2Mo, 1B, 1Si, Rem. Co
Metco 438 NS	11	WC/nickel aluminide, 75WC, 3Cr, 1.5Al, 0.8Fe, 0.8Si, 0.5B, 0.15C, Rem. Ni
Metco 439/439 NS-2	11	WC self-fusing, 50WC, 6Cr, 1.5Si, 1.5Fe, 1B, 0.7Al, 0.5C. Rem. Ni
Metco 439 NS	11	WC/nickel aluminide, 50WC, 6Cr, 3Al, 1.5Fe, 1.5Si, 1B, 0.5C, Rem. Ni
Metco 441	11	50 zirconia, 26Ni, 1.5 magnesia, 0.7 Cr, 2Al
Metco 1123	11	WC/Co self-fluxing blend, 75WC, 4.3Cr, 1Fe, 1Si, 1B, 0.2C, Rem. Ni
MT 1151M	5	65% magnesium zirconate, 35% Nichrome
MT 1153M	5	65% magnesium zirconate, 35% Ni/Al blend
MT 2010	5	WC/8%Co
MT 2020	5	WC/11%Co
MT 2030M	5	WC/12%Co
MT 2040M	5	WC/17%Co

TABLE A-5 (*Continued*)

Consumable tradename	Manufacturer code	Description and chemical composition
MT 2060 VF	5	90% chromium carbide, 10% Nichrome
MT 2070C	5	85% chromium carbide, 15% Nichrome
MT 2080	5	75% chromium carbide, 25% Nichrome
MT 2090	5	75% chromium carbide, 25% Nichrome
MT 2110	5	75% titanium carbide, 35% Inconel 600
MT 4081C	5	75% WC, 12% Co/Ni blend
438NS-1	11	WC self-fusing, 75WC, 3Cr, 0.8Fe, 0.8Si, 0.6B, 0.35Al, 0.2C, Rem. Ni
PAC 80	20	70% PAC 63, Rem. WC
PAC 86	20	65% PAC 60F, Rem. WC
PAC 87	20	80% WC, Rem. PAC 60F
PAC 89F	20	50% PAC 60F, 50% WC
PAC 125	20	WC/12% Co
PAC 129	20	85% chromium carbide, 15% Ni/Cr alloy
PAC 130	20	75% chromium carbide, 25% Ni/Cr alloy
PAC 131-1	20	93% chromium carbide, 7% Ni/Cr alloy
PAC 200S	20	WC/16Co
PAC 903	20	75% WC, Rem. PAC 128, 124, 900
Stellite No. 158	1	0.75C, 26Cr, 1.2Si, 1Mn, 1Mo, 2Fe, 3Ni, 0.7B, 5.5W, Rem. Co
Stoody 60T.G	1	1, C/Si/B/Fe/Ni/WC
Ticoat T-92	14	titanium carbide in iron-base alloy matrix
Ticoat T-94	14	titanium carbide in iron-base matrix
Ticoat T-916	14	titanium carbide in cobalt-base matrix
UCAR CrC-106	6	75% chromium carbide, 25% Nichrome
UCAR CrC-108	6	75% chromium carbide, 25% Nichrome
UCAR LC-1C	6	80% chromium oxide, 20% Ni/Cr alloy (d-gun)
UCAR LW-5	6	73WC, 20Cr, 7Ni (d-gun)
UCAR LW-11B	6	88WC, 12Co (PSP)
UCAR LW-15	6	86WC, 10Co, 4Cr (d-gun)
UCAR LW-1N30	6	87WC, 13Co (d-gun)
UCAR LW-1N40	6	85WC, 15Co (d-gun)
UCAR WC-104	6	88WC, 12Co
UCAR WC-106	6	88WC, 12Co
UCAR WC-107	6	22Cr, 5Ni, Rem. WC
UCAR WC-113	6	10Co, 4Cr, Rem. WC
UCAR WC-114	6	12Co, Rem. WC
UCAR WC-119	6	15Ni, Rem. TiC
UCAR WC-128	6	17Co, Rem. WC

TABLE A-5 (*Continued*)

Consumable tradename	Manufacturer code	Description and chemical composition
UCAR WC128–2	6	17Co, Rem. WC
UCAR WT 1	6	83(W, Ti)C, 17Ni (d-gun)
Wallex No. 55	7	50% Wallex No. 50, 50% WC
Wearclad W-516	14	WC in cobalt alloy matrix
Wearclad W-560	14	WC in nickel alloy matrix
	Nickel-Base Thermal Spray Consumables	
Amdry 315	9	20Cr, Rem. Ni.
Amdry 322	9	22Cr, 20Fe, 9Mo, Rem. Ni
Amdry 384	9	9Cr, 7Al, 5Mo, 5Fe, Rem. Ni
Amdry 721	9	6Cr, 3.4Fe, 3.25Si, 1.25B, 0.15C, Rem. Ni
Amdry 722	9	3.25Si, 1.25B, 1Fe, 0.05C, Rem. Ni
Amdry 723	9	2.5Si, 1B, 1Fe, 0.05C, Rem. Ni
Amdry 724	9	2.5Si, 1B, 1Fe, 0.05C, Rem. Ni
Amdry 754	9	10Cr, 3.65Fe, 3.5Si, 1.35B, 0.45C, Rem. Ni
Amdry 755	9	12Cr, 4/5Fe, 4Si, 2.4B, 0.55C, Rem. Ni
Amdry 756	9	13Cr, 4.4Fe, 4.25Si, 2.75B, 0.6C, Rem. Ni
Amdry 761	9	14Cr, 4.5Fe, 4.4Si, 3.15B, 0.65C, Rem. Ni
Amdry 769	9	14.5Cr. 4.5Fe, 4.4Si, 3.15B, 2.3Cu/Mo, 0.65C, Rem. Ni
Amdry 779	9	16Cr, 4Fe, 4Si, 4B, 2.4Cu, 2.4Mo, 2.4W, Rem. Ni
Amdry 961	9	17Cr, 6Al, 0.5Y, Rem. Ni
Amdry 962	9	22Cr, 10Al, 1Y, Rem. Ni
Amdry 963	9	23Cr, 6Al, 0.4Y, Rem. Ni
Amdry 964	9	31Cr, 11Al, 0.6Y, Rem. Ni
Amdry 997	9	23Co, 20Cr, 8.5Al, 4Ta, 0.6Y, Rem. Ni
CM No. 50	5	3.5Si, 1.9B, Rem. Ni, 30–40 HRC
CM No. 50B	5	2.5Si, 1.5B, Rem. Ni, 10–19 HRC
CM No. 50C	5	3Si, 1.8B, Rem. Ni, 20–29 HRC
CM No. 52	5	4.5Si, 3B, Rem. Ni, 55–60 HRC
CM No. 53	5	4.5Si, 3B, 7Cr, Rem. Ni
CM No. 54	5	3Si, 2B, 10Cr, Rem. Ni, 34–40 HRC
CM No. 55	5	4Si, 2.5B, 11Cr, Rem. Ni, 44–50 HRC
CM No. 56	5	4.5Si, 3B, 14Cr, Rem. Ni, 55–61 HRC
CM No. 58	5	4.5Si, 3.4B, 10Cr, 1.5Co, Rem. Ni, 56–62 HRC

TABLE A-5 (*Continued*)

Consumable tradename	Manufacturer code	Description and chemical composition
Cobex N-545	14	0.5B, 2C, 4V, 6Fe, 8Mo, 9W, 28Cr, Rem. Ni, 36–40 HRC
Cobex N-547	14	0.5B, 2.4C, 4V, 6Fe, 8Mo, 8W, 28Cr, Rem. Ni, 40–44 HRC
Cobex N-549	14	0.5B, 2.4C, 6Fe, 8Mo, 28Cr, Rem. Ni, 42–46 HRC
Colmonoy No. 8	7	chromium borides in a nickel matrix, 0.9C, 3.9Si, 26Cr, 0.5 max Fe, 3.1B, Rem. Ni, 53–58 HRC
Colmonoy No. 22	7	0.1C, 0.75Fe, 1.6B, 3.3Si, Rem. Ni, 28–33 HRC
Colmonoy Nos. 23A & 24	7	0.1C, 0.75Fe, 1.5B, 2.5Si, Rem. Ni, 16–23 HRC
Colmonoy No. 43	7	fusible Ni/Cr/B alloy, 0.4C, 2.3Si, 10Cr, 3Fe, 2.1B, Rem. Ni, 35–40 HRC
Colmonoy No. 53	7	fusible Ni/Cr/B alloy, 0.55C, 3.7Si, 12Cr, 3.7Fe, 2.5B, Rem. Ni, 45–50 HRC
Colmonoy No. 63	7	fusible Ni/Cr/B alloy, 0.7C, 4.35Si, 14Cr, 4Fe, 3.1B, Rem. Ni, 57–63 HRC
Colmonoy No. 69	7	fusible chromium carbide/nickel alloy, 0.7C, 4Si, 2.3Mo, 14.5Cr, 4Fe, 3.1B, 2.3Cu, Rem. Ni, 57–62 HRC
Colmonoy No. 77	7	WC composite, 82% Colmonoy No. 62SA, 18% WC, 58–63 HRC
Colmonoy No. 635	7	WC/nickel composite, 65% Colmonoy No. 6, 35% WC, 58–63 HRC
Colmonoy No. 705	7	WC/Ni composite, 50% Colmonoy No. 63, 50% WC, 58–63 HRC
Colmonoy No. 730	7	WC/Ni composite, 70% Colmonoy No. 72, 30% WC, 58–63 HRC
Colmonoy No. 750	7	WC/Ni composite, 50% Colmonoy No. 72, 50% WC, 58–63 HRC
Colmonoy No. 805	7	chromium borides in a nickel matrix, 75% Colmonoy No. 6, 25% chromium boride, 62–67 HRC
Colmonoy No. 42SA	7	chromium carbides in a nickel matrix, 0.4C, 2.3Si, 10Cr, 3Fe, 2.1B, Rem. Ni, 35–40 HRC
Colmonoy No. 52SA	7	chromium carbides in a nickel matrix, 0.55C, 37Si, 12Cr, 3.7Fe, 2.5B, Rem. Ni, 45–50 HRC
Colmonoy No. 62SA	7	chromium carbide/nickel matrix, 0.7C, 4.3Si, 14Cr, 4Fe, 3.1B, Rem. Ni, 57–62 HRC

TABLE A-5 (*Continued*)

Consumable tradename	Manufacturer code	Description and chemical composition
Colspray 80-20	7	Ni/Cr alloy, 90 HRB
Colspray 290	7	chromium boride/nickel matrix, 35 HRC
Colspray 395	7	chromium boride/nickel matrix, 45 HRC
Coltung No. 1	7	WC/nickel matrix, 60% Colmonoy No. 6, 40% WC, 58–62 HRC
Metco 12C	11	11Cr, 2.5B, 2.5Fe, 2.5Si, 0.15C, Rem. Ni
Metco 14E	11	14Cr, 4Fe, 3.5Si, 2.75B, 0.6C, Rem. Ni
Metco 15E	11	17Cr, 4Fe, 4Si, 3.5B, 2C, Rem. Ni
Metco 15F	11	17Cr, 4Fe, 4Si, 3.5B, 1C, Rem. Ni
Metco 16C	11	16Cr, 4Si, 4B, 3Cu, 3Mo, 2.5Fe, 0.5C, Rem. Ni
Metco 19E	11	16Cr, 4Si, 4B, 4Fe, 2.4Mo, 2.4W, 0.5Fe, Rem. Ni
Metco 20	11	16Cr, 4.5Si, 3.5B, 4Fe, Rem. Ni
Metco 440	11	14Cr, 7Fe, 6Al, 6Mo, Rem. Ni
Metco 447/447HS	11	self-bonding Mo/Al/Ni composite, 5.5Al, 5Mo, Rem. Ni
Metco 450/450NS	11	Ni/Al composite, 4.5Al, Rem. Ni
Metco 450P	11	Ni/Al composite, 4Al, Rem. Ni
Metco 451	11	Ni/Cr/Al, 9.5Cr, 2.5Si, 1.5B, 0.5Al, Rem. Ni
Metco 501	11	self-bonding Mo blend, 30Mo, 12Cr, 2.75Si, 2.75Fe, 2.5B, 0.75C, Rem. Ni
Metco 505	11	self-bonding Mo blend, 75Mo. 4.25Cr, 1Si, 1Fe, 0.8B, 0.2C, Rem. Ni
N-20	14	0.05C, 1.5B, 3.5Si, 1Cr, Rem. Ni, 15–20 HRC
N-40	14	0.25C, 2B, 3.5Si, 3.5Fe, 10Cr, Rem. Ni, 36–40 HRC
N-50	14	0.4C, 2.5B, 4Si, 4Fe, 12Cr, Rem. Ni, 47–53 HRC
N-60	14	0.55C, 3.1B, 4Si, 4.5Fe, 14Cr, Rem. Ni, 57–63 HRC
N-690	14	0.6C, 2.7Cu, 2.6Mo, 3.4B, 4Si, 1.5Fe, 15Cr, Rem. Ni, 57–64 HRC
PAC 60E	20	Ni/Cr/B/Si alloy, 60–65 HRC
PAC 65	20	Ni/Cr/B alloy, 30–35 HRC
PAC 69E	20	Ni/Cr/B alloy, 50–55 HRC
PAC 134	20	Ni/Cr/Al blend, 25–35 HRC
PAC 600	20	Si/Cr/B/Fe/Mo alloy, 58–62 HRC
PAC 912	20	Mo/Ni/Al blend, 70–80 HRB

TABLE A-5 (*Continued*)

Consumable tradename	Manufacturer code	Description and chemical composition
Stoody 63T.G	1	C/Si/Cr/B/Fe/Ni/WC, 60 HRC
Stoody 64T.G	1	C/Si/Cr/B/Fe/Ni/WC, 40 HRC
Stoody 65T.G	1	C/Si/Cr/B/WC/Fe/Ni, 50 HRC
Stoody 85T.G	1	C/Si/Cr/B/WC/Fe/Ni, 63 HRC
Stoody 86T.G	1	C/Si/Cr/B/WC/Fe/Ni, 63 HRC
Stoody 87T.G	1	C/Si/Cr/B/WC/Fe/Ni, 63 HRC
UCAR LN-2B	6	99% nickel
UCAR Ni-106	6	20Cr, Rem. Ni
UCAR Ni-107	6	20Cr, Rem. Ni
UCAR Ni-130	6	NiCoCrAlY
UCAR Ni-163	6	NiCrAlCoMoY
UCAR Ni-164	6	NiCrAlY
UCAR Ni-164–2	6	NiCrAlY
UCAR Ni-171	6	NiCoCrAlY
UCAR Ni-185	6	5Al, Rem. Ni (prealloyed)
UCAR Ni-191	6	NiCoCrAlY
UCAR Ni-192	6	NiCoCrAlHfSiY
UCAR Ni-211	6	NiCrAlY
UCAR Ni-220	6	MCrAlY
UCAR Ni-246–3	6	NiCrAlY (fine)
UCAR Ni-246–4	6	NiCrAlY (coarse)
UCAR Ni-278	6	NiCrAlY
UCAR Ni-292	6	NiCrAlY
UCAR Ni-332	6	Complex braze alloy
UCAR Ni-328	6	NiCrMoFe
UCAR Ni-334	6	complex braze alloy
	Cobalt-Base Thermal Spray Consumables	
Amdry 318	9	Co/Ni/Cr/W PTA powder
Amdry 326	9	30Co, 4W, 1Si, 1C, Rem. Co thermal spray and PTA powder
Amdry 557	9	10Ni, 25Cr, 3Al, 5Ta, 0.6Y, Rem. Co
Amdry 850	9	19Cr, 18Ni, 10W, 3.65Fe, 3.5Si, 3.25B, 0.7C, Rem. Co
Amdry 995	9	32Ni, 21Cr, 8Ni, 0.5Y, Rem. Co
Amdry 996	9	10Ni, 25Cr, 7Al, 5Ta, 0.6Y, Rem. Co
CM No. 4210	5	2.75Si, 1.75B, 13.1W, Rem. Co, 53–58 HRC
CM No. 8659	5	0.75C, 26Cr, 15Ni, 2.15Si, 1.75B, 9W, Rem. Co, 45–53 HRC

TABLE A-5 (*Continued*)

Consumable tradename	Manufacturer code	Description and chemical composition
Metco 18C	11	40Co, 18Cr, 6Mo, 3.5Si, 3B, 2.5Fe, 0.2C, Rem. Ni
Metco 66F-NS	11	62Co, 28Mo, 8Cr, 2Si
Metco 67F-NS	11	50Ni, 15Cr, 3Si, 32 Mo
Metco 68F-NS-1	11	52Co, 28Mo, 17Cr, 3Si
MT 4065C	5	Stellite 6
PAC 63	20	Co/Ni/Cr/Mo/B/Si alloy, 50–55 HRC
PAC 90C	20	Co/Cr/Ni/W alloy, 30–35 HRC
PAC 90VF	20	Cr/Co/W/Ni alloy, 30–35 HRC
S-1	14	2.6C, 12W, 30Cr, Rem. Co, 50–54 HRC
S-6/S-156	14	1.2/1.6C, 30Cr, 4.5W, Rem. Co
S-12	14	1.6C, 30Cr, 9W, Rem. Co, 40–46 HRC
S-21	14	0.2C, 5.4Mo, 27Cr, Rem. Co, 28–35 HRC
S-158	14	0.7B, 0.75C, 5.5W, 26Cr, Rem. Co
Stellite No. 156	1	1.6C, 28Cr, 1.1Si, 1Mn, 1Mo, 3Ni, 4W, Rem. Co, 43 HRC
Stellite No. 157	1	0.1C, 22Cr, 1.6Si, 1Mo, 2Fe, 2Ni, 2.4B, 4.5W, Rem. Co, 52–54 HRC
Stellite F	1	1.75C, 25Cr, 1.1Si, 0.5Mn, 0.6Mo, 3Fe, 22Ni, 12W, Rem. Co, 43 HRC
UCAR Co-103	6	25Cr, 10Ni, 7.5W, Rem. Co
UCAR Co-105	6	25Cr, 10Ni, 7.5W, Rem. Co
UCAR Co-109	6	28Mo, 8Cr, 2Si, Rem. Co
UCAR Co-110	6	CoCrAlY
UCAR Co-110–1	6	CoCrAlY
UCAR Co-123	6	CoCrAlY
UCAR Co-139	6	CoCrAlY
UCAR Co-170	6	CoCrAlPtHf
UCAR Co-174	6	CoCrAlHf
UCAR Co-210–1	6	CoNiCrAlY
UCAR Co-211	6	CoNiCrAlY
UCAR Co-222	6	CoCrNiTa
UCAR LDT-400	6	d-gun powder, 28Mo, 8Cr, 2Si, Rem. Co
Wallex No. 12	7	1.35C, 1.5Si, 29Cr, 8.5W, Rem. Co, 45–50 HRC
Wallex No. 40	7	0.5C, 7W, 2Fe, 24Ni, 2B, 16Cr, 1.5Si, Rem. Co, 41–46 HRC
Wallex No. 50	7	0.8C, 10W, 1Fe, 18Ni, 3.5B, 19Cr, 2.75Si, Rem. Co, 56–61 HRC

TABLE A-5 (*Continued*)

Consumable tradename	Manufacturer code	Description and chemical composition
		Iron-Base Thermal Spray Consumables
Amdry 970	9	24Cr, 8Al, 0.5Y, Rem. Fe
Amdry SS420	9	420 stainless steel thermal spray and PTA powder
Colspray 120	7	low-carbon steel
Colspray 400	7	martensitic stainless steel, 30 HRC
F-90	14	2.8C, 29Cr, Rem. Fe
F-410	14	0.12C, 12Cr, Rem. Fe, 36–42 HRC
Metco 91	11	low-carbon steel, 0.5Mn, 0.2C, Rem. Fe
Metco 97F	11	high-carbon steel, 3.5C, 0.35Mn, Rem. Fe
Metco 448	11	self-bonding low-carbon steel composite, 10Al, 1Mo, 0.2C, Rem. Fe
Metco 449	11	self-bonding high-carbon steel, 3Al, 3Mo, 3C, Rem. Fe
Metco 463	11	high-chromium iron/stainless steel composite, 2Mo, 6Al, 15Cr, Rem. Fe
Metco 465	11	high-chromium iron-base composite, 27.5Cr, 6Al, 2Mo, Rem. Fe
PAC 97	20	400 series stainless steel, 35–40 HRC
PAC 97 PTA	20	400 series stainless steel PTA powder
UCAR Fe-101	6	17Cr, 2.5Mo, 21Ni, Rem. Fe
UCAR Fe-123	6	31Cr, 14Mo, 2B, Rem. Fe
UCAR Fe-124	6	24Cr, 8Al, 0.5Y, Rem. Fe
UCAR Fe-132	6	19Cr, 4.5Al, 0.5Y, Rem. Fe
UCAR Fe-140	6	30Cr, 12.5Al, 12.5Co, 18Mn, 10Ni, 4.6Si, Rem. Fe
		Composite Thermal Spray Consumables
Amdry 951	9	60% Ni, 60% graphite
Amdry 952	9	70% Ni, 30% graphite
Amdry 953	9	85% Ni, 15% graphite
Amdry 954	9	80% Ni, 20% graphite
Metco 307NS	11	75% Ni, 25% graphite
Metco 308NS-2	11	85% Ni, 15% graphite
Metco 309NS-3	11	80% Ni, 20% graphite
Metco 311NS	11	55% Al, 7% Si, 26% graphite, 12% organic binder
Metco 313NS	11	40% Al, 45.5% graphite, 5.5% Si, 9% organic binder
Metco 601NS	11	12% Si/Al, Rem. Metco 600 polyester
Metco 605NS	11	90% aluminum bronze, 10% polyester

TABLE A-5 (*Continued*)

Consumable tradename	Manufacturer code	Description and chemical composition
Metco 625	11	plastic bondcoat, 95% stainless steel, 2.5% epoxy, 2.5% nylon
	Copper-Base Thermal Spray Consumables	
Amdry 1331	9	aluminum bronze
Colmonoy No. 15	7	copper buildup, 93% Cu, 7% P, 78–82 HRB
Colspray 500	7	aluminum bronze, 70 HRB
Metco 51	11	aluminum bronze, 9.5Al, 1Fe, Rem. Cu
Metco 51F-NS	11	aluminum bronze, 9.5Al, 1Fe, Rem. Fe
Metco 445	11	self-bonding aluminum bronze, 10Al, Rem. Cu
MT 4210C	5	aluminum bronze
PAC 16	20	10% Al, Rem. Cu, 50–58 HRB
PAC 940	20	aluminum bronze, 50–60 HRB
UCAR Cu-101	6	36Ni, 5In, Rem. Cu
UCAR Cu-102	6	36Ni, 5In, Rem. Cu
UCAR Cu-103	6	38Ni, Rem. Cu
UCAR Cu-104	6	10Al, 1Fe, Rem. Cu
	Pure Metal Thermal Spray Consumables	
Amdry 313X	9	molybdenum
Amdry 357	9	aluminum
Amdry 918	9	titanium
Amdry 919	9	tungsten
Metco 64	11	molybdenum
MT 3010	5	molybdenum
PAC 118	20	molybdenum 30–40 HRC
PAC 661	20	tungsten, 40–50 HRC
	Miscellaneous Thermal Spray Consumables	
Amdry 137	9	75% Mo, 25% self-fluxing alloy
Amdry 138	9	80% Mo, 20% self-fluxing alloy
Amdry 350	9	50% Ni, 50% Cr
Amdry 392C	9	12% silicon, Rem. Al
Amdry 500C	9	copper/nickel/indium blend
Amdry 956	9	95% nickel, 5% aluminum blend
Amdry 960	9	94% Ni/Cr, 6% Al blend
Colspray 955	7	nickel aluminide bond coat
Metco 350	11	high-carbon iron/Mo, 18Mo, 3C, 0.25Mn, Rem. Fe

TABLE A-5 (*Continued*)

Consumable tradename	Manufacturer code	Description and chemical composition
Metco 442	11	self-bonding SS composite, 8.5Cr, 7Al, 5Mo, 2Si, 2Fe, 2B, 3 titania, Rem. Ni
Metco 443/443NS	11	NiCr/Al composite 6Al, Rem. Ni/Cr alloy
Metco 444	11	self-bonding SS composite, 9Cr, 7Al, 5.5Mo, 5Fe, Rem. Ni
Metco 600/600NS	11	aromatic polyester
PAC 135	20	blend of Mo and Ni/Cr/B/Si alloy, 50–55 HRC
PAC 136	20	titanium 6Al, 4V alloy, 35–40 HRC
PAC 902	20	75% Mo, Rem. PAC 135, 35–40 HRC
PAC 1Co	20	Co/Cr/W alloy, 45–53 HRC
PAC 6Co	20	Co/Cr/W alloy, 35–40 HRC
PAC 12Co	20	Co/Cr/W alloy, 37–44 HRC
PAC 21Co	20	Co/Cr/Mo alloy, 26–31 HRC
PAC 156Co	20	Co/Cr/W alloy, 40–45 HRC
PAC 157Co	20	Co/Cr/W alloy, 48–54 HRC
SN-81	14	Co + Ni base, 2.5C, 5Fe, 14W, 28Cr, Rem. Co + Ni
SN-86	14	Co + Ni base, 2C, 29Cr, 14W, 9Fe, Rem. Co + Ni
SN-812	14	Co + Ni base, 2.2C, 9Fe, 9W, 27Cr, Rem. Co + Ni
SN-821	14	Co + Ni base, 0.4C, 8Fe, 7Mo, 28Cr, Rem. Ni + Co
SN-858	14	Co + Ni base, 0.9B, 1C, 8Fe, 6W, 28Cr, Rem. Ni + Co
Sweat-on Paste	7	chromium boride/self-fluxing, 82Cr, 18B, 68–72 HRC

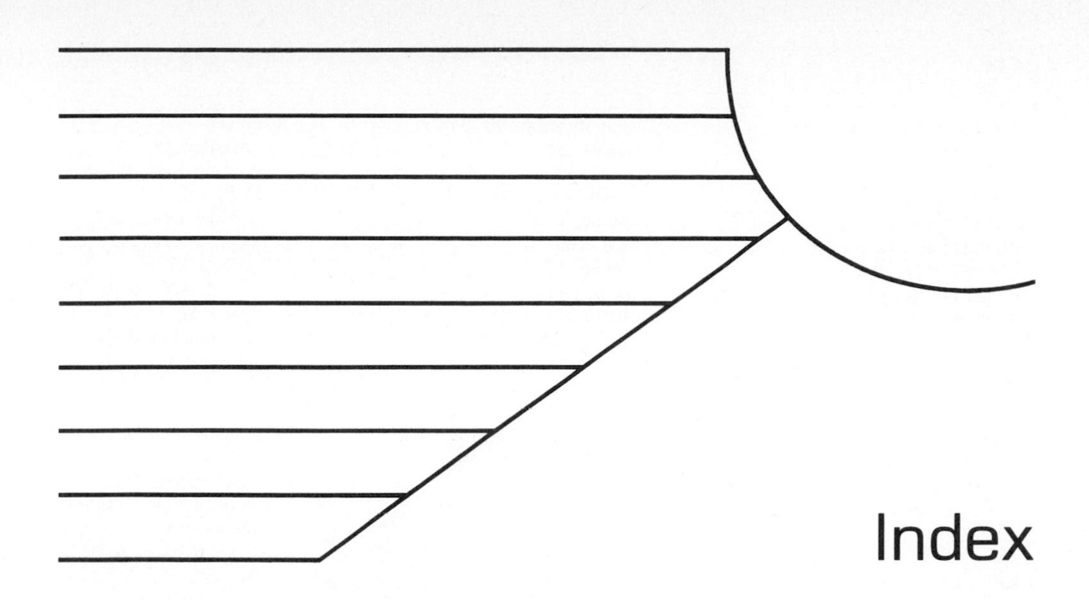

Index

A

B

C

D

E

T

U

V

W

X

Y

Z